PROTEIN–SOLVENT INTERACTIONS

PROTEIN–SOLVENT INTERACTIONS

edited by
Roger B. Gregory
Kent State University
Kent, Ohio

Marcel Dekker, Inc.   New York • Basel • Hong Kong

Library of Congress Cataloging-in-Publication Data

Protein–solvent interactions / edited by Roger B. Gregory.
p. cm.
Includes bibliographical references and index.
ISBN 0–8247–9239–4
1. Proteins. 2. Solvation. 3. Hydration. 4. Proteins—Biotechnology. I. Gregory, Roger B.
QP551.P697634 1994
574.19'245—dc20 94–24924
CIP

The publisher offers discounts on this book when ordered in bulk quantities. For more information, write to Special Sales/Professional Marketing at the address below.

This book is printed on acid-free paper.

Marcel Dekker, Inc.
270 Madison Avenue, New York, New York 10016

Current printing (last digit):
10 9 8 7 6 5 4 3 2 1

PRINTED IN THE UNITED STATES OF AMERICA

Preface

The interaction of water-soluble globular proteins with their solvent environment has long been recognized as an essential factor in determining their physical and functional properties. Protein stability, protein–protein association, ligand binding, and catalytic activity all depend on the structure and thermodynamic and dynamic properties of proteins, which are greatly influenced by the solvent. In addition to its fundamental interest, an understanding of protein–solvent interactions is also of considerable importance in the development of biotechnologies and industrial processes that make use of proteins, for example, in the application of enzymes in organic solvents, the use of immobilized proteins in chemical separations, and the development of strategies for the purification and storage of proteins, as well as in food processing and preservation.

Our understanding of protein–solvent interactions has greatly advanced over the last 15 years. Indeed, this period of research has been marked by a number of interesting and important discoveries. The contributions in this volume present some of these advances in our understanding of structural, thermodynamic and dynamic aspects of protein–solvent interactions. It is not intended to be a comprehensive survey of all that has happened in this broad field of research—such a volume would need considerably more space—but it is hoped that the chosen topics will give some sense of the real progress achieved and the outstanding questions that must be addressed.

As an introduction to the more specific descriptions of protein hydration and protein–solvent interactions, Rufus Lumry opens this volume by providing a detailed discussion of aspects of the structure, dynamics and thermodynamics of proteins as they relate to protein function and stability. This chapter discusses enthalpy–entropy compensation, protein activity coefficients, protein–protein association, and enzyme mechanisms, many of which are discussed further in later chapters. The "knot-matrix" construction principle, first described by Gregory and Lumry in 1985, and the protein "devices" that such a construction allows, form the central theme of this chapter. What emerges is a new paradigm for protein research. Much of the evidence for this emerging view of the modular construction of proteins is obtained from atomic Debye–Waller factors derived from x-ray crystallographic data, and it is to x-ray crystallography that we turn in Chapter 2, where Edward Baker describes the patterns and significance of solvent structure at the surface, in the interior, and at the active sites of proteins. The conservation of solvent sites and the role of bound solvent in stability and function are also discussed.

The next three chapters are concerned with protein hydration, and in particular with the effect of hydration on the dynamic properties of proteins. In chapter 3, Roger Gregory describes the sorption isotherm, hysteresis phenomena, and hydration-induced conformational changes, as well as the effects of hydration on protein dynamics and the glass transition temperature. The observation of a glasslike dynamical transition at 200 K in protein crystals and solutions and the recent description of the hydration dependence of its transition temperature, together with advances in our understanding of protein internal dynamics derived from hydrogen isotope exchange and crystallographic Debye–Waller factors, suggest a picture of proteins as heterogeneous glasses consisting of dynamically distinct domains in which free-volume rearrangement ("mobile defects") and the plasticizing action of water on the protein conformation provide the basis for conformational mobility. Chapters 4 and 5 focus on dielectric relaxation and Rayleigh scattering of Mossbauer radiation, two powerful techniques for studying the effects of hydration on protein dynamics. Ronald Pethig outlines dielectric theory and reviews the experimental results from both solution and solid state studies of proteins, including the evidence for the plasticizing effect of water and proton conduction of water-based hydrogen bond networks. Rayleigh scattering of Mossbauer radiation (RSMR), although not as familiar as Mossbauer absorption spectroscopy, is a far more versatile technique because it can be applied to proteins that do not possess Mossbauer nuclei. In addition, the broad range of correlation times (10^{-6}–10^{-13} s) that can be examined, and the possibility of defining correlated motions when coherent scattering effects are taken into account, make it an extremely powerful technique for exploring protein dynamic behavior. Vitalii Goldanskii and Yurii Krupyanskii have pioneered the application of RSMR to studies of proteins and in Chapter 5 outline the technique and describe the temperature, hydration, and cosolvent dependence of protein motions revealed by RSMR.

Catalytic activity has been demonstrated for a number of enzymes at hydrations well below the levels required for monolayer coverage and also in anhydrous organic solvents. In organic media, enzymes acquire a variety of novel properties, including enhanced stability, altered substrate specificity, and enantiomeric selectivity, and may also catalyze new reactions. The use of organic solvents also greatly expands the possibilities for biotechnological applications of enzymes in reactions involving nonpolar substrates. Darrell Williams, Igor Rapanovich, and Alan Russell review in Chapter 6 the current status of work on enzymes in organic solvents.

The remaining chapters are largely concerned with the dynamic and thermodynamic behavior of proteins in solution and the important solvent variables that influence protein behavior. Advances in our understanding of the dynamic properties of proteins have led to a fundamental reconsideration of the description of activated processes in proteins, in which the solvent viscosity enters as an important variable. In Chapter 7, Benjamin Gavish and Saul Yedgar outline Brownian dynamics and Kramers's theory and discuss the effect of solvent viscosity on protein dynamics, in particular, the effect of viscosity on volume-dependent relaxation processes in proteins as monitored by ultrasonic absorption. In Chapter 8, Wolfgang Doster and his colleagues describe recent flash photolysis results obtained for the well-known CO rebinding reaction in myoglobin in both water–cosolvent mixtures and hydrated protein films.

Issues of protein stability and protein–protein association are taken up by Arieh Ben-Naim in Chapter 9. The thermodynamics of these processes is examined from first principles, with particular emphasis on solvent effects involving hydrophobic and hydrophilic groups. A reconsideration of solvent-induced effects important in stabilizing proteins suggests that the conventional view of the dominance of hydrophobic hydration as a driving force for this process may be incorrect or, at best, that such a conclusion is premature. The full complexity of the contributions to biochemical processes from solvent-induced effects is revealed when the analysis is done correctly to separate direct (pairwise additive) interactions from indirect (solvent-induced) interactions. Ben-Naim provides an inventory of solvent-induced effects, and outlines approaches to obtain the missing thermodynamic data.

Enthalpy–entropy compensation is a ubiquitous property of proteins and aqueous solutions and there has been much discussion of the mechanisms that give rise to such behavior. (Rufus Lumry discusses several aspects of enthalpy–entropy compensation in Chapter 1.) In chapter 10, Ernest Grunwald and Lorrie Comeford describe a number of thermodynamic mechanisms for the phenomenon.

Thermodynamic and statistical mechanical descriptions of proteins in water–cosolvent mixtures have also been greatly developed. Studies of proteins in cosolvent mixtures, initially prompted by an interest in understanding the basis for the stabilization of proteins by polyols, such as glycerol and sucrose, have led to a new understanding of the underlying thermodynamic nonideality in terms of preferential interactions and excluded volume and a new appreciation of such fundamental concepts as ligand binding. In Chapter 11, Serge Timasheff describes the preferential interactions of water and cosolvents with proteins from first principles through the Wyman linkage relationship. The concept of "binding" and, in particular, the distinction between binding defined thermodynamically and in terms of site occupancy is discussed in detail. In Chapter 12, Peter Wills and Donald Winzor discuss the behavior of proteins in water–cosolvent mixtures in terms of excluded volume. They demonstrate the equivalence of the preferential solvation and excluded volume treatments of the thermodynamic nonideality. The second virial coefficients are the fundamental thermodynamic parameters of interest. Their determination and interpretation in terms of effective thermodynamic radii are discussed in detail.

The purification of proteins is often the most time-consuming and expensive process in the production of proteins for use in biotechnological applications or industrial processes. Large-scale purification of proteins from crude extracts can be achieved very efficiently and cost-effectively by precipitation methods. In the final chapter, Rex Lovrien, Mark Conroy, and Timothy Richardson discuss how protein–solvent interactions may be manipulated to form precipitates, coprecipitates, or cocrystals of proteins for use in protein separations.

Sincere thanks go to all the contributors for their time and effort in preparing their chapters, and to Andrew Berin and Anita Lekhwani at Marcel Dekker, Inc., for their work in the production of this volume and their support and advice throughout the editorial phases of its preparation. Thanks also to Jean Winsterman and Diane Baggesen for secretarial assistance in the preparation of the manuscript. Much of the preparation and editorial work for this volume was performed while I was on sabbatical leave at the University of Bordeaux and a special note of appreciation goes to Christian Courseille and the other faculty and staff of the Department of Crystallography for their very generous hospitality.

Roger B. Gregory

Contents

Contributors

Edward N. Baker Department of Chemistry and Biochemistry, Massey University, Palmerston North, New Zealand

Arieh Ben-Naim Department of Physical Chemistry, The Hebrew University, Jerusalem, Israel

Lorrie L. Comeford Department of Chemistry and Physics, Salem State College, Salem, Massachusetts

Mark J. Conroy Department of Radiology, University of Minnesota, St. Paul, Minnesota

Wolfgang Doster Department of Physics, Technische Universität München, Munich, Germany

Benjamin Gavish Department of Biochemistry, The Hebrew University—Hadassah Medical School, Jerusualem, Israel

Vitalii I. Goldanskii N. N. Semenov Institute of Chemical Physics, Russian Academy of Sciences, Moscow, Russia

Roger B. Gregory Department of Chemistry, Kent State University, Kent, Ohio

Ernest Grunwald Department of Chemistry, Brandeis University, Waltham, Massachusetts

Thomas Kleinert Department of Physics, Technische Universität München, Munich, Germany

Yurii F. Krupyanskii N. N. Semenov Institute of Chemical Physics, Russian Academy of Sciences, Moscow, Russia

Rex E. Lovrien Department of Biochemistry, College of Biological Sciences, University of Minnesota, St. Paul, Minnesota

Rufus Lumry Department of Chemistry, University of Minnesota, Minneapolis, Minnesota

Ronald Pethig School of Electronic Engineering and Computer Systems, University of Wales, Bangor, Gwynedd, Great Britain

Frank Post Department of Physics, Technische Universität München, Munich, Germany

Igor Rapanovich Department of Chemical Engineering and Center for Biotechnology and Bioengineering, University of Pittsburgh, Pittsburgh, Pennsylvania

Timothy I. Richardson Department of Chemistry, University of Minnesota, Minneapolis, Minnesota

Alan J. Russell Department of Chemical Engineering, University of Pittsburgh, Pittsburgh, Pennsylvania

Marcus Settles Department of Physics, Technische Universität München, Munich, Germany

Serge N. Timasheff Graduate Department of Biochemistry, Brandeis University, Waltham, Massachusetts

Darrell L. Williams, Jr. Department of Chemical Engineering and Center for Biotechnology and Bioengineering, University of Pittsburgh, Pittsburgh, Pennsylvania

Peter R. Wills Department of Physics, University of Auckland, Auckland, New Zealand

Donald J. Winzor Department of Biochemistry, The University of Queensland, Brisbane, Queensland, Australia

Saul Yedgar Department of Biochemistry, The Hebrew University—Hadassah Medical School, Jerusalem, Israel

1

The New Paradigm for Protein Research

RUFUS LUMRY
University of Minnesota, Minneapolis, Minnesota

I. INTRODUCTION

A. Purposes

This chapter is a preface to the following chapters, which deal with more specific topics to illuminate the interactions between proteins and water. Our goal, to provide an up-to-date discussion of the details of protein structure and function as they depend on dynamical behavior of protein conformations, has proved to be a major task but one of considerable interest to the author and, presumably, in time to physical chemists concerned with these protein topics. The appearance of new concepts has been remarkably rapid in recent years. This is largely a consequence of the discovery in 1982–1983 of protein substructures that revealed a much more powerful paradigm for protein research than that based on protein secondary structures. In this chapter we discuss many of the advancements in understanding protein behavior that have become obvious with this discovery. These focus attention on the gestalt features of protein behavior characteristic of most protein research rather than on x-ray–diffraction coordinates and site-directed mutagenesis (SDM). Under the new paradigm, many older ideas can now be seen to be naive, often incorrect. The old paradigm relates both structure and function to the "classic" secondary structures. It arose more by historical accident and the absence of an alternative than established utility. The old paradigm can be simply replaced by substituting the more recently discovered substructures for the secondary structures. Some of the discoveries that then become apparent have been discussed in papers written before 1989, but these, in turn, had consequences of even greater interest. Some of these appeared for the first time during the preparation of this chapter. All seem quite important, so all are discussed in at least preliminary fashion in this chapter. Just how free of error these discussions are remains to be seen.

The discovery of the substructures, just 10 years ago, was another serendipitous event in the long search by the author and his co-workers for the rules of protein structure, particularly those that make enzymic catalysis possible. The search begun in 1947 by Lumry and Kistiakowsky in an inconclusive examination of the sharp breaks reported in Arrhenius plots for enzymes finally became profitable in 1953 with the mechanical hypothesis of Eyring, Lumry, and Spikes, itself more a result of the exclusion of alternatives than a requirement of the experimental information then available. According to this hypothesis, the unique catalytic property of enzymes is their use of stress generated by conformation changes to enhance rate-limiting steps in catalysis. The idea that conformations change in function is implied by the early work of Pauling and Campbell on antibodies, but so far as we know, the concept under the name "configurational adaptability" was first given full expression by Karush, also in connection with the immune reaction. The applications to enzymic catalysis include the "passive" conformation change idea, termed by Koshland "induced fit," and the "active" conformation change, which formed the basis for mechanical mechanisms that Professor Eyring called "rack" mechanisms. In the former, the conformation changes are supposed to position the functional electron donors and acceptors in catalysis about the substrate in such a way as to enhance electron rearrangements in primary-bond rearrangements. In the latter, conformations are sources and sinks for mechanical free energy in free-energy rearrangements activated by conformation changes. Stress and strain effects are well known in catalysis of small-molecule processes, but the adjustability of protein conformations provides versatility in manipulation of mechanical free energy similar to that provided by electron push-pull adjustments in small-molecule chemistry. Catalytic mechanisms of proteins and presumably of RNA catalysts have been evolved by accidental exploitation of mechanical versatility, as well as the electronic trick familiar in small-molecule chemistry. The currently popular "transition-state stabilization" hypothesis of enzymic catalysis, as it is generally formulated, falls in neither the induced-fit nor the rack class since it requires a rigid conformation. When that restriction is removed, it becomes a rack mechanism in which the pretransition state is destabilized rather than the transition state stabilized. This fine distinction seems useful in describing the mechanism of catalytic antibodies, "abzymes," which are probably clever constructions for true "transition-state" catalysis, but it may be found to be nonsense for enzymes.

The transition-state hypothesis remains popular despite abundant evidence that conformations have soft parts which in function undergo conformation changes directly linked to mechanism. Hemoglobin is an honorary enzyme, and it is used in this chapter to illustrate a postulated mechanism that depends entirely on the dynamical consequences of fluctuations of conformation to effect linkage among the four sites for ligation, the classic example of "allosteric" coupling. This is a useful exercise to illustrate similarities and differences in mechanism that result from different construction principles, but it is also useful in suggesting new avenues for hemoglobin research. The first rack mechanism, which described the stress-strain relationship between heme iron and proximal histidine, has been the focus of most research on hemoglobin mechanism since its confirmation by Perutz et al. Despite the abundance and variety of high-quality data, the molecular basis of linkage, the Bohr effect, and some other well-known characteristics of this most frequently studied protein seem almost as obscure as they did 50 years ago. The data for ligand recombination by single-chain respiratory proteins obtained by Frauenfelder and co-workers and their description of the intense underlying dynamical activity indicate that conformational dynamics must be more fully factored into hypotheses of mechanism for the tetrameric hemoglobins. We do so with enthusiasm, but the result does not have the high reliability of the enzyme discoveries. It is well to point this out before proceeding.

In this chapter only homo-domain enzymes are considered in detail. Enzymes have two functional domains per catalytic function, and in homo-domain enzymes these are produced by an ancient gene-duplication event. Hetero-domain enzymes will be examined later. The general description of the rack for proteins such as hemoglobin and cytochrome c, which are complex ions, was given in 1961 [7] at which time it served as the stimulus for initiating the Gordon conferences on bioinorganic chemistry. Chymotrypsin is a good example of the broad class of homo-domain enzymes and perhaps the most frequently studied member of that class. Several features of its catalytic function can now be described in detail, as shown in Fig. 1, which illustrates the original rack mechanism for this protein. The central feature is the destabilization of the substrate just prior to transition-state formation followed by transient use of some mechanical free energy provided from the conformation as part of the free energy of activation to that transition state. Thus, mechanical free energy both destabilizes the pretransition state and supplements heat energy in transition-state formation. As depicted in Fig. 1, the molecular mechanism depends on the jaws formed by two domains. The jaws enclose the substrate and the chemically functional groups of the enzyme, forcing them toward and into the transition state for primary-bond rearrangement, here acylation of Ser 185 by an acid moiety of substrates. Only the primary groups, His 57 and Ser 185, were considered. The driving force for the closure process was

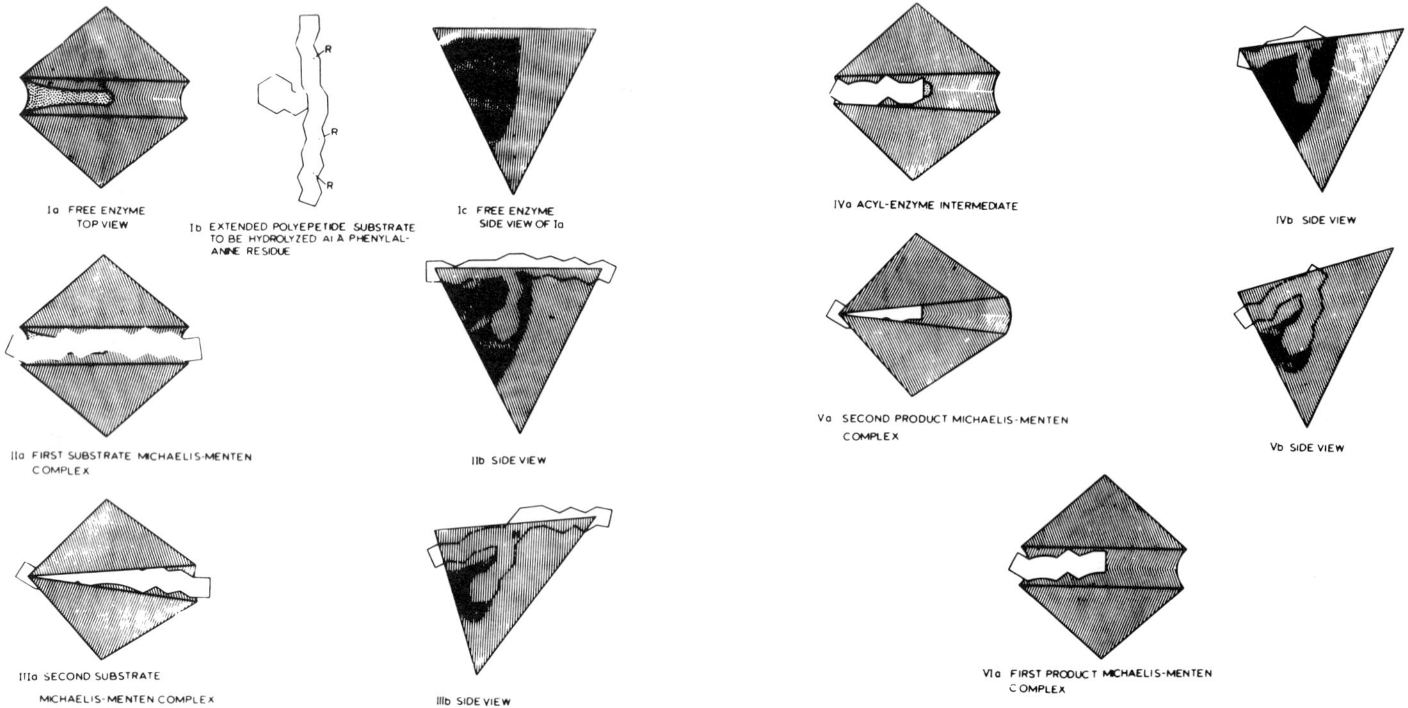

FIGURE 1 Rack mechanism for chymotrypsin as of 1967. The new features at that time were the domains, the functional domains, the pairing principle, the "subtle change" attributed to expansion-contraction of the total conformation, and the mechanical activation into transition state by compression of substrate into functional groups. (Modified from Refs. 9 and 10. Reproduced with permission of the copyright owner, CRC Press.)

then called a "subtle" conformation change and was suggested to be a general contraction of conformation. The other noteworthy feature was the movement of the substrate side chain deeper into its binding pocket as the closure process advanced. This detail was required to explain how free energy released on binding substrate can be used to increase the turnover rate [1]. At one time it appeared that all the increase might be due to this use of binding free energy, but many enzymes act on substrates which have no side chains or do not use them in binding, so there is little generality to the claim "the better the binding, the better the catalysis." The separation of the total mechanism into parts according to their contributions to free-energy management is detailed elsewhere [10].

This rack mechanism did not long attract enthusiastic support. It was soon eclipsed by claims, based on early interpretations of x-ray–diffraction data, that protein conformations are essentially static. It was not until the famous fluorescence-quenching studies of Lakowitz and Weber led to the overthrow of such rigidity that the conformational motility could be reintroduced. The history is particularly puzzling in the initial x-ray period because of inconsistencies between the expectations from static models and what was repeatedly observed starting with the first proton-exchange studies of Linderstrøm-Lang [2]. The *B*-factor data from x-ray studies could have been used to reduce the period of confusion, but no one, including ourselves, gave serious attention to those data. The fluctuational character of protein conformations is now given considerable attention, but as already mentioned, in the Western protein community conformational motility as a major feature of protein mechanism is neglected. The substrates and substrate transitions reported by Frauenfelder and his colleagues have not led to the revolution one might have expected. Rather, they seem often to be an embarrassment. Popular hypotheses such as the "induced-fit" and "transition-state" concepts, inconsistent with known motility characteristics, still dominate textbook discussions.

Information about conformational motility has come from many types of experiment. The accumulation is still insufficient for quantitative application to conformational dynamics. In the author's laboratory under Rosenberg's direction, the proton-exchange method was developed to detect and measure motility. Particularly important in this study was the discovery by Woodward and Rosenberg of the constancy of rank-order in proton-exchange rates [27]. The probability-density-distribution function for the rate constants essential for extracting physical significance from the data defied discovery until the early 1980s when Gregory successfully applied Provencher's numerical Laplace transform invention to the problem [5,138]. The substructures were immediately apparent in the first results, shown in Fig. 2, and progress has been steady since. Progress prior to 1989 has been reported in three papers but is in largest part recapitulated in this chapter as required to present newer findings. In particular, the distinctions among substructures establish a new paradigm for protein research to replace the historical assumption that secondary structures are the foundations of structure

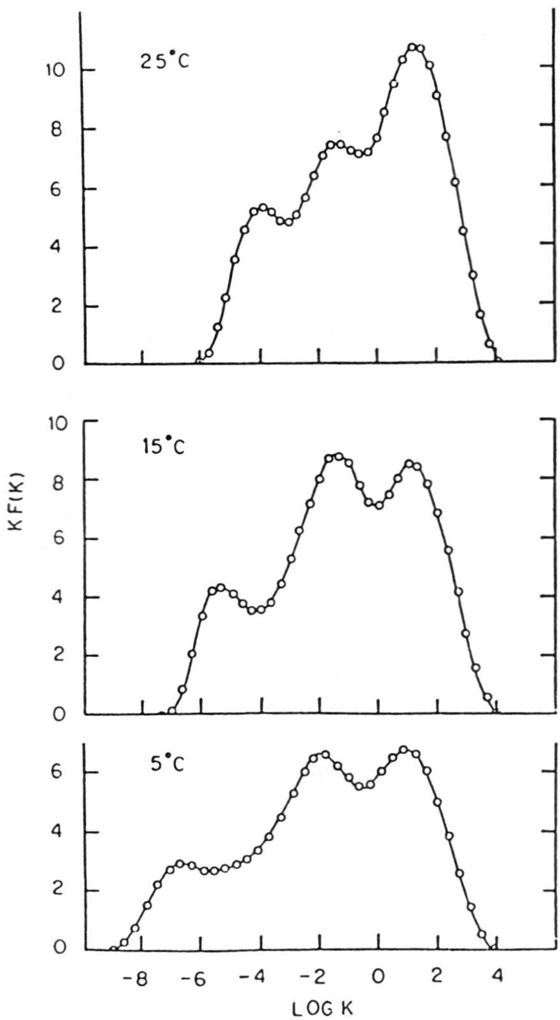

FIGURE 2 Probability-density distribution functions for proton-exchange rate constants at three temperatures. The rate constant is *K* and the pdf is *F*(*K*); protein is HEW lysozyme. The peaks correspond to group I (slow), group II (intermediate), and group III (fast), respectively. (Reproduced with permission of the copyright owner John Wiley and Sons, Inc. From Ref. 5.)

and function. The old paradigm was reasonable but has not been justified experimentally, so that in the long run its assumption has led to much waste of time and money. The new paradigm is sufficiently well developed to make obvious many of the devices discovered in evolution that make biology possible. These devices, in turn, suggest hypotheses likely to make protein research more efficient. Site-directed mutagenesis is a tool of great power, but it encourages only small questions and by its endless nature will continue to do so until the hypotheses on which its applications are based come much closer to addressing the true complexity of proteins. It is not clear that totally new hypotheses are required. Those arising from work by ourselves and our colleagues have been found to explain more and more protein processes, as illustrated by those discussed in this chapter. This is the fortieth anniversary of the discovery of the "rack" mechanism [3,4] and the tenth anniversary of the discovery of the "knot-matrix" construction principle common to enzymes and several other functional classes of proteins [5]. Both hypotheses are in excellent health and fully susceptible to further testing, modification, and extension. The rack mechanism for hemoglobin [6,9] and cytochrome c [7] provides the prototype for metal-protein interactions. Even without all details of conformation dynamics, the rack mechanism postulated for the serine proteases provides a much better model than the transition-state model for investigating other families of enzymes [8]. Gavish's model has similarities and differences with our rack model, but the two are both versions of the dynamic rack (see Ref. 143 and Chapter 7). His sophisticated examination of the physics of conformational dynamics complements the rack description given in this chapter. The two versions together are unrivaled in depth and detail.

Many topics in this chapter are old in the sense that they were discussed by Lumry and Biltonen in 1969 [9] but new in the sense that they have yet to attract attention from the current generation of protein biochemists. A recapitulation with some extension published in 1989 [10] but written 2 years earlier provides useful background specific to enzymic catalysis, and a paper in the *Journal of Molecular Liquids* written and published in 1989 [11] contains short introductions to several of the newest discoveries.

The most relevant references up to 1988 are given in the last mentioned paper, which is short enough to quickly relate topic and references. These are too numerous to be repeated here, so we give only a selection of these and of some new references which are particularly relevant to one or another of the topics in this chapter. Some topics, such as the role of mobile water in plasticizing conformations, are discussed in other chapters of this volume and so are covered only briefly here. Other topics are well developed in other fields. Thus the entropy-suppression effect known as the Helfrich effect or the "fluctuation force" effect, which has only recently become recognized as a major kinetic and thermodynamic factor in membranes and vesicles, is discussed in papers applying to membranes but is totally absent from papers on proteins. That effect is typical of phenomena from other fields that are included here because of the probability that they are important in protein systems. Our philosophy is that all such possibilities with potential importance in proteins can only be rejected on the basis of unfavorable qualitative evidence. The Helfrich effect is a particularly important example since it may be the major factor in the strong associations of proteins. Other effects little known in protein chemistry are discussed below, but it is more likely that we have missed still others than that we have been overinclusive.

B. Confusing Biology with Chemistry

This chapter addresses a serious problem frustrating progress toward understanding proteins and biological systems nearly as much now as in the earliest days of protein study. The problem arises from the confusion of chemistry with biology. The chemistry of the DNA double helix is that of its individual groups, all of which have been well studied by chemists and biochemists. The biological usefulness of the double helix lies in its structure, and this is an "invention" in which the intrinsic properties of these groups, i.e., their chemical and physical properties, have been utilized in evolution to produce a nonchemical result. To apply chemical knowledge to better understand biological mechanisms one must first discover the devices that make the mechanisms possible. The stable states of small molecules are determined by the intrinsic chemical properties of the molecules. Those of biological macromolecules individually and at higher and higher levels of cooperativity are intrinsically improbable, gaining their actual high probability only because of their coupling to the remainder of the biosphere. Chemistry deals with the cold earth and our attempt to breath life into cold molecules; biology deals with inventions not only unpredictable, but often so constructed as to subvert chemical expectations. The two disciplines require quite different orientations. Students of biology must be ever aware of the fact that chemistry is a poor guide to biology, a necessary foundation but not one extrapolatable to biological devices. Students of chemistry are adequately fortified by a general realization that in principle, often only in principle, chemical behavior can be traced back to the Schrödinger equation and its solutions. To say that this is also true of biology is more than a minor sophistry. Designs for research on biological mechanisms that are based on logical progression from chemistry are unlikely to produce real progress.

The use of conformation changes in enzymic mechanisms is another such device. This is not the chemists' chemistry, but rather a construct found in the biosphere that makes selection, specificity, and rate control possible and adjustable by DNA modification. Once such a device is discovered, it is likely that we can duplicate it, perhaps in some cases using classical chemical techniques. But one can neither understand nor duplicate something that is yet to be discovered. The problem is compounded by the fact that nature's solutions to many physiological problems depend on packaging several such devices in a single protein. Hemoglobin is an example in which progress continues to be impeded by inadequate models, and that protein is undoubtedly very simple compared with the glycolytic enzymes.

As used in this chapter the term "device" specifically refers to a discovery in evolution. There are many of these and few have counterparts in familiar chemistry.

C. Supporting Evidence

The unifying theme of this chapter is the construction principle for enzymes and many other functional classes of proteins. The reasons why the name "knot-matrix" principle is appropriate will become apparent in the following discussions of the discovery and consequences of this extremely useful device. Virtually all features of the construction and function of enzyme are consequences, so the principle is the single item in this chapter of most importance for protein chemistry.

The principle was established by the atomic temperature factors determined in x-ray–diffraction studies, genetic-conservation data and proton-exchange data. In the following section these sources are examined individually. It is fortuitous that the principle, like most other devices discussed in this chapter, can be confirmed by examination of the *B*-factor data for almost any protein in the Brookhaven Protein Database (PDB). A few proteins utilize other construction principles, but most appearing in the PDB have knot-matrix construction. We have found no exceptions in small enzymes, proteinases, redox proteins, or repressor proteins and suspect the principle is general, but large enzymes with synthetic or glycolytic functions have yet to be examined. The immunoglobulins and respiratory proteins are built using significantly different construction principles. The bases for this deduction can also be easily checked using data from the protein and sequence databases, but the deduction itself is not yet reliable for either class.

The palindromic patterns in *B* factor for enzymes arising from some primordial gene duplication can be detected in plots of *B* factors against atom number using a procedure given in the next section. The elegantly simple HIV-1 protease serves as an ideal model for proteins with two-domain palindromy.

Pictures constructed from atomic coordinates consolidate information in the tables and graphs and in doing so reveal the full beauty of the knot-matrix construct. All of these exercises are as easy for those just gaining familiarity with proteins as for the experts.

Some matters are not so easily confirmed by the reader. For the few requiring simple theoretical substantiation, the theory is given. Still others are less easily established because they require intricate integration of information presented in several papers. The deductions we have formed from the latter are sometimes conjectural, as is noted in the text, but most have some experimental support for their importance.

II. PROTEIN STRUCTURE

A. Information from *B* Factors

1. Reliability of Deductions from *B* Factors

The Debye-Waller factors, henceforth to be called *B* factors, determined in x-ray–diffraction studies of protein crystals, provides a wealth of information about the construction of proteins as well as invaluable information about conformational dynamics. They are calculated for each nonhydrogen atom from the dispersion of the x-ray scattering and can be simply converted to the mean-square displacement of an atom (the positional variance) from an ideal lattice point in the crystal lattice, μ^2. The quality of x-ray data for proteins is such that it is necessary to assume that the displacements are isotropic, in which case $B = 8\pi^2\mu^2$. The larger the *B* factor, the larger the displacement and the smaller the contribution of that atom to the coherent scattering. On the most frequently used scale, atoms with a *B* factor greater than 10 make a negligible contribution ($\mu = 0.35$ Å), and for this reason they are not included in refinement computations for small molecules. Protein structural determination is based on well-known information about the bond lengths and angles in the primary structure, and for this reason the *B* values have considerable reliability even when they are very large. When the atom *B* factors are refined with the coordinates atom by atom, the experimental contribution to the variances is negligible with respect to the other contributions: lattice disorder, local displacement, and global displacement. Many proteins have some *B* factors as low as 2, which suggests very low lattice disorder, but in general, the lower the factors, the less sensitive their values are to disorder. The first of these has discouraged use of the *B* factors because it was thought to dominate the other two. More recent evidence suggests that protein crystals have strong and specific interprotein contacts that minimize disorder. This is well supported by comparisons of families or proteins, the same protein in different crystals forms, and various tests of consistency with other kinds of data, particularly proton-exchange rate constants and genetic information. Consistency is the primary justification for our use of *B* factors to estimate the total extent of local and global disorder. Separate estimates of local and global disorder can only be made using other information to supplement that provided by *B* factors. This limits the reliability of interpretations of *B* factors. In addition, there is uncertainty due to crystal constraints. Proteins are in general much compressed on crystallization, there being whole-molecule compressions due to increase in protein activity coefficients and mechanical constraints by neighbors. These effects diminish the *B* factors relative to what is expected under the conditions of mild restraint that apply in free solution. Thus other things being equal, the crystal *B* values underestimate the true fluctuation spectra. Sheriff and co-workers [12] have attempted to assess the extent to which near-neighbor packing constraints reduce the *B* factors relative to free-solution motility. They assumed a rough empirical relationship between conventional estimates of surface accessibility and *B* factors to adjust the *B* factors. Their results for the single-subunit oxygen carrier myohemerythrin and the octameric form of the allosteric protein hemerythrin are shown in Fig. 3. These proteins correspond to myoglobin and hemoglobin, respectively, and, like the latter, have no low *B* factors and a high *B*-factor mean; actually the variance and the mean are larger for hemoglobin than for hemerythrin. The general reproducibility of the *B* factors is seen to be good. The *B*-factor adjustments on this assumption are often large and always positive, suggesting that crystallization reduces conformational

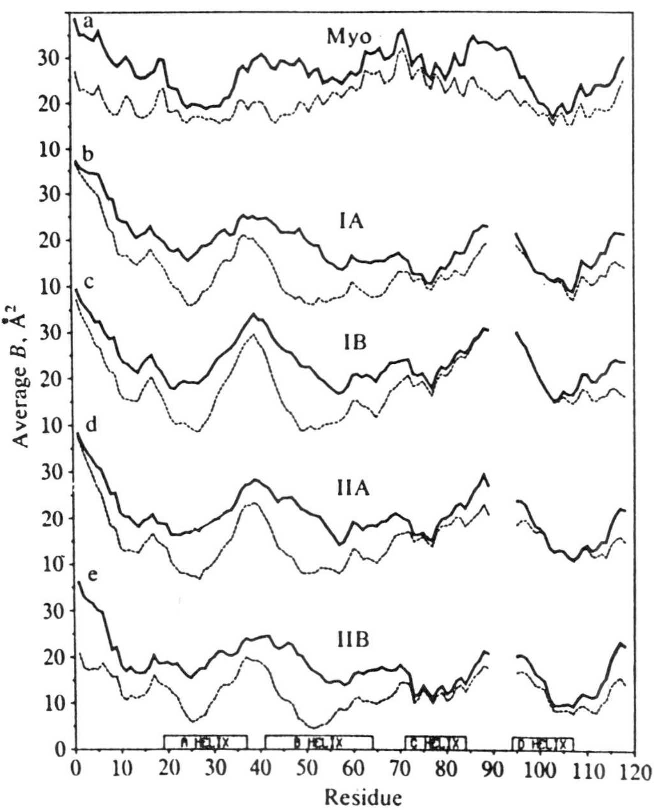

FIGURE 3 Residue-averaged *B* factors for myohemerythrin and octameric hemerythrin to show estimates of lattice suppression of *B* factors. Dashed line, uncorrected for lattice-contact effects; solid line, corrected for crystal-packing effects based on monotonic relationship between calculated solvent accessibility and *B* factors. The hemerythrins, like the hemoglobins and myoglobin, have no knot *B* factors or have such unusually high global fluctuations as to hide knots. They are in any event soft and highly deformable and typify one class of proteins that does not conform to the knot-matrix structural principles. (Reproduced with permission of the copyright owner, National Academy of Sciences, USA, From Ref. 12.)

motility. The motility reductions for the respiratory proteins are so large as to suggest that global fluctuations are the major source of high *B* values. The results cannot be generalized even for knot-matrix proteins but do indicate that the degree to which *B* factors underestimate the conformational motility of proteins in the dissolved state may be large.

The distance μ is the standard deviation of the distribution of the displacements if random, so at one standard deviation an oxygen atom with radius 0.75 Å and *B* factor of 10 sweeps out an average cavity of 1.77 $Å^3$ as compared with the atom van der Waals volume of 0.56 $Å^3$. This is an exaggerated estimate of the freedom of motion since each atom is constrained to a considerable extent to move in synchrony with its neighbors, but the order of magnitude of the comparison is correct. There can be considerable error, both random and systematic, in calculating the *B* factors. Most protein cystallographers use the same definition and the same method in computing *B* factors, but systematic errors due to crystal form, choice of dispersion cone angle, and type of refinement procedure used, if any, exclude absolute comparisons among *B*-factor data obtained by different groups or from different crystals. On the other hand, comparisons of relative values after a statistical adjustment of such scaling errors tend to be very good whether applied to the variances or the standard deviations. But *B* factors need not be known with high accuracy to be used in making deductions about the dynamics of protein conformations since the range of μ values is the same as the range of atom displacements in conformational fluctuations. This is in contrast to coordinate values determined in the same studies, which have absolute errors larger than these displacements. The several sources of error in determination and interpretation of *B* factors are very well discussed by Ringe and Petsko [13] in their review of the uses of x-ray–diffraction methods in the study of conformational dynamics.

B. Observations Based on *B* Factors

1. Substructures

Three groups of *B* factors can be distinguished by the three size ranges: group I, roughly from 2 to 8, values found in small-molecule crystals; group II, 8–14; and group III, all higher values [14]. The ranges and their midpoints vary somewhat from one type of protein to another, some differences being undoubtedly real and others due to differences in the method used for calculation and real effects of lattice disorder. There is considerable overlap of adjacent groups consistent with the more detailed three-peaked distribution function obtained from the rate constants for proton exchange. *B* factors distinguish the residues involved but give no reliable way to resolve the overlap into the families of substructures. This is demonstrated by the parent of the proteinase-inhibitor family, pancreatic trypsin inhibitor (BPTI), in Fig. 4. The group I *B* factors identify nine residues, but roughly half of

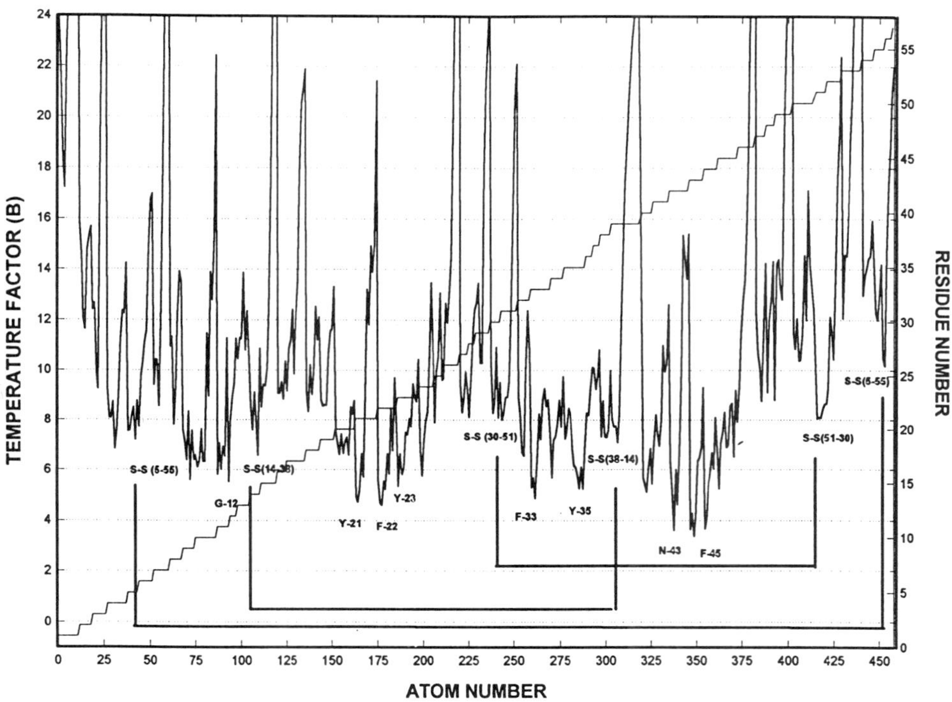

FIGURE 4 Single-atom *B* factors for bovine pancreatic trypsin inhibitor–*B* factor plot. All named residues are highly conserved. The six half-cystines form three disulfide bonds but nearby residues are not conserved and have variable *B* values depending on the crystal form. The three small clusters that combine into the single primary knot are each strongly stabilized by dispersion forces among aromatic residues and they combine by strong hydrogen bonds. The plot shows a distinct palindromic pattern heavily dependent on the placement of the disulfide bonds for matrix structure and properties. The palindrome is centered at residue 34 or 35, which might be said to be the keystone in function as well as structure. A diagram of the construction is given in Fig. 13. In the latter figure it can be seen that the structure of this protein has a palindromic organization which is also detectable in the *B* factors shown in this figure. The *B* factor palindrome does not appear to be as exact as those found with the gene-duplication enzymes but that may be a misleading consequence of a too strict definition of a functional domain. Whether or not palindromy or an approach to palindromy is characteristic of other families of single-domain proteins remains to be determined. (Data for 6PTI from the Brookhaven PDB).

the atoms of these residues have *B* factors in the group II range. Proton-exchange rates are in the very slow category for six of these residues. This family of inhibitors and venoms is called the Kunitz proteinase-inhibitor family. At least 47 members have been discovered using function as the test of family membership, but these all satisfy the more stringent test of genetic conservation [15]. They have the same three disulfide bridges and nine other residues (G12, K21, K22, K23, K33, L36, L37, K43, and K45) conserved. Conservation is not always exact at a few positions. Tyrosine, tryptophan, and phenylalanine are sometimes interchanged, and histidine replaces tyrosine or phenylalanine in two members. Genetic conservation has now been studied in many protein families; some, such as the mammalian phospholipases with more than 60 members, have already been compared. The Kunitz-family statistics and degree of conservation are characteristic.

In Fig. 5 all the atoms of the noncystine conserved residues of BPTI with low *B* factors except those of the cystines are shown in the complete molecule. The bonding making such structures stable is discussed in more detail later (see Figs. 11, 12, and 13). The study of BPTI has been very productive. One can single out in particular the study of BPTI substructures by Woodward and co-workers [16]. It is the first to be constructed using the new paradigm and an elegant, well-focused research model for future SDM studies (vide infra).

In addition to supplying specific information about a single protein family, *B* factors appear to be diagnostic for protein families and even more generally for different classes of proteins. As already mentioned, qualitative *B*-factor differences appear to establish the hemoglobin family and the Fab fragments as different from the knot-matrix class. Thus there are no atoms with group I *B* factors in the respiratory proteins. This seems to be the general situation for proteins with the "myoglobin fold," but there is also the real possibility that exaggerated helix motions in these predominantly helix proteins and the overlay of substrates wash out such detail. All atoms have unusually high *B* factors, the average for mammalian hemoglobins being about 50, which is twice the average usually found for enzymes. At present, the distinctions in motility data and in other kinds of comparisons of these proteins with enzymes require that hemoglobin, myoglobin, and so forth be considered members of another structural and functional class (see Sec. VI.C).

Our delineation of the immunoglobulins as a third class of protein is based on the extensive study by Constantine and co-workers of the antidigoxin Fab fragment [17]. One example is not an adequate basis for defining a new class, but the *B* factors have distinct differences from the other proteins we have studied. In contrast to the enzymes, group I substructures appear to be mobile, rearranging into new substructures on binding digoxin (see Sec. VI.F).

2. Variability Among Group I Clusters

Thus far three qualitatively different classes of group I clusters have appeared. The first is typified by the clusters of the trypsin proteases in which a large fraction of group I atoms are on residues located sequentially along the peptide. The helix

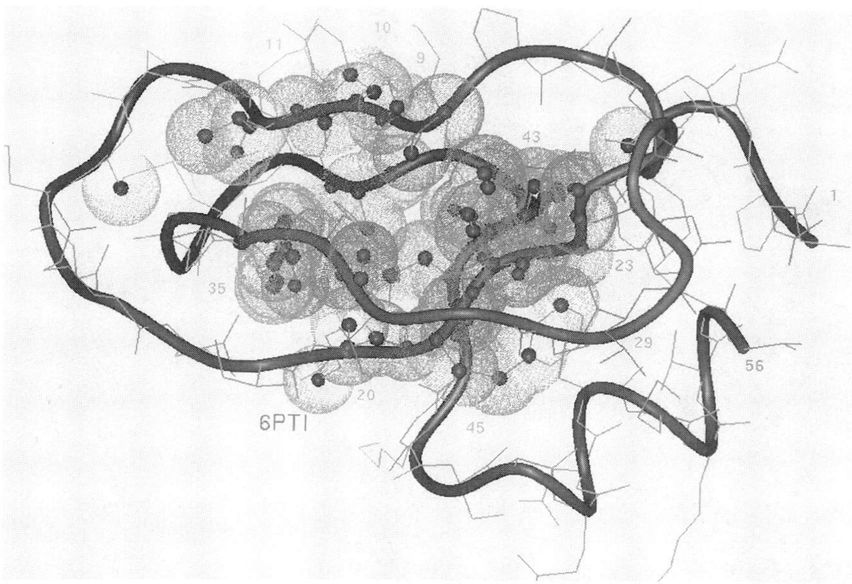

FIGURE 5 Atoms with group I *B* factors in bovine pancreatic trypsin inhibitor (pictorial presentation). Atoms with group I *B* factors are shown as van der Waals spheres. In addition to the conserved residues, which also are in group I for proton-exchange rates, residues 9, 10, and 11 are shown. The *B* values of the atoms in these residues and in group 52–54 show large variations in *B* values depending on the x-ray study. The variations occur in only one group in any study, demonstrating the coupling between these weakly connected groups. Coupling depends on the disulfide bonds and reflects features of the matrices important in function as proteinase inhibitors. (Data for 6PTI from the Brookhaven PDB.)

clusters of the members of the phospholipase A2 family are an extreme example; all the atoms of the two main group I clusters are backbone atoms in α-helices (Fig. 6). The second class is that in which the group I atoms lie in organized secondary structures but only a few group I atoms are in each piece of secondary structure, as in the subtilisins in which each cluster contains small parts from two or more layers of sheet. A third type, quite similar to the subtilisin situation, is typified by the single-chain acid proteases. The clusters contain blocks of residues from well-separated positions along the main chain. This produces a loop construction in the group II and group III residues in this case and in most others in which the group I cluster contains some noncontiguous subclusters.

Group I clusters as described using *B* factors are not always nicely clumped into single globular balls. In the subtilisin family one cluster appears to be split into two parts (Fig. 7). This appearance may be due more to the arbitrary choice

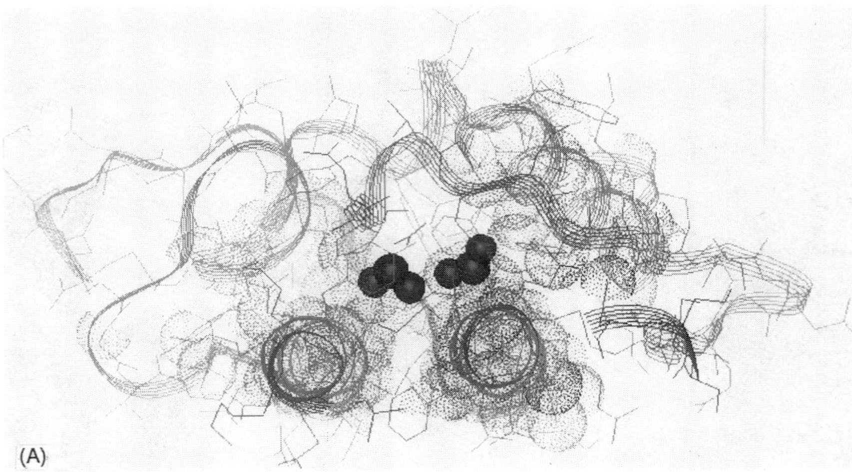

FIGURE 6 Knots of bovine mammalian phospholipase A2. The primary knots are the long helices shown in top view (A) and side view (B). Atoms with knot *B* values are shown as dotted spheres. Note the positions of the primary functional groups (solid balls for reacting groups of His 48 and Asp 99) and the absence of small *B* values in the side chains of the residues of the primary knots. With very few exceptions the knot atoms are main-chain atoms. Almost all side chains are small and usually nonpolar, so there are no "hydrophobic cores." The knots are very different from the PTI knot but quite common especially in heteroknot enzymes. The disulfide bond array is not shown but it is a major constraint on the matrices enforcing conformation and dynamics. It is difficult to see how knots of this type could control matrix properties without disulfide bonds. However, disulfide bonds seem essential for the same function in the PTI family (see Fig. 13). The secondary knot utilized in binding calcium ion is also shown. (Data from PDB 1BP2.)

of the lower cutoff for group II and the difficulty of showing the clusters in a single projection. The separation may be bridged by an H-bond skeleton as in the PTI proteins just discussed. However, there are occasionally small groups of atoms with group I *B* factors that are not parts of the main clusters and probably not a result of crystal constraints. The HIV-1 protease has such groups forming a partial shell around the catalytic assembly. These clusters probably mean that this kind of construction can produce more subtle refinements of structure. It will be seen that group I clusters come in several shades, each supporting a structural or functional role.

3. Functional Domains of Enzymes

The enzymes, proteinase inhibitors, and redox proteins so far examined consist of one or more of the composite substructure units discussed in the previous sec-

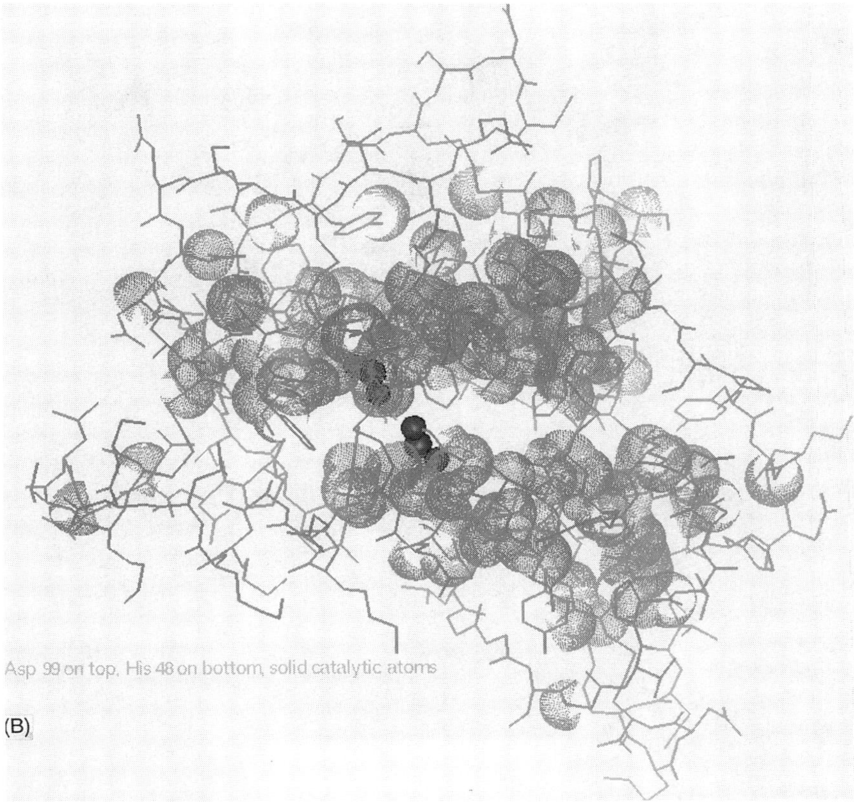

(B)

tion. To distinguish this "building block" assembly from the several other current uses of the term "domain," we have called them *functional domains.* These consist of one group I substructure to which is tethered most of the group II atoms of the domain. Group III atoms are restricted to surfaces, often the outer surfaces of loops that otherwise have group II *B* factors. Total surfaces of proteins usually contain many group II atoms as well and also frequently sides of group I clusters. Group III atoms have highly variable functions and achieve them in highly variable ways. They nevertheless form a third kind of substructure because their conformational characteristics and functions are different from those of group I and group II substructures.

The Kunitz PTI proteins consist of one functional domain. Each catalytic function of an enzyme requires two, and there may be more than one catalytic function in large enzymes. Regulatory molecules generally have their own functional domains with which they interact and influence function through perturba-

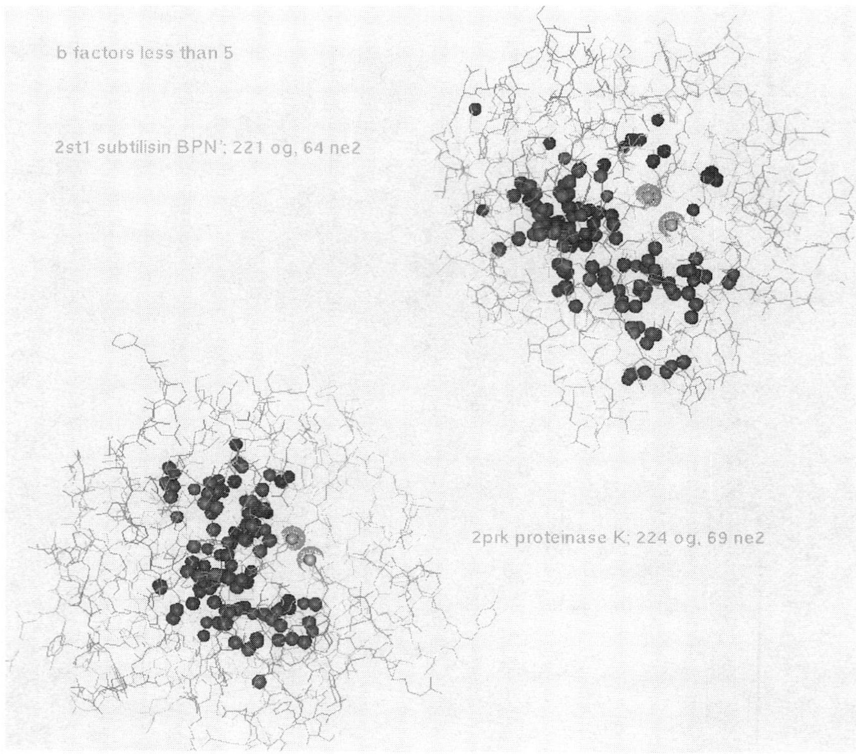

FIGURE 7 Knots of subtilisin BPN′ and proteinase K from *B* factors. The construction of the subtilisins is typical of the proteases. Proteinase K and subtilisin BPN′ are far apart on their evolutionary tree. Most knot residues are conserved but very few matrix and surface residues. The knots are similar as a result, as might be expected, but also the two knots in each protein approach good *B*-factor palindromy despite the absence of sequence palindromy (see Fig. 10). The two primary functional groups are shown as lightly shadowed spheres. In the absence of substrate they have high *B* factors (Data from 2ST1 and 2PRK in the Brookhaven PDB.)

tions of functional domains directly involved in reactions with substrates and so forth. For free-energy conservation, functional domains on the same or different proteins are often coupled. Simpler multienzyme systems are of this type. At a high level of function, and perhaps a higher level of evolution, enzymes with high-level functions as regulators, in synthesis, in glycolysis, and so forth, the functional domains can become integrated into large, complex, cooperative structures. In these the single functional domain is probably not a useful construct distinction.

Heat-stable proteins may also have complicated functional domain structure. Thermolysin is a much studied example [32].

The functional domain appears to be the smallest protein structural unit in physiological function. The proportions of group I and group II atoms vary among proteins of different families and different classes. In the enzyme class the group I substructures are generally small, roughly 10–20%. In thermophilic proteins the percentages can be higher. In the PTI family the percentage appears to be generally less than 10%.

The construction of catalytic and regulatory functional domains is modular. Zymogens such as pepsinogen and plasminogen package several functional domains, and on activation by cleavage yield several active proteins. The necessary cooperative competence among functional domains in each enzymic product occurs as a result of rearrangements that take place after initial activation.

Although the well-known *secondary structures* are usually used to describe protein structure, no obligatory relationship to the functional domains has appeared. It is unlikely that in enzymes with chemical function, secondary structures have any intrinsic functional importance. The reasons for this statement are given in some detail in several sections of this chapter.

As can be seen in Fig.8A, in the A2 phospholipases the atoms of the primary acid and base groups actually participating directly in catalysis punctuate the two cusps of the *B*-factor plots very near their midpoints. The low *B* factors at the bottom of the cusps are the group I atoms. The functional atoms in the free enzyme have at least group II *B* values. Although such groups in other proteins are usually not centered in the *B*-factor cluster, this freedom of motion is another invariant feature of enzyme construction. Thus, in the absence of substrates and most competitive inhibitors, the chemically functional groups are held on the first residue of the polypeptide chain as it emerges from the *B*-factor cluster and held in such a way that their average displacement amplitudes are high.

In Fig.8A, that part of the B-factor plot lying outside the cusps applies to the functional domain that binds calcium ion, a typical example of modular construction with functional domains. We can call the two catalytic functional domains the *primary* domains and the functional residues, here His 48 and Asp 99, the *primary* residues. As is well known, the catalytic chemistry can be additionally supported by other chemical groups. For example, the acid group of Asp 102 somehow supports the primary His 57 and Ser 185 in α-chymotrypsin. There is presumably some general rule about the positioning of these *secondary* chemical groups, but we have not yet sought it. Coenzymes often take over the role of polypeptide atoms as primary or secondary groups. In this chapter we cannot consider coenzymes in a general way, but the discussion of the protein-heme group interaction in hemoglobin has considerable generality in this connection (see Sec. VI. C.3).

The two functional domains of a catalytic function can be on separate proteins, as shown by Schachman et al. for aspartyltranscarbamolyase and more

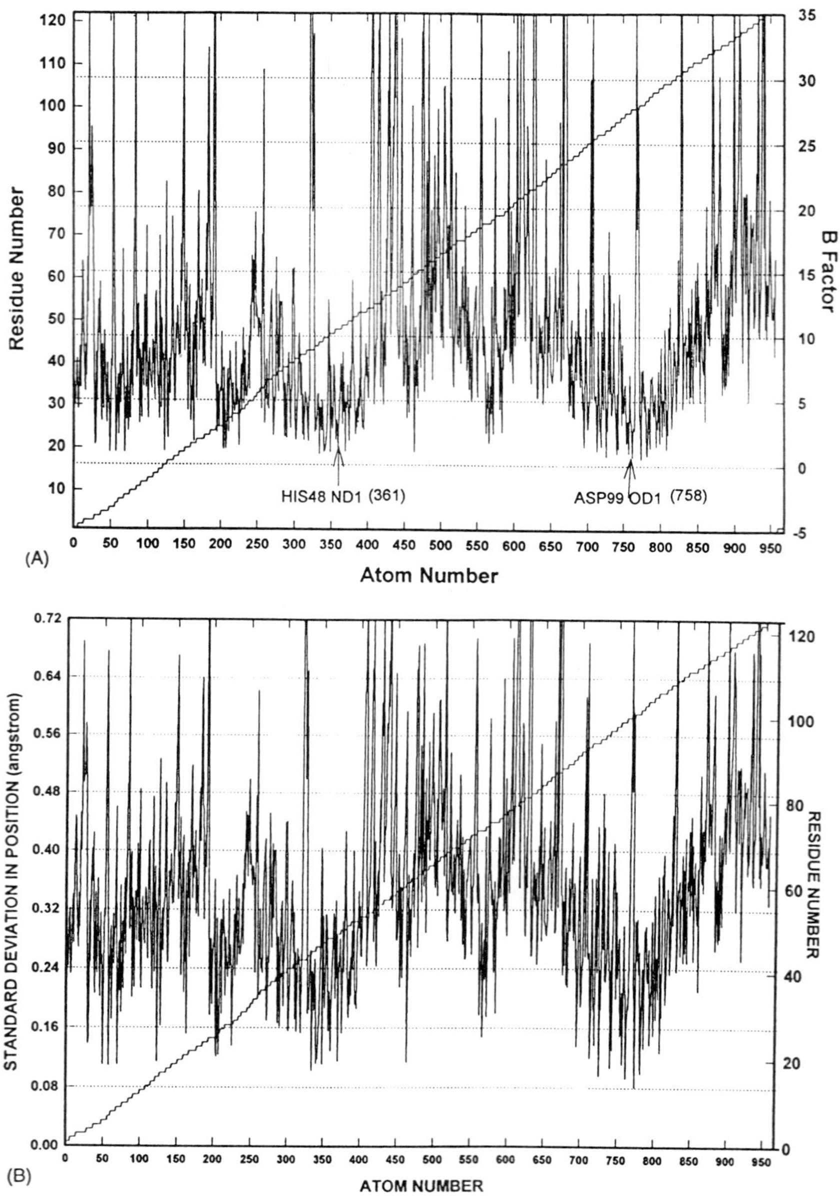

FIGURE 8 Atom B factors for mammalian phospholipases A2. *B* factors for 1BP2 (B) standard deviations of atom position for 1BP2 (isotropic model); (C) *B* factors for 4BP2; (D) *B* factors for 1PP2. In (A) the central atoms of the chemically functional groups are indicated. Their *B* factors are considerably lower than usually found for such participating atoms. The obvious palindromy pattern includes the

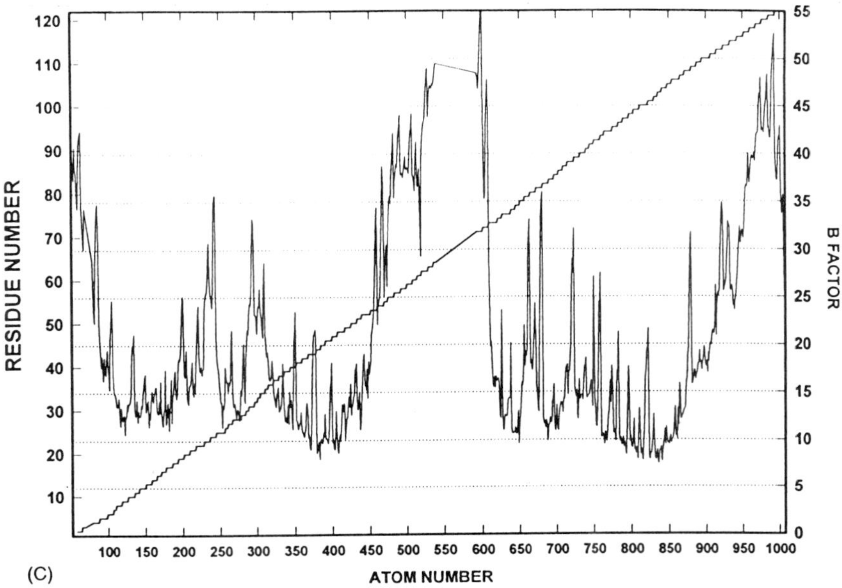

(C)

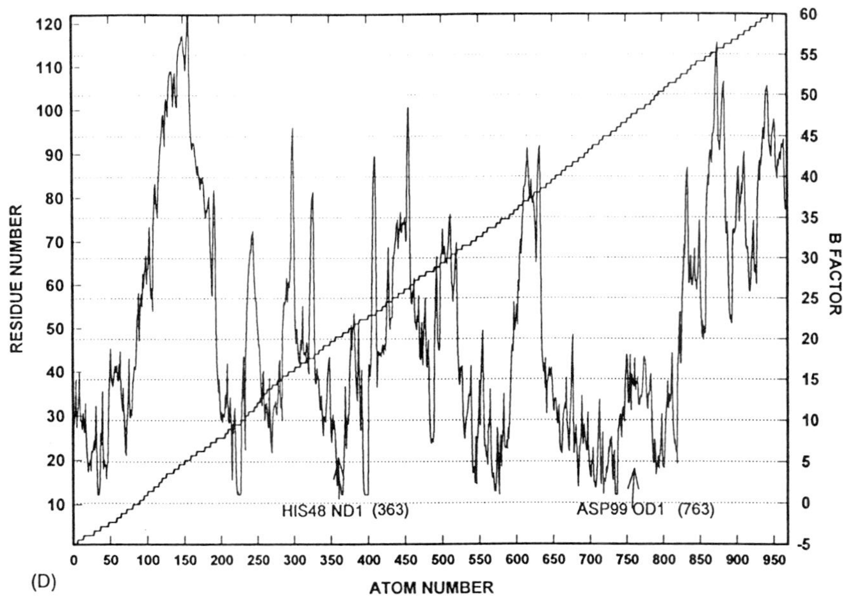

(D)

two primary knots and the “hinge” region connecting the two functional domains. The calcium-binding knot is on the extreme left. (C) and (D) are included to show the large experimental errors in *B* determination that can be found in the Brookhaven PDB. With such errors it is difficult to detect the palindrome. (Data are from the Bookhaven PDB.)

recently by others for other enzymes. Since there is little reason to suspect a different functional-domain construction in this type of enzyme, the two proteins must be strongly tied together at some point to establish the hinge. The HIV-1 protease is a simple prototype for this kind of two-subunit enzyme, and it shows how noncatalytic group I clusters are used to hold the separate proteins together in single functional entities (see Fig. 14).

Ancient enzymes may have been single functional domains. Some may be in existence today; e.g., myoglobin as an intracellular peroxidase? It seems probable that most lost the competition with their two-domain counterparts soon after the latter were discovered.

4. Palindromic Patterns in the Polypeptide Chains of Some Enzymes

The palindromic pattern is found in the *B* factors for those enzymes formed as a result of an ancient gene doubling. When the functional domains are separate protein units, their *B*-factor palindrome is a natural consequence of the residue palindrome since the two proteins always seem to be connected at either their C-terminal or their N-terminal regions. When both functional domains are on the same single polypeptide chain, there is no corresponding palindromic pattern in the residues, rather more nearly a repetition in the same sequence direction. Nevertheless, the *B*-factor palindrome pattern persists. Enzymes produced by zymogen activation are in the latter class, but so are enzymes like the subtilisins which do not have zymogen precursors. Figure 8 is a typical example.

There is a general procedure for finding these patterns in *B*-factor versus atom-number plots: Find the two primary functional groups. Find the residue lying halfway between them. The first atoms on the chain in each direction that are group I atoms form the first group I palindromic pair. Usually these are not at the same distance from the first-chosen central residue because the function groups are slightly displaced. One therefore finds the residue actually dividing the distance between the two members of the first *B*-factor pair. The new choice provides the reference to find the second pair of matching group I atoms, then the third pair, and so on. The asymmetry of the functional residues can be considerable. In the pepsins it is about four residues (Fig. 9). The functional atoms do not have group I *B* factors, as already noted, so they do not disturb the near perfection of the *B*-factor pattern.

Palindromy is also to be seen in the group II atoms because they tend to follow the group I atoms. It is usually detectable in the high-*B* atoms when they are in loops determined by the group I atoms. Variations from protein to protein in a family include residue insertions or deletions and even substitutions of helices for sheets in some loop conformations. These may cause slight changes in position of the group I pairs in the different members of a single enzyme family, but palindromy is preserved. Errors in *B*-factor determination are sometimes so large as to make the pattern search more subjective than objective. This is illustrated in Fig. 8C and 8D. Note that if perfect palindromy is desirable in evolution, it has been rarely achieved except in enzymes like HIV-1 protease, which combine identical protein subunits. Insofar as the *B* factors are reliable, we have not yet found perfect palindromy in a single-chain gene-duplication enzyme. Each gene-duplication enzyme family has its own palindromic *B*-factor pattern, as can be seen in the several *B*-factor plots. Two pepsins are compared in Fig. 9 and two subtilisins in Fig. 10. Additional features of the palindromic patterns are demonstrated particularly in the two subtilsins by comparisons of residue conservation.

The three-dimensional structures of enzymes with *B*-factor palindromes are not palindromic. Rather they are arranged about a noncrystallographic dyad axis running from the central residue or pair of residues through the separation between the primary functional groups. Shotton and Watson [18] noticed that the two functional domains of porcine elastase looked very similar but inverted in construction so as to show twofold rotational symmetry rather than mirror symmetry. As yet, there is no known mechanism for backward transcription to produce these patterns, but more than such a mechanism is required since the residues containing the group I atoms are highly conserved in the two domains, yet the domains are constructed to face each other along the chain. The HIV-1 protease satisfies this requirement easily since its two identical protein subunits are connected to each other by inversion and rotation so as to generate an accurate dyad axis. This is a logical consequence of *B*-factor palindromy when there is residue palindromy. It is not logical for single-chain enzymes, but evolution has succeeded, by many mutational accidents, in converting two initially identical sequences joined head to tail so as to now have head-to-head *B*-factor patterns. In the gene-duplication enzymes, the main chain crosses between functional domains only once and in doing so provides the hinge connecting the two functional domains. In some enzymes with different functional domains, the main chain crosses twice between the two functional domains. Double crossing is thus far not found to be common, but HEW lysozyme is an example. The double crossing suggests that both functional domains were developed together on a single chain. Again this must have required an extremely long voyage of evolutionary discovery much more tedious than combining functional domains from different original genetic events.

Although the two domains of gene-duplication enzymes often look rather alike in pictures, they are only superficially so. The highly conserved residues in the two domains, though very similar and aligned in the chain in the same direction, form two distinctly different structures, which nevertheless satisfy the linear *B*-factor palindromy requirement and thus produce in the three-dimensional structure the dyad axis, which in these proteins is somehow critical for function. Figure 9 illustrates this with two pepsins.

That the extraordinary protracted search for palindromy has been necessary to effect efficient enzymatic catalysis at its current high degree of efficiency and

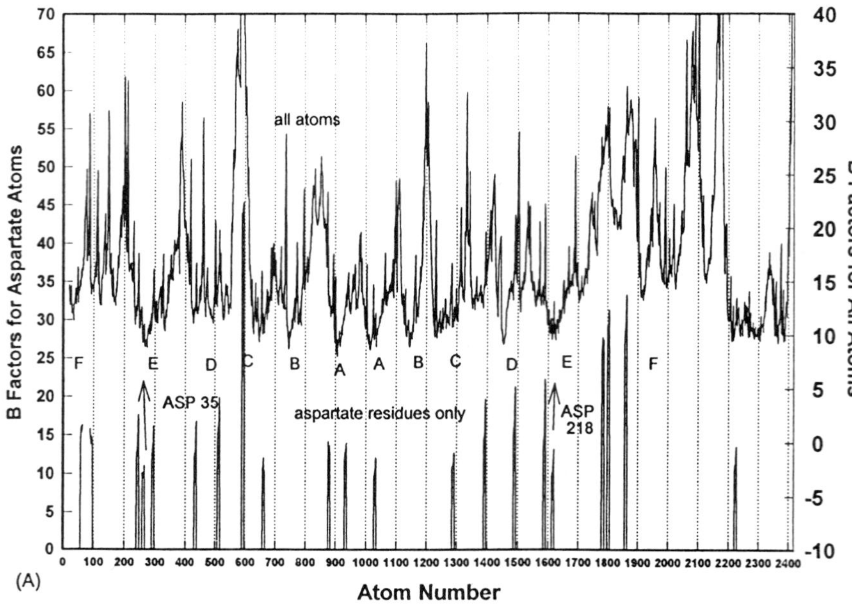

FIGURE 9 Atom *B* factors for rhizopepsin and penicillopepsin. (A) Rhizopepsin (2APR)showing the extensive *B*-factor palindrome, which extends well beyond the two primary functional residues. Note that the *B* factors of palindromic atom pairs are very similar, certainly the same within error. Also shown is the approximate palindromic pattern in distribution of the many aspartates of this protein. The latter is a consequence of the former required by the water environment. The outer loop positions provide for ionization. Pairs of corresponding minima are labeled with the same letter. The primary functional residues are Asp 218 and Asp 35. (B) Penicillopepsin and rhizopepsin compared. The molecules have similar *B* factors at all points but the palindromic patterns are centered between the two primary functional residues. This is a typical comparison of two members of the same family and it demonstrates that the *B* factors in fact are a reliable and reproducible source of information about the atoms. The palindromic patterns are not identical probably as much a result of experimental error as real difference. [The data for 2APR (rhizopepsin) and 3APP (*Penicillium* pepsin) are from the Brookhaven PDB.]

specificity should be an important clue as to how enzymes actually work (see Sec. VI.D.6).

The *B*-factor palindromy thus far found has been limited to enzymes, with the interesting exception of the Kunitz inhibitors. As shown in Fig. 4, there is a well-developed pattern of *B*-factor palindromy centered at residues 34 and/or 35 in this family. The protein is not an enzyme and the palindrome is not altogether similar to that found in gene-duplication enzymes. Some proteinase inhibitors of

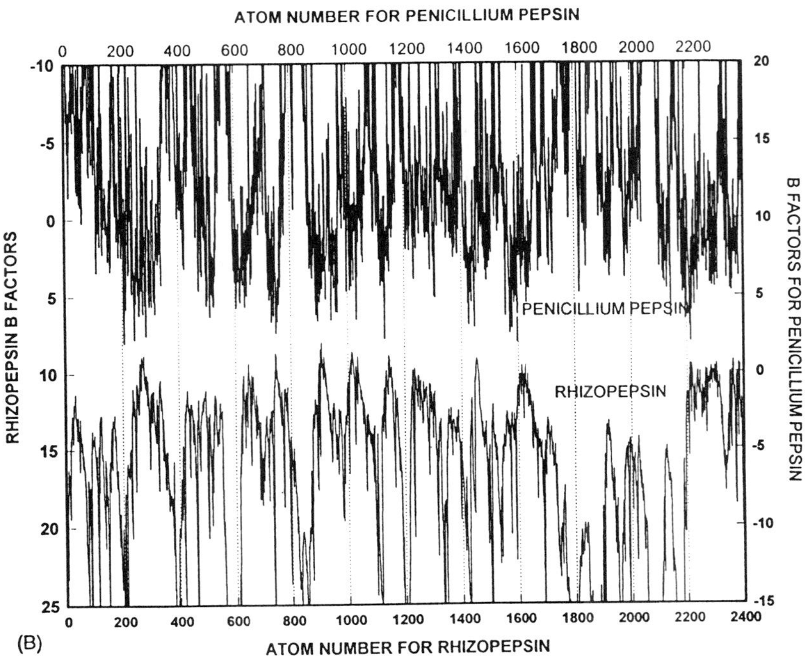

(B)

the Kunitz PTI family do acylate their target proteinases, but that is not general and the structural consequences of the rough palindromy appear to be of more interest than the chemical ones. In particular, note that the three disulfide groups fit well, although not exactly, into the structural palindrome (Fig. 13), where it seems obvious their major role is to help the knot control the matrix. Any arrangement of atoms or groups in a protein is likely to contribute to more than one important characteristic of structure and function. Many such proposals for disulfide bonds have been made and perhaps all are correct to greater or lesser degree. Their positioning in the Kunitz proteins and tight conservation seem to establish one such important role. In retrospect, it does not appear surprising that knots can have difficulty controlling their associated matrices. The phospholipase A2 problem of this kind apparently requires at least six disulfide bonds, again in quite a small protein, to control the matrices.

5. Palindromic Arrangement of Charged Groups

The pepsins have a very large number of aspartate residues, most of which are ionized and partially exposed to solvent, as indicated by their position just back from the outermost parts of the loops. The *B* factors are mostly in the group II size range and none are in group I. The residues are distributed in a pattern that is nearly iden-

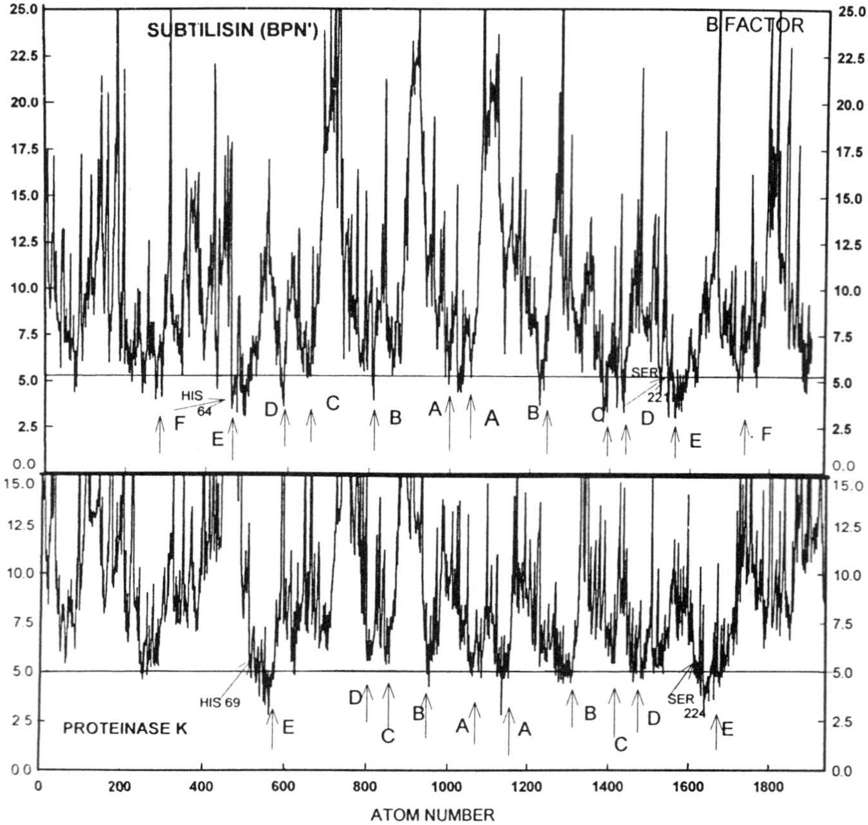

FIGURE 10 Atom *B* factors for subtilisin BPN′ and proteinase K. (Bottom) Proteinase K *B* factors and palindrome. The positions of the palindromic atom pairs are indicated by labels. The two proteins are well separated along their family tree and only about 70 residues of 325 constituting the knots are conserved in the two. Presumably it is because of knot conservation that the palindromes are very similar in positions and *B* values despite insertions and deletions along the evolutionary path from one to the other. The palindromes are fixed in position by the primary functional residues. The two knots in each protein conform to the requirements of the noncrystallographic twofold axes and each is very similar to its partner on the other protein. The difficulties and triumphs of evolution are thus particularly well demonstrated by this comparison. (The BPN′ data were supplied by Dr. R. Bott. The proteinase *K* data are from 2PKR of the Brookhaven PDB.)

tical to the *B*-factor palindrome of the protein. Of the 21 aspartates of rhizopepsin, only three do not have palindromic partners, as can be seen in Fig. 9A. *Penicillium* pepsin has a less complete aspartate arrangement, but that may be because glutamates are substituted for the missing aspartates. It is not clear whether these arrangements have importance in function, but like most other common details of structure, they probably do.

The acid stability and functional pH range of these acid proteases is due to their large number of aspartate residues, and one expects to find them at the ends of loops, so their pattern of arrangement is not surprising. Nevertheless, the electrostatic interactions between conformational electric moments, both local and global, must be quantitatively very significant so the arrangement of charges may be necessary to prevent distortion of the domain closure by an unsymmetrical field from the aspartates. In experiments with gross modifications in the charged groups of α-chymotrypsin, Shiao et al. [19] succinylated all amino groups of this protein and found only small changes in catalytic parameters and thermodynamic stability. Shiao and Sturtevant [20] found that nearly all acid and base groups of this protein had pK_a values and temperature dependencies identical within the error of their analysis with those of these groups in small molecules. Jolicoeur and co-workers [21] determined the heat capacity of several members of the chymotrypsin family as a function of pH and found results again similar to those obtained with the reference small molecules. Only hemoglobin demonstrates a significant involvement of charged cloud with oxygen-binding function. The ionizable groups of protein generally demonstrate linear enthalpy-entropy compensation behavior [83] (see Sec. VIII.B), and the compensation temperature is very close to the usual mean experimental temperature, 298 K. As a result, the free energy is often found not to vary significantly with pH. Much is hidden by this "accident." A classic example is the disappearance of the Bohr effect of ferrihemoglobins at 25°C. Experiments at 37.5° show a significant dependence of ligation equilibria on the degree of ionization of the many acid and base groups tethered to the surface. Since all compensation temperatures for the hemoglobin-myoglobin family are near 298 K, most of the potentially useful information is hidden in experiments carried out near this temperature. This same accident is quite general in other proteins. In part it explains the apparent simplicity of substrate and inhibitor reactions of the serine proteases. There are real puzzles in rationalizing the often simple pH dependencies of enzyme reactions with the complexity of the electrostatic interactions between the ionized groups and the gestalt field (Sec. VII.D).

Detailed examination of the placement of single kinds of residues and residues of classes such as all acid or all base promises to be profitable, particularly in understanding the residue arrangements necessary for global fluctuations.

The construction of phospholipase A2 is quite different and the knots do not so simply establish the loops. There are fewer of any single acid or base residue in this enzyme, so patterns of residue arrangement are likely to be more difficult to find.

6. Significance of Large *B* Factors

Group I *B* factors, although rarely as small as those in good small-molecule crystals, are nevertheless in the same range. Group II and group III *B* factors are much larger, and this reflects more the coordinated motions of groups of atoms, such as rotation of a section of peptide or *mobile-defect* rearrangements [22,23], than poor packing. There is considerable empty space in proteins even taking into account the internal fixed and mobile water molecules at normal hydration, but only a fraction of this space is available for local displacements. That part is the "free volume" and it plays the central role in all considerations of conformational dynamics. Its value in proteins is difficult to estimate but most estimates agree that the figure is 3–4% of the total volume. These estimates are based on homogeneous substances and may be somewhat high because of steric effects and the cohesive electrostatic force. HEW lysozyme has an increase in packing density of 4–6% on drying to 36 moles of water per mole of protein associated with many atom displacements of the order of 0.6 Å [24]. In homogeneous substances a free volume of a very few percent makes the difference between the liquid and solid phases. As in all liquid phases, the free volume in proteins is the determinant of mechanical and electrostatic properties and their fluctuations. The ice-water transition is a good model for protein matrices in that it demonstrates that free volume in such structured phases must be rearrangeable (mobile) to be useful. Ice is half empty space with virtually no mobility. In protein free volume it is responsible for the high *B* factors and thus for the machinery of protein function. Much of the remainder of this chapter is an examination of the ways this comes about and the magnitude of the effects.

C. Information from Proton-Exchange Studies

1. Permeability of Proteins

That there are only three *B*-factor groups, and thus three major substructures in the proteins we have examined, is not obvious from the *B* factors. Rather, it has been established by the finding in proton-exchange experiments that the site rate constants for exchange fall into three well-separated groups [5]. The *B*-factors and proton-exchange data give somewhat similar information but in different forms. The two are particularly powerful sources of information about conformational dynamics. *B* factors give average amplitude information directly and information about conformational dynamics indirectly. Proton exchange gives dynamical information directly.

The initial proton-exchange studies of Linderstrøm-Lang and co-workers established that proteins in solution are permeable to the catalysts for exchange;

protons and hydroxyl groups both carry at least one water molecule, there being no evidence that they combine chemically with protein groups, except in the proton-exchange process itself. Recent studies show that efficient exchange occurs in nearly dry proteins (0.08 g H_2O per g protein; called 0.08 *h*) and suggest that were sufficient catalyst ions present, mobility in entirely dry proteins is adequately high to support respectable exchange rates at all but group I sites. Hnojewyj and Reyerson [26] demonstrated water-catalyzed exchange at low degrees of hydration for most exchangeable sites but found roughly 20% of the sites in HEW lysozyme and similar small proteins did not exchange in many weeks. These findings indicate that catalyst molecules cannot get to some sites or that at these sites the transition state for exchange is difficult to form. Gregory examines this topic in Chapter 3 in connection with his demonstration that mobile water dissolved in proteins, though relatively small in amount, is critical for the gestalt behavior of conformations. Historically it has been accepted that proteins are hard, well-built solids with no internal space for mobile water. It is only in the last few years, due in major part to the work if not the interpretations of Klibanov and co-workers [25], that the existence and importance of mobile water have begun to be understood. Reexamination of x-ray–diffraction data, on the other hand, appears to reveal a large number of places to put water inside proteins. This is more consistent with a plasticizing effect of water indicated to be complete at about 0.15 *h* at 25°C. Although there is considerable variation among results reported for different proteins, most studies demonstrate catalytic activity comparable with that in free solution near 0.15 *h*. The long-range mobility necessary for catalysis arises either because of the plasticizing effect of dissolved water molecules or because the protein surface and consequently the protein interior is labilized by water held only at surface groups.

Internal hydration, plasticizing effects of water, glass transitions on a hydration-temperature diagram, and alternatives such as Careri's "proton percolation" proposal are reexamined with some reinterpretation in Chapter 3. Nearly all interpretations of hydration data ignore internal water and thus require revision. Many small molecules are known to diffuse into proteins, and there is no question about the ability of protein conformations to dissolve mobile water as well as the "fixed" water reported from x-ray–diffraction studies. What is uncertain is the amount of mobile water required for the onset of long-range conformational motility (vide infra). How much of this labilization is due to dissolved water and how much to the relaxation of surface constraints by water bound at the protein surfaces?

2. Preservation of Rank-Order in Proton Exchange

Woodward and Rosenberg [27] were the first to discover that the rate constants for slow and intermediate exchange sites maintain a fixed rank-order. This behavior is, of course, statistical, but it indicates that the fluctuational behavior of conformations is more nearly global than local (Sec. V.B.1). Woodward and Rosenberg employed the isotopic exchange method in free solution, measuring the tritium-to-hydrogen ratios through radioactive decomposition of tritium. Fortunately, NMR methods now greatly facilitate the acquisition of proton-exchange data by permitting detection of changes at individual exchange sites. The exploitation of such data to characterize the gestalt nature of conformational fluctuations is still primitive and still dependent on the constancy of rank-order preservation since it is the rank-order conservation that provides the basis for the enthalpy-entropy compensation behavior [5]. That is, both are consequences of the same fluctuations of conformation. In Sec. V.B.1 an explanation of the compensation behavior based on fluctuations along a single expansion-contraction coordinate is suggested.

3. The Discovery of Knots and Matrices

As already mentioned, in 1981 Gregory was able to extract the probability-density functions for the proton exchange rate constants by using Provencher's numerical method for Laplace transforms. With the isotopic exchange data for HEW lysozyme in free solution, he found the probability-density function (pdf) to have the three peaks [5] shown in Fig. 2. This characteristic of the pdf has been confirmed by proton-exchange data for some other proteins, but its generalization now depends primarily on the correlation of proton-exchange rates with the three *B*-factor groups discussed earlier and genetic-conservation data to fix group I residues.

The three-peaked exchange-rate distribution function immediately revealed the construction principle for this kind of protein. Before detailing the correlation among the three sources of information, the immediate deductions can be listed.

Although there is considerable overlap, the three peaks are clearly discernible since the positions of the peaks are well separated relative to the experimental standard deviation. In the absence of any exchange by actual melting (from states on the product side of the transition state for melting), the sites in the peak with smallest rate constant are characterized not only by their very small rate constants, but less subjectively by their manifestation of compensation behavior [5,28]. The compensation temperature for exchange of group I sites is 354 K ± 10 K. That for the sites in the group II peak is 450 ± 25 K. These same figures turn up in other kinds of studies of proteins, suggesting that they may be used to identify the kind of substructure altered in the study. However, the higher value found for group II exchange may not be an intrinsic characteristic of protein conformations. In particular, it is possible that the high value is due to the inclusion, in addition to the enthalpy and entropy changes in catalyst diffusion, of the activation enthalpy and activation entropy contributions from the chemical exchange process (see Fig. 16 legend for clarification). Evidence of other types suggests the true characteristic temperature is nearer 300 K (see Sec. VI.D.3.).

The fast-exchange sites, those of peak III, have high rate constants but have been successfully studied by Tuchsen et al. [29], who find them to be considerably smaller than expected from fully hydrated small models. We have called this class *surfaces* substructures (vide infra) and offer evidence for their importance as a basis of specificity in protein associations and tissue assembly in general (Sec. VI.B) as well as free-energy redistributions in multienzyme systems of all kinds.

The clear quantitative difference between the slow and intermediate sites indicated immediately that there were two distinctly different kinds of substructure in lysozyme, ribonuclease A, and BPTI, the three proteins for which the necessary data were then available. The slow substructures were called *knots* because the high activation energies indicated that they were responsible for the high kinetic stability of these proteins, a general property of proteins. The substructures are not knots in the topological sense; no such knots have yet been demonstrated in proteins. Rather the term was applied because knots must become "untied" in order for the protein to melt in denaturation. This use of the term has been much criticized but no satisfactory alternative has yet been suggested. Velcro is the best so far and it is actually in some respects a better term than knot.

Knots are solid structures, compact, hard, and difficult to distort. The cooperative contraction producing the knot is a single event characterized by a synergism of contraction and decreasing dielectric constant. The latter factor, as important as the former, is a consequence of the restriction of the motions of the polar groups in what is usually a preponderantly aliphatic and aromatic environment. The electrostatic potential energy of the knot is very low for both reasons. The aliphatic and aromatic groups lower the dielectric constant, just as does the oil in a power transformer. Attractive electrostatic interactions drop to lower potential energy as the dielectric constant drops, and repulsive electric interactions do just the opposite. The dispersion interactions in the oily clusters are always attractive and thus always strengthened by knot contraction. Interactions among the large dipoles of the peptide groups, especially via hydrogen bonding, make equal or greater contributions to knot stabilization. Since favorable orientation of these groups as in hydrogen bonding will result in low potential energy in a medium of low dielectric constant and unfavorable orientation will result in high potential energy, it is reasonable to conclude that the polar groups must be arrayed so as to interact attractively. Contraction of knots is limited by the van der Waals repulsive potential, which is minimized by very good packing so that there is little free volume at full contraction. Excellent packing is the hallmark of a strong knot although it may be more characteristic of parts of a knot than of the total. Again BPTI is the best example. The three "dispersion" pieces of the knot have very little free volume, but these are held together by the seven H-bond groups which, to achieve the best orientation, may require some free volume. But that free volume is not mobile free volume so it does not allow motions of polar groups that would increase the dielectric constant. The H-bond skeleton of the knot is exceptionally rigid and strong, but so are the dispersion pieces on full contraction. Although the exact atomic composition of a knot can sometimes be slightly obscure because not all atoms of a participating residue need be in a knot, it is clear that surprisingly small numbers of atoms are required to make very strong knots.

The sites with intermediate exchange rates (middle peak of Fig. 2) were classified as belonging to *matrices* in part because their low activation energies indicate higher motility and higher permeability. Again, however, the integrity of the class was considered to be established by their common compensation pattern. The preservation of rank-order requires linear enthalpy-entropy behavior, but the compensation temperatures for the two groups are well separated. Later *B* factors were found to support this assignment in terms of physical characteristics, and genetic conservation establishes that knot residues can be distinguished by a high degree of conservation. High conservation is sometimes a bit hidden by exchange of like-for-like residues, but as with BPTI, such exchanges are limited and clearly follow known similarities and small groups of residues. Atoms with group I (knot) *B* factors lying outside the knots on isolated residues can be attributed to crystal packing. The matrices are relatively soft, more like stiff liquids or rubbers than solids, but change with degree of contraction. They do not much resemble the conventional glassy state in fully hydrated proteins in their expanded forms but do so increasingly as the conformations contract in catalysis, for example (Sec. VI.D.3.a) or on large increase in protein activity coefficient. Because they lack the detailed packing of knots, on contraction they resemble porous glasses more than the solid, tightly packed knots. The motility characteristics are thus quite different. Matrices are deformable, and as a result knots can move as solid objects in matrices. In enzymes knots move as parts of functional domains, but the matrices of the latter adjust position to make closure possible. The softness of knots is a consequence of evolutionary choices for free-volume arrangement and rearrangement, but it is manifested thermodynamically by the equivalence of enthalpy and entropy. In knots, although the entropy is not zero, it is dominated by the large reduction in electrostatic potential energy that takes place in contraction.

The sources of kinetic and thermodynamic stability in knots have several molecular origins. The two knots of the mammalian phospholipase A2 enzymes (Fig. 6) are pure α-helices with little if any stabilization from interactions with the side chains, nearly all of which are small and none aromatic. The interhelix hydrogen bonds are of major importance in these proteins. The other helix knots we have found are complicated by the participation of the side chains, so this knot is especially important as a demonstration that the definition of a knot as a "hydrophobic core" is misleading. Clusters of hydrophobic groups are common in knots and have tended to obscure the at least equally important hydrogen-bond skeleton. In the phospholipases there are no oily subclusters, no hydrophobic cores. A pure van der Waals knot probably exists, but we have not yet found one. The many disulfide groups in this protein frustrate attempts to assign the importance of the factors responsible for thermodynamic stability. They may make major contributions to thermodynamic stability which are necessary because those provided by hydrophobic cores are absent or insufficient. That assumption would be in current fashion, but the positioning of the disulfide bonds in the Kunitz PTI proteins suggests caution, as is shown in the next paragraphs.

The knot of BPTI (Fig. 11) is quite different and much more like the knots found in the proteinases acid, for example. The primary knot consists of three

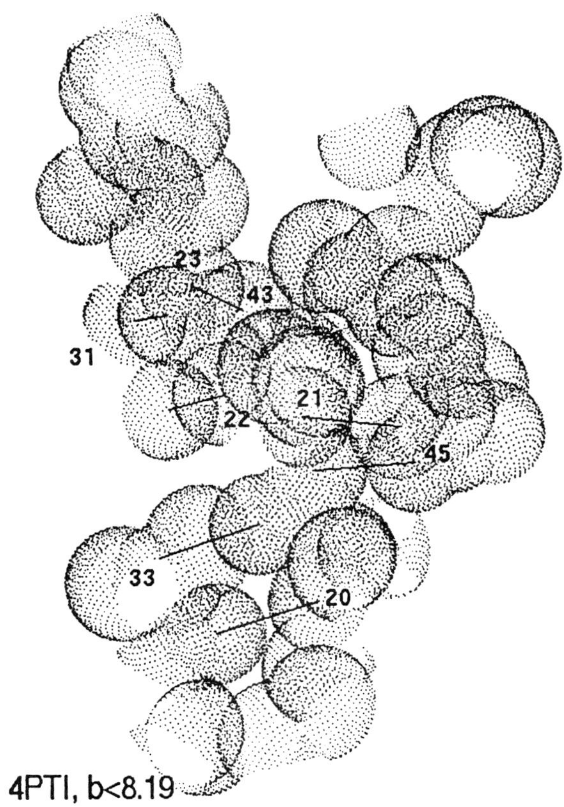

FIGURE 11 Details of knot for bovine pancreatic trypsin inhibitor. The knots of the Kunitz PTI family are exemplified by the bovine member, BPTI. The knot consists of three pieces with semi-independent folding integrity due to dispersion forces but achieving final integration by association through a double-H-bond structure. The latter are shown by the pairs of parallel lines. There is at least one other strong H bond (23–43) and there may be more. The H bonds derive their strength from the low dielectric constant. (The data are from 4PTI in the Brookhaven PDB.)

tightly constructed van der Waals pieces, largely aromatic, held together by a small number of exceptionally short, and thus strong, hydrogen bonds. These contain the highly conserved residues and, with the exception of three disulfide pairs, all those with group I proton-exchange rates. The protein has high thermodynamic stability and high kinetic stability although only about 62 non-H atoms of a total of 450 are in the knot. Judging from Levitt's [30] compilation of experimental information from the x-ray work of Diesenhofer and Steigmann [31], seven strong H bonds connect the pieces. The three disulfide bonds are themselves strictly conserved in the PTI family, and they form with the three pieces a palindromic assembly pattern in which the disulfides are incorporated as major determinants of matrix structures. The construction is diagrammed in Fig. 13 (p. 61). The palindrome is less exact than that found with the gene-duplication enzymes (vide infra). It is centered on residue 34 or 35 or both, and the two central disulfide bonds are not placed quite symmetrically. The construction makes it clear that although disulfide bonds may play some or all of the roles proposed for them, in this protein they are critical for matrix construction and thus for the inhibitory function. Removal of one disulfide by SDM was found to prevent stable folding, but considering the marginal thermodynamic stability of the matrix of these proteins, this is not unambiguous evidence that disulfide bonds are a major contributor to stability. The matter is discussed in more detail later. First a more detailed description of the folding process of the PTI proteins must be presented.

The knot parts reflect dispersion stabilization individually, but the parts are connected by a special kind of cyclic double H bond diagrammed in Fig. 12. This pairwise arrangement produces short, strong hydrogen bonds. It is another of the devices found in evolution, and it may be that all short, strong N-H-O bonds in proteins are of this type since short H bonds usually mean high proton polarizability. In this case, as in others elegantly discussed by Zundel [111], covalent character and bond strength increase as the N-to-O distance shortens. The double-

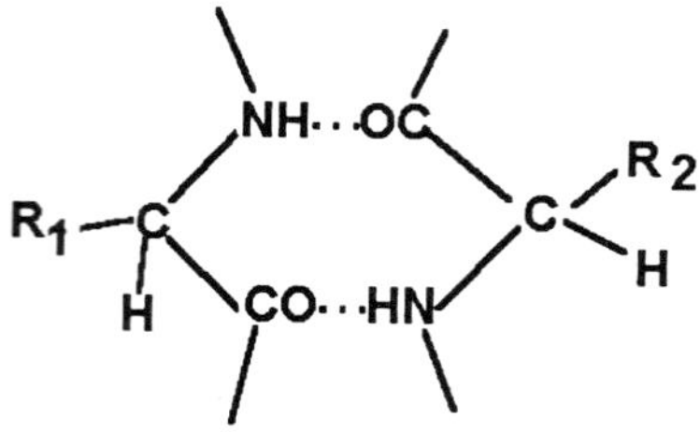

FIGURE 12 H-bond ring diagram, BPTI. This double-H-bond structure appears to be very strong in the PTI knots in contrast to its weakness in the normal antiparallel β-sheet. It appears that in a very low dielectric environment inductive delocalization of electrons through the α-carbon atoms produces considerable covalent character in the hydrogen bonds. It is probable that the ring is quite aromatic.

well potential progressively moves toward a single central well, reflecting an increase in pi bonding. In the ring structure the pi bonding produces some aromaticity. These conclusions based on short N-O distances and Zundel's rules require considerable inductive displacement of electrons through the α-carbon atoms and do not at first appear to be consistent with the H bonding in antiparallel sheet structures. The ring structures are the same but the bonds do not have any special strength. The only obvious possibility is the very low local dielectric constant at these bond pairs in the PTI knot. It is perhaps not surprising that we have thus far found no small-molecule examples of this special bonding situation, except perhaps the carboxylic-acid dimers. This is unfortunate since large delocalization through α-carbon atoms has major implications for secondary structures. In particular, it suggests that the strength and tightness of segments of such structure can be regulated by changing the effective dielectric constant. Thus helix segments in a low dielectric environment may have considerable electron cooperativity, yielding short, strong H bonds. In Sec. VII.C this reasoning is applied to explain the stability of the helix knots of the phospholipase A2 family.

The PTI knot construction illustrates the synergism of dispersion forces and hydrogen bonds mediated through the dielectric constant. In the compaction step of folding, the van der Waals pieces gain strength as they contract toward the repulsive potential and in so doing lower the prevailing dielectric constant not only inside the pieces, but in the regions between any pair of pieces where the H bonds form. As the latter become stronger, they draw the pieces together thus lowering the dielectric constant even more. Contraction continues until the contraction forces are balanced by the repulsive forces. There is also a limit on entropy reduction, which is probably set by electrostatic theory (vide infra). Real knots achieve varying degrees of compromise between fitting and strength, but very great strength is obviously possible with fortuitously exact assembly. Matrices are constructed to prevent extensive contraction so free volume is adjustable as required by function. Entropy is determined by free volume for matrices, and matrices need not be stable if knots are sufficiently stable to carry the burden.

The thermodynamic stability of knot-matrix proteins is a consequence of knot size and cooperativity. The aggregate knot is so built that it can decompose only as a unit, and this is, of course, the basis of the kinetic stability. To recapitulate: proteins functioning at ordinary temperature have small knots, between 10 and 20% of the non-H atoms. Proteins operating at higher temperatures can be expected to have larger knots and perhaps even less mobile free volume. Thermolysin has a large, complicated knot that has been extracted intact from its matrix [32]. There seems to be no reason why enzymes cannot be found that can function well above the boiling point of water. The keratins and some other strictly structural fibrous proteins probably define the upper limit reasonably well since they appear to be all knot. Proteins with physiological function have low thermodynamic stability because they must have soft, relatively weak matrices to support function. Kinetic stability makes this possible but it is the construction of the matrix to have high free volume and thus high entropy, which prevents the collapse of a matrix toward the contracted (low enthalpy) end of its coordinate. The latter is also partially a consequence of the strong knot, since the molecular and thermodynamic state of matrices is determined not only by fitting details worked into the polypeptide forming the matrix but also by the constraints placed on the matrices by the knots. Matrices are not in their intrinsic (lowest free-energy) states because of the constraints. The constraints arise because the whole protein is thermodynamically most stable when knots are strong and matrices weak. Some reliable estimates of the division of the standard free-energy change in folding into matrix and knot contributions are given in Sec. III.B This useful thermodynamic device may have been intrinsic to the knot-matrix construction when it was discovered, but that does not seem likely. Rather, one suspects that it was a later addition and thus to be considered a separate device.

D. Information About Groups from Evolution and Genetics

1. Genetic Conservation of Knots

Perhaps the strongest support for the knot concept comes from comparisons of the sequences of proteins from the same family. So long as the proteins have a common ancestry, their knots are the same or very nearly so since the few such residue differences we have so far found are like-for-like exchanges, such as tyrosine for phenyalanine. Insertions and deletions produce large changes in matrices and surfaces, but in enzymes demonstrating palindromy, the spacing of the knot residues along the peptide in the two functional domains must be quite exact so major differences in conformation of the two functional domains are in the outer regions of the loops.

Gregory and Lumry [5] noticed that knot residues were coded by the most highly conserved exons. The lowest *B* factors are in these residues, and all have proton-exchange rates that are in the group I exchange group.

The invariant residues are at present the best description of a knot and provide a basis for examining the details of construction and other important characteristics. Less superficial examination may require extension of this kind of description to include a few atoms from adjacent residues. This seems unlikely since no obvious candidates have thus far appeared, so for the present, the term "knot" refers to the substructure made up of the genetically conserved residues. Another example is the family of phospholipase A2 proteins. More than 60 of these have been sequenced and all demonstrate the same high degree of knot-residue conservation [33]. In still another example, Seizen et al. [34] found knot conservation in the 40 members of the subtilisin family that had been completely sequenced by 1991. Comparative sequencing has been popular for many years, so there is an abundance of information about knots only as far away as our computers. Of course, comparisons should be made with the *B* factors whenever possible, but in our limited experience the lower *B* factors are always found in the conserved residues. The pictures constructed using coordinates from x-ray–diffraction studies are also thus far entirely consistent with the group of conserved residues in that they show knot residues to be in closely integrated clusters. However, as in the subtilisins, satisfying the palindromy requirement may have resulted in splitting the knot in one or both functional domains into two or perhaps even three large pieces. Palindromy is not consistent with neat knots and probably as a result not always consistent with strong knots. The gene-duplication enzymes appear to establish that good palindromy is more important than very high knot stability.

A few examples of the same conserved residue set appearing in proteins with different function have already appeared. For example, the Protein Sequence Database reveals a few occurrences of the PTI knot in proteins not usually classified as proteinase inhibitors. The special conditions a knot sequence must have to be successful and the fact that knots are highly conserved and matrices are not seem to establish that the discovery of a new knot is a rare event in evolution. However, the term "rare" is definable only by teleology and thus not really defined at all. The test will be to find many instances of the same knot in proteins with different function and thus usually also with different matrices. The finding of a new combination of old knots from old functional domains to support a new function by new tailoring of the matrices of the old domains appears a more attractive probability than finding a new knot for that purpose. The supposition that only a few knots have thus far been discovered seems to be consistent with the suggestion of Dorit, Schoenbach, and Gilbert et al. [35] that the number of different nucleotide sequences in DNA is no more than one or a few thousand. Perhaps there are so few knots because there are so few DNA sequences or there are so few DNA sequences because so few knots have been discovered. We shall see. Probability considerations are particularly puzzling if the two knots of a new enzyme must be found more or less simultaneously, as is perhaps suggested by the fact that the main chain of the HEW lysozymes passes between the knots twice rather than the usual single passage.

2. Implications of the Palindromy for Enzyme Mechanisms

Finally we return to the problem that had to be solved in the subtilisins, which is in fact a general problem for single-chain enzymes with functional domains of common origin. The knot residues make two different knots with the same sequence in the same direction along the polypeptide in such a way as to generate the *B*-factor palindromic pattern. The functional domains are connected head to tail but the palindrome pattern has the symmetry of tail-to-tail associations. The A2 phospholipase (Fig. 6) knots take the easy path to this end. Apparently for such purposes as the palindrome the two helix knots are suitably palindromic even though they have the same direction in the two functional domains. Thus the major knots are both pure helix segments and, although not palindromic since both are right-hand helices, are sufficient for function. This protein appears to establish the minimum construction required for enzymic catalysis in gene-duplication enzymes. Antiparallel helices and similar antiparallel sheets satisfy the similarity requirement.

The subtilisins lack such simple knots so the symmetry requirement is satisfied only by awkward and imperfect knots (Fig. 7). BPN′ and proteinase K are very distant along their phylogenetic tree and have lost most sequence homology except in their knots. Only 70 of 325 residues are conserved, and these not always at exactly the corresponding positions in the two functional domains. The residue sequences of the two are not palindromic but quite close *B*-factor palindromy exists in both these representatives of the subtilisin family and the patterns of the palindromes are very similar. The *B*-factor data (Fig. 10) show small displacements at two positions.

Apparently an important aspect of symmetry is that the functional domains be able to associate in such a way as to generate a pseudo (noncrystallographic), twofold rotation axis running between them and bisecting the two primary functional groups. Because the knots are not structurally palindromic, these axes are not exact. The phospholipase A2 proteins and the HIV-1 type of enzyme have exact axes for two different reasons. The pattern thus far is ubiquitous in the gene-duplication enzymes, which suggests that it is the goal in evolution of modern enzymes and likely to be the central feature of their mechanisms. There appears to be a strong conservation requirement for form in matrices but not for residue composition. In knots residue conservation is more critical although a few enzymes in which conservation of knot residues is actually less than that of matrices have appeared. However, even in these, most deviations from conservation are due to exchange of similar residues. Conservation of primary functional groups appears tight, but there are enough exceptions to suggest that the requirement for specific catalytic residues in enzymes is weaker than that in conventional small-molecule catalysis.

It must have taken an extraordinary length of time for the palindrome to develop, and this is all the more remarkable since the HIV-1 protease satisfies the requirements by breaking the long peptide chain of the single-chain acid proteases such as pepsin at the hinge point. The two protein subunits are free to associate so as to satisfy *B*-factor palindromy (the twofold axis) without tedious mutation since the break allows residue palindromy. That is, the functional domains are identical, and because they are on separate chains, chain direction and the palindromic requirement are not in conflict. So far as is yet known, evolution is totally inefficient so there is some suggestion that the retrovirus proteins such as the HIV-1 protease and the Rous sarcoma protease, formed using the same principle, represent a long search for a simple solution to the palindrome requirement. But before one is tempted to suggest that some of the behavioral abnormalities of the retroviruses are a consequence of this kind of structure, it is well to note that many repressor proteins have identical subunits and use the same knot-hinge connecting device.

3. Genetic Conservation in Surfaces

Although it is already apparent that surface residues as defined by peak III in Fig. 2 are highly variable from one use of a functional domain to another, we have not yet explored this problem in any detail. However, the palindromic *B*-factor pattern of the aspartate residues in rhizopepsin suggests that a considerable degree of conservation of residues at the protein surface and at least rough palindromy patterns of charged residues may be essential for some proteins. Insofar as surfaces are in contact with the surroundings of the protein, they have been evolved as useful bridge structures. A special kind of surface is that between the two functional domains of enzymes. That is, we can distinguish surface residues that communicate with the world exterior to the knots and matrices, one and surface residues that allow functional domain of a domain pair to communicate with the other. These are probably very similar, if not identical. We suspect that both are critical in tailoring conformational fluctuations for specific functions from protein-protein association to catalysis, and in Section VI mechanisms for several such functions are proposed.

E. Information from Density Data

Several investigators, particularly Richards [36], have found higher densities in regions of proteins that can now be identified as knots. Hence density determination is another way to find knots and add descriptive information.

Because knots often have a high proportion of alkyl and aliphatic residues, they can have high number densities. The tighter these are packed together, the more effective they are in stabilizing knots. As a consequence the atom-number densities of knots should be high, and that appears to be the case.

III. HOW SUBSTRUCTURES DETERMINE GESTALT STRUCTURE AND PROPERTIES

A. Genetic Stability

Thermodynamics preserves the secondary structures of polypeptides; DNA preserves the successful results of evolution. The first obtains in a cold world, the second only in a biosphere. The "myoglobin fold" is a good and probably very ancient example of a major evolutionary discovery. Knots are an even better example since there are quite a few different ones and much more versatility in their modular assembly for new proteins with new or more efficient functions. By comparison with modern physiology, biological mechanisms must have been very crude or at least inefficient before some minimum threshold number of knots was discovered. It is likely that the transition from a knotless biology to a knot biology was revolutionary, the more so because it was accompanied by the discovery of the use of functional domains in pairs, what we call the "pairing principle."

The degree to which the conservation of matrix substructures is dependent on knot conservation is by no means clear. The BPTI class of knot obviously acts as a broad tether since it enforces association of several well-separated sections of polypeptide. The primary phospholipase A2 knots are formed from contiguous residues and as a result would appear to be poor at tethering a protein together. The palindromic pattern extends over many residues and is quite precise (Fig. 8), but the knots achieve intimate contact with matrices only through a network of disulfide bonds. As in BPTI, it is probable that the latter is essential to shape matrices and tether them to knots. Knots consisting entirely of helix are common, e.g., the N-terminal knot of RNase A, but these usually have low-*B* atoms in side chains. In this case the knot has an arrangement of charged groups exposed to solvent that may contribute to knot stability [37]. Most studies of the stability of helix knots are carried out on the excised knot. Richards' exploitation of this knot in separated state has been very productive, but the gestalt nature of protein electrostatic fields may be the major basis of stability (see Sec. VII.D).

It seems that given a preexisting knot, palindromy is a condition on matrix substitutions in enzymes associated with the preservation of function or the enhancement of efficiency in function. In contrast to knots, matrices are subject to evolutionary experiment. A reexamination of the history of protein evolution using knots rather than function to define the trees should improve our knowledge of the genealogy of the biosphere. This application of the knot-matrix paradigm may also reveal trends in evolution not yet detected.

B. Kinetic Stability

1. Knot Stability Determines Kinetic Stability

It is now generally appreciated that to support most physiological functions proteins must have relatively soft parts and thus have less than the maximum stability available only in highly structured proteins like the keratins. Judging from the small, very strong knots of PTI proteins and similar inhibitory or activator proteins, extremely strong and stable proteins are possible. Indeed the keratins have these properties. But for many proteins the thermodynamic stability is low, so the population of denatured species is not negligible. Aggregation and proteolysis are serious hazards in such a situation, but from the point of view of function there is a serious lifetime problem. It is necessary that an enzyme, for example, remain in its native active state long enough for its catalytic function to be completed. Evolution found a way to minimize all these problems by making the activation free energy for denaturation very large, much larger than the overall free-energy change, thus making the lifetime in the native state very long. This result is termed high *kinetic stability*. Biltonen found that the rate and equilibrium data obtained for denaturation of chymotrypsinogen B by Eisenberg and Schwert could only be explained in terms of a transition state very similar to the native species conformation [9,38]. The heat capacity of activation was negligible, the entropy of activation a small fraction of the expected total configurational entropy change in equilibrium melting, but the enthalpy of activation was very large. The activation process is not clearly shown in Fig. 1 to be an expansion-contraction process. At the time the data did in fact suggest this, but not unambiguously, so the activating conformational process was originally called a "subtle change." The data did not reveal how much of the protein is involved. The discovery of knots has removed some of the uncertainties shown in Fig. 1.

The discovery of the knot has also made it clear that the transition state for melting of most knot-matrix proteins is the expansion of knots to the point that their cooperativity is lost. A variety of data supporting this important conclusion are available. Most are based on more extensive studies of the rate parameters themselves. The most extensive studies of the kinetics of denaturation are those of Pohl, and much of his work was reported only in his dissertation [39]. Pohl's study of denaturant effects on several members of the trypsin-trypsinogen family and their acyl derivatives was the first of this kind. In particular, he found large decreases in activation enthalpy and entropy in 8 *M* urea but no activation heat capacity, results indicating that the effect of urea on the melting rate is due to modifications of the native state and these are maintained in the transition state. Pohl's extensive studies of this kind [40] confirmed Biltonen's deductions for denaturation processes that conform to the two-state model. Pohl also confirmed the three-state model for chymotrypsin [41] suggested by Brandts et al. to accommodate *cis*-proline isomers. The three-state model is often required for melting rate processes in which molten-globule intermediates appear. The latter are usually enzymes with one functional domain melted (Sec. III.D). However, since the latter add complexity without much enlightenment, only two-state denaturation processes are discussed in the remainder of this section.

More recently, Segawa et al. [42] carried out studies of HEW lysozyme that gave good agreement with those of Biltonen and Pohl. They found the rate parameters to be independent of pH, and in subsequent studies Segawa and Sugihara established that neither strong denaturing cosolvents nor cross-linking in the crystal altered these quantities to any large degree [43]. They found that amphiphiles known to diffuse easily into the protein somewhat decreased the activation free energy as their concentration was increased. Propanol was particularly effective and apparently diffuses into many proteins easily and in large concentration, behavior reminiscent of that of ethanol. These results establish that for HEW lysozyme the position of the transition state along the reaction coordinate for melting is independent of temperature, cosolvent, and pH, but the degree of relaxation that occurs at the transition point is decreased because denaturants in effect swell the matrices of the native state. Knots should be impermeable and little altered by the several effects studied by Pohl and Segawa et al. Chen et al. [44] carried the identification with knot disruption one step further in demonstrating that the kinetics of melting of a T-4 lysozyme was the same in low-temperature melting as in high-temperature melting. Thus the activation free energy was constant and the entropy of activation nearly zero and only weakly temperature-dependent. The heat capacity of activation was 20% of the overall value and not zero, but the experiments had to be carried out in 3 *M* guanidine HCl so the significance is not clear. However, the most important point is that the activation enthalpy remained positive and large even though the overall enthalpy change became large and negative at low temperatures. With HEW lysozyme in the higher-temperature thermal denaturation, Segawa and Sugihara [45], using Pohl's three-state model to correct for a maximum of 10% of the third species, found for unfolding a constant activation energy of 47.8 kcal/mole and a negligible heat capacity in pure water and in several denaturing solutions, including guanidine HCl up to 3.2 *M*. The activation free energy is a good estimate of the cost of conversion of knot to matrix. The constant activation enthalpy is a useful number since it is a close estimate of the potential energy of stabilization of the two knots and thus a reference point on which to base estimates of the complete set of activation free energy, activation enthalpy, and activation heat capacity for the unfolding and folding rate constants. Since the data show that matrices do not become relaxed until the transition state is passed, the values of these quantities for the unfolding rate constant give information about the knots and the values for the folding rate constant give information for the matrices.

There are several other studies of the kinetics of protein denaturation, e.g., that of Gekko and Timasheff on chymotrypsin in glycerol solutions [46], but the total number of studies is small and the results often so different as to clearly establish that different proteins respond in different ways to denaturants, stabilizers, pH changes, and so on. Although preliminary attempts to compare results have been reported [14], not much will be clear until this kind of study, perhaps the most profitable that can be made on the denaturation process, becomes routine.

Segawa and Sugihara [47] also demonstrated that the binding site for the much studied competitive inhibitor, $(NAG)_3$, to HEW lysozyme does not exist in the transition state. This molecule is bound by the equilibrium native species, which is then stabilized with a decrease in the unfolding rate. Thus the binding ability of the protein exists only on the native side of the transition after the system on refolding moves past the transition state. Similar experiments on other proteins show the appearance of native properties only in the segment of reaction coordinate between native and transition state. Recently the study of "protected" proton-exchange sites provides the same sort of information although the logic is somewhat different in that the relaxation process between native and denatured states is approached from the denatured side [48]. With careful timing the sites that exchange most slowly in the native species can be selectively labeled. Protected-exchange studies can locate knots in this way and show the dual role of the transition state but only with considerable difficulty. In most applications the method is an elaborate substitute for direct measurement of the forward rate constant in melting, but it can provide more detail relating knot construction to its melting process.

Woodward and co-workers have used measurements of the fine details of hydrogen-exchange processes of knots to confirm for BPTI that the knot, in current parlance, the "hydrophobic slow-exchange core," is the source of the thermodynamic stability of the native state [49]. Specifically they replaced all knot residues one by one with alanine and found that the pattern of change in the activation free energies for exchange was paralleled by the standard free-energy changes in equilibrium melting. The two kinds of experiment must reflect a common origin in either the native state, the transition state for melting, or some combination. We can suggest further resolution by noting that although the matrices of the mutants can be somewhat perturbed by residue changes in the knot, the effects are not large and certainly not thermodynamically large. As one consequence, the transition-state position along the melting coordinate is nearly the same for the wild type and its mutants. This being so, the change in overall free energy of folding must be attributed to changes in the formation free energy of the native species. The experiments are rounded out by the somewhat earlier reported finding of Segawa and Kume that with HEW lysozyme there is no correlation between matrix dynamics measured by proton exchange of matrix sites and protein stability [50]. Gregory et al. [52] suggested that exchange at knot sites of HEW lysozyme in the presence of glycerol added as a cosolvent is due to compaction of the matrix about the knots, a consequence of increased protein activity coefficient and reduced free volume. Knots can be perturbed in a few other ways (vide infra).

The large overall heat-capacity change on product formation and most other peculiarities of the overall denaturation reaction take place on the product side of the reaction coordinate. Many disulfide rearrangements occur in this part of the reaction coordinate but some may take place on the "native" side of the transition state (vide infra). The latter would be a sort of "permanent waving" of matrices as they come into equilibrium with the newly formed knot. The disulfide bonds between knots and between knots and matrices of the phospholipase A2 enzymes are so simply structured as to suggest that they are formed after the knots. Such adjustments can be confusing since their formation, even if it makes only a small contribution to overall folded stability, will complicate interpretations of rate data. The principle of microscopic reversibility in such a case requires that the disulfide bond system of the native state be disrupted before the knot. This will give the impression that the disulfide bonds are the major source of stability. The disulfide system of the phospholipase A2 proteins presents some hard problems for nature. On the other hand, that of BPTI, already discussed, suggests that final disulfide rearrangements occur between the formation of the pieces of the knot and the fusion of the pieces into the knot. Itah and Haas [51] and Woodward et al. [49] show that the pieces are formed very early in the folding of this protein. In view of the magnitude of the dispersion interactions in these pieces (vide supra), it is not surprising that they have considerable independent stability.

The "thermodynamic principle" as the basis for protein folding was elegantly established by Anfinsen and seems to be in no disarray. The early history of this principle is interesting and no longer well known since it does not appear in textbooks. We describe it here briefly.

The exchange reaction between disulfide groups and sulfhydryl groups was discovered and characterized at least by 1933 [6]. In 1936, Mirsky and Anson had demonstrated that heat-coagulated proteins could be "uncooked," perhaps the most important experiments in the history of protein chemistry. Anson [53] seems to have understood, on the basis of the reversibility of protein denaturation, that minimization of the global free energy determines folded conformation, but he did not state this explicitly perhaps because the placement of disulfide bonds may not have appeared to follow from this "thermodynamic principle." Whatever the case, an explicit statement of the thermodynamic basis of protein folding was not made until 1953 [4], and any uncertainties about the ease of disulfide-sulfhydryl exchange processes in proteins were finally eliminated by Kolthoff and co-workers [54] in a series of papers beginning about 1958 showing, among several important but little appreciated discoveries, that disulfide bonds rearrange rapidly as they are reformed by oxidation or bisulfite-catalyzed exchange. Working with bovine serum albumin, they were able to reform all 17 of the disulfide bonds of this protein. A variety of products with varying numbers of disulfide bonds was formed, but the amount of native protein so regenerated could not be determined because it was not found possible to make an antibody for that protein. This crucial uncertainty was later resolved by Anfinsen and White using RNase A, but the work of Kolthoff et al. established that even with complete disulfide-bond formation, rearrangements to new partners occur rapidly and without much change in free energy. Once the disulfide arrangement problem was understood, there was little doubt that free energy minimization determines folding. Chaperones catalyze the folding inside cells to replace water as catalyst, but thus far there is no evidence that folding to the native species is other than spontaneous.

Because of the discovery of knots and matrices, "Anfinsen's hypothesis" needs to be restated in more detail. We do so in Sec. III.C.3.

2. The Thermodynamic Phase Relationship

The rate of change of enthalpy with advancement along the reaction pathway for denaturation exceeds that of T times the entropy change until the transition state is reached and then equals it at the transition state and from thereon to product falls. The pattern is not always quite so simple because the contributions to changes in H and S from the features of the actual melting process vary from protein to protein and with experimental conditions. The ratio of rate of enthalpy change to rate of entropy change has been called *the thermodynamic phase relationship* [9]. It has had some useful application in attempts to relate thermodynamic changes in denaturation to molecular information. It could be used more effectively to explain deviations from two-state behavior. Generally protein denaturation processes are closely approximated by the behavior of a two-state model, an appropriate model for weak first-order phase changes. When deviations from that model are found, the thermodynamic phase relationship can often distinguish between real and only apparent deviations. In some cases it may be able to distinguish between knot-matrix proteins and proteins constructed using another construction principle. This becomes somewhat more obvious when one notes that the changes in the phase relationship listed above are an immediate consequence of knot-matrix construction.

It has just been observed that so long as knot cooperativity determines the transition-state position and does so because of the removal of constraints on matrices, the positions of the transition states along the reaction pathway for mutant families of a single kind of protein will tend to be the same or very similar when only knot residues are exchanged. Variations in activation free energy for melting, which then measure changes in the ground state of the native protein, can then be large, often so large as to preclude stable folding. Matrix mutations, in contrast, will have a small effect on the activation free energy but can have large effects on thermodynamic stability. Thus, as already mentioned, the determination of the activation free energies for mutant series not only provides generally reliable information about knot mutations, it also provides quantitative information about the construction of the knots.

C. Thermodynamic Stability

1. Relationship Between Kinetic Stability and Thermodynamic Stability

The material in the previous section not only establishes the central role of knot melting in kinetic stability, but rather thoroughly establishes knots as the principal source of thermodynamic stability in the many proteins constructed using the knot-matrix principle. In both cases the unit of stability with these proteins is the functional domain. Whenever the thermal denaturation of a functional domain conforms closely to the two-state model, and the transition state is that species in which knot cooperativity has just been lost, the species appearing after the transition state are thermodynamically unstable. This is a consequence of the fact that the model does not allow any species past the transition state to be more stable than the transition state. Only one product state is permissible in the model, and if the model is an accurate fit to the data, only one product state is possible, although that single macrostate may be very complicated by its composition of microstates. Intermediates important in the kinetics and thermodynamics can only be important if their formation free energies are virtually identical with that of the major transition state. This is rare in small-molecule reactions but not so rare in proteins because solvent conditions, residue exchanges, and even temperature, pH, and pressure can change relative formation free energies. Multiple transition states are rare in protein folding under normal conditions, that is, physiological conditions and no denaturants or stabilizers. Although great amounts of time and money are being spent on the detection and characterization of such intermediates, they add little to the understanding of protein stability. A possible exception is some forms of what is called the "molten-globule" state. These are the multidomain proteins with at least one intact and one melted functional domain (Sec. III.D). The *cis-trans* isomerization phenomena of Brandts often further complicate the situation. Incorrect isomer incorporation in folding produces metastable states that can be very long-lived and thus are of some use to nature in the regulation of protein activities. Proline residues are usually found in matrices and surfaces, and these are sometimes sufficiently soft to accommodate the wrong isomer in a folded species. Matrices and surfaces appear to be generally quite tolerant of residue alterations made by nature or by humans insofar as they find ways to allow knot formation. Knots are much less tolerant in this respect, but results such as those of Feng Tao and K. Sun et al. (see BPTI discussions) show that some kinds of changes do not destroy knots but only reduce their stability.

In contrast to the stable knots, large variations in residue composition and conformational properties are required to tailor matrices for specific functions. Such developments would have been much more limited had matrices been required to be thermodynamically stable. It is a frequent assumption that energy or free-energy minimization in all parts of a protein is necessary to establish folded stability. This is a reasonable deduction if one assumes that protein conformations and construction are homogeneous and that stability is marginal and possible only with near total maximization of the stabilizing interactions. The net folded stabilities of most familiar proteins are of course not great, roughly 10 kcal/mole, but this low stability is a result of natural selection rather than intrinsic to polypeptides. A major reason for this selection becomes apparent in the knot-matrix construction. It provides a way to achieve the adjustability required of matrices to achieve necessary physiological functions without limiting choices by requiring that matrices be thermodynamically stable. The knots take care of overall stability, and once knot constraints are removed, matrices can collapse into their own *intrinsic* states of minimum free energy quite different from their state in the folded protein. At least the intrinsic state is a respectable virtual state even though it may be actually established only at low temperature or not at all. To be useful in function, matrices have to have conformational properties inimical to maximum stability. The knots carry the thermodynamic responsibility, and that is a major reason for the original choice of the term "knot."

This division of labor between knots and matrices is illuminated particularly well by data from studies of dehydrated native (see Chapter 3) and denatured protein films. For example, the rates of proton exchange from matrix sites in dry native proteins are apparently catalyst-concentration limited rather than catalyst-diffusion limited. Those in dry films of denatured proteins appear to be very slow, although the data are sparse. Dry native proteins are apparently somewhat porous rubbers with considerable short-range motility. At about 0.15 h long-range motility sets in and with it catalytic function in the case of enzymes. It is almost certain that the onset of long-range motility is a result of the plasticizing action of water molecules dissolved in the protein. It is an unfortunate theme in protein chemistry that the assumption that there is negligible mobile water dissolved in proteins has been used to provide inaccurate guidelines in drawing inferences from experimental data rather than being tested by the latter for its own reliability. But even then, so much subjectivity is allowed by protein data that the assumption probably could not be rejected, as, of course, it was not.

In any event, the number of dissolved water molecules required for long-range motility is very difficult to determine. Gregory and Pethig give examples in Chapters 3 and 4. The short-range motility is conserved on drying of native proteins and supports diffusion for proton exchange but cannot support normal functions such as catalysis. Dry denatured proteins, on the other hand, are amorphous solids rather than aperiodic crystals. They certainly lack long-range motility, and short-range motility seems also to be limited since proton exchange in nearly dry films is very slow.

2. Limitations on Rate Studies of Folding Mechanisms

When the equilibrium denaturation process for a protein conforms closely to the expectations of the two-state model, only three species of reactant or product have practical importance. Knot-matrix proteins generally follow the two-state model, and the transition state is usually the point at which knot cooperativity is lost. In such cases kinetic stability is easily explained on the basis of the quantitative differences in free energy between equilibrium native species and the transition state. There do not appear to be major intermediates in the activation processes, so sophisticated questions about thermodynamic and kinetic stability as they depend on these two states can be asked and answered. The problems restricting full understanding of thermodynamic stability lie on the product side of the transition state. The many metastable intermediates along the many paths from denatured state to transition state will remain a source of ambiguity. Fortunately such detail is irrelevant once the transition state has been found and characterized, as it has in this chapter for some knot-matrix proteins. The variety and variability of the product state are the sole source of the remaining confusion. To ask and answer questions about thermodynamic stability the products must be much better characterized under experimental conditions of interest than they are now. Fortunately measurements of the pure rate constant for formation of the transition state from the native state provide most of the information needed or even useful in the absence of detailed knowledge of the product state. Molten-globule states (vide infra) are of major concern when they have functional utility or complicate the analysis of stability data, but they and other major metastable species, primarily because they are sufficiently long-lived for thermodynamic studies, provide opportunities for better quantitative description of protein states and thus for better molecular understanding of the folding process.

Most equilibrium and virtually all quantitative rate studies of protein denaturation have utilized knot-matrix proteins. The hemoglobins are an exception, but the data are likely to become much more illuminating when they are not forced into the same mold as the knot-matrix proteins. The immunoglobulins as representatives of still another protein class are in the initial states of such investigations.

3. Further Characterization of Denaturation

Many old and new viscosity data, somewhat newer Raman data [55] and, more recently, chromatographic [56] and NMR data establish that the soluble denatured states of some proteins, with or without disulfide crosslinks in water in the absence of denaturants, have about twice the volume of the native state. However, these apply to ambient temperatures, and even more recent studies [68] suggest that expansion will generally occur as temperature is lowered. Migration of water into the protein, which allows matrices and knots to expand, involves removal of water from the bulk phase, the positive enthalpy of interaction between internal hydrophobic groups and water, and replacement of peptide-peptide interactions by water-peptide interactions. The first and last tend to cancel each other and the second makes a significant positive enthalpy contribution. The major entropy contributions to the free energy are from mixing of water and the melted protein and an increase in the configuration entropy of the polypeptide, both positive. Although the values of the several contributions have as yet been given no reliable estimates, the aggregate change in free energy is probably small because of considerable cancellation of positive and some negative contributions. What remains in the total expansion process is the increased exposure of hydrophobic groups to bulk water. The contribution to the free energy is then the positive reaction-field enthalpy change as discussed by Lumry and co-workers [103]. The remaining part of the enthalpy change and the large entropy change, both negative, tend to compensate each other but do so to a smaller and smaller degree as temperature drops. Then insofar as the exposure of hydrophobic groups determines the sign of the free energy, the netlike denatured state will tend to expand as temperature is lowered and contract as it is raised. The dominant factor appears to be the negative entropy change in the hydrophobe-water interaction, but the transfer of peptide groups from water to protein stabilizes folding at physiological temperatures. The data from small-molecule models in transfer reactions suggest that protein stability is considerably dependent on the poor solubility of the peptide bond in water, perhaps as much as the poor solubility of hydrophobic side chains in water.

In recent years peptide-group solubility has received less attention than hydrophobic hydration. Interesting recent evidence of its importance is the study of the effects of osmolytes on protein stability. Cosolvents like sarcosine and sucrose prevent freezing but also stabilize proteins at subzero winter temperatures, thus contributing to the preservation of life at such temperatures. The data obtained by Liu and Bolen [57] for transfers of peptide-group models and models of hydrophobic side chains from water to osmolyte solutions showed much diminished solubility of peptide-group models but not hydrophobic side chains. Although there are still no reliable ways to apply model data quantitatively to protein stability problems, studies of transfer models for peptide groups have always suggested that protein folded stability may depend as much on the poor interactions between water and peptide groups as on those between water and hyrophobic side chains (see Sec. VII.C). In any event, the peculiarities of protein denaturation at lower temperatures are largely those expected from the melting of matrices in a bulk-water environment. Matrix effects in denaturation are mostly solvation related and dominate the overall free-energy change in denaturation only at temperatures generally below the TMD of water. Nevertheless, at higher temperatures, their contributions to the enthalpy, entropy, and heat-capacity changes are large and obscure the true basis for free-energy change. In the formation of the transition state for two-state denaturation, the knots are converted to matrices, and this process is much simpler, having relatively little dependence on temperature and solvent composition. "Cold denaturation" near the TMD is easily rationalized using the abnormal features of hydrophobic hydration in the cold region, but the true story is probably more complicated since, as Grunwald demonstrates, the abnormal properties cannot contribute to the free energy of solvation of solutes unless a new species of solvent is produced by the solute. Grunwald has treated hydrophobic hydration on the basis of the two macrostates that exist in pure water so there is no free-energy contribution from perturbation of the populations of these states. The extended treatment will be given in Chapter 9.

Cold denaturation near 50°C has been found in at least one protein found in thermophilic bacteria at vents in the ocean bottom [58]. This is not so easily explained and may require increased sophistication in our models for proteins. Explanation of peculiarities of the familiar "hot" denaturation process at high temperatures may also require a more elaborate or at least a more correct model.

There are two phenomena in thermal denaturation reactions near the boiling point. The first is the well-known convergence temperature for the standard specific heat capacities of denaturation discovered by Privalov and co-workers and much discussed since then. The second is equally puzzling but not at all well known. It appears in the compensation plots of the activation enthalpy versus activation entropy reported for the unfolding rate constant in denaturation. Rather, one should say it appears from the fact that these data demonstrate linear compensation behavior (Sec. V.B). Pohl plotted activation enthalpies and entropies for melting of trypsin, chymotrypsin, their zymogens, and a variety of their acyl enzymes obtained in water and in 6 *M* urea. He also gave data for RNase A, and we have added data from Hopkins and co-workers for other acyl chymotrypsins, the HEW lysozyme data of Segawa et al. [42], and the T4 lysozyme data of Chen et al. [44] (Fig. 17 of Ref. 14 includes most of these points). The data ranges are large and the plot is linear by regression testing but may not be reliably so using χ^2 since the variances in the several methods used are large. The T_c value is about 355 K and the plot has an enthalpy at zero entropy of about 15 kcal/mole. One implication of this plot is that all these proteins have the same melting rate at about 355 K. The same deduction arises from the compensation plot for unfolding rates of several mutants of amycin nucleotidyltransferase [59] in which residue exchanges have been made in the knot. Finally, recall that 354 K is the T_c for proton-exchange rates of knot sites in proton exchange. More and better data may establish that the phenomenon is only a reflection of low precision and high coincidence, but for the present it must be taken seriously.

A second use of Pohl's compensation plot is to estimate the relative stabilities of the knots and matrices of the protein that lie on the plot. HEW lysozyme is an example. The equation for the plot is $\Delta H_{uf}^{\pm} = 15 \text{ kcal } M^{-1} + 350\ \Delta S_{uf}^{\pm}$ and the activation heat capacity for the unfolding process is zero independent of pH. So using 47.8 kcal/mole for $\Delta H_{uf}^{\pm}$ (Segawa and Sugihara; pH 2.6) and $\Delta S_{uf}^{\pm}$ = 94 cal/K *M* and $\Delta G_{uf}^{\pm}$ at 300 K is found to be 20 kcal/mole. This is the work required to reach the transition state, and it is an estimate of the work required to convert the knots to a matrix-like state. The overall standard free-energy change for denaturation in pure water at this temperature and pH 7 is 14.2 kcal/mole [60], so the transition state is unstable relative to the product state by about 6 kcal/mole. Assuming that 20% of the latter change is due to denaturation of the knot as matrix, the instability of the matrix in its nonintrinsic state is roughly 4.5 kcal/mole. Then in folding $-14.2 = (1.2 - 20)$ from knots $+4.5$ from matrices.

The standard free-energy change in converting a product under these experimental conditions to the transition state of 6 kcal/mole means that 1 in 10,000 denatured lysozyme molecules is in the transition state at equilibrium. Whatever the explanation of this number, it is a low probability easily overcome by the 20-

kcal drop in free energy that occurs in knot contraction from its matrix-like condition near the transition state to its native state conformation.

The pattern of unstable matrices supported by very stable knots is common to the proteins falling on Pohl's plot and probably obtains for most enzymes since this division of labor makes matrices easily adjustable for specific tasks. For short-lived processes it is the knot stability which limits the amount of stress that can be placed on the protein. For longer-lived processes the overall stability is limiting. In neither case is the limit high, but it is apparent that the mechanical uses of protein conformations increase in number and in amount of allowed transient increase of free energy as the event time decreases.

These considerations of matrix instability provide a basis for the previous comments about the secondary importance of disulfide bonds in protein thermodynamic stability. The thermodynamic problem involved may be unusual. In syllogistic style we can write: (1) The placement of disulfide bonds in native proteins is that which minimizes the free energy of formation of the native state. (2) The rearrangements effecting the minimization occur as the knots are formed. (3) Matrices as constrained and configured by knots are not in their intrinsic states of lowest free energy in the most stable state of the total molecule. (4) The thermodynamically determined positioning of disulfide bonds takes place in an element of structure that is not itself in equilibrium. As a result, these bonds cannot add stability to the metastable matrices but can help determine the structure of the matrices. The alternative to this argument is that the disulfide groups determine final protein conformation by favoring knot formation; that is, knots may form only when proper disulfide positioning has taken place.

A possible illustration is provided by the three disulfide bonds that connect the two primary functional domains of phospholipase A2 at the knots. The knots are pure helices, and a priori it seems unlikely that the disulfide links form before the helices. They may form as the helices do but they can at most help helix formation but not determine it. This has been discussed using BPTI as the example. Weissman and Kim [61] show that in the spontaneous folding of BPTI the native pairing of half-cystines occurs early in the sequence of folding events. It is hardly likely that this occurs before the three pieces of the PTI knot form, just as it seems unlikely that the interknot disulfide bonds form before major pieces of the knot form in phospholipase A2. As with BPTI, the remaining disulfide bonds in this disulfide-rich small protein appear to provide constraints necessary for structure and function of matrices.

Because of the very high degree of conservation of residues contributing atoms to a particular kind of knot, it has seemed reasonable that any change in such residues or even simple rearrangements would make a knot unstable. However, Kim et al. [49] have exchanged one by one all knot residues in BPTI with alanine residues to find small to large decreases in thermodynamic stability. The knot is the source of folded stability, but its formation can be sensitive to residue changes. Klemm and co-workers [62] had in fact established the existence of this tolerant behavior toward matrix changes in phage T-4 lysozyme. Among the 28 mutants, 21 produced at least a 2 kcal/mole decrease in thermal stability and an increase in melting rate. The reductions in the activation free energies were as much as 35% of the standard free-energy change under the fixed set of experimental conditions. Not surprisingly, since the 21 mutants had at least marginal folded stability, the *B* factors of atoms in the wild type indicate that all mutations occurred in matrix or surface residues. The probable exception is M6 in the wild type. The rate increase suggests considerable destabilization of the knot by the increased stresses from the matrix. The additional thermodynamic instability is then attributable to the higher formation free-energies of the matrices. Knots have been shown to undergo changes in stability following residue exchanges in matrices and in aqueous solvent mixtures having primary effects on matrices. Examples of the latter are urea and glycerol. These have opposite effects on both the transition state and knot proton-exchange rate. This sensitivity, which we can call second-order, becomes less surprising as a result of the finding of Kim et al. that knots can tolerate some changes in their own residues without total loss of their stability. Lin and Sauer [63] rearranged by SDM three knot residues of a lac repressor and found only one rearrangement entirely eliminated folded stability. Changes in knot side chains can perturb the H-bond skeletons of knots but apparently rarely destroy them. Both sets of experiments are consistent with the idea that the skeleton of the knot is a major source of knot stability (see Sec. VII.B). The experiments of Klemm et al. show that the knots are sufficiently weak to respond to the mechanical stresses placed on them by the matrices. The increased stabilization of the protein, the reduction in exchange rates for knot sites, and the increases in activation energies for the exchange reported by Gregory et al. [52] for HEW lysozyme in glycerol solutions (vide supra) seem to require a mechanical explanation.

In the previous section, the practical advantage of measuring rate constants in melting processes rather than equilibrium thermodynamic changes was emphasized because pure forward rate constants in the denaturation knot-matrix proteins are not confused by inadequately described and easily altered properties of the product state. Of course, equilibrium data are of major importance when they can be interpreted unambiguously.

However, Bolen and co-workers [64] appear to be uniquely successful in devising a method for obtaining uncomplicated thermodynamic information about the differences between two native forms of a single protein, e.g., residue substitution, acyl-enzyme formation, pH, and so on. They eliminate the complication presented by the melted state in water by using high concentrations of strong denaturants, which force all forms of a protein into a common extensively unfolded product state. The latter provides a reliable reference state not only for changes in native states of proteins, but also for studying the complexities of states produced in ordinary denaturation.

D. "Molten-Globule" Conformation States

The "molten-globule state," so titled by Ptitsyn and co-workers [66] and now well known, is a collection of macrostates that demonstrate characteristics of both native and denatured states. Many examples are undoubtedly due to weak coupling between functional domains of enzymes. Brandts et al. have shown that quite weak coupling can cause simultaneous melting of the two functional domains of simple enzymes [65]. However, Privalov and co-workers have demonstrated step-by-step melting of functional domains in many proteins. Whenever there is step-by-step melting, the intermediates whether stable or metastable will show properties that define them as molten-globule states [66]. Yeast phosphoglycerate kinase is one such protein [67]. Recently, Damaschun et al. extended this study using a variety of methods to measure hydrodynamic properties [68]. In the presence of 0.7 *M* guanidine HCl, in cold denaturation the two steps were well separated. The hydrodynamical structure was determined as a function of temperature to show expansion of the product state to nearly random-coil dimensions at 274 K. The last finding has somewhat limited generality because of the guanidine hydrochloride even though present in small amount. Nevertheless, the observation strongly supports the expectation already discussed, that the denatured net will tend to expand on cooling and, in doing so, again illustrates the hazards in studies of denaturation carried out without control or knowledge of the product state.

The molten-globule state of α-lactalbumin has been as extensively studied as any other representative of this class. The study has been especially complicated by the involvement of calcium ion. Alpha-lactalbumin is a member of the general HEW lysozyme family converted to a regulatory role by adding or strengthening the metal-ion binding site. Muranker and co-workers have shown that the two primary domains of HEW lysozyme itself can be made to fold independently [69]. Calcium ion on binding to alpha-lactalbumin effects a reintegration of part of the protein into a state that appears to be different from those available in its absence. There has been considerable confusion among the results obtained by the many who have studied this protein, but Baum et al. now appear to have eliminated most confusion by establishing that the molten-globule state has one of its two lysozyme-like domains melted [70]. Calcium ion controls molten-globule formation and in this way provides a regulating device for the enzyme supported by its lactalbumin companion.

Another state of mixed native-denatured character has been discovered by Kim and co-workers with BPTI [71]. By replacing a single tyrosine residue, they destroyed the native conformation of the matrix with little, if any, effect on the knot in this single-domain protein. Partial matrix melting may be possible by one or another modification in any knot-matrix protein in which the knots do not have complete control of the matrices. The PTI family is a good example, not only because of the disulfide system in the matrix, but also because inhibition of trypsin by these proteins depends on the ability of the inhibitor to match the severe requirements of the protease to achieve the necessary tight binding (see Sec. VI.E).

Once the knot of a functional domain is untied, the matrices are free to move into their (intrinsic) lowest-free-energy state. At low hydration this state is favored. At the usual hydration levels in aqueous solution there is competition between formation of this state and the hydration process in which the latter usually wins. Under some circumstances it is not unlikely that the probability of the intrinsic state is sufficiently high to yield complex kinetics and complex distribution functions, including features not usually found in either native or fully hydrated species. A rough counterpart of this is the species with high helix content formed when sufficient amphipathic cosolvent is present. The small alkanols and short fatty acids, all destabilizers of the native state under some conditions and stabilizers under others, apparently form a sort of mixed micelle in which matrices and possibly also knots are much altered. It is known that small alkanols, especially propanol, weaken knots [42]. The bipolar nature of these molecules provides an effective glue, competing with water successfully and taking advantage of the poor packing of matrices to produce new conformations.

The molten-globule phenomenon is often related to the *completing the knot* phenomenon (Sec. IV.E).

E. Structural Dependence of Common Experimental Observables

1. Phosphorescence, Fluorescence, and Circular Dichroism

Some proteins demonstrate high phosphorescence yields, usually indole phosphorescence, when fully hydrated and at ordinary temperatures. This is unusual behavior in an ostensibly nonrigid physical system since internal conversion of triplet states is very sensitive to motion and to paramagnetic quenchers such as oxygen. In the absence of such quenchers, a variety of shorter-lived triplet states are demonstrated by new phosphorescence emission. As a probe of both local dynamics and the diffusion rates of quenchers through the protein, phosphorescence is a powerful tool. Its virtues are discussed by Strambini in some detail. He was the first to realize that the extraordinary rigidity of the local structure necessary for room-temperature phosophorescence is that of knot chromophores [72]. With the shorter-lived emission sources the conformational dynamics determining the rate for oxygen diffusion through matrices can be determined. The phosphorescence method cleanly separated knot dynamics from matrix dynamics and may be the most sensitive method now available for measuring small changes in conformational fluctuations.

A similar explanation appears to account for several aspects of the circular dichroism (CD) of proteins. Manning and Woody have shown that specific CD effects in BPTI are those of knot residues [73]. There has long been a question as to why many residues do not give CD effects in the low ultraviolet (near 200 nm). One explanation is that most CD-active side-chain groups do contribute specific effects, but these average to give a nearly featureless spectrum. However, judging from models, the average values are not consistent with this explanation, indicating rather that there must be some local effects. The probable explanation for what is observed is suggested by the high *B* factors of nonknot atoms and the low *B* factors of the knots. The motility of matrix atoms tends to average their contribution individually and collectively to zero. The contributions from tightly held knot residues are then responsible for the CD spectrum. Fluorescence is an especially useful tool for the study of proteins, since the fluorescence lifetimes of phenol and indole groups lies in a time region characteristic of many conformational fluctuations. Major conformational processes in enzymic catalysis appear to fall in the time range 0.1–10 ns. It has nevertheless been difficult to separate the wheat from the chaff in fluorescence studies [74] primarily because emission from these groups is very sensitive to near neighbors and their relationship to their neighbors reflects conformational fluctuations. Exceptions are those chromophoric groups actually held in knots. These are invariant in most experiments and as a result provide more or less unambiguous information about knots and their sensitivity to changes in matrices and surface. This is a limited use and may be better replaced by studies of phosphorescence.

2. Compressibility and Thermal Expansion

Although the values of protein thermal expansion coefficients appear to be reasonable in relation to those of comparable polymers and small molecules of similar nature, protein compressibilities are much smaller than would be expected if proteins are impermeable to solvent [75]. Because of their hardness, knots can be expected to have low compressibilities, but the considerable empty space in matrices and surfaces should produce high values if the interior of proteins were mechanically isolated from the environment through which the pressure is applied. Thus far it is found that the hemoglobins have isothermal or adiabatic compressibilities in the normal range for liquids, but most other proteins studied had values characteristic of soft solids. Apparent compressibilities would be consistent with 100% knot and no free volume. The explanation appears to be that although the true low knot compressibilities contribute correctly to the apparent compressibilities, the matrices are permeable to water and thus mechanically transparent. The matrices contribute only relaxation contributions from conformational readjustments under changing pressure and the flow of water between bulk phase and protein interior. The few relevant experiments indicate that application of pressure forces water into protein matrices. The interior of the protein is in mechanical equilibrium with the solvent medium so it is the primary-bonded structure that provides the true phase boundary. Examining the x-ray results for lysozyme at pressures of 1000 bar determined by Kundrow and Richards [76] shows that only the knots are undistorted.

IV. SOME DEVICES THAT BECAME POSSIBLE AFTER THE DISCOVERY OF THE KNOT-MATRIX CONSTRUCTION PRINCIPLE

A. Modular Construction of Knot-Matrix Proteins

It is now apparent that proteins in this class are constructed by combining substructures. Evolution has a family of knots to use in new experiments and also a broader and more variable family of matrices. The elementary unit is the functional domain defined in a broad way by its knot and in a most specific way by its matrix and surface. Much of evolution consists of experiments in which matrix and surface modifications are tested for acceptance or rejection, but the experiments are roughly of two kinds. In one, new knots are found and thus potentially new functions. In the other, new functions are found by combining old functional domains and tailoring matrices and surfaces for the new function. A variety of combinations of knots and matrices are possible and all seem to have known representatives. All-knot proteins are of the structural class, almost perfect H-bonded secondary structure (e.g., keratins). Semiknot proteins have generally low *B* factors but variable knot-forming abilities. The knot regions are not fixed in the sequence order but vary in position and size with functional state. The immunoglobulins are discussed in Sec. VI.F as probable examples of this type. Knot-matrix proteins have knots and matrices in almost any proportion and the knots are fixed in sequence position. For high thermal stability with retention of physiological function, knots must be either large or exceptionally stable or both. Enzymes have two knots per catalytic function but may also have structural knots such as hinge knots to strengthen and establish the association of the functional domains. When the two functional domains are on different proteins, they are usually held together to form the catalytic unit by knots formed from parts of both proteins. Many proteins consist of a single functional domain, e.g., most proteinase inhibitors. By current tests and with current data the hemoglobins have no knots. Proteins like the serum albumins and insulin have not yet been examined for their classification.

B. Expansion-Contraction Processes

Knots are the true examples of Schrödinger's "aperiodic crystals"; matrices have properties that vary from rubbery or liquid-like to porous glass in native proteins and amorphous nonporous polymer in the denatured state at low hydration. The mechanical characteristics of matrices depend on free volume and dissolved water molecules, small changes in the amounts of either having large consequences for function. Matrix characteristics are heavily determined by free volume. In contrast to knots, matrices have considerable free volume much of which is mobile. Thus in enzymes free-volume changes in large parts of the matrices probably provide the mechanism by which catalysis is effected (Sec. VI.D.4). Specifically, local fluctuations are somehow coordinated to produce gestalt matrix fluctuations in which there are major changes in amounts and distribution of free volume. The characteristic thermodynamic feature of these "functional" fluctuations is the correlation of enthalpy and entropy changes to produce at temperatures close to, but below physiological, a near-isoergonic condition along the fluctuation coordinate (Sec. VI.D.4). Functional domains have some fluctuational autonomy, but the two catalytic domains of enzymes are at least coordinated in catalytic events in a symmetrical scissor-like closing process that we and some others call "domain closure" and still others call "hinge bending." Domain closure is driven by the expansion-contraction behavior of the matrices in which there are large changes in free volume due not only to changes in total volume, but also to changes in the amount of dissolved water (Sec. VI.D.4).

The large free-volume fluctuations have a rough analogy to a harmonic spring. The energy is fixed and, of course, also the Helmholtz work, but cycles between potential and kinetic energy. Expansion-contraction cycles of the conformation interconvert entropy and enthalpy so as to maintain near-constant free energy. If the process is isoergonic, it is quite like a spring, but the total energy is not constant, so the fluctuation behavior depends on the environment and its temperature. If it is not isoergonic, motions along the expansion-contraction coordinate must be coupled to a source and sink for free energy. Such coupling is usually disadvantageous because it decreases the probability of the major fluctuation, but in multienzyme systems it may facilitate free-energy management. Catalytic rates are found to be high at low hydration where protein aggregates serve as the thermal reservoir. The thermostatting problem appears to present no difficulties.

C. Free Volume and Dielectric Constant

Protein conformations and their properties are, of course, determined by the electrostatic fields and their changes as much as by entropy. The adjustability of local dielectric constants is one of the major benefits of the discovery of proteins. All electrostatic interactions are increased by lowering the effective dielectric constant, and the lower the latter, the larger the change. The dielectric constant decreases because the mobility of the groups with permanent polarization is reduced. So both interatomic distances and dielectric constant, already small, decrease until some repulsion limit, which prevents further loss of free volume, is reached. The lower the free volume, the lower the electrostatic potential energy.

Contraction of the knot is limited by repulsion forces of which the short-range repulsion is the more important, but unfavorable dipole and quadrupole interaction enforced by local structure can also be important. Because knots are very stable relative to their matrix-like properties just before the transition state for melting, it is unlikely that there is more than one pathway to the knot. Once the soft system has found this pathway, it drops like a rock to the final native state. Alternate knots with their own formation pathways would produce major metastability. There is as yet no evidence for such long-lived alternative "native" structures in knot-matrix proteins. This is probably a result of evolutionary selection, but improbable in any event. That does not mean that evolution has failed to find this kind of switching device.

The contraction process occurs in matrices as well as knots but the relatively loose structure of matrices establishes that the potential-energy limits that their various clusters reach are much less negative. The entropy losses are also much less, so the entropy and enthalpy are roughly equal in importance and ordinary temperatures. The knots and matrices are qualitatively different because of their thermodynamic differences. This is the major virtue of the knot-matrix discovery since it allows a division of labor that makes the functions of modern knot-matrix proteins possible. Since the knots determine thermodynamic stability, matrices need not play an important role in stability. They need not be and usually are not in their intrinsic lowest free-energy conformations but rather in conformations dictated by the knots. They must assume conformations that further favor stability of the knots if such conformations are available and if the net result is a sufficiently low free energy to maintain folded stability for the total protein. Then the matrices can be adjusted to carry out the specific function of the protein with only minor restrictions on their own instability.

D. The "Pairing Principle"

The discovery in evolution of the utility of pairing functional domains was not dissimilar in importance to the later discovery of the opposed thumb. The catalytic mechanism that now obtains depends on the *pairing principle* as manifested in the drawing together of two functional domains. Its precursors must have used a simpler invention. Survivors of the old order probably remain because they can carry out some function better than their knot-matrix competitors. Hemoglobin is, of course, our candidate. But survival success in such a competition is all or none, so it is likely that all modern enzymic mechanisms are manifestations of a single device, even though the details vary greatly from enzyme family to enzyme family.

In addition to the acceleration and specificity advantages of the domain-closure mechanism for catalysis (Sec. VIII.A), it also makes possible control of rate processes along the entire reaction coordinate. The thermal activation of small-molecule reactions is generally an improbable event as well as an uncontrollable one. The properties of the reactants determine the reaction path and the products, and the chemical free energy released can be conserved only by coupling to another reacting system usually by the transfer of atoms or groups. In domain closure the uncertainty principle may be less restrictive because less heat is used in forming the transition state, but transient use of PV fluctuations of conformation for this purpose is subject to the same restrictions. Destabilization of the pretransition state by free-energy rearrangements in the conformation is not so limited.

A major virtue of domain-closure mechanisms is that what happens after the transition state is nearly as controllable as what happens before. A special advantage is the conservation of free energy discussed in past papers as *free-energy complementarity* [9,10] and illustrated particularly well by Fisher and co-workers using glutamic dehydrogenase [7]. It is a general basis for the evolution of free-energy management toward the thermodynamic limit without loss of speed. By refining through evolution the details of domain closure, less and less of the free energy released in a given protein process is lost as heat. Specialists in multienzyme processes have always been attracted to complementarity for obvious reasons. It seems reasonable to expect that such improvements are open to any evolving system. The latter conjecture becomes even more reasonable with the finding of an enzyme mechanism ideally suited to such improvements.

Examples of the use of the pairing process are so common they are taken for granted. Thus, for example, receptor sites such as that for the human growth hormone are constructed using a very sophisticated pairing-principle device that transmits information about the separation of partners to and through membranes. Another interesting example is provided by some families of repressor proteins that appear to derive their specificity in selecting sites on DNA helices by pairing of subunits so as to be sensitive to groove geometry and more specifically to local dynamical behavior (Sec. VI.B.2). The so-called "second genetic code" appears to depend on matching of the fluctuations of repressor proteins and other proteins regulating DNA expression to local dynamical behavior of DNA. Complementarity in protein-protein association depends on matching of the fluctuations of surfaces, and the example of hemoglobin given in Sec. VI.C.4 describes one possible way such matching can be tailored to effect useful exchange of free energy.

E. "Completing the Knot"

The high degree of cooperativity of knots is the basis for many useful devices. One of the best known of these is the "leucine zipper" [78] in which the separate knots and associated matrices of two proteins are so constructed that on association they interdigitate and contract to form a single common knot. This device appears to be widely used to effect strong association between proteins. It is another example of modular construction since the functional domains that form the zipper do little more than join separate proteins. Intercalation of the two proteins allows formation of a single knot joining them, thus "completing the knot." The occurrence of the latter is signaled by a decrease in the *B* factors for those atoms that lose their matrix status.

A more common example of completing the knot is production of a knot from matrix where no knot had appeared before. Practically this is found by comparison of the *B* factors assuming such are available. Such conversions signal a drop in enthalpy and a lesser drop in entropy to produce the lower free energy accompanying knot formation. The latter is the thermodynamic signature of completing the knot as it has been defined. It is less specific than it needs to be because matrix contraction has the same thermodynamic characteristics but must rarely produce true knot properties. Were it to do so, the stable state of the protein would tend to be all knot. The determining factor is the balance of enthalpy against entropy, the "thermodynamic phase rule" considered later.

Successful knot completion is, of course, essential for native-state folding in any protein constructed using the knot-matrix principle. An early demonstration of this now familiar principle was given by Goldenberg [79]. These authors generated many mutants of BPTI and found that substitutions in what they defined as the hydrophobic core and we define as the knot either destroyed folded stability or greatly decreased it as indicated by rapid denaturation rates. The matrix mutants in which all the knot residues were unchanged folded successfully. Kim, Fuchs, and Woodward have extended this kind of investigation and made it quantitative (Sec. III.C.3).

HIV-1 protease and the gene V protein [80] are formed from two protein subunits and will not fold except in a bimolecular process. Some repressor protein are dimers and only stable as dimers, e.g., the arc repressor [81]. The dimers are connected by a hinge device referred to, for obvious pictorial reasons, as a "fireman's grip." Each chain of this dimer provides a phenyl group near its C terminus, which fits into a hole available in the native conformation of the other chain. This may be accompanied by what might be called a strap-hinge knot. It is a secondary knot consisting of a short section of antiparallel sheet structure constructed using identical residues of the two proteins. The two structures together actually form the hinge. Figure 14 depicts the HIV-1 protease, which has the same hinge structure. Both proteins fold into native form in a bimolecular process, which seems to demonstrate that completing the hinge knot is critical to folded stability. In the case of the λ repressor, Pakuta and Sauer confirmed this supposition by completing the knot using a hydrophobic peptide grafted onto the monomer. The single protein subunit then folded spontaneously using the peptide to complete the knot [82]. This experiment provides an alternative strategy for inactivating the HIV-1 protease.

Completing the knot is a powerful and much used device. The advantages are twofold both arising from the special characteristics of the knot produced in this way. The entire free energy associated with contraction into the knot

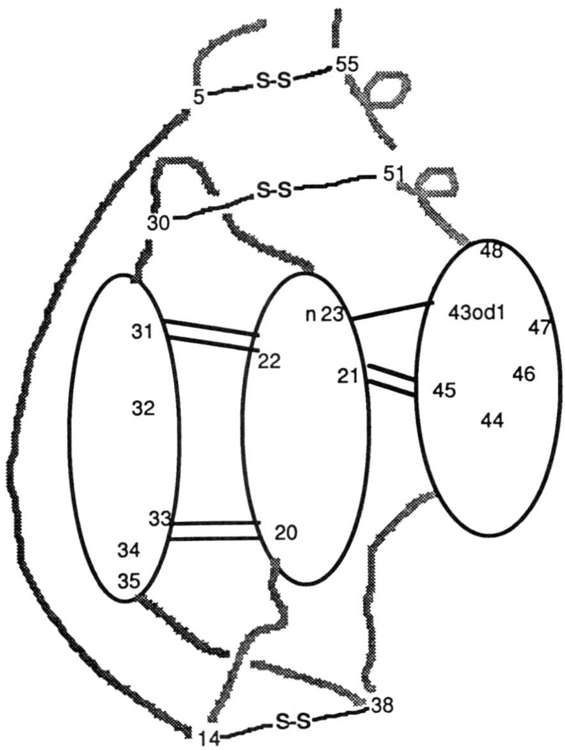

FIGURE 13 Construction diagram for the PTIs. Assembly diagram for the Kunitz family of inhibitors showing the integration of the strong double-hydrogen-bonded rings and the three highly conserved disulfide groups. The knot consists of three pieces of which the 22–23 piece is central, being connected to the 33–35 piece by two of the double-H-bond structures (Fig. 12) and to the 43–45 piece by one of the latter and one very short single H bond. A few residues numbered in the pieces have *B* factors outside the group I range and no group I proton-exchange sites nor are they strongly conserved. They lie between or next to knot residues but should not be classified as knot residues. The disulfide bonds are also arranged symmetrically and are seen to provide firm constraints on the matrix conformation and fluctuation amplitudes. They may also improve the thermodynamic stability of the matrices but very likely have little effect on kinetic stability. Although they are not shown in Fig. 6, the disulfide bonds in the phospholipase A2 proteins provide essential constraints on the matrices. In both families the disulfide bonds supplement the knots in determining matrix construction and dynamical properties.

conformation becomes coupled to the process. It is much larger than any amount of free energy released in the contact association of trigger group and polypeptide. The second advantage is the strict specificity requirements on groups that are to be part of the knot. Very small differences in structure and polarization separate a successful trigger molecule from an unsuccessful one since the binding of the former is amplified by the free energy released in knot formation. Although the free energy released in knot formation, roughly equal, we believe, to the measured standard free-energy change in thermal denaturation of the functional domain so completed, is relatively small by comparison with changes in primary bonds, it is nevertheless large on the scale of biological processes. It is not unlikely that the binding of small molecules to proteins with association constants much larger than $10^7 M^{-1}$ sometimes depends on completing the knot, but matrices may also function in a somewhat similar way. If the substrate or inhibitor can favor a general change in matrix conformation that reduces the total free energy of binding, this will be the case. Emphasis in such binding has usually been placed on direct contact of substrate or inhibitor with binding region of the protein to take advantage of their high potential energy and low entropy when unoccupied. However, studies of enzyme *B* factors in these processes although still limited in number, suggest a general trend toward lower average *B* factors for matrix atoms at intermediate states of the catalytic process (Sec. VI.D.3). This kind of change is consistent with the idea that binding to matrix sites displaces the equilibrium position of the expansion-contraction process more often toward contraction rather than toward expansion.

F. Protein Activity Coefficients: Gibbs-Duhem Consequences

The chemical potentials of proteins depend on surface area, A; total protein volume, V; free volume, v; amount of dissolved water per protein molecule, α; T, and P. These dependencies can be approximately expressed using the Gibbs-Duhem (G-D) expression as a sum of linear terms, but the variables interact, often strongly, because of geometrical relationships and dependence on free volume. This is not a disadvantage to the investigator because the cross-relations can be determined from experiment and contain much of the most valuable information about protein systems obtainable with thermodynamics. The G-D derivation as given by Lumry and Rajender [83] is correct in terms of the chosen extensive variables but does not take water dissolved in protein into account correctly. The corrected version given next is still inadequate to take into account all the ways solvent additives can alter the G-D balance.

We fix the total moles of protein as n_p and the total water as $W = n_e + n_i$ in which the water external to the protein is n_e and the internal water is n_i. Changes in these totals can be taken into account if needed. For the present, one might think of the protein concentration as being a chosen standard concentration. The total protein-solvent interface is $n_p a$ and the total protein volume is $n_p v$. The intensive quantities σ and π are the partial derivatives of the standard chemical potential of the protein with respect to area and volume, respectively.

The internal-energy expression is Eq. (1),

$$dU = TdS - PdV + \sigma n_p da + \pi n_p dv + \mu_e dn_e + \mu_i dn_i \tag{1}$$

The G-D expression is Eq. (2),

$$VdP - SdT + n_p ad\sigma + n_p vd\pi + Wd\mu_e = 0 \tag{2}$$

The last term on the left is the result of the equality of the chemical potentials of internal and external water at equilibrium. The chemical potential of the internal water depends on the protein variables, but since we do not know its concentration, it is better to use the chemical potential of the external water, which includes its activity coefficient.

The cratic terms for protein and bulk water can be neglected so long as the total amounts of protein and water can be assumed constant, but there is an entropy of mixing term for water dissolved in the protein. All mobile water contributes to this term. The term may not be large, but proteins are large with respect to water molecules and a large part of the protein volume is accessible to this water. Effects of protein charge or changes in solvent composition can be included as new variables, or more safely perhaps through σ and π. Donnan effects can be significant and nonideality large. Just as in more rigorous small-molecule studies, thermodynamic quantities are related to activities and not concentrations. Nonideality in processes of large and small molecules in aqueous mixtures arising from interactions between nonionic solutes and cosolvents can be as large as that arising from ionic solutes. Primarily because water is not a good solvent for proteins and is substantially altered by changes in solution composition not the least of which is the protein concentration, protein activity coefficients are usually far from unity even when the protein reference state is infinite dilution in the solvent system.

The chemical potential of the protein can be written as Eq. (3):

$$\mu_p = \mu^\circ(P, T, v, A, n_p) + RT \ln n_p \tag{3}$$

and this defines the protein activity and activity coefficient.

Protein crystals are a poor reference for solution chemical potentials. Protein vapor pressures have been measured with isotopes, but the perfect protein gas is an impractical reference state. Bolen's random-coil denatured state in principle would be the most reliable activity reference, but because of the labor involved in its determination it is not a practical choice in the current climate of funding. Usually one can settle for differences in protein activities rather than absolute values, but these are tedious and expensive to measure.

Most extensive variables appearing in Eq. (1) are related through a common dependence on protein free volume, surface volume and geometry, hydrostatic pressure and other quantities of the protein system under examination and the level of rigor desired. Water "freely" dissolved in the protein is closely related to free volume on the one hand and to the activity of solvent water on the other. Because of the constancy of the chemical potential of water throughout any system at equilibrium, dissolved water is quantitatively important roughly in inverse proportion to its concentration. The analogy is with osmotic pressure, but it is a complicated one because of the deformability of matrices and the several other cross-dependencies. Since Linderstrøm-Lang's demonstration of the catalysis of proton exchange by water ions, one could only escape the implications of the G-D equation for dissolved water by assuming that exchange occurs only on unfolding, and that possibility was laid to rest by the discovery that helix-coil rates are very fast [84].

Equation (2) is essential in any consideration of the dependence of proteins on external variables as in melting, response of catalytic parameters to solvent composition, any dependencies on protein charge, and so on. Effects attributed to changing viscosity fall outside thermodynamics and thus outside this discussion except insofar as they are not true viscosity effects but rather reflections of solvent effects on the variables of Eq. (2). An example from Timasheff and co-workers makes this clear. They have measured the variations in protein activities caused by characteristic cosolvents much used in protein studies. Thus they find that glycerol, a common example of the polyhydroxy family, generally raises the protein activity coefficient but does so more for denatured than native species. In practice, stabilization is a consequence of selective destabilization. Just the reverse was found with the denaturants urea and guanidinium ion, which decrease the activity coefficients of both native and denatured forms. The effects of glycerol and polyhydroxy compounds have been explained by Timasheff in terms of preferential hydration of the protein and by Winzor in terms of excluded volume. These important matters are discussed by these authors in Chapters 12 and 13. Winzor has now established that the two explanations are equivalent. The size and hydration structure of glycerol diminish the surface entropy and probably the fluctuational entropy of the matrices when in close contact with the protein surface. The increase in free energy is minimized by reducing the glycerol concentration in the surface region, but the protein activity coefficient is nevertheless raised and the protein contracts. This is easily accommodated by adding a glycerol term to the G-D equation. Changes in water activity are minor. Timasheff and Winzor give less casual arguments.

Polyethyleneglycol (PEG) is very frequently used in crystallizing proteins for x-ray and neutron diffraction studies. PEG raises protein activity coefficients and forces contraction and thus constrains conformational dynamics. Although the thermodynamic explanation is thought to be complicated by protein-solute direct contacts and the unique effects of PEG on water structure, the consequences for the G-D relationship are the same. Usually one expects the facilitation of crystallization to be due to an increase in solute activity in the liquid versus the crystal state, but in proteins even polymers like 5000-molecular-weight PEG can be taken up by the mother liquor in the crystal. The effect on the activity coefficient of the protein in the crystal may be about the same as in the liquid, but both are raised by these cosolvents under conditions used for crystallization. The consequence of most practical importance is that the protein in the crystal is compressed thermodynamically as well as by the original restraints in the crystal and thus resembles the free-solution state even less. In particular, the *B* factors will be lowered even further by the precipitating agents. The efficient approach to this matter is likely to be *B*-factor changes rather than coordinate changes.

A second consequence of the protein volume, surface, and geometry on polyhydroxy solvent additives is the confusion generated in experiments seeking to find important effects of viscosity on enzymic function. An up-to-date discussion and review of the relative importance of viscosity and thermody-namics in glycerol solutions is that of Gonnelli and Strambini [85]. Cosolvents used to increase viscosity are usually polyhydroxy compounds, and these are among the most effective cosolvents in raising protein activity coefficients. There is clearly a connection between the two characteristics of such solutions, but the mechanism of their effects on conformational dynamics is totally different. The studies that most closely demonstrate agreement with expectations from the frictional form of Kramers's theory are on myoglobin and single-chain hemoglobins. Judging from the high *B* factors for all atoms of myoglobin, these proteins might be expected to display viscosity-dependent conformational dynamics, and it is true that agreement between Kramers's theory predictions and observations is poorer with knot-containing proteins. Glycerol, for example, by causing contraction reduces matrix motility thermodynamically and by its increase in solvent viscosity does the same thing. One must distinguish then between the thermodynamic effect and the viscosity effect of the additive. Kramers's theory has not proved to be an unambiguous aid in making this distinction. Unfortunately, additives that are experimentally useful in increasing the viscosity of water solutions are also very effective in raising protein activity coefficients. The sulfate ion may be the exceptional additive with which to resolve this problem, but the problem may not exist since enzymic activity in the absence of a bulk solvent phase probably precludes an essential role for solvent viscosity. The elegant solution of physical chemist Timasheff and his students has provided much clarity in these matters. It has also revealed a number of additional subtleties in the effects of protein stabilizing and destabilizing agents that require explanations (see Fig. 14 in Ref. 10 for an especially interesting example of polyhydroxy compounds as exemplified by glycerol). Compensation behavior is very useful in revealing and categorizing the several families of stabilizing and destabilizing solvent additives (Sec. V.B) [86]. Aqueous mixtures of water and hydrazine and pure hydrazine considerably simplified the understanding of solvent effects in micelle formation by systematically removing the

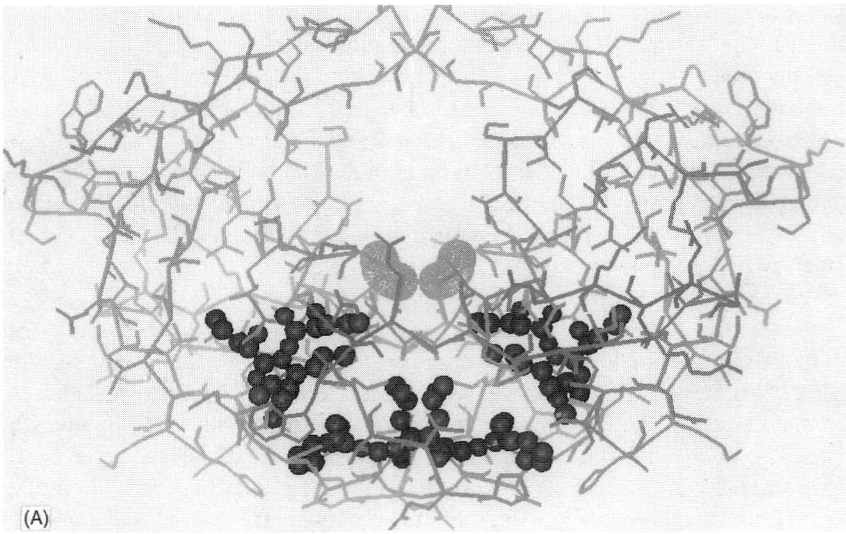

(A)

FIGURE 14 Construction of HIV-1 protease—a model of a perfect enzyme. (A) The two primary functional groups are large blobs in the center. The major parts of the two knots are the small spheres below them, and the strap-hinge arrangement holding the two protein subunits (functional domains) together is at the bottom. The substrate-binding region is the empty space at the top. The noncrystallographic twofold axis lies in the plane of the figure and it is in this plane that domain closure occurs, as described in the text. (B) Detail of the "strap hinge" viewed from the rear. The molecule has been rotated forward 90° from the view in (A) so as to show the beveled method by which the two subunits combine and the locking hinge knot. The hinge consists of two short sections of main chain arranged to form a very small antiparallel β-sheet. In addition, each subunit has a terminal phenyl group inserted in the other subunit to complete its knot. This construction has been called a "fireman's grip." The electrostatic factors responsible for the stability of the strap hinge are in detail highly obscure. (Data from 3HVP in the Brookhaven PDB.) (C) View from the end opposed to that shown in (B). The solid balls have high *B* factors in the absence of substrates or inhibitors and so do not form a knot. The construction is otherwise identical with that shown in (B). One suspects that with natural substrates the pieces combine to complete a knot very similar to the strap hinge. If so, a good part of the thermodynamic stability in substrate binding is derived from "completing the knot." This interesting artifact may be quite common. The inverted juxtaposition of knots occurs in the vertical plane. It is suggested in the text that this maximizes the force on the reacting assembly. In the plane of (A) the molecule is a perfect pair of scissors. All gene-duplication enzymes so far examined approach this pattern.

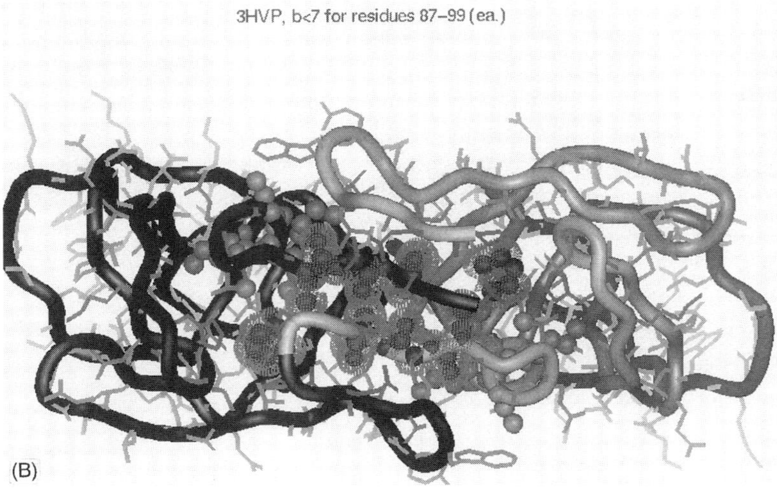

(B)

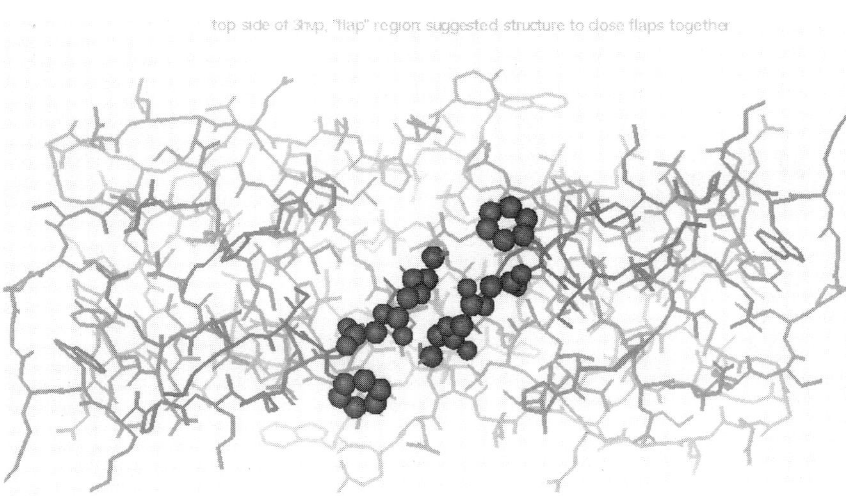

(C)

abnormalities of water. Unfortunately, similar studies have not been carried out with proteins.

G. Intermolecular Communication Through Surfaces

Surfaces have large *B* factors and, as a result, high sensitivity to fluctuations of conformation. Contractions and expansions of all or large parts of the protein produce the largest changes in the *B* factors of atoms at the surface, very often those at the end of loops (see Fig. 15). Surfaces provide the contacts with the world outside the protein and thus play major roles in associations with other macromolecules, in response to solvent changes, ionization state of the acid and base groups in the surface region, and so forth. These interactions have a large dynamical component, and since that is modulated by the fluctuations of the remainder of the protein, the expansion-contraction changes associated with function are strongly reflected in corresponding changes in surface dynamics. In turn, external conditions and changes in those conditions modulate the surface and, through the surface, modulate the internal fluctuations and thus the function. This device is an important vehicle for specificity in association and free-energy exchange between the protein and its surroundings and may indeed be the major mechanism for such exchange.

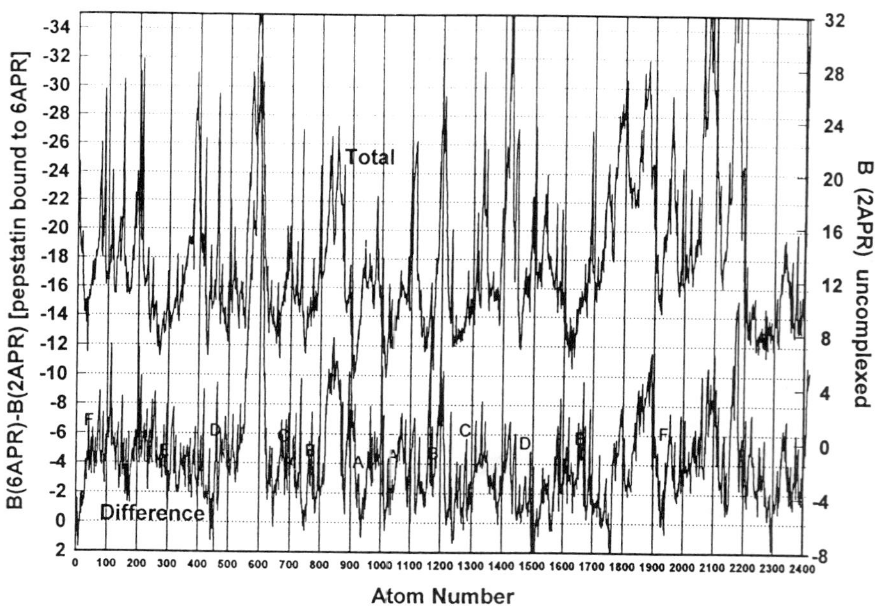

FIGURE 15 Rhizopepsin with and without pepstatin. The small but remarkably tightly bound peptide inhibitor produces major contraction of the matrices. The larger the *B* factor of an atom in the free enzyme, the greater the reduction in its *B* factor with pepstatin bound. This is shown by comparing the difference profile at the bottom with the total profile for the free protein. The difference profile has the same peaks and troughs as the total *B* profile of the free protein. The *B* factors in these two studies by the same authors are so well determined that many details of the contraction process can be made apparent [135]. (Data from 6APR and 2APR in the PDB.)

V. SOME THERMODYNAMIC TOPICS OF SPECIAL IMPORTANCE FOR BIOLOGY

A. Weak Relationship Between Free Energy and Its Temperature and Pressure Derivatives

1. Benzinger's Discovery

In the 1960s T. H. Benzinger discovered an omission in the conventional presentation of the thermodynamic relationships of isothermal processes. Errors in interpretations of enthalpy, entropy, and volume data that result, although often small, are likely to be large for protein and other macromolecular systems including many polymers. In fact, any system with a large number of vibrational and librational modes will have quantitative errors in relating free energies of formation or free-energy changes in processes to enthalpy, entropy and volume data. In 1971 Benzinger published a Carnot-cycle argument explaining the problem [87]. In the same paper he estimated that only about 6% of $\Delta H°$ for thermal denaturation of RNase A contributed to $\Delta G°$. Although the correct percentage is not yet known and the estimate undoubtedly too low, his argument is correct. The very important consequence is that several of the most important uses of thermodynamic data for isothermal processes are quantitatively invalid. None of the physical-chemistry textbooks we have examined give the argument and thus have incorrect thermodynamic derivations of the relationship between G and H, S, and V. The way this occurs is shown next using statistical mechanics rather than classical thermodynamics. Guggenheim once observed that one finds difficulties using thermodynamics and explains them using statistical mechanics. That is certainly true for Benzinger's discovery.

Let the eigenvalues of the energies at constant, T, V, and N be E_0, E_1, ... so the cannonical partition function in the classic limit is Eq. (4) and the Helmholtz work function $A(T, V, N)$ is Eq. (5). The temperature derivatives of A at constant V and N give the expressions for the internal energy $U(T, V, N)$ and the entropy $S(T, V, N)$ [Eqs. (6) and (7), respectively].

$$Z = \sum_i \omega_i e^{-E_i/\kappa T} \quad \text{(sum over total absolute energies, } \omega \text{ are the degeneracies)} \tag{4}$$

$$A = -\kappa T \ln Z \tag{5}$$

$$U(T, V, N) = E_0 + \frac{\sum_i (E_i - E_0)\omega_i e^{-(E_i - E_0)/\kappa T}}{\sum_i \omega_i e^{-(E_i - E_0)/\kappa T}} \tag{6}$$

$$S(T, V, N) = \kappa \ln \sum_i \omega_i e^{-(E_i - E_0)/\kappa T} + T^{-1} \frac{\sum_i \omega_i (E_i - E_0) e^{-(E_i - E_0)/\kappa T}}{\sum_i \omega_i e^{-(E_i - E_0)/\kappa T}} \tag{7}$$

The last equation is written to show that only the energy differences appear in S. This third-law requirement is well known but generally neglected in the derivation of A (vide infra). The cancellation of terms that this fact requires is one of two fundamental bases for Benzinger's discovery. The other follows from the second law (vide infra). The cancellation takes place as follows.

In terms of the "absolute" partition function A is given using its definition, $A = U - TS$, by Eq. (5a). This returns us to Eq. (5) but in doing so reveals a cancellation of terms from the internal energy and entropy.

$$A(T, V, N) = E_0 - \kappa T \ln \sum_i \omega_i e^{-(E_i - E_0)/\kappa T} \tag{5a}$$

The familiar defining statement of Boltzmann, $dS = dQ/T$ at equilibrium, applies to the exchange of entropy between a closed system and its environment but says nothing about how the entropy change in a system undergoing an isothermal process is divided into thermal and work parts. This is perhaps the most common source of confusion.

The second term in Eq. (6) is the energy required to change the system from its ground state with energy E_0 to its equilibrium state. To avoid confusion with some uses of the term *heat* we call it the *thermal energy*. It is canceled in A by the second term in Eq. (7) and that quantity will be called the *thermal entropy*. In some early papers the term *compensation* has been used. Rhodes uses that term in his discussion of Benzinger (see below).

Only one term in S and one term in U contribute to A, and thus to any work changes. We will continue to use the terms *motive energy* and *motive entropy* for these. The Gibbs free energy G, and enthalpy H, and the isopiestic entropy S are similar expressions derived using the Guggenheim partition function. The motive energy, E_0 includes the energy of the lowest electronic level plus the vibrational zero-point energies. The motive enthalpy contains in addition a pressure-volume term, $PV(0)$.

For a reversible isochoric process of a closed system between an initial state a and final state b the Helmholtz work is given by

$$\begin{aligned} \Delta A &= \Delta U_m + \Delta U_t - T\,\Delta S_m - T\,\Delta S_t \\ &= \Delta E_0(T) - T\,\Delta S_m \end{aligned} \tag{8}$$

The thermal parts of ΔU and ΔS are in the ratio of T and their mutual cancellation is a second-law requirement on an isothermal system. Specifically, in any part of a total system (system plus reservoirs plus work storage) at thermal equilibrium any change in ΔS must include a part $\Delta S_t = \Delta U_t/T$. Were this not true, the system would spontaneously depart from thermal equilibrium. Note, however, that this is rigorous since in general there is a change in the total heat of the system plus reservoir, which would effect a change in T unless the reservoir has infinite heat capacity. An infinite-capacity thermal reservoir is always assumed for isothermal processes. The considerations of this section then apply to all isothermal processes.

The thermodynamic equivalents of the motive and thermal terms are obtained by integrating the temperature integral over the entropy change by parts and rearranging, Eq.(9).

$$\Delta S(T) = T^{-1}\int_0^T \Delta S(T)\,dT + T^{-1}\int_0^T \Delta C_v(T)\,dT \tag{9}$$

The motive entropy change, first term on the right, can be written as a double integral over the heat-capacity change. The corresponding expression for ΔA, Eq. (10), is different because the reference temperature is 0 K rather than the temperature of interest. The only experimental method to obtain $\Delta U_0(T)$ requires calorimetric measurements of the heat capacities of the reactants and products from T to 0 K, yielding $\Delta E(0)$ and the equation of state. As is discussed in the next section, this introduces a practical uncertainty principle in an otherwise rigorous discipline.

$$\Delta A(T) = \Delta E(0) - \int_0^T \Delta S(T)\,dT \tag{10}$$

To illustrate the magnitude of the errors in entropy and the quantum-mechanical effects the values of the total formation entropies and their heat-entropy parts are given for some metallic solids in Table 1.

2. Consequences of the Discovery

a. The Motive-Thermal Separation of U, H, S, and V The motive internal energy, or enthalpy $E_0(T)$, is the lowest electronic level plus the zero-point vibrational energies for a system at the temperature of interest. In primary-bond-rearrangement processes it is large compared with the thermal energy and, as Benzinger suggested, this may be the reason why the separation was not quantitatively obvious in early chemistry. The situation is entirely different for biological and polymer processes that involve changes in many secondary interactions. However, as shown in Table 1, even strong solids have large thermal entropies with respect to motive entropies. The behavior manifested in the table is characteristic of systems of oscillators. All the average kinetic energy of such systems is heat and thus neither it nor the thermal entropy contributes to the free energy. For an Einstein crystal with $h\nu/\kappa T = 1$ the motive entropy is 44% of the total entropy. If the frequency is doubled at the same temperature, the motive entropy is 32% of the total. Doubling of that frequency further decreases the motive entropy to 20% of the total. Not only is the total entropy of formation at all these frequencies a poor es-

TABLE 1 Motive and Heat Entropies of Some Metals[a]

Metal	S (total)	S (thermal)	S (motive)	S (the.)/S (tot.)
Cesium	20.2	6.2	13.9	0.31
Lead	15.2	5.4	9.8	0.36
Gold	11.7	4.8	6.8	0.42
Copper	8.1	4	4.1	0.49
Chromium	5.7	3.3	2.4	0.58

[a]Entropies of formation at 298.15 K in cal/*M*/K [88].

timate of the motive entropy, but processes in which the frequency changes, e.g., contraction processes of solids and proteins, have further confusion in the change in fraction of motive entropy and energy with frequency change.

For systems that remain in the same phase between 0 K and *T*, e.g., the metals in Table 1, calorimetry can provide values of the motive and thermal parts as illustrated by the table since it is experimentally possible to determine the free-energy change at 0 K. Dry proteins may be similarly tractable, but even that is in doubt and the data have no useful application to processes of wet proteins at physiological temperatures. Most systems change state before they reach room temperature, so the calorimetric method is no longer applicable, there being no experimental way to make the separation for the phase transition. The only recourse is then to solve the appropriate system Schrödinger equation at *T*, a task presently impossible for systems like proteins that have large thermal terms and present particularly large uncertainty as to the meaning of *U*, *H*, and *S*. Experimental values of the first-derivative quantities *U*, *H*, *S*, and *V*, or their changes thus are unreliable sources of information about changes in *G* or *A*, all motive, but also are themselves difficult to interpret without some good idea as to the values of the motive parts. For example, the changes in motive and thermal parts may have opposite signs.

Hutchens et al. [89] measured the enthalpy, entropy, and heat capacity of dry and damp crystalline insulin and chymotrypsinogen A as a function of temperature. Had there been no changes in state over the range from 0 to 300 K, the motive and thermal parts of the enthalpy and entropy of formation at 300 K could have been determined. Even though protein crystals are very soft relative to the crystalline metals in Table 1, the computed motive/thermal fractions are very similar to those shown by these metals. This result shows that Benzinger's errors depend more on the kinds of degrees of freedom than on the hardness of the solids. The motive/thermal ratio for S and H depends on the relationship between mean vibrational potential energy and mean vibrational kinetic energy. In a vibrational-librational system, the former is the motive part and the latter the thermal part. The smaller quantitative differences among the metals and the protein reflect the quantum behavior of the onset of heat capacity with increasing temperature. The higher the crystal frequencies, the smaller the motive/thermal ratio. The more rapidly the degrees of freedom approach the classical limit, the larger their contribution to S at low temperatures, so the larger their motive terms at higher temperatures.

The most generally used method for extracting molecular information from thermodynamic data is by comparison with the same kind of data obtained for other systems. There is no rigor in this procedure but it is nevertheless the major use of thermodynamics by chemists and biochemists. It is apparent from Benzinger's discovery that there is even less rigor and less reliability in the procedure than has been believed.

3. The Hierarchy of Thermodynamics

The enforced separation of H, U, S, and V into thermal and motive requires a revision of the old hierarchy of thermodynamic quantities relevant to data for isothermal processes in which those quantities were grouped with G and A. The new hierarchy is:

i. Gibbs and Helmholtz free energies (always related to reversible work changes and thus always reliable).
ii. H and U, S and V: These first temperature and pressure derivatives of A and G contain work-related parts that contribute to free-energy changes, their motive parts, and heat parts that do not, their thermal parts. The heat parts are due to equilibrium thermal and mechanical fluctuations and cannot contribute to work.
iii. The heat capacity (constant pressure and constant volume) and all its temperature and pressure derivatives: These are pure measures of thermal and pressure fluctuations at equilibrium.

Quantities in level ii as mixtures of motive and thermal quantities have no rigorous connection with the other levels. The hierarchy can be extended to apply to isothermal processes of open systems.

It should be noted that although all equilibrium fluctuations contribute to H and S, so long as they do not involve changes in hamiltonians, i.e., there are no transitions among substates, the contributions are in the thermal parts and will not appear in G. Changes in potential energy as occur in transitions among substates may or may not appear in G. Though it may seem surprising, they generally will not. If these reflect only fluctuations, they complicate the effective hamiltonian and partition function but do not appear in the motive parts. Some details of these unexpected results are given in Ref. 14.

4. "Quantum Thermodynamics"

As can be seen in Table 1, the total entropies of formation and their motive parts are inversely related to the strength of the metal lattice. This is a consequence of the quantum nature of vibrational excitation, which requires use of the correct partition function rather than that for the classic limit. In applying Benzinger's discovery to a model system of harmonic oscillators, Rhodes discusses in detail this quantum feature of thermodynamics [90]. It can certainly be important for processes in which many vibrational and librational modes change either mode type or force constant. The partition function for a single harmonic mode is $(1/1 - q)$ with $q = e^{-h\nu/\kappa T}$, and ν is the frequency. Rhodes defines a "classic" region in which the mode is excited and a "quantum-aberration" region in which the mode is either unexcited or only partially so (*frozen mode*). He distinguishes these regions according to whether the quantum is greater than $\kappa T/h$ or smaller. At 298 K the frequency equal to this quantity is $\nu = 6 \times 10^{12}\ \mathrm{s}^{-1}$ so that by his reasonable definition those modes with lower frequencies are active and those higher are frozen. Although there is little reliable information about the frequency distribution in proteins, undoubtedly some modes in addition to those of the primary bonds fall in the quantum region. These stronger modes may become weaker or weak modes become stronger as a result of changes in conformation. Thus frozen modes may be thawed by denaturation or even frozen during function. Conversion of frozen modes to thawed modes makes positive contributions to the changes in H, S, and C_p. Just how big are these contributions likely to be in proteins at room temperature? In a rough way we can identify the low B factor values, group I, as indicating behavior found in strong molecular crystals. All other atoms are presumably like those in stiff rubbers. In neither knots nor matrices will the stretching and bending modes of the primary bonds of the polypeptide be much affected by the inclusion of the latter in the protein. Many backbone modes are, of course, not fully excited at ambient temperatures. The behavior of the heat capacity of protein crystals with temperature and the glass transition already discussed suggests that most secondary-interaction modes of matrices become classic at about 200 K. Knots retain their peculiar semiglass state at physiological temperatures, but only the modes in which atoms with the group I B factors participate are likely to be nonclassic. These are few in number but they will become classic in denaturation and the resulting contributions to entropy and heat capacity may be important in the overall thermodynamic bookkeeping. It is unlikely that the contractions associated with function as in enzymic catalysis will produce a significant number of frozen modes. The B factors of matrix atoms drop in the examples of this process for which data are available but are still in group II (see Secs. VI.D.3, VI.D.4).

B. Enthalpy-Entropy Compensation Behavior

1. General Basis of the Phenomenon

Because of the mutual cancellation of thermal terms in free-energy changes, enthalpy and entropy changes tend to have the same sign and same proportionate changes. This cancellation is exact and a rigorous consequence of thermodynamics but since thermal terms are only parts of H or U and S, the totals of the latter do not demonstrate the tendency exactly. This important consequence of Benzinger's discovery is not experimentally obvious in gas-phase primary bond rearrangements because the thermal parts are negligible. In processes in which there are changes in large numbers of secondary interactions, the thermal terms can be significant with respect to the motive terms. As noted in the previous section, any significant changes in nonfrozen vibrational and librational modes make thermal contributions roughly equal to their motive contributions. These effects can be large in the reactions of proteins and many other polymers but yet be overwhelmed by compensation of motive enthalpy by motive entropy. The latter has no rigorous or fundamental basis in thermodynamics and thus is a well-known subdivision of "extrathermodynamic" phenomena. It is shown below that most such phenomena also come under the heading of scaling. Scaling, to be accurate or even observable, will produce near-linear relationships between enthalpy changes and entropy changes in related series of reactions. As a consequence, most extrathermodynamic compensation plots are linear unless experimental precision is unusually high.

Linear plots of enthalpy change versus entropy change in a related series of reactions are extrathermodynamic. Such plots are extremely common in biology, almost ubiquitous, and they are frequently found in studies of liquid mixtures, especially aqueous mixtures. Extrathermodynamic compensation behavior is part of the well-known phenomenon called *isokinetic relationships* by Leffler and Grunwald in their classic monograph on the subject [91] and *linear enthalpy-entropy compensation relationships* by others including Lumry and Rajender [83]. Linear free-energy relationships, *LFE* relationships, and linear compensation behavior have led to the production of a vast, often contradictory and confusing literature. Possible molecular bases of the phenomenon and their relationship to thermodynamics have been suggested by many authors. In lieu of an exhaustive discussion, which may not be possible for many classes of the phenomenon, there are some general comments that are worth making, and these are presented in the following paragraphs using the simplest model, which has a further virtue in that it has some general applicability.

Linear *compensation plots* are plots of ΔH_i° versus ΔS_i° for a related series of reactions indexed by i. The series might be solubility of nonelectrolytes in aqueous solutions of systematically varied composition, rates of enzyme-catalyzed processes in congener series, exchange of residues as altering stability or reactivity of a single protein, pH variation of protein processes, and so on. With some rare exceptions, compensation behavior of this kind has a common basis. There are several ways to plot the data. The least likely to conceal error is the $\Delta G_i^\circ / \Delta S_i^\circ$ plot, but the $\Delta H_i^\circ/\Delta S_i^\circ$ plot is better for obtaining parameter values and for detecting LFE behavior. If there is linear compensation behavior, there is always an LFE behavior though it may not be linear in an arbitrarily chosen index and thus not obvious from an LFE plot based on that index.

The important parameter is the slope of the $\Delta H_i^\circ/\Delta S_i^\circ$ plot, called the *compensation temperature*, T_c.

The cancellation of thermal terms discussed in previous sections cannot explain the linearity observed nor a temperature-independent T_c, but it does provide a basis for explaining why compensation behavior arises from changes in secondary interactions.

The explication starts from a general LFE relationship and the corresponding enthalpy, entropy, heat-capacity, and volume relationships:

$$\Delta G_i^\circ = \Delta G_o^\circ + f(i)g \tag{11}$$

$$\Delta H_i^\circ = \Delta H_o^\circ + f(i)h - \Delta G_o^\circ T\left(\frac{\partial f(i)}{\partial T}\right)_P \tag{12}$$

$$\Delta S_i^\circ = \Delta S_o^\circ + f(i)s - \Delta G_o^\circ\left(\frac{\partial f(i)}{\partial T}\right)_P \tag{13}$$

The step quantities g, h, and s and the motive and thermal parts of h and s arise from the coupling to other processes. Assuming the temperature derivative of $f(i)$ is independent of i, the compensation relationship is given by Eq. (14).

$$\Delta H_i^\circ = \Delta H_o^\circ - \frac{h}{s}\Delta S_o^\circ + \frac{h}{s}\Delta S_i^\circ \tag{14}$$

$$\text{Here } T_c = \left(\frac{h_m + h_t}{s_m + s_t}\right) \tag{15}$$

The temperature dependence of T_c is generally small:

$$\left(\frac{\partial T_c}{\partial T}\right)_P = \frac{c_P}{s}\left(1 - \frac{T_c}{T}\right)$$

In these processes as in all isothermal processes of closed systems $h_t = Ts_t$ (in fact $h_t \equiv Ts_t$) so the constancy of T_c in any example requires that all the four quantities in the bracket in Eq. (15) change with the series index by the same factor. Accidental simplification includes thermal parts negligible relative to motive parts or accidental choice of temperature and other experimental variables that yield a ratio of motive enthalpy to motive entropy near the mean experimental temperature. Gas-phase primary bond reactions have negligible thermal parts so any compensation behavior observed reflects a linear relationship between motive entropy change and motive enthalpy change. The argument based on scaling is trivial in this case.

But in the general case, the motive and thermal parts of each $\Delta H_i°$ and $\Delta S_i°$ are scaled by the same factor for each i so the bracketed quantity in Eq. (15) is the unit step for all i. Using Grunwald's terminology [93], in each case all thermal and motive parts are said to be *scaled* by the same factor and the factor is constant within experimental error for the series and the temperature. This is not a very interesting explanation for compensation and it is not exactly consistent with the extrathermodynamic nature of the phenomenon. The solubility of a set of alkanols in water is a well-known example of compensation. The enthalpy and entropy changes are found to be ordered along a single line in a way that can be rationalized on the basis of changing mass or of changing number of methylene hydrogen atoms. Methanol falls off the line because its alkyl group is free to rotate about the HOC axis. The other linear alkanols have different moments of inertia, which is another way they can be considered to be scaled. With only moderately precise data one can see that the branched alkanols really do not belong on the same line with the linear alkanols. Scaling only produces reasonable precision in linearity when there is only a single quantity that has to be scaled. In the alkanol example that would probably be the number of water molecules actually involved in accommodating the solutes. But that is not easy to determine although it may be the objective of the study that disclosed the compensation behavior. A very large family of protein compensation examples arises because a measured process and a driving one are coupled in a nonobligatory way through protein conformation. Change in one process perturbs the conformation and the conformation then perturbs the other process. The series then might be congener families of reactants or products, pH changes, or residue exchanges that perturb the conformation. In these examples it is the conformation change that is scaled. So long as the perturbations are not so large as to change the mechanical character of the conformation, linear compensation behavior results.

Grunwald has developed the scaling explanation without using the thermal-term cancellation [93]. The compensation of the thermal parts of the unit stepwise enthalpy and entropy changes as included using Eq. (15) provides a better, rigorous basis. It is now quite general that regardless of the specific source of the compensation phenomenon, a process demonstrating linear compensation must contain one feature that introduces additivity. Excluding methanol, the linear alkanols are so near perfection in additivity that even the highly precise heat-capacity changes obtained by Wadsö and co-workers do not depart from additivity [94]. Congener series, such as the alkane series, on transfer from gas to water, scale the same way by mass and number of methylene protons. All have essentially the same polarizability. If $f(\text{mass}) = \gamma f(\text{methylene proton number})$, γ being a constant and other characteristics of the members of the series are not quantitatively significant, compensation behavior is linear. But if *f*(mass) and *f*(polarizability) are not strictly linearly related, and both features are important in the process studied, compensation is not linear. The noble gases provide an example [95,96]. Because the latter condition is not satisfied for the noble gases in transfer from gas to water, the experimental transfer enthalpy and entropy changes yield nonlinear compensation behavior. One can use either the mass or the polarizability but each gives a different slope. One would expect scaling either by mass or by polarizability to produce the linear behavior, and both undoubtedly exist, but the scaling factors for the two characteristics do not vary through the family in the same way. The atom itself is not a suitable index for linear compensation.

Systematic changes in chemical reactivity, usually in rates, are the most familiar source in organic chemistry. The prototype for most is the famous Hammett "sigma-rho" relationship. Compensation behavior is a consequence of systematic small changes in electron distributions. These too can be considered examples of scaling, but less profitably. The scaling explanation for compensation does not make the phenomenon a particularly useful source of information, but by being more specific about the things scaled, the phenomenon is potentially a major tool in converting thermodynamic and rate information to molecular information. Thus the examples arising from linkage of subprocesses of a protein through conformation can provide unique information. For example, the fact that the first three steps of the four-step ligation processes of ferrous and ferrichemoglobins demonstrate compensation of this type establishes unique limitations on the mechanism responsible for linkage.

Linkage compensation, tests for statistical reliability of observed compensation behavior, conversion to an activity-coefficient formalism, and the uses and misuses of temperature and compensation temperature are discussed most correctly and in most detail in Ref. 14, but a complete and correct discussion of the broad and often bewildering LFE and compensation phenomena has not been published.

When enthalpy-entropy compensation behavior is a result of simple scaling, there must also be linear relationships for each pair of scaled thermodynamic quantities (vide infra). That between standard entropy change and standard volume change is well known, and the linear entropy–heat capacity relationship has attracted attention in data for processes in aqueous-solutions and studies of protein processes mediated by conformation readjustments. Usually the latter include conformation changes to link primary-bond rearrangement reactions. The latter are correctly written as stoichiometric relationships. The conformation changes are not stoichiometric but are rather changes in the distribution functions for conformers even though they may appear to reflect exchanges between two distinct, narrow distribution functions. Frauenfelder et al. have shown that they are often nearly continuous. Compensation behavior, often over a major part of the total advancement of the linked processes, must in general be attributed to the distribution feature of conformation states. Real stoichiometry can produce linear compensation only over a very limited range of advancement [98]. There are several approximations that yield experimental linearity over large degrees of advancement. The explanation we have usually given is based on weak coupling among subprocesses either directly coupled or coupled through a common subprocess of the system such as change in protein conformation. Weak coupling allows the use of the activity-coefficient formalism and thus the linkage formalism of Wyman and co-workers. It is because of the latter connection that this kind of enthalpy-entropy compensation behavior has been called *linkage compensation* [14].

Proteins may provide a special situation in which the weak-coupling formalism remains a good approximation even though quite strong coupling actually exists. Characteristic biological free-energy differences in a single enzymic process, for example, are of the order of 10 kcal/mole and the activation process somewhat more. These are well above the weak-coupling limit, but Gregory has shown that if the potential-energy function for coupling can be approximated by a mean field, which changes only slowly with advancement, linear behavior can occur even though large free energies and enthalpies of coupling are involved [98]. In view of the nearly ubiquitous appearance of linear compensation behavior in protein processes, this is an important deduction. Linear compensation behavior in linked systems makes them demonstrate *linear-response* behavior, and its basis of the mean-field approximation makes the latter appear even under conditions of moderately strong coupling. The advantages for nature and the experimenter that arise from the validity of this approximation are discussed in Sec. VI.A.

In conclusion, it is worth noting explicitly that every linear compensation example implies the existence of an LFE process and thus provides a basis for the linear relationships between any pair of the following: ΔA_i, ΔU_i, ΔS_i, ΔV_i, $\Delta C_{v,i}$, and so forth (also Gibbs energy change, enthalpy, etc.) mentioned above. Because the scaling factor is explicit in LFE expressions, it is not always easy to find the scaling function $f(i)$ and thus the LFE itself. Compensation behavior is a better source of information than LFE behavior simply because the scaling factor cancels out the compensation expression. Every LFE has a corresponding compensation relationship. More data are required to establish the latter, but it provides the correct scaling function and the most useful parameter for both expressions, which is the compensation temperature.

2. An Example: The Preservation of Rank-Order in Proton Exchange of Proteins

An interesting example of protein compensation behavior is the preservation of rank-order in proton exchange discussed in Sec. II.C.2. In fact, the preservation of rank-order requires that there be linear compensation behavior for the pairs of activation enthalpy and activation entropy changes for the exchange sites [5,99]. A clue to the molecular basis of both phenomena is provided by the ring-flipping events observed in NMR signals. These indicate large stochastic fluctuations, which briefly release the spatial restrictions on phenyl and phenolic groups to the degree that they can flip around an in-plane axis. Lesser fluctuations must occur more frequently and very extensively and thus provide, at least in part, a mechanism for the diffusion of exchange catalysts [92]. Hence we can postulate that the exchange rate constants are the product of a preequilibrium catalyst-diffusion term dependent on stochastic free-volume rearrangements and an intrinsic rate constant, k, for the actual exchange reaction between catalyst and amide group. Proceeding from this assumption, the average path to the ith site is $p^{j(i)}C$, in which C is the catalyst concentration in the solution. The rate of exchange at the ith matrix site is then $R_i = p^{j(i)}kC$. The apparent activation free energy for the ith site is Eq. (16).

$$\Delta G_i^{\pm} = \Delta G_k^{\pm} - j(i)RT \ln p \tag{16a}$$

$$\Delta G_i^{\pm} = \Delta G_k^{\pm} - j(i)\, \Delta G_p \tag{16b}$$

Since $f(i) = j(i)$, this is another example of simple compensation behavior discussed in the previous section. On eliminating $j(i)$ from the equations for the apparent activation enthalpy and apparent activation entropy, the compensation relationship [Eq. (17)] is obtained. The T_c is given by Eq. (18).

$$\Delta H_i^{\pm} = \Delta H_k^{\pm} - \left(\frac{\Delta H_p^{\pm}}{\Delta S_p^{\pm}}\right)\Delta S_k^{\pm} + \left(\frac{\Delta H_p^{\pm}}{\Delta S_p^{\pm}}\right)\Delta S_i^{\ddagger} \tag{17}$$

$$\left(\frac{\Delta H_p^{\pm}}{\Delta S_p^{\pm}}\right) = T_c \tag{18}$$

The compensation temperature is written as the ratio of an enthalpy change characteristic of the jump step to the corresponding entropy change. An alternative mechanism might depend on the equilibrium distribution of catalyst at each exchange site, and the equations are also appropriate for that alternative. However, it seems unlikely that the equilibrium catalyst concentrations would be determined by a single parameter.

The apparent activation enthalpies and entropies extrapolated to the most rapidly exchanging matrix site taken as an estimate of $j(1) = 0$ are those for k. The adequacy of this simple model, regardless of the mechanism that finally proves to be correct, implies that matrices are highly homogeneous. It suggests that p is determined by fluctuations of a matrix, perhaps a single major fluctuation such as the expansion-contraction fluctuation discussed above. If so, proton-exchange provides a means for measuring the response of these fluctuations to temperature, pH, and some other variables. The probability-density distribution of exchange rate constants gives some information about the density of matrix sites relative to the protein "surface," but not much. Back exchange to sites that have undergone first exchange has not been included in the development given above, and the experimental rate-constant distribution will be considerably distorted by this process.

The chemical-exchange constant, k, is apparently the same at all matrix sites since the compensation behavior is linear within relatively small experimental errors. From this we have deduced for the derivation above that rank-order conservation implies conservation of rank order in the diffusion pathways and in the relative times for a catalyst molecule to reach a site. The latter suggest that the free-volume rearrangements required for catalyst diffusion are nearly always produced by a single matrix fluctuation mode. This possibility receives considerable support from existing literature, and as noted earlier, the most promising candidate is the more or less gestalt expansion-contraction of matrices. That this may have important implications for enzymic catalysis will become clear in Sec. VI.D.3.

The model does not apply to knot exchange via melting since the significance of i is lost. On the other hand, the model can be applied to the compensation behavior observed in exchange from knot sites in the absence of melting, but only if it is modified to take account of the distortion of the knot itself. This follows from the fact that the probability of the necessary distortion varies with the site. Thus in addition to catalyst diffusion (no exchange by unfolding), there is an additional probability factor due to the distortion. But in fact we know the process is quite simple since the T_c values for the forward constant in denaturation and for the exchange process of knots are the same within rather small errors. This identifies the dominant process in both cases to be knot expansion, varying in amount for exchange at each knot site. Distortion then is simply expansion toward the transition state. In this way the exchange sites are exposed in order of decreasing constraint in the knot. Exchange is not limited by diffusion of catalyst since that process is fast compared to transient knot expansion. It is not limited by the recontraction of knots since that is very fast. First exposure of a knot exchange site may be sufficient to accommodate the formation of the chemical-exchange complex and transition state with the catalyst so exchange occurs. Single exposure may not be sufficient, but the mechanism is otherwise the same, and the probability factor for failed exchange on each exposure is probably small and very similar for all knot sites. This is just the sort of behavior Kim et al. [49] have found with the knot mutants of BPTI (see Sec. III.C).

These deductions return us to the puzzle of the constant T_c for the forward constant in denaturation. It will be recalled that all available data say that near 353 K proteins have the same value of this constant and the same intercept on the activation-enthalpy axis. The only obvious explanation seems to be the correct one, unreasonable though that may seem, that all knot-matrix proteins that lie on this compensation plot have knots that are very similar. At their high degree of contraction, the atomic groupings being of the same kind and organized in the same way, their conformations are dominated by the same potential-energy function. The additivity feature, that is, the position of their points on the compensation plot, arises from differences in degree of contraction and knot size, but the ratio of enthalpy advancement to entropy advancement is the same for all and quite possibly a requirement of the applicable basic electrostatic theory. Local distortion for exchange at a site is a function of the whole knot rather than local structure. Again we note this is in excellent agreement with the studies of BPTI mutant knots [49]. What is not yet clear is how cosolvents like propanol and urea alter the thermodynamic changes in expansion and contraction. They both reduce the activation enthalpy and entropy but urea maintains T_c near 355 K.

Because the increase in potential energy during expansion of one of these knots is purely electrostatic, there are few ways to change the nonelectrostatic part of the entropy. It is possible then that the observed activation entropy is primarily a manifestation of the unfreezing of librational and vibrational degrees of freedom due to the secondary forces holding the knot in contraction. The configurational entropy increase, that associated with the translational degrees of freedom, is probably very small until the transition state is passed. With any of these sources the activation entropy and enthalpy are entirely motive, which means their total values contribute to the activation free energy. The activation heat capacity is zero, so T_c is independent of temperature. This is an unusually simple situation and it makes T_c a reliable source of information about motive quantities. Thus 353 K is a close estimate of the ratio of motive enthalpy change per unit of motive entropy change. When a series of obviously related processes manifest linear compensation behavior, the members are related by changes in amount, not in quality. The particular chemical basis is the same and the members differ only in degree of advancement of this common feature.

VI. CONFORMATIONAL DYNAMICS AND "DYNAMIC MATCHING"

A. The Facts

1. Theoretical Tools

a. Some History For some years it has been reasonable to suspect that the unique features of enzyme function, high rates and specificity, are manifestations of dynamical machines built using conformational fluctuations. Nevertheless, until recently most considerations of such mechanisms have continued to be based on conventional "cold-earth" chemistry. The "transition-state" hypothesis does little to redress this error since it is still based on a hard protein and thus inconsistent with the data. Fortunately, the shear weight of evidence for an essential dependence on conformational fluctuations is slowly clearing the deck for more profitable investigation. In this section "fluctuation" hypotheses advanced for the function of enzymes and hemoglobin are reexamined in light of the structural information presented earlier. Features not previously discussed in detail, such as "dynamic matching" of conformational fluctuations, are also introduced to accommodate the new information in the old hypotheses. There are two important ground rules in formulating these hypotheses. The first is that the entire mechanism is to be found in the protein itself. Earlier Lumry and Rajender proposed that bulk water was coupled into the catalytic process not necessarily as a source of activation energy, but rather as a device for interchanging entropy and enthalpy using the two-state process of water [83]. There may be some truth in this, but studies of proteins at low hydration establish that full catalytic activity arises well below the vapor pressure of bulk water. The second rule is that the conventional kind of transition-state formation by thermal fluctuations is at best only part of the activation mechanism. All "rack" hypotheses have been developed with the postulate that mechanical work is an important source of activation free energy. This necessarily requires consideration of the equilibrium aspects of stress and strain, but now these can be examined as consequences of dynamical behavior.

The motile nature of protein conformations was established by Linderstrøm-Lang and co-workers in their demonstration that all protons could be exchanged with water protons. Eyring et al. reasoned by an exclusion process that protein functions must depend on conformational fluctuations because the latter make possible the substitution of mechanical energy for heat. However, early interpretations of protein x-ray–diffraction results appeared to establish that proteins were quite rigid and had nearly all H-bonding possibilities satisfied. This is in sharp contrast to the implications of Linderstrøm-Lang's results and the claims of Lumry and Eyring. Modern interest in conformational dynamics begins with the paper by Lakowicz and Weber [100] on oxygen diffusion into proteins measured by quenching of tryptophan fluorescence. This single observation forced a reconsideration of x-ray–diffraction interpretations and the thermodynamics of protein folding. The discovery of "mobile defects" using proton-exchange data [22] and their description in terms of transitions among substates developed by Frauenfelder and co-workers [97] using myoglobin ligand-recombination rate data have in recent years been supplemented by data from a wide variety of studies of protein conformational dynamics. The discovery of knots and matrices has revealed much more detail about the mechanical heterogeneity of proteins and in this way has provided a basis for understanding the large amount of information on conformational dynamics now available. The understanding requires use of several theoretical tools. These are discussed briefly in general application before we describe the specific molecular hypotheses of mechanisms for several functional classes of proteins.

b. Probability-Density Distribution Functions In a system at equilibrium, the fluctuations in enthalpy or entropy are fully described by its temperature and pressure derivatives, which give the moments of the probability density distribution function. In principle these can be used to construct the enthalpy and entropy probability density functions *pdf* [14]. This relationship, due to Gibbs, is now reasonably well known but almost never applied because too few moments are experimentally available to provide any detail. However, the first few moments are sometimes quite informative. The moments are the expectation values of the powers of H or H-$\langle H \rangle$ determined by integrating the product of the nth power of H times the probability of H over the range of allowed H values. This is carried out for any T, but the joint PT moments of H can also be determined by a similar procedure [14]. For $n = 1$, the corresponding moment is $\langle H \rangle$. The second moment of H is the variance of the distribution function; the third is its skewness, and so on. The moments are thus related to the derivatives of H with respect to T. The pdf is the normalized partition function and as such provides a description of the probabilities of occurrence of the accessible states. The latter is the normalized pdf for enthalpy fluctuations at thermal and mechanical equilibrium. Cooper's application to proteins is faulted by the assumption that the protein has a special construction which makes the unit for thermal equilibration a large fraction of the protein [101]. As applied to the enhancement of enzymic catalysis, it is the old "contributing degrees of freedom" hypothesis long since laid to rest. Conformational fluctuations at constant P produce volume fluctuations as well as fluctuations in energy independent of V, which might be called true heat in rough consistency with the behavior of isochoric systems. As regards true heat, each small region of the protein is independently in equilibrium with the thermostat and thus with the other local regions of the protein. There is no way to prevent this in a soft system since almost by definition of the term "soft" as applied here, the characteristic times are very long compared with thermal-equilibration times. Even the knots appear to be soft in this respect. Cooper's argument would have been quite correct had he applied it to PV fluctuations since the ability of these to be controlled by protein construction appears to be the basis for protein machines (see Sec. VI.D.3).

C-H. Chen [102] has applied the method of moments to data for pure water using a four-parameter approximation to the partition function. More correctly stated, he has used the temperature derivatives of V to evaluate the normalized partition function. The absolute partition function cannot be estimated in this way but the shape information provides the most rigorous evidence possible to substantiate the mixtures model of water. His procedure is applicable to proteins so that with the heat capacity, its first temperature derivative and a reasonably reliable estimate of the second equilibrium distribution of heat, and thus the dynamical spectrum at equilibrium, can be described [105]. The latter is particularly useful if the protein is a "linear-response" machine (vide infra). If it is not, the theory of nonlinear coupling, called "chaos" theory, is required (Sec. VI.A.1.g).

c. ***Equilibrium Dynamics and Linear-Response Behavior*** We discuss the distribution functions and their uses in the study of conformational fluctuations because it is not improbable that the fluctuations important for the physiological mechanisms of many proteins are predominantly the fluctuations at equilibrium. Insofar as that is true, thermodynamics provides the essential theoretical basis, which is much simpler than time-dependent theories. This is Onsager rather than Eyring, chemical-relaxation theory and reversible thermodynamics rather than unrestricted transition-state theory.

For purposes of this chapter, linear response is described by such equations as $\Delta Q = C_p \Delta T$ with $C_p = \sigma_H{}^2/RT^2$ in which $\sigma_H{}^2$ is the variance of equilibrium probability density distribution function for the enthalpy. The response of Q to a change in T is dependent on the existing fluctuation spectrum of H and is linear so long as ΔQ and ΔT are not too large. There are many such linear relationships that connect changes linearly through one or another parameter of the fluctuation distribution. Thus $\Delta V = -[\sigma_v{}^2/RT]\Delta P$ relates the response of V to a pressure perturbation through the variance of the volume. Diamond has an unexpectedly large change in V for a small change in P because of its high volume variance. These are examples of what is called linear-response behavior. The response of a system to a perturbation is governed by the fluctuations at equilibrium, which are the same in the absence of the perturbation as in its presence. The hypothesis of interest then is that the conformational fluctuations with which protein mechanisms such as those of enzymes are constructed are those occurring at equilibrium. If this is in fact true or a close approximation to the truth, most fluctuation information needed to describe and understand these mechanisms exists in native proteins at the temperature and pressure at which studies of mechanism are made and is contained in the higher derivatives of the partition function. This has important experimental consequences since it makes thermodynamic quantities determined at equilibrium powerful sources of information about rates as well as equilibrium, which depend on conformational dynamics. Heat-capacity data obtained at constant T and P as a function of advancement of a protein process sample the macrostate population changes, usually chemical events or changes in the protein substates but also the microstate changes reflecting conformational dynamics in these states. The oxygenation of hemoglobin studied in this way by Lumry et al. [103] is a good example. Later in this chapter we argue that linear-response behavior has advantages in evolution as well as the practical advantages it provides investigators.

d. Nature Substitutes Mechanical Work for Heat Although thermodynamic data cannot provide temporal information directly, they can be useful in interpreting time-dependent quantities. Transition states for difficult processes such as primary-bond rearrangement have very low probability of occurrence because their formation requires large thermal fluctuations. As a result, they do not significantly influence the microstate distribution functions. Rate processes are reasonably well treated by the theory of absolute reaction rates which uses thermodynamic concepts but introduces time by an ad hoc assumption. If all states of an enzyme along its reaction coordinate are well-populated states at equilibrium, the formation of the low-probability transition states can be treated as a conventional rate effect involving no more atoms directly than in small-molecule reactions. This is necessary for application of absolute rate theory. When passage through the transition state is rapid relative to changes among conformational states, the rate process is adiabatic and an enzyme process can be treated by combining a true thermodynamic theory of equilibrium fluctuations with conventional ("small-molecule") rate theory. The conformation fluctuates into states that lie close in free energy to the transition state for the rate process, and the transition from one or more of these pretransition states to the transition state is a "small-molecule" process. In principle this does not deviate from the assumptions made in the development of absolute rate theory. In practice it divides the enzymic or other protein rate process in such a way that the special features of the protein as catalyst are to be found in the equilibrium fluctuation behavior of the protein. The conformation provides mechanical work and the small-molecule process makes up the difference by transient conversion of heat to work. If the mechanical work is supplied in an isoergonic process, the probability of the transition state occurrence is proportional to $e[(\Delta G^{\pm} - \Delta W)/RT]$ in which $\Delta G^{\pm}$ is the Eyring free energy of activation and $(\Delta G^{\pm} - \Delta W)$ is the apparent activation free energy. The latter may or may not govern the uncertainty lifetime. If free energy of protein and substrate are simply redistributed (destabilization of the pretransition states), there is no such restriction. If conformational PV fluctuations contribute to the activation free energy (transition-state stabilization), the lifetime of the latter is limited by the uncertainty principle just as in the case of total activation by heat.

Nature can experiment with different electronic situations that lower $\Delta G^{\pm}$ and these have preoccupied most scholars of enzyme chemistry. However, a more fruitful focus of evolution is ΔW. A ΔW of 14 kcal/mole is equivalent to 10 orders of magnitude in a rate constant. Progress in evolution toward maximizing rate is then made by finding mechanisms to provide ΔW which more and more closely approach the isoergonic condition. In general, although the original idea of stress has sometimes been incorporated in mechanisms of enzymic catalysis, the assumption has persisted that enzymes only minimize $\Delta G^{\pm}$. Rack mechanisms focus on maximizing ΔW but nature carries out experiments with both, so one expects enzymes to depend on hybrid mechanisms. We can expect to have difficulty in distinguishing the parts.

e. Mean-Field Approximations for Electronic Potential Energy Accurate solutions to the exact hamiltonians for many-atom problems in which secondary interactions are important are a very long way away. Polarization, multibody interactions, the free-energy problem, and its averaging over coordinate fluctuations are all too difficult for the best of modern computers. Treatments of critical behavior are forbidden for some of the same reasons, of which a wide distribution of highly populated macrostates is most important. This complication has limited theoretical treatments of critical phenomena to mean-field theories. The mean-field concept is that an atom or group is subject to a potential that is the static average of the individual potentials it experiences from its neighbors and usually also the time average. If the linear-response behavior of protein conformations is sufficient to support the physiological mechanism, mean-field theories are applicable to these problems and in fact have been so exploited for enzymic catalysis [104]. More exact theories may be useless because the average values of coordinates are in considerable error relative to van der Waals distance dependencies and also because the large B factors tell us that atoms spend little time at their true lattice points.

The several sources of quantitative and thus qualitative uncertainty require that the use of explicit but necessarily curtailed potential-energy functions developed with small molecules be abandoned in favor of the inexact but tractable *mean-field* methodology to provide physical insight even if not prediction of experimental numbers. Applications of mean—field approximations to the understanding of enthalpy-entropy compensation will be made in later papers [105].

f. Resonance and the "Fluctuations-Dissipation" Theorem In a system of oscillators, those with the same frequency communicate even with very weak coupling. This classic physical concept and its consequences are described by an elegant theory called "fluctuation-dissipation" phenomenon or theorem. Careri has explored the possibility that some of the energy of activation in enzymic processes is provided in this way from other processes, particularly ionization and counterion binding, coupled to the protein. Experimental attempts to support the idea do not appear to have been successful, and like Careri's concept of "proton percolation," the use of energy transfer to effect enzyme catalysis via the fluctuation-dissipation theorem requires reconsideration and probably revision in light of the new ideas about protein hydration. (See Chapter 3.)

The term "resonance," arising initially from considerations of pure energy in isolated small samples has many applications in chemistry in its original form but it has also come to be applied to thermodynamic systems with two or more macrostates. Specifically, two macrostates of such a system are said to be in resonance (at constant T, P, and total mass) when they have the same free energy. The chemical potentials of the species of any component are equal so the species populations are equal. In simple two-state systems this is the condition of maximum heat capacity. By analogy with strong first-order phase changes, the temperature at which this occurs in protein denaturation reactions is often called the melting temperature. The extension is much used in thermodynamics and statistical mechanics and applies only to systems with large numbers of particles. Since only mean values have any significance, deductions based on thermodynamic quantities always apply to hypothetical average (most probable), chemical species, and differences in those quantities to differences between macrostate averages.

As is very well known, in a two-(macro)state system the "between-states" heat-capacity term is $\alpha(1 - \alpha)$ $[(\Delta H^\circ)^2/RT^2]$, in which α is the mole fraction of one species and ΔH° is the standard enthalpy change per mole of chemical species or more generally per mole of cooperativity unit, the total number of degrees of freedom that change as a unit in the transition from one macrostate to another. The cooperative unit determines the separation and volumes in phase space. The between-state heat-capacity is largest at resonance so it is sometimes called the "resonance heat capacity."

A large heat capacity means that the temperature change per increment of heat intake is small. Thus the higher the heat capacity, the larger the range of linear response to a temperature perturbation. Protein systems poised at the resonance condition exhibit linear response behavior with respect to temperature and pressure perturbations but also to any perturbation of their population ratio. Acid-base buffering is in this respect the same as temperature buffering. The relevance of this matter for protein processes is better illustrated with the ligation of ferrihemoglobins as a function of pH, residue composition, variation in ligand kind, and so on. Because of its compensation characteristic the system is almost perfectly balanced at the resonance point with respect to both temperature and pH perturbation at 295 K and as a result has no detectable Bohr effect near this temperature nor does it display sensitivity to residue composition and so forth. The standard enthalpy for binding four ligands decreases as pH increases to a minimum near the isoelectric point and then rises at alkaline pH values to its low-pH value. The total change is −8 kcal/mole. Lumry and Rajender [83] gave many examples of this compensation from the study by Beetlestone and co-workers [105], and they suggested that the basis for all manifestations of compensation by hemoglobin systems was the two-state behavior of water. That water does have two or more macrostates seems now to be established by Chen [102], but it has become less and less certain that water is the major source of compensation in proteins. Unfortunately, a critical or even detailed consideration of this situation is precluded in this chapter by space limitations. Compensation phenomena are definitely correlated with the expansion-contraction processes of proteins, but some of the data necessary to untangle that process from the water process as sources of protein compensation are still missing. The compensation pattern is manifested by pH, residue exchange, and some other variables used to produce compensation in the ferri- and ferrohemoglobin family. Figure 16 shows compensation behavior of chymotrypsin in inhibitor binding and in normal steady-state kinetics. This figure and the phenomenon it may reflect are discussed briefly later.

Since heat capacity as the variance of the enthalpy distribution measures the width of the probability-density function of the enthalpy, a large heat capacity means large fluctuations in enthalpy. The width narrows with any change that makes the free-energy difference between the two states different from zero. An important example is the association of two proteins. If the two are dynamically

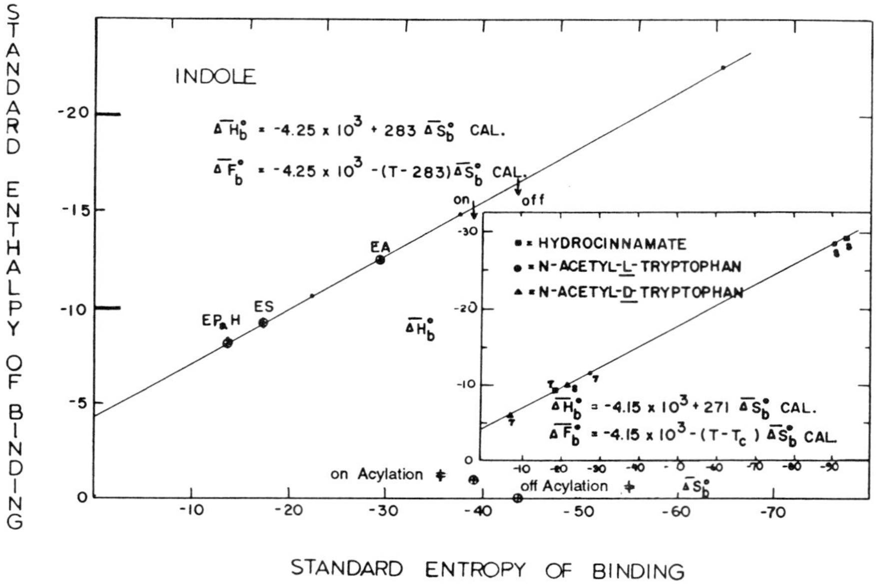

FIGURE 16 Enthalpy-entropy compensation patterns from α-chymotrypsin. The compensation line is from Yapel's direct measurements of the binding of indole as a function of pH [136]. The points EPH, ES, and EA shown on this line are from the steady-state rate studies of hydrolysis of *N*-acetyl-L-tryptophan methyl ester [137] at pH 7.5 as are the crossed circles at the bottom. The downward-pointing arrows on the line refer to the latter and estimate the contributions from the domain-closure process to activation enthalpy and entropy changes in the "on-acylation" and "off-acylation" rate processes as described in detail in Refs. 9 and 54. The compensation line in the insert is for Yapel's result in measuring the equilibrium binding of hydrocinnamate, *N*-acetyl-L-tryptophan, and *N*-acetyl-D-tryptophan. The variable used to detect these compensation patterns was pH. The errors in the steady-state parameters are significant, so the appropriate compensation temperature may be in error by as much as 10°. The compensation temperatures for the inhibitors and virtual substrates have an experimental error of only a few degrees, but the simplicity of the results shown may be deceptive since there is no established mechanism for inhibition of the serine proteases. The latter confusion arises in part because the effect of pH on equilibrium and rate constants at temperatures near 298 K is minimized by enthalpy-entropy compensation. (Redrawn from Ref. 9.)

matched over an otherwise suitable contact region, their total heat capacity will remain about the same in association as in separate proteins. It may even increase. On the other hand, if dynamic matching is poor, the proteins can only suppress each other's fluctuations on association. This reduces the heat capacity, narrows the enthalpy distribution, and depresses the thermal and motive parts of the entropy. The entropy depression is called the Helfrich effect (vide infra). When the dynamical spectra are similar in the region of contact, which in turn reflects to some extent the entire matrix regions, we say the proteins are *dynamically matched* or *tuned* to each other. Other things being equal, proteins tuned to each other will associate much more strongly than those pairs that are not tuned. Although the quantitative consequences of the Helfrich effect are difficult to estimate for proteins, it may be more important in proteins than in membranes, in which case tuning is likely to be the major basis for specificity in protein association. Similarly, tuning and detuning of associated proteins or subunits of a complex protein can provide a mechanism for free-energy redistribution. This is discussed below as a basis for linkage among ligand-binding sites in hemoglobin. Its application to other multisubunit proteins, multienzyme systems, and tissue assembly has been considered elsewhere [11].

g. Theory of Nonlinear Systems: Chaos The fluctuation-dissipation theorem, resonance, and the entire theoretical structure of molecular dynamics developed for small molecules are important to understand the dynamical structure of conformations. How complicated are the modes? What is special about modes able to participate in physiological functions? Can normal-mode analysis of proteins be simplified enough to be useful? Are normal modes useful in understanding systems with substates? These questions are about time averages and unlikely to be easily answered or even very useful in understanding the dynamical behavior of protein matrices since the latter are consequences of very nonlinear equations of motion. The behavior then lies in the realm of the theory of chaotic processes. Havsteen is a pioneer in the application of the theory to protein processes [106]. The major input in such applications is the *B*-factor data and from these Havsteen has exposed an internally consistent picture of dynamic behavior in chymotryptic catalysis, which considerably exceeds in sophistication other treatments of enzyme dynamics. The jump from the latter to chaos is mathematically complicated but now well known and well covered in the literature, e.g., Gleick [107]. Havsteen's summary of the concepts is an entirely adequate preface to his applications to chymotrypsin. The latter, presented briefly in Sec. VI.D.9, introduce the possibility that any realistic discussion of protein functions even at an elementary level may have to be based on nonlinear theory. The *B* factors appear to be the only kind of observation generally useful for the applications, but more extensive exploitation of relaxation methods such as ultrasound is needed to confirm and extend mode structure extracted using chaos theory.

B. Protein-Protein Association

1. Surfaces and Contact Types

Elegant x-ray–diffraction studies such as those of Perona, Tsu, Craik, and Fletterich [108] on the complex of BPTI and trypsin display a wealth of structural detail of interactions between proteins in protein-protein association, but there is little reliable thermodynamic information about the association. Accurate data for protein-protein association processes are in surprisingly short supply. The estimates extracted from models have proved unreliable data, and the models are more complicated than suspected in traditional chemistry: e.g., Zundel's elaboration of the great variability of the hydrogen bond [109]. It is generally assumed that a detailed examination of the time-average structure in the region of contact will lead to quantitative understanding of the free-energy changes. But there are macroscopic effects, primarily dynamic ones, that cannot be explained, even detected, with this approach. The Helfrich effect and a time-average enthalpy effect discussed in the following sections are two dynamical factors in association that nevertheless have thermodynamic foundation at both levels ii and iii of the hierarchy. Mutual three-dimensional distortion usually occurs even when micron-sized particles of solids associate. This static effect should be increasingly significant as particle size decreases [11]. Proteins may have been evolved to accommodate to this effect, but more likely evolution has exploited it. The mysteries of cell assembly imply the existence of still more subtle evolutionary discoveries of critical importance to the association processes of biological macromolecules.

Previously we grouped protein-protein association processes into three classes: *wet*, weak association between protein surfaces separated by nearly complete layers of water; *damp*, association dependent on a smaller number of water molecules filling holes, tying off H-bond valences, and so forth, at the contact surface; and *dry*, association in which water may be involved but most contacts are direct interactions between protein groups. Since the strong and highly specific associations are probably always dry, intercalation and mutual refolding are possible for the dry class, and it is the class which is most sensitive to dynamical aspects of protein conformations.

2. Dynamic Matching: The Helfrich Effect

The large *B* factors of most matrix and surface atoms in knot-matrix proteins demonstrate average displacements from ideal atom lattice sites of the order of 0.2–0.5 Å, and these may be considerably larger when the protein is freely dissolved. These are large relative to the distance dependence of Keesom, London, and hydrogen bond forces. The last applies to real hydrogen bonds. Ion-dipole interactions are stronger and have a weaker distance dependence. These displacements measure the mean amplitudes of fluctuations, and regardless of their source in long-range versus local fluctuations, they provide some ground rules for association between proteins that are nonspecific in contrast with the tailored interactions. The bases for two such rules are quite simple and only partially independent. The enthalpy rule for association of macromolecules and motile surfaces is well known. Short-range interactions among the groups that make up the two contact surfaces will produce a large time-average decrease in the enthalpy only if their motions are coordinated. Although the degree of such matching required for stable association in any case depends on the contributions from other factors, in general, some minimum of dynamical matching is necessary to make the enthalpy contribution favorable to association.

In the limit of perfect matching, the dynamics of the associated proteins over at least the contact region are unchanged on association. Some gain in entropy is possible because of an increased cooperative unit, but the dominant feature in the thermodynamics is the second rule. The second rule, entirely due to entropy, states that association of proteins dynamically mismatched over their contact surface must reduce the total entropy. In application the two rules are really one. Mismatched surfaces of contact do not decrease the enthalpy and, when forced to be matched, decrease the entropy.

The entropy rule is a consequence of the Helfrich effect [110], recently established as important in membrane chemistry [111], where it is known as the "fluctuation force" [111]. The effect can be illustrated by the act of stilling a vibrating drumhead with the force of one's hand. The hand acts as a heat reservoir for the enthalpy released but of at least equal importance is the work done against entropy. Lowering quantum numbers in this way releases vibrational energy. Changing force constants in this way forces the entropy to decrease. Both are hard on the hand, but only the second actually involves decrease in the partition function of the drumhead. Poor matching in protein-protein association introduces a negative entropy contribution to the free-energy change in association. Hence the better the dynamic matching, the less the negative entropic contribution to the free-energy change and the more negative or less unfavorable the enthalpy contribution.

The presence of the Helfrich effect in protein associations is clearly signaled by the high *B* factors of the surface substructures. Estimates of its quantitative importance for membrane systems are simpler to make than those for spherical proteins, not only because semi-infinite models can be used for the former, but more so because the dynamical spectra of proteins are still very inadequately characterized by experiment. Presumably, *B* factors will once again come to the rescue but available information provides no confidence that the Helfrich effect in proteins will turn out to be minor. It may prove to have less quantitative importance than the long-range interactions of groups chosen by natural selection specifically to favor association. This class contains charged groups and hydrogen-bonding groups all of which interact with water when the proteins are separated, to present as much quantitative confusion as the Helfrich effect itself. Specific associations of aromatic side chains as in the $\alpha_1\beta_1$ interface of the tetrameric hemoglobins is a more likely example of specific tailoring for association.

Despite much discussion and hypothesis, the true explanation for protein-protein associations remains shrouded in conjecture about water displacement and so on. The Helfrich effect is a good example of such conjecture but one well worth exploring, not only as a factor in strong association, but also because together with the enthalpy consequences of good dynamic matching it provides a basis for selectivity in association and strength of association not available in static states. Since dynamic matching is adjustable by DNA changes, it must be included in the list of devices. In particular, it is the major device available to adjust the way a protein interacts with the rest of the world. This is achieved by increasing or decreasing cooperativitly among particles consisting of associated proteins, membranes, DNA, etc. One very important consequence of the negentropy driving force in evolution is that evolutionary experiments will tend to increase the complexity in new or more successful organisms. Complexity is achieved by increasing the size of the cooperative unit and there is a hierarchy of such units starting with the single functional domain of proteins and extending to that which consists of the total organism. The cell is one such level. Its size, growth, and function are undoubtedly a consequence of the dynamical matching of its components. Only dynamical matching provides the sensitivity required for the assembly of tissues to produce organelles, cells, and organs with gestalt identity and individuality.

Elsewhere, the role of dynamic matching in immune reactions has been discussed as an example of specificity in selection through dynamic matching of antibody to antigen. In this chapter it is shown how a supplementary device related to completing the knot can supplement or even override dynamic matching (Sec. VI.F.1).

3. Operating Temperatures and Fine Adjustments for Resonance

The very wide range of opportunities for evolutionary experiments provided by conformational fluctuations is utilizable only with considerably more stringent controls on the values of independent variables than would be necessary in a static system. Mammals, for example, "pay" more for temperature control than muscle function, and temperature control is a necessity in systems that depend on dynamics for their operation or have very large cooperative units, probably a consequence of dynamical factors enhanced by natural selection. It is perhaps not surprising that even individual degradative enzymes like the trypsin family have resonance temperatures near but below physiological temperatures, generally in the 285–295 K range. Apparently the difference between 20 and 37.4° C is sufficient for control and fine tuning of most functions. The coordination of functions, which is to say the management of the free energy transfers, is controlled by the operating temperature of the organism or often, as in many insects, by local prevailing temperatures. A linkage diagram based on compensation temperatures is a good way to describe the free-energy structure. The flow of free energy in a linked system, whether in a single protein or in a multienzyme system, can best be described by a diagram arraying the T_c values of the subprocesses above and below the operating temperature. In principle, the individual T_c values can usually be determined, but in practice, the work required is formidable [14]. In such a diagram the subprocesses with T_c values higher than the operating temperature receive free energy lost from processes with T_c values below the operating temperature. The detailed distribution depends on the separations among the compensation temperatures. All such quantitative relationships change when the operating temperature is changed. This kind of diagram is a good basis, perhaps the best, to use in formulating explicit thermodynamic and steady-state descriptions of living machines.

C. The Oxygen-Binding Mechanism of Hemoglobins

1. Slow Progress in Explaining Linkage in Hemoglobin

High B values for all atoms, rapid proton exchange at all sites, high motility by other tests, the range and complexity of mobile-defect states, rapid mutation time, lack of severe genetic restrictions, and insensitivity of oxygenation to changes in residue composition all suggest that hemoglobin and myoglobins, in fact all proteins with the myoglobin fold, are not members of the knot-matrix class of proteins. These data are not entirely convincing since they may reflect global fluctuations more than local fluctuations and thus conceal knots, but the absence of genetic conservation, rapid proton exchange, and the many substates of myoglobin provide sufficiently strong evidence for the assumption.

The absence of knots means that some other mechanism is responsible for the thermodynamic stability of these proteins, perhaps the long-range fields of the helix dipoles (Sec. VII.D). These proteins nevertheless manifest the pairing principle and a sophisticated use of mechanical work and strain in their ligand-binding functions. The latter makes them potentially useful models for the mechanical systems of enzymes, but in fact, after nearly a century of study, there is very little about this most studied of proteins that is understood. There are probably more and better experimental data for hemoglobin than any other protein, but the framework on which these data have been organized is incomplete. Without adequately detailed hypotheses even the best data will not be useful. Two deficiencies in particular deserve presentation in this connection.

2. The Possibility that Spectrophotometric Isotherms Are in Error

It is not impossible, but certainly extremely difficult, to obtain for these proteins oxygen-binding isotherms which have both high precision and high accuracy—in particular, to achieve the precision and accuracy sufficient to test the nearly universally accepted hypothesis that such isotherms obtained by spectral change are the same as those obtained by direct binding. A consequence is that nearly all studies of these proteins rely on spectrophotometric determination of the degree of oxygen binding. Historically, the easy acceptance of this assumption without reliable testing is understandable as a consequence of inadequate apparatus, but that excuse only partially applies now and what is known about mechanism makes the assumption not only an inadequate basis but also a dangerous basis on which to base hemoglobin research until such time as it has been tested.

Undoubtedly the most reliable oxygen isotherms obtained by direct measurement of oxygen uptake are those obtained by Roughton and co-workers using the van Slyke method, but these studies were carried out more with physiology in mind than hemoglobin. Hemoglobin from lysed cells without further purification is now known to be complicated by effectors such as disphosphoglyceric acid and by partial oxidation. The only experiments in which the isotherm was determined by direct binding and spectrophotometry applied simultaneously to the same purified hemoglobin solution appear to be the very few that Rifkind and Lumry [112] were able to complete using the Hersh coulometric cell for oxygen determination. They used only purified hemoglobins, presumably reasonably well stripped in the preparation, although the importance of DPG was not appreciated at the time. The isotherms determined by the two methods were always found to be in poor agreement. The direct-binding isotherms, which may be assumed to give the correct measure of oxygen binding, rose more rapidly at low oxygen pressure and fell more rapidly at high. The differences were smaller in phosphate buffer than in 0.1 *M* NaCl but all are so large as to preclude the use of the spectrophotometric isotherms for computing the binding constants. The total number of studies is nevertheless inadequate to the potential importance of the problem. Only a few pairs of accurately comparable isotherms were determined. The undertaking was further limited by experimental problems, and 20 years later there are still no instruments capable of providing the direct-binding data needed to complete the testing.

Errors in the spectrophotometric isotherm are produced by true differences in the spectra of the α and β chains but these are small. Prior to these experiments Rifkind and Lumry, on the basis of the hemoglobin rack mechanism, suggested that under some conditions HbO_2 is a mixture of low-spin and high-spin species and some of HbO_6 is totally low-spin. It is well known from the work of Pauling and Coryell that the spectral changes in the Soret and visible region correlate linearly with the magnetism and thus with the spin state, but no such correlation has been established between oxygenation and magnetism. A non-linear correlation might be expected as a consequence of the mechanical mechanism that regulates oxygen affinity by control of the position of the iron ion in relation to the porphyrin plane. This mechanism has been generalized for complex ions formed with either a pure metal ion and protein ligands or a metal coenzyme with some protein ligands [7]. Its application to hemoglobin was the first example of the mechanical feature in the function of a small, dissolved globular protein. Specifically, in hemoglobin, attachment of oxygen to heme iron draws the iron ion toward the mean plane of the prophyrin group to the extent limited by the proximal imidazole group as a result of its constraint by the protein conformation. The iron ion is thus stretched on a rack, hence providing a name for mechanical mechanisms as "rack" mechanisms. This tug of war can have any number of averaged positions, as is now well known. The range of variability provides an explanation for errors in spectrophotometric isotherms in the following way: In the first step of binding of dioxygen the protein pulls too hard. Oxygen binds relatively weakly to the high-spin form but nevertheless binds, so HbO_2 is a mixture of the high-spin and low-spin magnetic species. After three dioxygen molecules are bound, the protein conformation provides little mechanical resistance to iron displacement, so the unliganded heme group oscillates between low-spin and high-spin states. The spectral change measures spin state, not oxygenation. It does not necessarily follow from the data of Rifkind et al. that spectrophotometric isotherms are always incorrect. Rather it means that they are unreliable and will remain so until verified by direct-binding isotherms.

3. Hemoglobin as a Knot-Free Matrix Protein

The mixed character of the first and third oxygenation states satisfies the requirements for molecular resonance; the protein fluctuates between the two spin states of equal free energy. Takashima and Lumry [114] did indeed find resonance effects at 25% and 75% oxygenation in dielectric-relaxation studies, but their results have been challenged. Battistel et al. [115] found a large heat capacity spike consistent with two-state behavior near 25% oxygenation on varying oxygenation at constant temperature. A similar relaxation heat-capacity peak near 75%, if present, was buried in baseline uncertainties. The hemoglobins and myoglobins in both ferrous and ferric forms demonstrate compensation behavior under a wide variety of variations in independent variables: e.g., pH, oxygenation changes in ferrous hemoglobin, change of ligand type in ferric hemoglobin, species variation and thus residue exchange, solvent composition, and so on [14,83]. The compensation temperatures for ferric forms are all near 290 K within experimental error with the exception of some myoglobin experiments that lie as low as 275 K. Of course, the errors in compensation temperatures even in these experiments are 5–15°. With ferrous forms, most compensation studies have utilized the several steps of oxygenation to demonstrate compensation behavior. The standard enthalpy and entropy changes in the four steps of oxygenation show compensation behavior with T_c values near 265 K for Roughton's gasometrically determined single-step binding constants and in the range 300–305 K for those computed from spectrophotometric isotherm. The points for HbO_6 and HbO_8 formation lie on top of each other in all the experiments. Replacement of iron by cobalt changes the affinity but not the compensation temperature. The existence of compensation behavior with slope independent of complex ion or species implies that progress along the oxygenation coordinate is a single process with a single initial and a single terminal end as discussed in Sec. IV.C.3. This suggests a two-state conformational process and one more or less uncomplicated by the many substates and activated transitions observed in the single-chain respiratory proteins by Frauenfelder et al. [97]

These proteins thus appear to display most of the conformational characteristics deduced thus far for matrices. The compensation temperatures are independent of temperature over the experimental ranges, so compensation is of the linkage class, that is, determined by the linkage between ligand binding and conformation, and the T_c values are thus "accidentally" near the usual mean experimental temperature 298 K. It will be recalled that when the experimental temperature equals T_c the free-energy change in the process responsible for compensation is zero. This is the resonance condition, so the conformational process not only undergoes maximum spontaneous fluctuations, but also at this temperature makes no contribution to the single-step standard free-energy changes in ligand binding. The corresponding enthalpy and entropy changes then reflect motion along the conformational coordinate at near-constant total-system free energy. The enthalpy and entropy changes cannot be separated into their thermal and motive parts, but they probably provide a rough index of progress along the conformation coordinate.

4. Dynamical Basis of Linkage

Data for a variety of types of hemoglobin studies can often be fitted with considerable precision to the two-state model. That this is so despite the well-established existence of many macrostates of the protein is an old paradox that for many years dominated hemoglobin research. Much effort went into developing alternative mechanisms for "allosteria" in this protein and more effort into experiments designed to test for the supremacy of one or another of the alternative hypotheses. The experimental undertakings, often among the best yet carried out on protein systems, did not resolve the problem almost certainly because the hypotheses did not include the right device or the right devices. The matching of the fluctuations of two proteins over their contact surface provides a totally different hypothesis, one taking into account the complex dynamical nature of protein conformations. It has been presented elsewhere [11] but needs some recapitulation as a potentially better model for hemoglobin and as an introduction to other mechanisms for function and energy transfer that involve dynamic matching. Hemoglobin provides a basis for describing the potential virtues of multi-subunit systems that function through changes in dynamical matching. This is the second of the potentially important deficiencies in older hypotheses of hemoglobin function. As noted in the Introduction, dynamic tuning is a very promising idea if the Helfrich effect is a major factor in the association of proteins, a possibility that in turn receives support from the high *B* factors of matrix atoms and especially surface atoms. The development is neither supported nor denied by existing data as we interpret those data and so rests heavily on logical deductions from those topics in this chapter better supported by experiment rather than on directly relevant experimental information.

It is assumed in this discussion that all subunits of deoxy hemoglobin are dynamically matched at their contact surfaces. In view of the small size and large *B* factors of matrix atoms in the subunits, it is likely that all the subunits are dynamically matched to one another in the ligand-free protein. This balance is upset by the first oxygenation process, which destroys the dynamic matching of the oxygenated subunit to the other three. Initially this raises the free energy at the dimer-dimer interface because the two dimers have become mismatched. The dimers must then dissociate or perhaps only greatly reduce their contact surface or the protein must be promoted into a higher free-energy state in which the matching is so much improved that the dimer-dimer association has again become tight. The first alternative would produce a destabilized protein and the second a considerable loss of the new free energy. Promotion, on the other hand, traps much of that free energy in the tetramer. The promotion step is critical to the argument because in this step the free energy is distributed among the four subunits of which the non-liganded three can then use their increased free energy to increase their oxygen affinity. Promotion in this description is the vehicle for "heme-heme interaction." Ligation at the second and third sites releases free energy, which is redistributed to the subunits in exactly the same way. The coupling among the subunits varies from step to step, so the phasing of this free-energy redistribution system changes. The tertiary conformation provides the coupling and is perturbed along its normal conformational coordinate whether that is simple distortion or an expansion-contraction process. The compensation temperature then remains unchanged with advancement toward total ligation again in agreement with experiment. The interfaces and the subunits go through a progressive series of changes quantitatively different but all focused on the increases and decreases of free energy at the subunit contacts. Such changes at the $\alpha_1 \beta_2$ interfaces are well known. Those at the $\alpha_1\beta_1$ interfaces generally thought to be minor because of the strong interdimer interaction at that interface have only recently been established and described by Rifkind and Ackers and their respective co-workers. Retuning reduces the stress as it develops at the interfaces and does so with a minimum of useless heat production. Retuning after the third step of ligation is apparently less dramatic as judged by oxygen affinity and more direct measures of the protein differences of HbO_6 and HbO_8 . Since the quaternary change reduces the contact areas, it may reflect an inability to store all the free energy released in some step by promotion. This would make the transition from the second to third step a likely point for the quaternary change, but the free and ligated species are not single chemical species but rather broad distributions of macro-and microstates. Our argument must refer to average species named HbO_2, HbO_4, and so forth. Slight differences in experimental conditions, major differences like DPG binding, changes in residue composition, and so on will favor a particular pathway of changes in the distribution functions but with a large amount of overlap. Transient behavior can be very complicated because measurements of the kinetics will disclose unrelaxed states, by which we mean various transient species during promotion, which have negligible concentrations at equilibrium. For example, the stress on the bond from proximal histidine to heme iron appears large in kinetics experiments and often small in equilibrium states. This completes the argument at its bare-bones level. The relationship between dynamic tuning and strong protein-protein association is another device discovered in evolution, as is *conformation promotion* on which the association free-energy redistribution depends. In hemoglobin these are inevitable consequences of the knot-free, high-*B*-factor structure. What is not at all clear is whether they work as described in the linkage mechanism. Are there static coupling devices among the subunits that are quantitatively more important?

It is necessary to explain why processes based on these devices demonstrate two-state behavior. Somehow all the substate and microstate variations are entrained in a single major process. This may be no more than a consequence of averaging over the substate populations or of a more complicated coordination of atom motion, but the simplest idea is probably the right one, and that is the dynamical properties of the subunits are intrinsically identical and their conformation changes are not large enough to destroy the identity, at least not in normal function. Their conformations then have the same enthalpy and entropy properties and the same compensation temperatures. The aggregate changes will then generate a single compensation temperature. The two-state behavior is an artifact reflecting the intrinsic similarities of the subunits rather than a true gestalt behavior of the tetramer. The two-state characteristic is itself an artifact though a useful one. The two states are simply the limits of conformational distortion that take place in the series of ligation steps. Since the pH dependence and the many other variables that produce compensation behavior are all linked to one another through conformation rather than by direct stoichiometric or nonstoichiometric coupling, all the subprocesses will also yield a common T_c value. This is an example of perfect linkage compensation [14], and for the present we shall be content with this last rationalization and move on to consider a few more aspects of compensation in this protein family.

All those changes in the hemoglobin-solvent system that produce compensation behavior in ligand binding under change in any of the several variables already mentioned in previous sections displace the two-state conformational process in each subunit along its coordinate. In this protein as in many others, the compensation coordinate is the conformational coordinate. The thermodynamic changes that occur in the individual chains depend on the interchain interactions and their phase relations due to the fact that all are coupled to a single conformational coordinate. To the extent that the intrinsic free energy of binding at the heme group of each chain depends on stress on the iron ion through proximal and distal connections to the conformation, the intrinsic or, perhaps better, the mean affinity of a given species of hemoglobin as well as the degree of advancement along the conformational coordinate in each chain for an advancement in any one depends on how the various subprocesses are phased. Phasing is determined by the residue composition and thus adjustable through DNA changes. A finer detail of such control and perhaps one actually to be found in nature appears when one or more heme groups exist in the mixed-spin state. The oxygen affinity of HbO_2 increases as the proportion of high-spin form increases. The proportions, like all other aspects of hemoglobin behavior, depend on phasing and thus more fundamentally on differences in residues. The hemoglobin fold is extremely tolerant of such changes and thus does not denature readily. Most changes nevertheless have quantitative consequences more often detectable in enthalpy, entropy, and volume than in free energy.

Coordination among the subunits and thus the flow of free energy in the course of oxygenation is mediated by their interactions, which certainly include those at intersubunit contacts but may also be consequences of the special construction of the myoglobin fold. Thus there may be direct coupling among the long-range electric fields of the helices since the dipole vectors of these helices are oriented for favorable interaction not only within the chain, but also between chains. Just how these fields might be constructed is discussed in Section VII.D.

The compensation temperatures in these knot-free proteins give the enthalpy-entropy balance point, which is nearly constant along the "conformational coordinate" in oxygen binding and indeed in ligand binding in general. But the T_c value(s) are not entirely dependent on tertiary construction since they appear to be different in the ferrohemoglobin family from their well-determined value for the ferrihemoglobins. T_c values obtained by oxygen binding for ferrohemoglobins do not agree from study to study. They are near 310 K for HbA when binding is measured spectrophotometrically and near 265 K from the gasometric data of Roughton et al. This is interesting in light of the discussion of the two kinds of isotherm but does not help much in gaining better understanding of the function of this protein and evidence for or against the dynamic-matching mechanism. At least that mechanism illustrates the virtues of surface dynamics as a means for free-energy transfer and redistribution.

D. Enzyme Mechanisms: Updating the Rack Mechanism

1. Some History

Our search for enzyme mechanisms has been marked by discoveries that pointed new research in unfashionable directions. The recently discovered devices of this chapter continue this tradition, and their application to enzymic catalysis in light of textbook history may seem unnecessary and unjustified. The history of our research correctly connects the new to what went before so the new material in this section gains necessary perspective when seen as a logical continuation of that history. The relevant facts of that history are discussed in the following paragraph.

Carboxypeptidase A then (1948) and now was too complicated and enzyme kinetics proved to be of very limited utility in the absence of an accurate description of protein construction. Progress with hemoglobin was more successful, leading almost immediately to the rack hypothesis, which with the appearance of x-ray data for myoglobin could be described in some molecular detail. The search for conformational changes in catalysis began to produce encouraging results in this laboratory and that of Prof. G. Hess at Cornell, so an attempt was made to apply the rack concept to the serine proteases. There were many contributors to the experimental effort and sufficient consistency in the results to suggest even in the early 1960s, the mechanism shown in Fig. 1. A turning point in this exploratory research period was our discovery of the enthalpy-entropy compensation phenomenon. It was found quite by accident that a variety of data for the binding of inhibitors and substrates fell on straight lines with only slightly different slopes. Thus as shown in Fig. 16, data from steady-state studies of the esterase action of chymotrypsin by Rajender and Han gave the same compensation line as that found with binding data for the inhibitor indole determined as a function of pH. The steady-state experiments carried out over the same pH range produced the same line although the precision in the slope determination has since proved to be no greater than 5%.

Once investigators were alerted to the existence of the phenomenon, the straight line with slope near 290 K was found in a wide variety of protein studies [83]. This compensation pattern was found to be associated with catalysis and inhibitor binding and soon became very useful as a simple phenomenological device to relate protein structure and stability with the catalytic process. Its utility in this use and its value in suggesting different ways to think about protein processes are amply demonstrated in this chapter. In particular, it led to the idea of a single conformation coordinate in the catalytic process because it was initially thought to be a characteristic of bulk water so that its changes reflected the protein-water interaction. This still popular idea has only recently been found to be erroneous.

Figure 1 owes much to our interest in compensation behavior. Its publication in 1969 brought to the end the period of exploration. Despite the lack of x-ray–diffraction data, several of the most important deductions about structure were included in the mechanism of Fig. 1. The most important were the recognition of *functional domains* and the *pairing principle* for enzymic catalysis, *domain closure* for the application of force, and the *subtle-change* process driving domain closure. All these were clearly suggested, some even required, by experiment. Expansion-contraction involving a large part of the protein appeared then to be the best guess as to the subtle-change process, which is related to the concept of protein "breathing" popularized later by Go and many others. Since such conformational motility seemed to be counterindicated by early x-ray–diffraction results, the molecular details began to emerge only in 1971 with a more realistic interpretation of x-ray data and the realization that proteins have considerable mobile free volume, *mobile defects.* These made it impossible to take most hypotheses of mechanism seriously since none could accommodate these experimentally established characteristics of the enzyme. However, our hypothesis was not much better until the knot-matrix construction principle was discovered in 1982. It revealed structure in such a way as to make most details of function obvious. Progress has been rapid so that even the last updating, prepared in 1985 [10], now appears primitive and incomplete. This chapter is reasonably comprehensive in covering discoveries that have been made on the knot-matrix foundation. It is less successful in utilizing the discoveries to actually establish an improved molecular description of enzyme catalysis.

2. The Anatomy of an Enzyme

The HIV-1 protease is a good model enzyme structure since in this elegant molecule the catalytic mechanism is virtually obvious from the construction of the protein. The HIV-1 protease consists of two functional domains, which in this enzyme are two identical proteins. The several structural features required for the catalytic mechanism and their organization are shown in Fig. 14.

The two proteins are joined at the base by residues at the C-terminal ends, which produce the strong hinge by the "completing the knot" procedure using insertion of a terminal phenyl group into a pocket in the protein partner and a short strap hinge consisting of antiparallel association of short backbone segments near the C termini. The strong association thus produced in formation of this secondary knot is probably due to release of electrostatic potential energy, but it nevertheless poses some puzzles, discussed in Sec. VII. This is an example of "completing the knot" (Sec. IV.E) to form a hinge.

In single-chain enzymes the two functional domains are coupled by the peptide chain, which provides the necessary strong hinge. The two proteins so joined lie in a mean plane containing a noncrystallographic, twofold axis lying in the plane, perpendicular to this "hinge" and passing between the two aspartate functional groups. This geometry is also common to all the "gene-duplication enzymes" so far studied; it is less precisely twofold in the other enzymes because the palindromic *B*-factor patterns are less perfect. Heteroknot enzymes like HEW lysozyme and ribonuclease A have the same organization of functional domains but neither the twofold axis nor the palindromy.

The primary knots are face to face in the mean plane with twofold symmetry of the noncrystallographic axis. The knots are neither spherically nor cylindrically symmetrical, but this arrangement guarantees that the mean plane passing between them contains the twofold axis. In domain closure the knots approach one another in this plane. In the absence of substrate, the knots are not in tight contact at their interface. In the presence of substrate or inhibitor competitive with substrates, the average degree of closure depends on the bound species.

The primary functional groups are positioned with their carboxylate groups separated by a few tenths of an angstrom in the free enzyme. In the free protein, the atoms of these groups making chemical contact with substrates in the catalytic process have matrix *B* factors and thus are not parts of the knots. In some cases they may be parts of knot residues, but judging from the pictures and the *B* factors of these residues, the functional groups are on the first residues of the polypeptide chains as they emerge from the knots, as is usually obvious.

The long arms engage the substrate, positioning it so that its reaction element is in contact with the functional groups, or very nearly so. This complex is a Michaelis complex. As shown in Fig. 14C the long arms contain at their N-termini structures nearly identical to those forming the hinge knot but slightly separated. It is not unlikely that these associate to complete a new knot confining the substrate. The particular latch construction is not general for enzymes, but it is part of a large family of conformational processes that enforce tight binding between substrates and inhibitors to the protein. The appearance of two examples of this construction in the same enzyme appears to be coincidental but it nevertheless suggests some parsimony in enzyme construction.

3. The Conformation Coordinate for Catalysis

Figure 16 led first to the recognition of a single conformational process underlying the catalytic function of chymotrypsin and trypsin, but the description of this process as an expansion-contraction of conformation has become well established only recently using *B*-factor data to compare free enzymes with their substrate and inhibitor complexes. Several kinds of intermediate states are possible in these complexes and it is not always easy to tell the pertinent class of an example. Those produced by inhibitor binding are not necessarily the same as those produced by substrates. The *B* factors do much to reduce the confusion, and *B* factors determined for many enzyme-inhibitor complexes and for a number of the "acyl-enzymes" in such proteins as the serine and asparate proteinases can provide molecular information not otherwise available even from x-ray coordinate values. A typical example—a superior one because of the high quality of the data—is that of rhizopepsin with and without pepstatin. The *B* values obtained by Suguna and co-workers [116] have an unusually small experimental standard deviation after full refinement. Pepstatin is thought to be a typical pepsin inhibitor but atypical only in its very-high-affinity constant for the protein. The decrease in *B* factors are large but not larger than found in some other comparisons of this kind.

In Fig. 15 the *B*-factor differences given as that of the inhibited protein minus that of the free protein are compared with the total *B* factors for free enzyme. The difference for the atoms in the knot residues is zero within error. Otherwise the difference is positive, indicating that the matrix regions all undergo contraction along radii centered in the primary knots and specifically on the small number of atoms with lowest *B* values. Since the free-enzyme factors increase on moving out toward the periphery, contraction is larger the larger the *B* factor in the free protein. This yields a pattern in which the (negative) spikes in the difference profile project upward into the peaks in the profile of the free protein. Thus the largest differences are those in surface atoms, which in this protein are at the ends of the main-chain loops.

The reductions in *B* factors are thus roughly in the same ratio to the total *B* factors for all atoms once the baseline has been subtracted. This ratio is large, indicative of loss of much free volume and thus considerable hardening of the conformation. Matrices now look a bit more like knots but still have higher *B* factors and greater motility.

It is of much interest that all matrix and surface atoms experience changing *B* factors in the contraction process. In this case the *B*-factor data are so good that accurate molecular descriptions of the contraction process can be constructed.

There is considerable additional information of value in Fig. 15, especially when comparisons are made with the positions of various kinds of residue such as the aspartates. The accumulation of such information is likely to be one of the more successful areas of protein research.

In general, we have found that enzymes and other chemically functional proteins have lower average *B* factors when they are trapped in intermediate states. This is also true for liganded forms of some nonenzymes we have examined, such as hemoglobin. A particularly important study of is that of Bone et al. [117]. They determined the average *B* factors of a series of derivatives of α-lytic protease, a typical member of the trypsin family of serine proteases, with several alkyl boronic acids. The latter closely resemble the acyl enzymes formed from analogous ester substrates. The average *B* factors were found to decrease in a way strictly correlated with the increase in the catalytic rate constant for the corresponding substrates, that is, ester substrates with these side chains. The average *B* values drop by more than 30% through the series. We find that the *B*-factor changes are very similar to those discussed above for the pepsin comparison except that with the two derivatives corresponding to very good substrates even the knot atoms with lowest *B* factors experience some reduction in the acyl enzymes. It is unlikely that this is an artifact of systematic error in these closely integrated series of x-ray studies. Instead it suggests that sometimes primary knots of enzymes may undergo important changes in physical properties as part of the catalytic process. This is as yet as isolated observation, but more intensive study of knots will almost certainly show new complexities that may or may not be common.

As in the pepsin study, *B*-factor changes are widespread rather than concentrated in a small number of regions. All matrix atoms participate in the contraction process, again demonstrating the gestalt rather than local nature of that process.

Mangel found an even larger decrease in average *B* values on acyl-enzyme formation with trypsin using a different kind of acylating molecule [118]. The strong acylating agents for the serine proteases such as di-isopropylfluorophosphate work by forcing the contraction process and thus preventing domain opening or entry of water for hydrolysis. The *B*-factor results obtained with the more effective acylating agents may lead to overemphasis on the degree of contraction since in previous studies even with simpler, nonchemical inhibitors, contraction along the conformational reaction coordinate past this critical juncture for transition-state formation has been observed. Virtual substrates during domain closure do not always catch the serine group in the serine proteases and so slip past this critical point to produce greater contraction than that occurring in normal catalysis. The reversal is often very slow even with the simpler inhibitors.

As the total volume shrinks, free volume is lost. Lüscher et al. [119] found in comparing the hydration isotherms for chymotrypsin in free form and as the tosylated acyl enzyme that the latter has a much reduced number of tightly bound water molecules but many more in the less tightly held classes. The first finding suggests loss of internal water molecules either from the region between the functional domains or from the populations of fixed and mobile water molecules within the domains. X-ray–diffraction studies often locate high-occupancy sites for water between the domains, so the former loss is to be expected, but the loss of interior water molecules seems just as likely. The increase in weakly bound water remains to be explained. The loss of tightly held water suggests a relationship between the conformational process in function and the glass transition (vide infra).

Matrix enthalpy and matrix conformational entropy (as opposed to configurational entropy) increase with free volume and thus with expansion, so expansion is driven by the entropy increase. As a result, with decreasing temperature the mean position along the conformational coordinate moves toward the state of maximum contraction, and that may have characteristics of a glass as the knots do. This raises the question of the relationship of the conformational process in catalysis and the glass transition produced by dehydration or lowering the temperature. In fully hydrated proteins the free volume, mechanical properties, and dynamical behavior of protein matrices are glasslike below 180–200 K and rubbery above that temperature. At low hydration the transition temperature range rises, and in fact the dry protein does not demonstrate any transition. These observations seem to require that the two processes are the same although not all their manifestations need be identical.

Glass formation as facilitated by temperature and hydration changes is discussed in Chapters 3 and 4. Because it is totally at variance with the popular opinion, the most important topic is the existence of mobile dissolved water in proteins and its role in plasticizing conformation to establish the motility required for function. As Gregory discusses in detail, the hydration behavior is that to be expected of a glass transition that is dependent on the amount of plasticizer. Catalytic function is lost in the transition. At 298 K it is absent at zero hydration increasing to almost full activity (that in bulk water) near 0.15 g water per gram dry protein (0.15 *h*) or slightly higher. Similar results were found in the behavior of the magnetic susceptibility of oxidized horse-heat cytochrome c [7]. At 300 K the oxidized protein demonstrated highly cooperative transitions at 0.07 *h* and 0.19 *h*. With reduced protein only one transition at 0.03 *h* was observed. The number of plasticizing water molecules for any protein can only be roughly estimated as between 15 and 60 depending on molecular weight, and the hydration processes, though cooperative, do not demonstrate a high order in water activity. Since the limiting state of contraction and the glassy state studied by hydration and temperature at this time appear to be the same, changes in internal water content such as were found by Lüscher et al. on tosylation of α-chymotrypsin are consistent with the unified picture. The domain-closure process in enzymes, cytochromes, and so forth may have sites for water when open that disappear in closure. This is suggested by many x-ray–diffraction studies. With insufficient water available, domains can open only with difficulty if at all. The increase in hydration leading to catalytic activity then includes interdomain water in the open conformation, water going into substrate sites, side-chain sites in particular, which are expensive to build but less so with some water, and the water diffusing through the matrices. Obviously, thermodynamic bookkeeping is a formidable undertaking for realistic models.

4. A Summary of Hypotheses

There are three obvious ways in which an enzyme can facilitate a chemical process, that is, a change in primary bonds. The first involves arrangement of polar groups about the substrate so as to favor the electronic distribution in the activated complex. The protein may be rigid or flexible but the specific mechanism depends on electron borrowing and lending, what is often called "electron banking." The second is to destabilize the system just prior to the transition state using either distortion of substrate on binding to a rigid protein or some source of mechanical energy to produce distortion in a flexible enzyme-substrate complex. The third is to use some mechanical work for activation free energy thus diminishing the amount of transient heat required for this purpose. Obviously a clean separation between the second and third mechanisms is a bit difficult to make in practice. Stress-strain mechanisms of catalysis are known in small-molecule chemistry and they have been suggested for rigid enzymes. The use of mechanical force derived from protein conformation has no small-molecule counterpart. Enzymes use mechanical and electronic mechanisms of catalysis. Electronic mechanisms constructed using familiar models from physical organic or inorganic chemistry dominate the literature, and propinquity effects and other general thermodynamic consequences of substrate-protein association common to all possible mechanisms are described in detail. The enzyme discussion in this chapter can thus be limited to the basic mechanical hypotheses utilizing protein conformations with realistic softness and motility.

Mechanical hypotheses can emphasize destabilization of the enzyme-substrate complex prior to the formation of a rate-limiting chemical bond rearrangement or stabilization of the transition state. There are possibilities for semantic confusion here because the free energy of the transition state can be raised or lowered and both probably occur. A second problem is that mechanical mechanisms said to stabilize the transition state cannot actually do so. Rather they can provide a source of mechanical free energy to supplement the heat source in the very short-lived transition state. The free-energy of formation of the transition state is the same as in the absence of the mechanical reservoir, but the apparent free energy of activation will usually be less by the mechanical contribution. This follows because the activation free energy is defined in absolute rate theory in terms of the experimental rate constant. This is a nontrivial detail, which needs some restating of that rate theory [138].

The mechanical contribution to the free energy of activation if short-lived arises from *PV* fluctuations of the protein that as a form of the total heat (at constant *P*) are subject to uncertainty-principle restrictions. Again the distinction is semantic because stabilization using a long-lived loan from the protein is an ambiguous way to describe what is really destabilization prior to transition-state formation. The distinction is important because it is critical to understanding enzyme mechanisms.

The unique possibilities for mechanical catalysis that proteins have can depend on simple redistributions of conformational free energy or the special ways they can be constructed to manage the *PV* fluctuations. Note first that proteins are no better thermal reservoirs than liquids or other soft polymers and thus cannot be more efficient than any collection of small molecules in providing thermal energy for transition-state formation. Knots may turn out to be so hard they provide a few contributing degrees of freedom, but that is both unlikely and irrelevant in this discussion. Each local region of a protein is in independent thermal and mechanical equilibrium with its surroundings, and proteins provide no new ways to collect and channel the heat from many such regions to the point where it is needed. On the other hand, proteins do have a unique way to channel *PV* fluctuations which are an intrinsic feature of protein construction and one that can be adjusted in evolution. This property distinguishes proteins from other molecules and it must be this property that makes most protein processes possible.

The manifestation of the special property that has become most obvious is that revealed by the compensation behavior observed with most enzyme reactions. Compensation plots are the signature of fluctuations along the conformation coordinate. This is not to say that the molecular description of that coordinate is simple, but rather that its fluctuation behavior is simple. It is simple fluctuation behavior that reveals the unique management of *PV* fluctuations. The protein construction maximizes fluctuations along that coordinate and limits most others. Just how it entrains the many local positional fluctuations of the atoms and groups into a single mode is obscure.

One kind of rack hypothesis includes the use of these allowed *PV* fluctuations as a transient source of mechanical free energy for the actual transition of the system from its equilibrium states to the transition state. Then, as already observed, the *PV* free energy thus used is subject to the same uncertainty-principle restriction on lifetime as true heat (the partial derivative of internal energy with respect to *T* at constant *P*). The advantage is nevertheless considerable because the many degrees of freedom of the protein are coordinated by the construction to provide large amounts of mechanical free energy in single fluctuations. To recapitulate, the protein as heat reservoir has a very narrow heat pdf, which means large fluctuations in heat such as those required to produce many kilocalories of free energy for transition-state formation are rare. The mechanical situation is much different. Because conformations are well integrated, there are long-range coordinated *PV* fluctuations that greatly broaden the *PV* pdf. Detailed construction of the conformation may have been chosen in evolution to enhance this behavior, but of more interest is the fact that the pdf, which includes only the scalar parts of the *PV* fluctuations, can be given some directional constraints. Choice of conformational details in evolution makes possible the conversion of a scalar into a vector in the sense that long-range fluctuations will tend to have a characteristic direction. The matrix expansion-contraction processes we find in enzymes can be viewed as large *PV* fluctuations with a roughly approximate radial direction. They actually consist of simultaneous contractions or expansions in the two primary functional domains much like the motions of the blades in a pair of scissors. Thus, in contrast to the true heat parts of the enthalpy, for *PV* fluctuations a large part of the protein acts like a single isolated molecule with a very wide *PV* distribution. This comparison is somewhat oversimplified, but a given large *PV* fluctuation has a probability about $\sqrt{N}$ greater than the same fluctuation using heat [138]. *N* is the number of vibrational and librational modes in the protein matrix.

5. The Rack Mechanism for the Serine Proteases

The essential mechanism is well illustrated in Fig. 1. As the matrices contract after substrate binding, the two functional domains move toward each other forcing the knots together and thus driving the assembly of functional groups and the reaction element of the substrate into a state of elevated free energy from which the jump to the transition state for primary-bond rearrangement requires much less thermal energy than conventional catalysis.

The domain-closure process appears as a single major conformational mode, but in detail it is rather a manifestation of a large set of smaller modes that are so constrained as to be in phase. However, it is unlikely that phase correlation is exact and thus that the major mode is strictly cyclical. Either the micromodes themselves or their phase relations are sensitive to the specific substrate or inhibitor bound and the degree of advancement of the catalytic process. Thus both the position of the transition states and the detail of the most common reaction pathway in phase space varies from substrate to substrate, so to some usually small degree each substrate for a particular enzyme dictates its own catalytic pathway.

Domain closure is driven by a contraction of the matrices. Compensation behavior is a ubiquitous feature of protein conformations, and the compensation temperatures suggest that at or slightly below physiological temperature the contraction-expansion process may be poised close to its isoergonic state. This near-resonance condition constantly cycles the conformational free energy between the high entropy of the expanded state and the low enthalpy of the contracted state.

The acid product of a typical "good" substrate for chymotrypsin is an inhibitor when unprotonated and a virtual substrate otherwise [120]. An example is the pair *N*-acetyl-L tryptophan and *N*-acetyl-L-tryptophan methyl ester. The two parents, *N*-acetyl-glycine and *N*-acetyl-glycine methyl ester, are less well bound but have the same ability to block the domain-closure process by combining with the functional groups of the protein to form a physical entity restricting contraction. The unprotonated acids cannot make this connection and produce additional contraction past the transition state. The indole side chains of the tryptophanyl compounds stimulate domain closure by contributing some of the free energy released in interaction with the protein but they cannot restrict domain closure. They favor the contraction process as does indole itself, but the side-chain effect is large only if the chemical link between the indole side chain and *N*-acetyl-glycine methyl ester exists. The acid enantiomer, *N*-acetyl-D-tryptophan, is bound like a side chain and unable to restrict contraction nor does it produce the deep contraction found with the anion of *N*-acetyl-L-tryptophan. The so-called "transition-state analogs" of substrates act like unprotonated virtual substrates and other good competitive inhibitors because they do not form a bond to the protein which blocks further motion along the reaction coordinate. They are thus better bound than substrates. The strong binding of these analogs is the major experimental support offered for the "transition-state" hypothesis of mechanism, but appears to be equally well rationalized by rack mechanisms.

It has been observed earlier in this chapter that for many congener series of substrates, e.g., the methyl esters of the *N*-acetyl-L-amino acids for chymotrypsin, there is a parallelism between activation free energy measured from the catalytic rate constant and the standard free-energy change in substrate binding computed from the Michaelis-Menten constant. This demonstrates that some of the latter is utilized to increase the rate constant, and virtually all hypotheses of enzyme mechanism since that time have contained features attempting to explain the obviously important connection. This phenomenon is the basis for the statement that the better the binding, the better (faster) the catalysis. For chymotrypsin substrates better binding means, in general, stronger interaction between protein and the α-carbon side-chain group in small substrates and for large, more normal polypeptide substrates, multiple interactions between substrate and conformation. The most complete examination of this phenomenon in chymotrypsin is given in Ref. 10, pages 39–45, based on data for the *N*-acetyl-L-amino acid methylester family of substrates provided by Dorovska-Taran and co-workers [121] and James [122]. Steady-state kinetics data begin to be very useful when obtained for families of substrates demonstrating compensation behavior. Only in this way do the thermodynamic patterns characteristic of the general as opposed to the special features of mechanism become apparent and quantifiable. Many enzymes have substrates and inhibitor families with compensation behavior, but only data for chymotryptic catalysis appear to have been analyzed using compensation theory. The half-dozen major findings are described in the references and will not be repeated here. Among the more interesting phenomenological patterns revealed by the data is the qualitative difference between effect of main substrate body and side chain in the way they enhance the rate. The side-chain effect is a manifestation of the use of its binding free energy, but through increasing the entropy of the enzyme-substrate complex. This peculiarity probably means that the side chain also influences the domain-closure process directly as well as through its connection with the total substrate.

Less ambiguous is the finding that the enzyme-substrate (ES) compound formed with the side-chain-free parent, the glycine substrate, is isoergonic at 298 K. The corresponding enthalpy and entropy changes are very large and negative, demonstrating that a large degree of domain closure takes place in this isoergonic process. About five orders of the total of 10 orders of magnitude in rate enhancement obtained with the best substrates in this family are a consequence of the domain-closure effected by binding the glycine substrate. The contribution of side chain is by no means minor, but it appears to have been added on in evolution to the unique mechanism put into play by side-chain-free substrates. Either the ES compound is raised in free energy by compression in its formation or a transient compression occurs in ES to facilitate its conversion to the transition state. Both probably occur, so we should distinguish a minimum of four factors responsible for catalysis: static and dynamic electrostatic factors, including related inductive and resonance perturbations, side-chain contribution of free energy, domain closure in ES formation, and transient use of mechanical free energy in transition-state formation. The latter two remain to be explained but these are likely to have more general significance since many enzymes do not have substrates with side chains or other features that contribute to binding of free energy to catalysis. Consider carbonic anhydrase, the enzyme with the highest known turnover number.

The small conformational adjustments that occur in the contraction process will in general interact with the substrate somewhat differently at each point along the conformational coordinate. The most efficiently catalyzed substrates will then be those for which the conformational coordinate was evolved. This condition provides a device that can greatly enhance specificity in substrate choice by basing it on the details of dynamical changes along the domain-closure coordinate. If this is true, the static fitting of small molecules to x-ray pictures, which is now currently the basis for most pharmacological computer research, is neither a reliable nor an efficient way to find new drugs to solve physiological problems. Although there is a close connection between this "kinetic specificity" and the survival advantages of efficient retention of free energy [free-energy complementarity (7)], the two reflect different aspects of successful evolution. As already discussed, the rack mechanism makes it possible to tailor the reaction coordinate and thus the reaction from beginning to end, an impossibility in small-molecule reactions. By such tailoring, high degrees of free-energy conservation without loss of reaction velocity become possible. This is effective only if the substrate and conformation are able to be tightly complementary along the reaction path. That path then forms a maze that a substrate, to be successful, must be able to solve. The more accurate the solution, the better it satisfies the intrinsic specificity requirements of the protein. Other potential substrates are rejected for even minor failures.

6. Do the *B*-Factor Palindromic Patterns Reveal the Catalytic Mechanism?

Most updating of the rack mechanism for gene-duplication proteins like chymotrypsin follows immediately and obviously from better descriptions of the construction of the enzyme. Confidence in the mechanism increases because the subtle change of Lumry and Biltonen is now positively identified and easily detected using *B* factors. Contraction of matrices produces domain closure. This is a very well-coordinated process, which is very poorly understood. It may be very simple or very complicated, but enzymic catalysis will not be respectably understood until that driving force for contraction is clearly described.

It does not yet seem possible to satisfy this requirement, but the following attempts to do so are instructive since any such attempt must come to grips with the palindromic pattern so far omnipresent in gene-duplication enzymes.

Contraction or at least domain closure appears to take place in two steps the first of which is an equilibrium process either making ES unstable or preparing it for the second step, which drives ES into the transition state. With many enzymes the first may be missing. The second may also be missing in some cases, but a somewhat dangerous use of the principle of parsimony suggests that it will appear in the mechanisms of all enzymes. By what mechanism does it occur? In this subsection the first question is considered and in the second, the second question.

In the gene-duplication enzymes domain closure is always associated with a *B*-factor palindrome. For the several reasons already discussed, e.g., residue conservation, invariance, and cost in evolutionary time, the palindromic pattern now appears to be the most important construction feature and a signature of the mechanism in this class of enzyme. Apparently domain closure to be most effective must be symmetrical in the two functional domains or as nearly so as evolution has been able to achieve. Palindromy is the signature of that symmetry but it gives considerable fine detail about the conformational changes making it possible. It probably also indicates that the primary functional domains are dynamically matched as the latter has already been defined. The great similarity of the domains means that if not exactly tuned, they are close to this condition. Were they not, it is probable that contraction would occur in one domain before the other or in one only. Since they are in contact at their mutual interface and come into closer contact as closure advances, the conditions of dynamic tuning must be met or the free energy will rise as closure advances. Thus the lowest path of free-energy change is that along which the domains remain dynamically matched to each other.

The next step of sophistication in domain closure may well depend on substrate as a tuning device. If the domains are not well tuned prior to binding of substrate and become so afterward, the substrate gains some control over the process, which may or may not be useful. In fact, something like this must happen because substrate modifies conformation and must then also modify tuning. That is, the protein is untuned before a good substrate is bound and more appropriately tuned for domain closure after such binding. This novel new branch of chemistry requires serious consideration for its possible roles in protein reactions at least until we are certain that it is quantitatively negligible.

The thermodynamic stability of the protein as reflected in its mechanical strength is probably not a factor in the activation process since the lifetime of the application of maximum force is much shorter than the lifetime of the native state of the protein. The matrices are the weak link mechanically but must become less weak as contraction proceeds since there are more and stronger secondary interactions as contraction advances.

It is interesting that just about the time we recognized that protein conformations can and probably do manage *PV* fluctuations, the palindromic patterns appear as one such mechanism for this purpose. The palindromic construction is obviously of major importance for the catalytic process. There may be several bases for this importance, but the only obvious one is as a device to regulate the application of force to catalysis. The device can be illustrated with the HIV-1 protease (Fig. 14). The noncrystallographic twofold axis requires that the primary functional domains face each other across this axis. In the plane of the axis as shown in Fig. 14, corresponding ends face each other. The second plane containing the dyad axis is vertical to the plane of the figure. In this plane the knots face each other in an inverted way. The rationalization suggested by this organization is that it regulates the application of force in two ways: As the domains move together, close matching in the first plane favors dynamic matching. More or less exact mismatching of the knots in the second plane is unfavorable to dynamic matching, at least so it appears, but favorable to maximizing the force exerted by the knots on the reacting assembly of substrates and functional groups. Because of the inverted orientation as the domains close, the two force vectors are always constrained to lie in the first plane. If the two domains are identical as in the HIV-1 protease, the force vectors always lie exactly in this plane, and although they are almost oppositely directed, they are not colinear. The angle at which they meet in this first plane is determined by the way the knots approach each other, but in HIV-1 protease, the vectors have mirror symmetry in the second plane. In this protein no matter how unsymmetrical the knots, the exact palindromy keeps the force vectors in the first plane. With less perfect palindromy these vectors move out of that plane, although rather little unless the palindrome is very imperfect. The second consequence of imperfect palindromy is that the two force vectors in the first plane lose perfect mirror symmetry in the second plane.

With poorly developed palindromy this effect is distorted so that only part of the the total force of contraction is directed in a useful way on the reacting assembly. The remainder is not balanced and so distorts the conformation in ways that are probably useless. Nature may have been able to so improve the already elaborate construction that the force vectors are directed so as to achieve the maximum possible rate for a given kind of enzyme.

The high degree with which a once discovered palindromy has been conserved also seems to imply that if it is the signature of a mechanical-force device, that device then has about reached the maximum a knot-matrix protein can produce. In single-chain proteins the patterns are less perfect than in HIV-1 protease, for obvious reasons, but the patterns are nevertheless very well developed. In more complex proteins in which several matrices and their knot cooperate in catalysis and control, there may be more force available, as perhaps also in large enzymes. The mechanism is a triumph of engineering able to beat the best catalytic rates of humans by as much as 10 orders of magnitude and to generate delicacy in specificity and control almost exceeding comprehension. Added to all this is the essential element of adjustability for experiment and improvement via DNA experiments. Recall all the other virtues, such as free-energy conservation, which are already suggested by experimental findings.

The palindromic patterns may reflect some totally different kind of mechanism. The fact that we cannot think of an alternative does not mean that one will not be found. At this time these patterns seem to be the strongest support for rack mechanisms. The strongest confirmation will come if and when it can be shown that the construction of enzymes with different domains, such as the lysozymes, is functionally equivalent to that of the homodomain enzymes. One does not expect to find palindromy in the heterodomain enzymes and none has yet been suggested in the data so far examined. However, the researcher must be prepared to find different mechanisms. Examination of the enzyme-like behavior of the immunoglobulins in a following section presents one such possibility.

The vibrational-librational structure of protein matrices is extremely complicated in the "open" (soft, high S) end of the coordinate because there is no fixed geometry. That is, interatomic distances are determined more by free-volume distribution than by vibrational amplitudes. A normal-mode analysis is a highly inadequate description of a mode structure in which the modes seldom get fully developed before their parameters change. Even the breathing mode is a superposition simplified by large-scale averaging and only superficially considered a vibration. On contraction the situation approaches the greater simplicity of the knot, which does have a fixed, well-defined vibrational structure possibly constructed so well as to provide a few contributing degrees of freedom. However, thermal equilibrium in the best built knot can be no longer than a few picoseconds. Large amounts of photoexcitation energy in proteins, as in smaller molecules, become equilibrium heat at such times.

In sum, it seems that nature uses matrices to produce catalytic mechanisms using the *PV* part of the enthalpy, and this is possible because random *PV* fluctuations can be sorted with minor loss in entropy to act in ways that are not totally random. Such constructions are probably possible in many kinds of polymer, but the discovery of polypeptides has allowed nature to raise this kind of design to a high art. The conservation of the structural elements necessary for such constructions—functional domain structure, palindromic *B*-factor pattern, and hard knots—tends to confirm the supposition that this is the fundamental discovery that has made protein-based life possible. If this is true, it is no longer a mystery as to why proteins are large.

7. Compression and Transition-State Formation

The profiles of thermodynamic changes along the reaction coordinate for the *N*-acetyl-amino acid esters given in Ref. 7 do not provide any explanation of the mechanism by which mechanical free energy from conformational processes actually produces the very large reductions in activation free energy. The free-energy change in the transition from separated substrate and protein to the ES compound, though small, often even negligible, greatly labilizes ES for bond rearrangement. As previously mentioned, the enthalpy and entropy changes are ambiguous because their separation into motive and thermal parts is not possible. The heat changes in contraction are large, and it is quite possible that the motive parts actually have positive signs but are overwhelmed by the very large negative thermal contributions. Too much dependence on *S*, *H*, and *V* in protein processes is especially dangerous.

The contraction process must focus forces on the reacting assembly presumably distorting the substrate bonds directly or through their interaction with the participating polar groups. If this is the primary basis for the low-activation free energy, then we can conclude that the free energy of the ground state from which the transition state is formed has been raised in ES. This is possible with negligible free-energy change only if that of conformation drops and that of substrate increases, typical complementarity behavior. But that introduces some problems of free-energy conservation that seem too special to be general. Again, enzymes like carbonic anhydrase without such elaborate possibilities come to mind, and once again it is necessary to postulate a unique, more general conformational process participating only in transition-state formation.

8. What Drives Domain Closure?

A major virtue of the rack hypothesis is that it explains the involvement of domain closure in terms of changes in mechanical work. In contrast to the "transition-state" hypothesis, the ground state is raised rather than the transition state lowered, but one might alternatively describe the process as a provision of some of the activation free energy from domain closure. The two descriptions are not equivalent and it is probable that domain closure works in both ways. The matter turns on the mechanism by which domain closure is effected. When a substrate side chain has a binding site, interaction between site and side chain would be expected to favor domain closure because it reduces the availability and probably the amount of free volume in the matrix. This entropy loss must be paid for by the improved interaction between substrate and protein, but the enthalpy also decreases on contraction, so the free-energy problem is not difficult to solve.

But this is equilibrium thermodynamics and what seems to be missing is a direct contribution of mechanical work from conformation to pay in part for the activation free energy. Rack mechanisms postulate spontaneous additional contraction arising from large *PV* fluctuations. For many years Gavish has championed this idea (see Chapter 7) using sophisticated arguments but depending not on *PV* fluctuations of the conformation itself but rather on those of the surrounding water. As noted here and especially in Chapter 3, studies of catalysis in the absence of any bulk-phase water indicate that only the plasticized protein matrix is essential for catalysis. This suggests that bulk water does not play this role or does so in a supplementary way. Indirect approaches may give necessary details rather more simply. For example, Levashov et al. used the adjustable degree of hydration of enzymes trapped in micelles to advance domain closure and employed spin labels to measure the rigidity so produced [123]. They found that the slower the spin-label motion, the more rapid the catalysis. It is quite interesting that by advancing domain closure using dehydration (with mixtures of water and *p*-dioxane), they were able to increase the hydrolysis rates of poor substrates in pure water to be equal to those of very good substrates. The argument advanced to explain these observations is that the enzyme has an active and an inactive state, the former hard and the latter soft. This is consistent with the domain-closure mechanism but does not include, at least explicitly, that the transition from inactive to active occurs after substrate binding. The activation of the poor-substrate mechanism and not the good-substrate mechanism by solvent and hydration changes suggests that the free-energy of the contracted state is reduced by these changes. Micelle-based studies of proteins have become very informative. Recently published papers of similar importance by Dorovska-Taran et al. [124] and Mao and

Walde [136] reporting delicate control of conformation by hydration adjustments using micelles extend and explain some findings by Levashov.

9. Chaotic Behavior, Domain Closure, and Transition-State Formation

That the expansion-contraction process of the protein occurs spontaneously and at random in the absence of substrate is reasonably well established by the random occurrence of ring flips and by the isotropic and gestalt behavior of conformation in proton exchange (Secs. II.C.3 and V.B.1). At temperatures near its compensation temperature, roughly 285 K, the process is nearly isoergonic, so fluctuations from one end of the coordinate to the other are not unexpected. The process manifests simple oscillatory behavior but it is actually the complicated result of a high degree of coordination among the many motions allowed in matrices made complicated by random free-volume rearrangements. Were there no knots, large degrees of expansion and contraction such as occur in catalysis would be stochastic and rare, as though there were no nonlinear couplings. The bed of complex oscillators would not reflect chaotic behavior despite the fact that most coupling is nonlinear. In his extraordinary paper on chaos [109], Havsteen has applied modern theories of systems with nonlinear coupling to chymotrypsin and its tosyl acylenzyme using *B*-factor data from Tsukada and Blow. The dynamical behavior can be parameterized in several ways, and when the several results are combined, most details of the rack mechanism become apparent. Havsteen not only extracted most of the new devices discussed in this chapter, but also provided a reasonable explanation for the coordination manifested in the contraction-expansion process. In the latter he finds that trajectories arising from the conservative chaotic behavior of the protein matrix are constrained by two "strange attractors." The latter in this case, as in most, are two separated valleys in phase space. In the region of one, the domains are open; in the other, closed. The trajectories are constrained by these attractors to be near the attractors, so they mimic normal oscillator behavior as they circle through one region back into the other. The time sequence of these transitions is chaotic rather than stochastic. Palindromy is not explicit but may be fully reflected in the results because the *B*-factor palindromy is well developed in the *B*-factor data that Havsteen used. Knot, matrix, and surface regions are clearly delineated even with this abbreviated model, but the trajectories in time rather than the substructures are extracted. Thus the step from B factors to trajectories is direct and the dynamical characteristics are revealed in much more detail. An example is the perturbation of the trajectories by disulfide bonds, important because it suggests that their presence and placement may be determined by dynamical requirements.

Just as the new paradigm based on knot-matrix construction displaces the old secondary-structure paradigm, it is quite likely that theories of nonlinear dynamics will eventually direct protein research. The eventuality is probably far away but that may be unfortunate if enzyme function can be explained correctly only as a manifestation of chaotic behavior.

10. Convergent Evolution?

Enzymes with essentially the same sets of substrates and products and with the same x-ray–diffraction description of active site geometry are defined as examples of convergent evolution when the proteins are otherwise distinctly different. This definition can be improved by stating that one or more of their functional domains are different. The three families of the lysozymes appear to be a good example. One can broaden the definition to include enzymes with the same qualitative behavior as catalysts but with totally different construction, the plant and animal enolases for example. Insofar as the domain-closure mechanism is general, these convergencies may not be remarkable since different domain-closure machines may be tailored for many specific catalytic mechanisms. Catalytic sites require only a fairly nonspecific pair of functional residues attached in the free enzymes quite loosely to their knots. In the enzymes examined thus far the functional groups are acidic and basic groups and the specificity may not be high. Examples of glutamic acid residues replacing aspartates can be found in the sequence literature without much trouble. Such changes apparently can be accommodated in the catalytic site during the functional contraction with quantitative but not qualitative consequences for the catalytic process. The fine details of the catalytic process itself are determined more by gestalt characteristics of matrices than by their local changes in geometry. The construction of active-site regions is tailored to a specific process and essential to that process, but active-site geometry can give only a limited insight into the total catalytic mechanism. Enzymes may have very similar active-site geometries but detailed catalytic mechanisms that are quite different because their matrices are otherwise different. The subtilisins and the serine proteases have very similar active sites but very different matrices and very different knots. Their *B*-factor palindromy patterns are thus very different, and it is to these pattern differences in the "gene-duplication" enzymes that we must look for the subtle differences in catalysis. Differences in binding sites such as the charge difference in the side-chain binding sites of trypsin and chymotrypsin provide important information about differences in substrate specificity because the two proteins have the same knots and the same palindromy. Differences in specificity between the subtilisins and the trypsin family of proteases are likely to be correctly understood only when the consequences of their palindrome differences are understood.

E. The Kunitz Proteinase Inhibitors

Of the several ways to account for conserved disulfide groups discussed in this chapter, as well as by others for the Kunitz inhibitors, the appearance of these groups in the *B*-factor palindrome and in matrix positions suggests that all three participate in function. By this is meant that they satisfy a specific requirement in function rather than a more general requirement for protein stability. This follows from the fact that matrices are associated closely with function and knots with stability considerations. There is ample evidence that they can and perhaps generally do favor folded stability, but this function appears to be secondary. A more restrictive statement to the effect that they play a major role in the stability of matrices is consistent with their guywire role in matrices. We have not yet examined disulfide data for other families of nonenzyme proteins to see if disulfide participation in rough structural palindromy is common, or to determine whether or not such palindromic construction is common. A case can be made that the structural palindromy of the PTI proteins is necessary because their targets, the serine proteinases, have domain-closure mechanisms that reflect *B*-factor palindromy. A very efficient inhibitor must not only accommodate to domain closure in its target, but take advantage of it to improve inhibition efficiency. The *B* factors for the trypsin-BPTI complex determined by Huber and Deisenhofer (2 PTC in the PDB) show few significant changes in atom *B* factors of either protein. A small but general increase in all the PTI *B* factors on complex formation may be a crystallographic artifact. In any event, complex formation does not force trypsin into a contracted, glass-like state. The absence of *B*-factor change is consistent with the idea that association is due to pretuning of the conformational dynamics of the inhibitor to match those of the enzyme at least in the contact region. The considerations of tuning and dynamic matching suggest that this is the only way to achieve the limiting strengths of protein-protein association (bimolecular association constants of the order 10^{14}).

F. The Immune Reaction

1. High Specificity and Strong Association

We have previously suggested that the very tight association of some antigens with their corresponding immunoglobulins, specifically γ-globulins, requires dynamic matching of antigen to antibody. The arguments remain valid but the more detailed examination of the x-ray data for the antidigoxin Fab fragment and its combination with digoxin suggests an additional factor associated with the knots rather than the surfaces. The structural comparison of antidigoxin with and without digoxin shown in Figure 18 reveals a redistribution of the low *B* factors. The lowest *B* factors tend to crowd around the contact site formed from one end of the 740-molecular-weight antigen and disappear from the other parts of the heavy and light chains although there is a considerable increase in the number of low-*B* factor atoms (Fig. 17). None of the enzymes and enzyme inhibitors we have examined show anything like this. Their knots are fixed in place. Without genetic information it is not possible to identify knots, but the migratory behavior of the lowest *B*

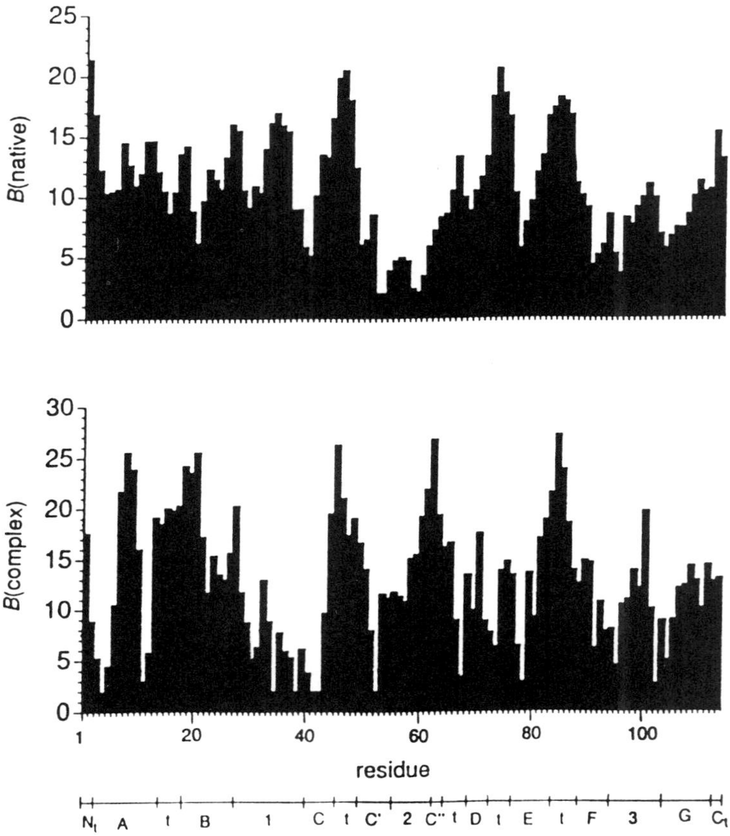

FIGURE 17 The *B* factors for antidigoxin Fab with and without digoxin bound. (Top) Antigen-free antidigoxin Fab fragment. (Bottom) Complexation with digoxin. The residue-averaged *B* factors give a description of the conformation changes in these proteins, which are very different from those observed with enzymes and other knot-matrix proteins. The *B* factors are on average considerably smaller and no groups of low-*B* residues persist on binding digoxin. If Fab can be described by the knot-matrix model, the knots are few in number until digoxin is bound and those few are not fixed. Digoxin produces many large changes in *B* factors most of which reflect a generalized stiffening more or less uniformly spread throughout the L and H subunits. The localized knot characteristic of knot-matrix proteins is replaced by a general increase in knot-like character. Fab thus appears to be a different kind of protein in which even the higher-*B* factor regions have physical characteristics more like knots than matrices. The large contractions, a rich source of free energy, are possible at least in this isolated example only when antigen is bound. (From Constantine et al. (1993), *Struct. Funct. Genet.* *15*:290 and reproduced with permission of Wiley-Liss.)

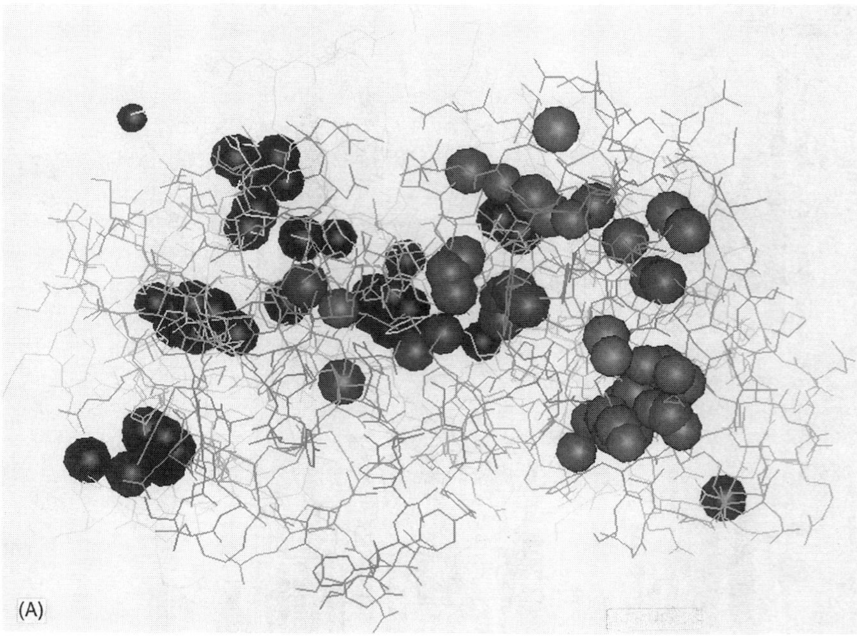

FIGURE 18 *B*-factor change produced in antidigoxin on binding digoxin. The darker spheres are atoms with *B* factors at or below 2. The protein is shown end-on to display the changes around the digoxin binding site. Digoxin (the ball of string in Fig. 18B) is a long, thin molecule of molecular weight 740. Only one end is in contact with Fab. The clustering of low-*B* factor atoms in Fab produced by digoxin is significantly higher near the contract region than elsewhere, and to some extent this suggests that the immune process is one of completing the knot. However, it is obvious that for this concept to be useful in this case, it must be extended to include the entire Fab molecule. (The *B* factor values are from 1IGI and 1IGJ in the Brookhaven PDB).

factors suggests a construction very different from the knot-matrix construction and one in which knots cannot be identified in a unique way. Although we have examined only a small sample of the *B*-factor data for the immunoglobulins, the differences between ordinary knot-matrix proteins and antidigoxin are so marked that they suggest a totally different use of the same principles to produce structures with qualitatively different characteristics and physiological function.

In Fig. 17 the average *B* factors for the residues are plotted against residue number for the free antibody and the immune complex. It can be seen that groups of residues lose their low *B* factors on complex formation while other groups drop

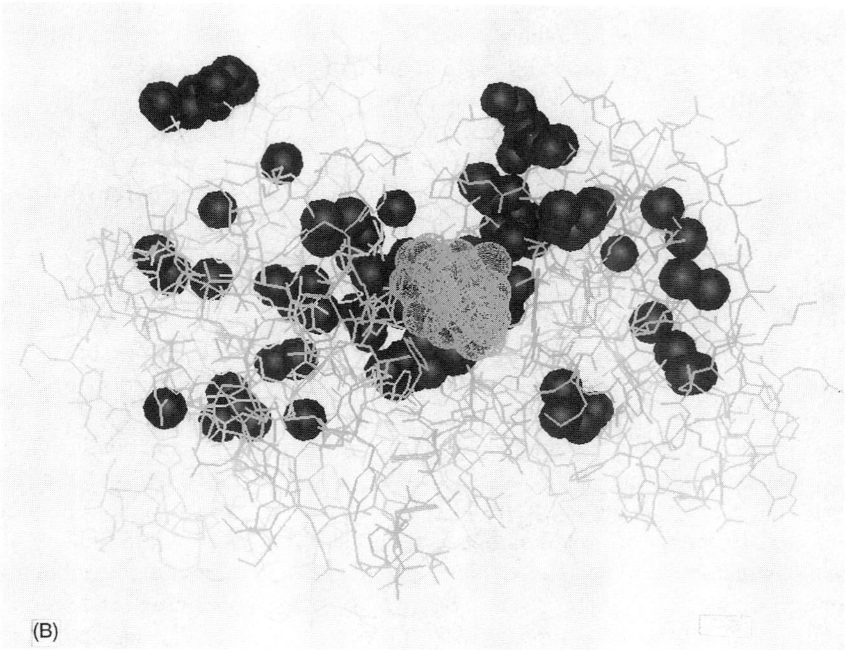

to low *B* values. Knots change in number and position on binding digoxin. Not only are these *B*-factor rearrangements not found in enzymes, but the lowest *B* factors are the very low and the *B*-factor average falls well below those found in enzymes, whether in complexed or free state. The *B*-factor variance is very low relative to that of enzymes and furthermore is higher in the complex than in the free fragment, just the opposite of what has so far been found for enzymes. These observations suggest that the construction principle is such that all regions can have the characteristics of a knot or a matrix and can shift from one to the other in binding functions. All four domains, two to each heavy chain and two to each light chain, are similar in appearance and *B* factors. All change to about the same extent when digoxin is bound.

Although the construction falls in the knot-matrix class, as indicated by the large differences between the lowest *B* values and the average, it is so different from that of enzymes and other proteins with fixed knots as to define a new class or subclass. The single knot of antigen-free antidigoxin is replaced by several on binding digoxin, but the number of atoms in the contact surface is a very small fraction of the total. The lack of organization of the very low *B* atoms into fixed knot-like structures makes attempts to predict thermodynamic stability dangerous. The low *B* average suggests high thermal stability, but the free antibody may be constructed to have a high free energy in the absence of antigen. One can only guess that the folded stability is low without digoxin and much increased by digoxin.

The simple "completing the knot" process described in Sec. IV.E would apply if some atoms of the antigen and some of the protein combined to form a knot. This does not seem to be the case. Digoxin binding does increase the number of low *B* atoms near the combining site, but most of these are separated from that site by atoms of higher *B*. A more accurate description arising from the limited data is that combination with digoxin shifts most *B* factors both near and well removed. Such preliminary deductions and the useful information supplied in the publications of Constantine et al. [17] suggest that the Fab fragment is constructed using principles similar to those involved in matrix manipulation of enzymes, but using knots or a general conformational state more like knots than matrices to provide the function and flexibility similar to those of matrices in enzymes.

Two major features of immune reactions are high specificity in binding antigen and strong binding of the latter. Another is multiple valency, by which is meant many combination reactions ranging from complement binding (although not to Fab), binding of protein antigens at constant regions [123], to binding of small unstructured polypeptides as much at constant regions as at the variable regions. Since all four domains seem to be involved in any combining reaction, the coupling and thus cooperation among such processes must be considerable. It is as though the entire molecule is poised to contract toward one or another state of lower free energy but is restricted by its construction from doing so unless certain arrangements of atoms are available in foreign molecules for combination. This is in close analogy with other examples of completing the knot and thus thermodynamically allowed. It is perhaps consistent with the imperfections that prevent accurate, stable β-sheet structure in the domains.

These observations suggest the device by which IgG achieves high specificity and great strength in its most fundamental physiological function. IgG suggests a lock in which each potential knot is a tumbler. A good specific antibody must accommodate its geometry and dynamics to its antigen and does so by changing to a specific set of knots. The more knots, the greater the specificity, just as occurs when the number of tumblers of a Yale lock is increased, and also the more stable the complex. This lock-and-key mechanism is a considerable elaboration of the ancient lock-and-key analogy for enzymic catalysis.

2. Abzymes and Enzymes: Similarities and Differences

The transition-state hypothesis of enzyme mechanisms has been cleverly utilized to explain and exploit the catalytic efficiency of specific antibodies. Antibodies (*abzymes*) specific for binding of a molecule thought to resemble the transition state for the chemical conversion of another molecule are found to catalyze the reaction of the latter. This is pure "transition-state-stabilization" reasoning and probably correct. The catalytic efficiencies where comparable are low relative to those of enzymes, which is not surprising since the mechanisms are considerably different and one expects evolution for catalytic efficiency to be different since it is unlikely that antibodies were developed in evolution for catalytic purposes. The transition-state hypothesis of catalytic mechanism is at best only a partial explanation for catalysis by true enzymes, and its adequacy in explaining abzyme catalysis makes conjectures about one kind of mechanism from the other somewhat hazardous. In contrast to enzymes, the immunoglobulins are constructed to make very strong bonds in a very select way. The mechanism proposed in the previous section for strong association suggests the protein has a built-in store of conformational free energy, which is released on association with antigen. Now the antigen is the transition-state analog become abzyme substrate. The gradient of free-energy change acting on the transition-state-analog substrate produces a large mechanical force on the substrate, which in many cases results in catalysis. The requirement is that the substrate is so constructed that force will facilitate reaction. This is a brute-force mechanism, which does not seem to have found much use in evolution though it is used a good deal in biotechnology. Enzymes have no such gradients, or at least none of comparable magnitude, and so demonstrate more subtle ways of accomplishing their function.

VII. DYNAMICAL ASPECTS OF PROTEIN ELECTROSTATIC POTENTIALS

A. The Next Level of Complexity

Several of the simplifications that have been made to facilitate understanding of the knot-matrix construction principle and its consequences can be removed or justified by somewhat tedious use of existing data. Most, however, cannot be put on a rigorous quantitative foundation because all such features of proteins reflect the behavior of very complex potential-energy functions. Some fundamental questions about the substructures can be posed on the basis of the knots used as illustrations for this chapter. Thus we ask first, "What is the atomic description of a knot?" and then "What factors are responsible for knot stability?"

B. What Is the Atomic Description of a Knot?

In this chapter the position has been taken that the real differences between knots and matrices are the differences in mobile free volume, but from the practical viewpoint it is very difficult to define a knot accurately in this way. Structural information lacks the precision required to measure free volume and to determine its mobile part. A careful study of density variations can be of some aid but is inferior to *B*-factor and genetic information. The last can be confusing because there is considerable latitude in the replacement of knot residues by similar residues, but the aromatic residues are the major subclass in this connection and they are very common as efficient van der Waals partners in knots. Exchanges among aromatic residues are more common than among the subfamily of long-chain alkyl residues.

The information about the construction of a knot from genetic, *B*-factor, and proton-exchange data generally agrees quite well in defining the residues but not so well in establishing which atoms of the residues are actually in the knots.

The conservation of knot residues down protein family trees is very high when the exchanges of like-for-like are taken into account. As a consequence, conservation of palindromic *B*-factor patterns in enzyme families appears to be high with small changes in details consistent with small changes in knot construction. The PDB is useful but still inadequate in providing highly-refined *B*-factor data for more than a few members of each protein family. Fortunately, as shown in Fig. 8, unrefined *B*-factor sets from relatively low-resolution studies can be used to confirm the existence of a palindromic pattern even though precise comparisons of those patterns among members of the same family are not possible. At present, comparisons between results obtained with the same protein in different laboratories are not much more accurate than those between two proteins of the same family. As a result, the degree of conservation of palindromy is still uncertain. It may not be particularly high to be effective in whatever its role since palindromy in single-chain enzymes is achieved only by many mutations and is thus always in progress and rarely completed. One gets the impression that although perfect palindromy may be required to yield the maximum catalytic efficiency possible for an enzyme family, nature does quite well with something less than perfection.

C. What Factors Are Responsible for the Stability of Knots?

The term "stability" as used here refers to the work required to expand a knot to the point at which its cooperativity is lost. In contrast to the thermodynamic stability of the total protein, the stability of knots can be treated to a first-order approximation as independent of the matrices. Specifically, since the configurational entropies of knots are low relative to the configuration entropy of any melted states, stability is primarily a potential-energy consequence. The two major potential-energy factors in protein folding are associated with the transfer of peptide dipoles and side-chain groups without permanent polarization, from a highly aqueous medium to the protein interior. These transfer processes are very important for the total thermodynamic stability of proteins in vivo only when referred to the denatured state determined by the conditions in the organism. The only good reference state is the true random coil and this corresponds to the theta-point condition so useful in most polymer chemistry. Bolen establishes this condition with very high concentrations of some denaturants but, as remarked earlier, the undertaking is laborious and thus not used much.

For those investigators interested in proteins as chemical species, knot stability or instability is generally of most interest. The four major aspects of these interactions are: (1) the van der Waals interactions, (2) packing so as to reduce free volume and thus reduce motility, (3) interactions of groups having permanent polarization, and (4) participation in cooperative contraction producing lower free volume, thus lower motility in the knot. An important consequence of item 4 is the reduction of the effective dielectric constant since this is the major contribution of the groups with low or zero polarization to knot stability.

Knots do not appear to be successful simply because of their strength. All knot-matrix proteins so far examined have much lower thermal stability than the keratins and most other strictly fibrous proteins. Knots of enzymes may be softer than those of proteins, like the pancreatic trypsin inhibitors, and there is some evidence, already discussed in connection with the acyl derivatives of α-lytic protease, that even the knots may undergo slight contraction if domain closure exceeds some as yet to be determined limit. *B*-factors for hetero-domain enzymes have not yet been examined for knot contraction during domain closure. The domain-closure construction is apparent in all families of such enzymes so far examined but the expansion-contraction process is yet to be studied. In any event, the similarities in knot construction for all classes of enzymes indicates that knots for almost any purpose can be found in evolutionary search.

It is difficult to estimate the contributions to the free energy of the knot since the potential-energy functions are sensitive to changes in distance and angle parameters so small as to exceed the errors in determination. The free volume on which the matrix entropy depends is similarly sensitive. Average values will not be of much use if protein functions are consequences of conformational fluctuations. Nor for that matter do average coordinate values as contrasted with *B* factors give much information about the probability-density distributions of proteins and associated dynamical behavior.

The quantitative description of protein stability is to a major extent determinable from experimental data on knots, at least this applies to proteins that demonstrate accurate two-state behavior in thermal denaturation. For these the measurement of the pure melting rate constant makes possible as extensive an assessment of substructure and total protein stability as is possible and, for that matter, necessary given that the variability of the denatured state, the distribution of conformers of the matrices, and the general problem of protein activity coefficients have not been determined for the class of proteins under study. Just about all this information is necessary if small-model data are to contribute reliably to an understanding of proteins. But the relevant small-model data are themselves in a state of considerable confusion. A good example of the latter is the uncertainty about the interactions of water with polar groups of peptide and nucleosides. On reexamining the solubility data for thymine in pure water, Alvarez and Biltonen [124] found that about 25% of the hydrogen-bonding groups of this molecule do not make hydrogen bonds with water.

With the knots so far examined, not a great many, two qualitative classes of knot behavior appear to be identifiable. Major cooperativity seems to be an essential feature of the hydrogen-bonded skeleton, in which case loss of a residue or chemical change destroying a strong polar interaction is likely to destroy the stability of the knot. The effects of outright changes in number of residues in knots by SDM experiments in which whole residues are added or subtracted would appear to be more important at this stage of knowledge than side-chain permutations. The reasoning here is that very few SDM experiments have supplied information about the H-bonded skeleton of knots. The latter are the most important in knot construction and stability but the most difficult to study.

D. Gestalt Versus Local Fields

There is a general belief that the best way to attenuate the strong fields of the dipoles is through pairwise hydrogen bonding. This is very difficult to establish since the point-dipole approximation, although roughly applicable to arrays of well-separated dipolar groups, is inapplicable at the short distances of separation of such groups in folded proteins. However, Miller, Flory, and Brant [125] did simple calculations on large arrays of peptide groups which suggested that the reduction in electrostatic free energy of these groups might be lower in three-dimensional arrays without pairwise H bonding but with optimum orientation of the dipole vectors. Hydrogen bonding is rarely extensive except in secondary structures, and we have already discussed the fact that such structures do not in general make special contributions to folded stability. These considerations suggest that array interactions among dipoles may be an important gestalt feature of protein conformations and individual hydrogen bonds may be less important than now assumed.

The primary phospholipase A2 knots demonstrate that α-helices can be very stable in knots, but they do not explain why this is so. One suggestion is that the small side-chain groups do not restrict contraction of these helices to very low local dielectric constant. A possible supplement is that when considerable contraction is possible, inductive electron rearrangements through α-carbon atoms become large. The covalency that results strengthens the interhelix peptide bonds, knitting them to a greater or lesser degree into single electronic structures with some aromatic character. This effect seems to be manifested by the H-bonded ring structures found in the BPTI knots (Fig. 12). Helices and sheets do not allow the same division into single six-membered rings, but in the helix, for example, the 1-4 hydrogen-bonded pairs may be so strongly coupled by induction as to confer gestalt electronic properties on sections of helices. Such extensive and longer-range electronic rearrangement, if it occurs, is likely to be important especially in determining dynamical behavior. All this would be in keeping with the H-bond polarization results of Zundel [109], but even when the study of homopolymers was very popular, not much information on this effect appears to have been accumulated.

An opposite example is provided by the helices of myoglobin and hemoglobin the atoms of which have high *B* factors and thus high motility and little contraction. These have the usual distribution of side chains rather than the collection of small side chains found in the primary phospholipase A2 knots. The larger side chains may restrict contraction of the helices, thus reducing the strength of interchain H bonding. This does not seem very promising, but alternative explanations for soft helices and sheets that depend on the long-range electric field structure are equally unattractive although probably closer to the mark. At present it is as difficult to explain the hard helix knots of the phospholipases as to explain the soft knots of myoglobin and most other secondary structures. In the studies of the S and C helices of the N-terminal knot of RNase A, already mentioned, some research groups seem to have established a special basis in the organization of nearby charged groups. Removed from the protein these helices have unusual stability in helix form, so local groups rather than gestalt protein fields may be the explanation. Local fields play important roles in structure and function, but because of everyday familiarity with small molecules, they tend to divert attention from the gestalt fields which distinguish a protein from a collection of local groupings. The latter is illustrated in the next few paragraphs with the few relevant examples so far to appear.

In phospholipase A2 primary knots, the fields of the peptide dipoles are completely or nearly completely attenuated within the helix. By comparison, we see that this cannot be the case for the myoglobin helices. The dipole fields leak out of the helices to be attenuated by groups outside the helices. The regular array guarantees some interhelix cooperativity and thus some coherency in the escaping fields. The result will be a tendency for helix segments to behave like single dipoles with a large moment and a long range. This explains the previous suggestion that the interactions among the escaping fields of the helices are important in folded stability. It has already been suggested that the alternative basis for the construction and stability of proteins containing the "myoglobin-fold" is the long-range interactions among these favorably oriented helix "dipolar" fields. The myoglobin fold has proven to be a major achievement in evolution, undergoing very little change despite the fact that only two residue positions have been unchanged in existing proteins of this structure. The residue composition is very poorly conserved, but form is very strongly conserved.

The escaping fields can be attenuated in several ways. A popular idea is that the whole-helix field is capped by ions, which seems a bit unlikely in view of the maximum amount of polarization along the helix that is possible. In any event, the fields must exist and they present splendid opportunities for nature and great challenges for chemists not only in stability matters, but also in dynamics during fluctuations. Thus, for example, contraction of the helices will tend to draw the fields back inside the helices. This behavior must also undoubtedly occur in these soft proteins with high motility and in these proteins it must be a major determinant of conformational dynamics.

Discovery of a knot in evolution may sometimes depend on a single fortuitous mutation, perhaps always if H-bonded skeletons are as critical as we believe. On the other hand, tailoring a knot to improve the function of a given kind of protein or to generate a new function in a new protein is likely to be through saltatory and uncertain progression. That is, knots may have some threshold of appearance in evolution but proteins themselves reflect quantitative advances. This matter is also well illustrated by the results with the knot mutants of BPTI [49], which need not again be recapitulated except to note that the loss in thermodynamic stability ranged from one to four orders of magnitude in the equilibrium constant for unmelting and that these stability losses reflect primarily differences between ground states of the folded species and a single invariant transition state for unfolding. The latter follows from the fact that the transition states are determined by loss of knot cooperativity. The matrices are little changed, so their absolute free energies of formation must be about the same. To that approximation the thermodynamic differences among the mutants are differences in their ground-state free energies.

Three other examples round out this section. The first of these is the "strap hinge" formed of the penultimate four-residues block at the C-terminal ends of the two subunits of the arc and cro repressor proteins and also found in the similarly structured HIV-1 protease (Fig. 14). The structure is clearly a secondary knot, but nevertheless a structural one and a strong one since without it the necessary coordination of functional domains would not be possible. The two-strand antiparallel structures have knot *B* factors and look like knots. As with the primary helices of the mammalian phospholipase A2 proteins, nearly all atoms with knot *B* factors are backbone atoms. The stability of this small isolated structure presents the same puzzle as the phospholipase knots. Are these strengths due to interactions with nearby atoms or are they consequences of whole molecule fields?

The second example concerns the similarity and dissimilarity of the Fab fragments and the keratins. While both seem to depend on secondary structure, the keratins provide a reference state of highly ordered and highly stable secondary folding, which is quite different from the extensive but very disordered piles of β sheets in both domains of the heavy and light chains of digoxin-free Fab. Keratin is strictly structural, apparently almost a continuous knot. Antidigoxin has a chemical function, like the enzymes, but differs from such "chemical" proteins not only in having much more sheet structure than usual but also in having other more notable differences, some of which have already been discussed. Particularly noteworthy are the low-*B* factors throughout and the large number of very low-*B* factors found in free antidigoxin and even more prevalent in the complex with digoxin. If the *B* factors of matrix atoms in enzymes are the reference, it is difficult to decide whether most atoms of Fab are hard matrix atoms or soft knot atoms. As shown in Fig. 18, most peptide groups are integrated into strong three-dimensional structures but with less than perfect secondary organization. The changes on binding digoxin are widespread, implying that function depends on the delicate balance between knots and no knots in this protein. More accurately, the construction suggests a frustration of sheet structure as though nature started with the keratins containing β structure and then upset their ability to fold in normal, stable, ordered ways: Changes in antidigoxin Fab occur throughout the four domains but the impression this gives is one of propagation of short-range interactions rather than domination by long-range fields or by fixed knots. The antidigoxin Fab is constructed so differently from the typical knot-matrix proteins as to require a new classification. In particular, the absence of a fixed knot system in itself requires a new classification and the low-*B* factors establish that it is not a member of the hemoglobin family.

The last example arises from a very different kind of stability peculiarity. Kiefer and co-workers [126] have studied human carbonic anhydrase with special attention to the complex-ion structure in the protein at which the zinc ion is bound. As is not unusual, they find the chelation of this ion by the protein is thermodynamically considerably more stable than in most small-molecule chelates for zinc ion. The ligands have been previously shown to be three imidazole groups and one water molecule, but Kiefer et al. find the strength of the direct interaction of these with the zinc ion is strongly dependent on the shell about the chelate complex. The changes they have made in these neighbors reduce the stability of the complex to that of small-molecule zinc complexes. Several explanations for this effect can be postulated, but sorting them out is more or less guesswork until the gestalt aspects of the dipole arrays in proteins are described in some detail.

VIII. SUMMARY

A. Thermodynamics in the Biosphere

In *What Is Life?* published in 1944, Schrödinger explained the apparent violations of the second law of thermodynamics that make life possible in terms of a series of heat engines operating between the characteristic temperature of the light from the sun absorbed by the top members of the food chain, the plants, and progressively lower and lower temperatures down to that of the lowest-temperature reservoir, space. All systems participating in the biosphere are either parts of such engines or are coupled to them for food, the food being, of course, free energy.

This discovery was first reported by Ludwig Boltzmann in 1886. Boltzmann observed that the struggle for existence lies not in the energy taken up from the sun but rather in the negentropy taken up with the energy. The sun's energy has very nearly zero entropy so the negentropy flux is very large. The second law is, of course, obeyed because there is an irreversible production of entropy due to the "frictional" characteristics of real processes. In his biography of Boltzmann, Broda presents the most important quotations from Boltzmann and gives the major references [127]. Lumry and Biltonen particularized the concept for series of enzymic processes [9] to point out that the only general means available to all organisms to improve their survival value is to maximize their efficiency in the use of the free energy they are able to get and keep. This is done by minimizing the irreversible entropy production in their processes without loss of high reaction rates and control. Usually a compromise is necessary, but there is an underlying pressure in evolution to approach the ideal by finding and keeping mutational changes improving the ability of an organism to carry out physiologically important reactions at useful rates and do so in such a way as to increasingly approach the thermodynamic efficiency limit that would obtain in a frictionless biosphere. Enzymes are the primary vehicle for these improvements since in catalytic functions, both simple and complex, the enzyme and successful substrate are tightly coupled so as to conserve free energy along the reaction path from reactants to *products.* This kind of coupling in single proteins is called *free-energy complementarity* and in systems of coupled proteins *supercomplementarity.* The most experimentally based illustration of how far this tendency has become advanced in evolution is that of Fisher and co-workers [77], but it is probable that the entire glycolytic system has been evolved as a unit so as to satisfy the efficiency requirement [128].

Complementarity need not be exact, or even necessary, in proteins with a purely degradative function such as the proteases, and it is likely often to be secondary in importance to the chemical conservation of free energy in second-order, primary-bond rearrangement processes. It nevertheless establishes a general requirement on enzymic mechanisms that must be taken into account in any hypothesis of mechanism that is offered for serious consideration. Thus far only rack hypotheses attempt to accommodate to this requirement.

Mechanical stress and strain and complementarity, which must usually be achieved by mechanical means, are devices developed in evolution. They reflect more fundamental devices: the stable folded conformations of polypeptides, the adjustment of protein potential energy functions, and enthalpy-entropy balance. And since all these are adjusted by mutation, they also reflect DNA, a device without which none of the others would have been discovered. The more-recently detected devices discussed in this chapter arise from these same fundamental discoveries in a roughly linear progression, most being enhancements of previous discoveries with branching and combination.

B. The Evolution of Devices

On the basis of the still limited information now available, one can conjecture about the progression of discoveries of devices in evolution after the discovery of the knot and thus the knot-matrix construction principle. Thus, the pairing principle as expressed in the domain-closure process of enzymes probably came soon after. Its manifestation in the antidigoxin Fab fragment, clearly a slender reed on which to lean, appears to be much more complicated. Thus most of the surface seems to form a complex set of allosteric sites with a wide range of possible responses. The qualitative difference between the immune proteins and the knot-matrix proteins again demonstrates the remarkable variety available in proteins from nothing more than sequence choices. The only even approximate analogy with familiar chemistry is that of carbon.

Functional domains are direct consequences of knot-matrix construction but their pairing for catalytic function is a subsequent discovery. If single-domain catalysis ever existed, it was unlikely to withstand the competition provided by modern bidomain enzymes, but a few relics may yet be discovered. The reasoning here is that the use of the natural expansion-contraction (free-volume) process of matrices as a source of mechanical energy in catalysis is more effective when it is mediated through domain closure than through simple expansion and contraction in a single domain. The pairing principle is exemplified by the cooperation of the two domains in enzymes and thus is similar to the discovery of the opposed thumb much farther up the line. We suspect the major gain in pairing is the mechanical advantage even in nonenzymic proteins. One might even call the rack construction of hemoglobin an example of pairing.

Proteins with inhibitory or other regulating function, e.g., the BPTI family of protease inhibitors, get by with a single functional domain. "Completing the knot" is then a useful discovery for regulating the regulators in addition to being a very useful way to supply free energy by formation of knots.

The use of conformational dynamics to improve the operation and versatility of the devices so far mentioned and as the basis for still other devices may be recent or very old. The evidence for the device we call dynamic matching and tuning is quite impressive, even if much of it rests on logical arguments rather than immediate experimental evidence. Despite its novelty, the hypothesis of dynamic matching as the basis for heme-heme interaction in the multichain hemoglobins (Sec. VI.C) is at least as sound as the more conventional explanations insofar as dynamic matching must occur. It may prove quantitatively inadequate, but the point is that so long as mechanisms depending on conformational dynamics can be advanced, there is no way to reject them out of hand.

C. Function Follows Form?

The devices discussed in this chapter do not explain why the evolution of enzymes followed the path it did but they do suggest how they may work. The devices generating the most insecurity are those involving dynamic matching, yet there has not appeared any other explanation for the existence of the palindromic *B*-factor patterns in the single-chain, gene-duplication enzymes and the considerable expenditure of time in evolution to find and improve them. These patterns in the sequence *B* factors reveal at least one level of complexity in protein construction and function, which goes well beyond those under current consideration. They occur in the gene-duplication enzymes, but there are similarities and differences in the patterns, which suggest a reason for these patterns not only in these enzymes but in all enzymes. The core of the argument is that despite considerable differences in the way palindromic patterns arise in the gene-duplication enzymes, one structural feature is common to all we have studied. This is the approximate dyad axis lying in the plane separating the two functional domains and passing between the two primary functional groups. The only use in catalysis for this kind of construction that we have been able to imagine is as a source of mechanical work and strain to supplement the normal heat activation of transition states for primary-bond reactions. The dyad arrangement giving the protein vectorial control over the conservation of force and thus the use of the mechanical energy ceases to seem improbable when one searches for alternatives and fails to find them. So one searches for similarities and differences among the enzymes.

One can compare the construction of phospholipase A2 (Figs. 6 and 8) with that of the subtilisin protein (Figs. 7 and 10) to see that although both show *B*-factor palindromy, in detail the construction is quite different. In the latter the minima in the *B* factors outline the palindrome so the knots are themselves constructed in palindromic fashion. In the former the knots are not constructed in this way except in the not-so-trivial sense that so far as the needs of function are concerned, a helix is a helix despite the chain direction. Similarly, although the HIV-1 protease and the single-chain aspartate proteases, the pepsins, have some similarities in their palindromes, the former is constructed by reversing chain direction. In the comparison of subtilisin BPN′ with proteinase K (Fig. 10) it was shown that the palindrome pattern is the same despite very few similarities in the matrix sequences. The sequences of knots are highly conserved in the individual families, but the two knots themselves have nothing in common except their common palindrome. Phospholipase A2 is a good example and also a useful reminder that there is so much variation among knots that construction classes have to be recognized. Nevertheless, the features necessary for the mechanical machine are to be found in all the classes so far examined. What emerges from such considerations is the suggestion that it is the form of the conformation rather than any local details that supports function. Then form determines function. This is known to be true for proteins with the myoglobin fold. The extreme residue variations that occur in the hemoglobin family destroy neither the fold nor the function. The knots in knot-matrix proteins are highly conserved. The matrices are weakly conserved as is consistent with their role in catalysis and their necessary adjustability for experiments in evolution. Knots determine palindromic patterns and matrices put them to work. In the gene-duplication enzymes the knots in the two functional domains need to be similar only to the extent required to establish the palindrome and thus the form. So the palindromic patterns appear as a direct manifestation of the fact that function follows form.

The conservation of form is most easily seen when it is revealed by a palindrome. Heterodomain enzymes probably do not have such patterns. We have not found them, and after due consideration we do not expect to. But they have some important similarities with the homodomain enzymes used as examples in this chapter. Thus they have knots and matrices so that the functional domains are organized internally in what appears to be exactly the same way and the two functional domains are organized like a stapler or a pair of scissors, not so neatly but nevertheless as attractive for function.

If indeed function follows form, there must be some essential similarity in the forms of the gene-duplication enzymes and the heterodomain enzymes. The fundamental basis of the catalytic mechanisms may be the same, but if the basis is the mechanical one, it would not generate palindromic *B*-value patterns. A useful consequence of this is that heterodomain enzymes provide a means for testing the mechanical hypothesis. There should be an alternative construction feature in the heterodomain enzymes playing the same mechanical role but not demonstrating palindromy. This assumes, of course, that all enzymes have the same mechanism. That too is a question that can be answered by such comparisons.

One can suggest, for example, for the heterodomain enzymes such as lysozymes and ribonucleases that the alignment of the two functional domains during domain closure is enforced by the substrates. The latter are held in a trough between the domains, and thus held perpendicular to the direction of domain closure. The situtaion is similar in the serine proteases, but in the latter the similarity of domains provides the basis for palindromy. This example is not very satisfying but it illustrates the search to be made.

D. Consequences for the Immediate Future of Protein Chemistry

Of the several devices discussed in this chapter, the one most important for protein research at this time is the knot-matrix construction principle itself. It provides just the right framework at this time for organizing the data obtained with site-directed mutagenesis methodologies. That it and such related features as the control of protein folding by knots have not become generally known has made much of recent research on protein structure and function rather more tedious than it need be. The old paradigm, based on Linderstrøm-Lang's classification of structures persists, but the devices of biology have not recognized it. They make very little distinction between secondary and tertiary structures. The new paradigm will greatly improve research efficiency since so many properties of proteins are consequences of knot-matrix construction. A major factor in accelerating the use of the new paradigm is the abundance of *B*-factor data and sequence data in the several databases. Atomic coordinates are always invaluable, but it is likely that they will take a backseat in usefulness and importance to the *B* factors as the dynamical bases of protein functions emerge in greater and greater detail.

E. Hypotheses Based on the Knot-Matrix Principle

Kinetic stability is due to knots. It is a consequence of the low potential energy of knots and their cooperativity, both of which arise from a symbiosis of excellent fitting and enhanced electrostatic stabilization. Woodward and co-workers have established that for BPTI knot disruption is the activation process for melting and "folding." This is in agreement with deductions made using several other proteins but it is an experimentally more impressive proof. Nevertheless, the degree of generality of this mechanism needs to be better established. It is not always correct for a few knot-matrix proteins under normal conditions, and for many it can become incorrect under special conditions.

Thermodynamic stability is primarily due to knots for the same reasons although the all-or-none characteristic of thermal denaturation is due to kinetic stability. Strength and other physical characteristics, thermal stability, and the temperature range of physiological function are adjustable by changes in size, composition, and packing of knots and matrices. Even under the constraints arising from the intrinsic chemical properties of the amino acids, DNA mutation provides virtually unlimited opportunities for evolutionary experimentation.

Matrices and surfaces can make contributions to the free-energy change in folding which are negative, but in general such contributions are positive. Enthalpy dominates knots, but matrices must be soft to support function and thus have entropy values roughly balancing enthalpy values of the same sign. Their free energy is higher than that in their state in the collapsed denatured protein. The high entropy values reflect free volume, which in turn is responsible for the soft-rubber physical characteristic of the free protein and the stiffer contracted states in catalysis. Matrices and surfaces are held at high free energy by the knots. In this way it has become possible to overwhelm the natural tendency of matrices to collapse, and this makes them useful in specific physiological functions.

The discovery of a new knot is probably a rare event. If so, genetic stability depends on tight conservation of knots. A few changes in knot atoms may be tolerated but these are like for like. A given kind of knot can be associated with several, probably many, matrices and surfaces to produce different functional domains and thus different proteins with different functions. Functional characteristics such as specificity and the description of transition states are due to matrices and surfaces.

Functional domains are the smallest structural elements of the catalytic machine and can be mixed and combined through matrix mutation to produce new proteins for new or old functions. Functional domains with secondary functions (rate enhancement, regulation, etc.) can be added or subtracted. The two primary functional domains can be provided by different proteins.

Two functional domains are combined to effect each catalytic function. They are organized like an office stapler and, like a stapler, enhance rate processes mechanically by contraction of the jaws (domain closure). The transient mechanical free energy used in transition-state formation appears to be supplied by expansion-contraction processes of matrices, enthalpy-entropy interconversions at near-constant free energy. The basis by which the expansion-contraction process is effected has not been established and may require chaos theory for its explanation.

The palindromic *B*-factor patterns of the gene-duplication enzymes are difficult to explain on any other basis than the dynamic matching of the fluctuations of the matrices of the functional domains. The changes in this tuning that result on substrate binding would provide control of the matrix contraction process by substrate and product. Mechanical force on substrate-enzyme reaction assemblies is applied directly by the knots driven by contraction of the matrices in which they are embedded. The knots position the two primary functional groups of the enzyme in the "active-site region" prior to contraction and otherwise set limits on the adjustability of this region during contraction into the transition state. Thus, although the functional groups themselves are not in the knots but only held at knot surfaces, the region is always to some extent bounded by knot surfaces.

The supplementation of heat by mechanical work in transition-state formation provides a new kind of chemistry modifiable by evolutionary experiments and probably not resembling very closely the familiar chemistry of small molecules. The requirements on conservation of protein functional groups may be much less than in catalytic processes of primary bonds in small molecules.

Surfaces provide contact between protein and its environment and are adjusted as separate structural units to effect such communication. Insofar as this communication, which includes both association and free-energy exchange with other macromolecules, depends on adjustment of conformational dynamics, matrices and surfaces are not functionally independent. The dynamics of surface and matrices are adjustable by mutations in the same way as the other properties of these substructures. Charged groups may have to be considered as forming a separate substructure. They are attached near and on surfaces with varying degrees of contact with bulk solvent but have long-range fields that can have important interactions directly with the electrostatic fields of matrices.

Additional hypotheses having to do with genetics as much as structure and function will arise as existing data are examined. We have thus far found no exceptions to the knot-matrix principle, but our sampling is far from complete and the exploration of *B*-factor data is just beginning.

1. Final Questions

How variable are the matrices found combined with a given knot? What corresponds to the *B*-factor palindromes in heteroknot proteins? Do the palindromes reveal fine detail associated with specific residues, groups, or kinds of atoms? How do the family trees of proteins determined by knot and matrix conservation compare with those determined by other methods? Are there enzymes that are built with a different principle or are functional with a single functional domain? How many different kinds of knots have thus far been discovered in evolution? What are the details of the interplay in evolution of a knot and its matrix? How many different matrix functions, e.g., catalysis, can a single kind of knot support?

But the most interesting puzzle arises from the palindromic patterns and the implication that what they reflect is a universal catalytic mechanism. What is there about that mechanism which is so much more efficient than the alternatives that all enzyme families independently find it to be the road to success in evolution [138]? Can someone think of another basis for *B*-factor palindromy?

These questions can all be answered to at least some degree with data now available.

ACKNOWLEDGMENTS

The conformational basis of protein function has been studied in the Laboratory for Biophysical Chemistry of the University of Minnesota since 1953. Since 1990, support has been provided by the Lumry Family Foundation of Hunts Point, Washington. During most of this period Professor Andreas Rosenberg directed the search in this laboratory for the knot-matrix concept using proton-exchange methodologies. In recent years, Professors Roger Gregory and Clare Woodward and Dr. Joseph Rifkind have provided particularly valuable information and support. Frequent discussions with Professor Gregory have catalyzed both the nucleation and growth of the many ideas discussed in this chapter. We are grateful to Dr. Keith Constantine for bringing the digoxin-Fab problem to our attention and for transmitting the x-ray–diffraction results. Professor A. Wlodower and Dr. R. Bott provided x-ray–diffaction data, for which we are also grateful. Professor Woodward made graphics computer facilities available, which Mr. Olav Jaren used to construct the molecular-model figures Ms. C. Makkyla did much of the editing. Nearly all coordinate and *B*-factor data came from the Brookhaven Protein Database. This work would not have been possible without that information. So many different sets of protein data were used that listing detailed references to all the sets is not possible. We are, of course, very much indebted to all the contributors.

REFERENCES

1. Lumry, R. (1959). *The Enzymes*, 2nd ed., pp. 1, 186.
2. Linderstrøm-Lang, K. and Schellman, J. (1959). *The Enzymes*, 2nd ed., pp. 1, 443.
3. Eyring, H., Lumry, R., and Spikcs J. D. (1954). *Mechanisms of Enzyme Action* (W. McElroy and B. Glass, eds.), Johns Hopkins Univ. Press, Baltimore, p. 123.
4. Lumry, R. and Eyring, H. (1954). *J. Phys. Chem.*, *58*:110.
5. Gregory, R. and Lumry R. (1985). *Biopolymers 24*:301.

6. Lumry, R. (1961). *Biophysics* (Jpn) *1*:3.
7. Lumry, R, Solbakken, A, Sullivan, J., and Reyerson, L. (1962). *J. Am. Chem. Soc. 84*:142. See also Lumry, R. (1971). *Electron and Coupled Energy Transfer in Biological Systems* (T. King and M. Klingenberg, eds.), Marcel Dekker, New York, Vol. 1, Chapt. 1.
8. Lolis, E. and Petsko, G. (1990). *Ann. Rev. Biochem. 59*:597.
9. Lumry, R. and Biltonen, R. (1969). *Structure and Stability of Biological Macromolecules* (S. Timasheff and G. Fasman, eds.), Marcel Dekker, New York, Chapt. 1.
10. Lumry, R. (1991). *Study of Enzymes* (S. Kuby, ed.) CRC, Boca Raton, FL, Vol. 2, Chapt. 1.
11. Lumry R. and Gregory R. (1989). *J. Mol. Liq. 42*:113.
12. Sheriff, S., Hendrickson, W., Stenkamp, R., Sieker, L., and Jensen, L. (1985). *Proc. Nat. Acad. Sci. USA 82*:1104.
13. Ringe, D. and Petsko, G. (1985). *Prog. Biophys. Mol. Biol. 45*:197.
14. Lumry R. and Gregory, R. (1986). *The Fluctuating Enzyme*, (G. R. Welch, ed.), Wiley Interscience, New York, Chapt. 1.
15. Creighton, T. and Kemmink, J. (1993). *TIBS*, *18*:484.
16. Woodward, C. (1994). *Curr. Opin. Struct. Biol. 4*:112.
17. Constantine, K., Friedrichs, M., Goldfarb, V., Jeffrey, P., Sheriff, S., and Mueller, L. (1993). *Prot.: Struct. Funct. Gen.*, *15*:290; Constantine, K., Friedrichs, M., Metzler, W., Wittekind, M., Hensley, P., and Meuller, L. (1994). *J. Mol. Biol. 236*:310; Constantine, K., Goldfarb, V., Wittekind, M., Anothony, J. Ng, S-C., and Meuller, L. (1992). *Biochemistry 31*:5033.
18. Shotton, D. and Watson, H. (1970). *Nature 225*:811.
19. Shiao, D., Lumry, R., and Rajender, S. (1972). *Eur. J. Biochem. 29*:377.
20. Shiao, D. (1969). *Biochemistry 8*:4910; Shaio, D. and Sturtevant, J. (1969). *Biochemistry 9*:1083.
21. Jolicoeur, C. Pers. commun.
22. Lumry, R. and Rosenberg, A. (1975). *Colloq. Int. CNRS 246*:53.
23. Frauenfelder, H., Paral, F., and Young, R. D. (1988). *Ann. Rev. Biophys. Biochem. 17*:451.
24. Klibanov, A. (1983). *The Biological Basis of New Developments in Biotechnology* (A. L. Hollaender and P. Rogers, eds.), Plenum, New York, pp. 497–518: Zak, P. and Klibanov A. (1984), *Science 224*:1249, *J. Biol. Chem. 263*:8017; Klibanov, A. (1988). *Protein Engineering* (M. Inouye and R. Sarma, eds.), Academic Press, Orlando, FL, pp. 341–349.
25. Kachalova, G., Morozov, V., Morozova, T., Myachin, E., Vagin, A., Strokopytov, B., and Nekrasov, Yu. (1991). *FEBS 284*:91.
26. Hnojewyj, O. and Reyerson, L. (1961). *J. Phys. Chem. 68*:1694;*65*:1694.
27. Woodward, C. and Rosenberg, A. (1971). *J. Biol. Chem.246*:105.
28. Gregory, R., Knoz, D., Percy, A., and Rosenberg, A. (1982). *Biochemistry 24*:6523.
29. Tuchsen, E. and Woodward, C. (1985). *J. Mol. Biol. 185*:405, 430.
30. Levitt, M. (1981). *Nature 294*:379.
31. Diesenhofer, J. and Steigmann, W. (1974). *Acta Crystallogr. B-31*:238.
32. Dalzoppo, D., Vita, C., and Fontana, Jr., A. (1985). *J. Mol. Biol. 182*:33.
33. Davidson, F. and Dennis, E. (1990). *J. Mol. Evol. 31*:228.
34. Seizen, R., de Vos, W., Leunissen, J., and Dijkstra, B. (1991). *Prot. Eng. 4*:719.

35. Dorit, R., Schoenbach, L., and Gilbert, W. (1990). *Science 250*:1377.
36. Richards, F. and Kundrot, C. (1988). *Prot.: Struct. Funct. Gen. 3*:71; Richards, F. (1979). *Carlsberg Res. Commun. 44*:2; *J. Mol. Biol.* (1974). *82*:1.
37. Sundaralingam, M., Sckharudu, Y., Yathindra, N., and Vichandran, V. (1987). *Prot.: Struct. Funct. Gen. 2*:64; (1987). *Intern. J. Quant. Chem.: Quant. Biol. Symp. 14*:2S9; Creighton, T. (1987). *Nature 386*:547.
38. Lumry, R. and Biltonen, R. (1965). *Abstracts*, 150th Am. Chem. Soc. Meeting, Atlantic City, NJ, Sept., Abstr. C197.
39. Pohl, F. (1969). Habilitationsschrift: Kinetik der reversiblen kn oformationsänderung globulär proteine, Univs. Göttingen and Konstanz.
40. Pohl, F. (1969). *FEBS Lett. 3*:60; (1968). *Eur. J. Biochem. 4*:373; *7*:146.
41. Pohl, F. (1969). *FEBS Lett. 65*:293.
42. Segawa, S-I., Husimi, Y., and Wada, A. (1973). *Biopolymers, 12*:2521.
43. Segawa, S-I. and Sugihara, M. (1984). *Biopolymers 23*:2473, 2489.
44. Chen, B-I. Baase, W., and Schellman, J. (1989). *Biochemistry 28*:691.
45. Segawa, S-I. and Sugihara, M. (1984). *Biopolymers 23*:2473.
46. Gekko, K. and Timasheff, S. I. (1981). *Biochemistry 20*:4677.
47. Segawa, S-I. and Sugihara, M. (1984). *Biopolymers 23*:2489.
48. Radford, S., Dobson, C., and Evans, P. (1992). *Nature*, 358.
49. Kim, K-S., Fuchs, J., and Woodward, C. (1993). *Biochemistry 32*:9600; Kim, K-S. and Woodward, C., p. 9600. Extended in Tao, F., Fuchs, J., and Woodward, C. (1993). *Techniques in Proteins* (R. Angelette, ed.) pp. 4, 449.
50. Segawa, S-I. and Kume, K. (1986). *Biopolymers 25*:1981.
51. Itah, V. and Haas, E. (1994). *Biophys. J. 66*:A180.
52. Gregory, R. (1988). *Biopolymers 27*:1699.
53. Anson, M. (1945). *Adv. Prot. Chem. 2*:361.
54. Kolthoff, I., Anastasi, A., and Tan, B. (1960). *J. Am. Chem. Soc. 82*:4147 (1965). Kolthoff, I. and Tan, B., *87*:2717.
55. Koenig, J. and Frushour, B. (1972). *Biopolymers 11*:2505.
56. Corbett, R. and Roche, R. (1984). *Biochemistry 23*:1888.
57. Lim, Y. and Bolen, D. W. (1993). Pers. commun.
58. Anfinsen, C. (1991). Pers. commun.
59. Data supplied by M. Matsumura and plotted in Fig. 9 of Ref. 7.
60. Aune, F. and Tanford, C. (1969). *Biochemistry 8*:4586.
61. Weissman, J. and Kim, P. (1991). *Science 253*:1386.
62. Klemm, J., Wozniak, J., Alber, T., and Goldenberg, D. (1991). *Biochemistry 30*:589.
63. Lin, W. and Sauer, R. (1989). *Nature 339*:31.
64. Izbicka, E. and Bolen, D. (1978). *J. Am. Chem. Soc. 100*:7625; (1981) *Bioorg. Chem. 10*;118; Huang, Y. and Bolen, D. (1993). *Biochemistry 32*:9329.
65. Brandts, J., Hu, C., and Lin, L-N. (1989). *Biochemistry 28*:89.
66. Ptitsyn, O. (1987). *J. Prot. Chem. 8*:272.
67. Griko, Yu., Venyaminov, S., and Privalov, P. (1989). *FEBS Lett. 244*:276.
68. Damaschun, G., Hilde, D., Gast, K., Misselwitz, R., Müller, J., Pfeil, W., and Zirwer, D. (1993). *Biochemistry 32*:7739.
69. Muranker, A., Radford, S., Karplus, M., and Dobson, C. (1991). *Nature 349*: 633.

70. Baum, J., Dobson, C., Evans, P., and Hanley, C. (1989). *Biochemistry 28*:7.
71. Kim., K-S., Fuchs, J., and Woodward, C. (1993). *Prot. Sci. 2*:588.
72. Strambini, G. (1989). *J. Mol. Liq. 42*:155; Strambini, G. and Gonnelli, M. (1986). *Biochemistry 25*:2471.
73. Manning, M. and Woody, R. (1989). *Biochemistry 28*:8609.
74. Hershberger, M. and Lumry, R. (1976). *Photochem. Photobiol. 23*:391.
75. Gavish, B., Gratton, E., and Hardy, C. (1983). *Proc. Natl. Acad. Sci. USA 80*:750.
76. Kundrow C. and Richards, F. (1987). *J. Mol. Biol. 193*;197.
77. Fisher, H. and Singh, N. (1991). *FEBS Lett. 294*:1.
78. Johnson, P. and McKnight, S. (1988). *Ann. Rev. Biochem. 58*:799.
79. Goldenberg, D. (1989). *Protein Structure: A Practical Approach* (R. Creighton, ed.), IRL, Oxford, pp. 225–250.
81. Ohlendorf, D. et al. (1983). *Br. J. Mol. Biol. 169*:757.
82. Pakuta, A. and Sauer, R. (1990). *Nature 344*:363.
83. Lumry, R. and Rajender, S. (1970). *Biopolymers 9*:1123.
84. Legare, R., Lumry, R., and Miller, W. (1964). **2**:84.
85. Gonnelli, M. and Strambini, G. (1993). *Biophys. J. 65*:131.
86. Ramadan, M., Evans, D. F., Lumry, R., and Philson, S. (1985). *J. Phys. Chem. 89*:3405. (1983). Ramadan, M., Evans, D. F., and Lumry, R., *87*:5020; Evans, D. F. and Wightman, P. (1982). *J. Coll. Interface Sci. 86*:515.
87. Benzinger, T. (1971). *Nature 229*:100; see also Benzinger, H., and Hammer, C. (1981). *Curr. Top. Cell. Reg. 18*:475.
88. Data from Lewis, G. and Randall, M. (1961). *Thermodynamics, 2nd ed.*, revised by Pitzer, K. and Brewer, L., McGraw-Hill, New York, Table A7-1, p. 671.
89. Hutchens, J., Cole, A., and Stout, J. (1969). *J. Biol. Chem. 244*:26.
90. Rhodes, W. (1991). *J. Phys. Chem. 95*:10246.
91. Leffler, J. and Grunwald, E. (1963). *Rates and Equilibria of Organic Reactions*, Wiley, New York.
92. Feitelson, J. and Mclendon, G. (1991). *Biochemistry 30*:5051.
93. Grunwald, E. (1984). *J. Am. Chem. Soc. 106*:3414; see for alternative definition of scaling: (1985). *107*:125.
94. Hallen, D., Niellson, S., Rothschild, W., and Wadso, I. (1986). *Chem. Therm. 18*:429; Gill, S., Dec, S., Oloffson, G., and Wadso, I. (1985). *J. Phys. Chem. 89*:3758.
95. Benson, B. and Krause, D. (1989). *J. Sol. Chem. 18*:803.
96. Huot, J-V., Battistel, E., Lumry R., Villeneuve, G., Lavallee, J-F., Anusiem, A., and Jolicoeur, C. (1988). *J. Sol. Chem. 17*:601.
97. Frauenfelder, H., Parak, F., and Young, R. (1988). *Ann. Rev. Biophys. Biochem. 17*:451.
98. Eftink, M., Anusiem, A., and Biltoneun, R. (1983). *Biochemistry 22*:3884.
99. Gregory, R. (1983). *Biopolymers 22*:895.
100. Lakowicz, J. and Weber, G. (1973). *Biochemistry 12*:4171; see also Eftink M. and Ghiron, C. (1985). *Biophys. Chem. 32*:173.
101. Cooper, A. (1984). *Prog. Biophys. Mol. Biol. 44*:181.
102. Chan, C-H. (1994). *J. Phys. Chem.*, in press.
103. Lumry, R., Jolecoeur, C., and Battistel, E. R. (1980). *Biophys. J. 32*:648.
104. Thompson, C. and Kotz, I. (1971). *Arch. Biochem. Biophys. 147*:178; Sturgill, T. and Biltonen, R. (1976). *Biopolymers 15*:337.

105. Beetlestone, J., Adeosun, O., Goddard, J., Kuchimo, J., Ogunlesi, M., Ogunmola, G., Okongo, K., and Seamonds, B. (1976). *J. Chem. Soc.* (Dalton), 1251.
106. Havsteen, B. (1989). *J. Theor. Biol. 140*:101; see also (1991). *151*:557.
107. Gleick, J. (1987). *Chaos*, Viking-Penguin, New York.
108. Perona, J. et al. (1993). *J. Mol. Biol. 230*:919.
109. Zundel, G. (1986). Enzymol. *127*:31; Zundel, G. and Fritsch, J. (1984). *J. Phys. Chem. 88*:6295; Merz, H. and Zundel, G. (1981). *Biochem. Biophys. Res. Commun. 101*:540; Lindermann, R. and Zundel, G. (1978). *Biopolymers 17*:1285; Zundel G. and Fritsch, J. (1980). *Chemical Physics of Solution* (R. Dogonadze, R. Kolman, A. Kornysheve, and J. Ulstrop, eds.), Elsevier, Amsterdam, Vol. 2.
110. Helfrich, W. (1973). *Z. Naturforsch. 27*:693.
111. Tsao, Y-h., Evans, D. F., Rand, R., and Parsegian, V. (1993). *Langmuir 9*:233; Evans, D. F. and Parsegian, V. (1986). *Proc. Nat. Acad. Sci. USA 83*:7123; Dubois, M. and Zemb, T. (1991). *Langmuir 7*:1352.
112. Rifkind, J. Lumry, R., and Abugo, O. (1994). National Institute of Aging, to be published; See Rifkind, J. and Lumry, R. (1967). *Fed. Proc. 26*:2328.
113. Rifkind, J. and Lumry R. (1967). *Fed. Proc. 26*:2328.
114. Takashima, S. and Lumry, R. (1958). *J. Am. Chem. Soc. 80*:4238.
115. Battistel, E., Jolicoeur, C., and Lumry, R. (1994). *Biophys. J. 41*:9a. to be published.
116. Suguna, K, Bott, R., Padlan, A. Subramanian, E., Sheriff, S., Cohen, H., and Davies, D. (1987). *J. Mol. Biol. 196*:877; Bott, R., Subramanian, E., and Davies, D. (1982). *Biochemistry 21*:6956.
117. Bone, R., Frank, D., Kettner, C., and Agard, D. (1989). *Biochemistry 28*:5925, 7600; see also Bone, R., Fujishige, A., Kettner, C., and Agard, D. (1991). *30*:10388.
118. Mangel W. et al. (1990) *Biochemistry 29*:8351.
119. Lüscher, M., Schindler, P., Ruegg, M., and Rottenberg, M. (1979). *Biopolymers 18*:1775.
120. Vaslow, F. and Doherty, (1953). *J. Am. Chem. Soc. 7S*:928.
121. Dorovska-Taran, V., Momtcheva, V., Gulubova, N., and Martinek, K. (1982). *Biochim. Biophys. Acta 702*:37.
122. James, D. (1981). Dissertation, University of Minnesota.
123. Levashov, A., Klyachko, N., Bogdanova, N., and Martinek, K. (1990). *FEBS Lett. 268*:238.
124. Dorovska-Taran, V., Veeger, C., and Visser, W. (1993). *Eur. J. Biochem. 211*:47; in *Biocatalysis in Non-Conventional Media* (J. Tramper, ed.) Elsevier, pp. 697–703.
124a. Derrick, J. and Wigley, D. (1992). *Nature 359*:752.
124b. Alvarez, J. and Biltonen, R. (1973). *Biopolymers 12*:1815.
125. Brant, D. A. and Flory, P. (1968) *J. Chem. Soc. 87*:663; ibid., 2791; Flory, P., and Miller, J. W. (1966). *J. Mol. Biol.* 6284; Miller, W., Brant, D., and Flory, P. (1993). *J. Mol. Biol. 23*:67.
126. Alexander, R., Kiefer, L., Fierke, C., and Christianson, D. (1993). *Biochemistry 32:*1510.
127. Broda, E. (1983). *Boltzmann: Man, Physicist and Philosopher*, Oxbow, Woodbridge, CN.

128. Friedrich, P. (1984). *Supermolecular Enzyme Organization*, Pergamon, Oxford.
129. Lumry, R. and Jaren, O. (1994). To be published.
130. Yapel, A. (1967). Dissertation, Vol. 2, University of Minnesota.
131. Lumry, R. and Rajender, S. (1971). *J. Phys. Chem. 75*:1387.
132. Bowie, J. and Sauer, R. (1989). *Biochemistry 28*:7139; Bowie, J., Reidhaar-Olson, J., Lim, W., and Sauer, R. (1990). *Science 247*:1306.
133. Liang, H. and Terwilliger, T. (1991). *Biochemistry 30*:2772.
134. Renetseder, R., Brunies, S., Dijkstra, B., Drenth, J., and Sigler, P. (1985). *J. Biol. Chem. 260*:11267.
135. Mao, Q. and Walde, P. (1991). *Biochem. Biophys. Res. Comm. 178*:1105.
136. Gavish, B. (1986). *The Fluctuating Enzyme* (G. Welch, ed.), Wiley, New York, p. 263.
137. Sosnick, T. and Trewhella, J. (1992). *Biochemistry 31*:8329.
138. Lumry, R., in: *Methods in Enzymology Cenergetics of Biological Macromolecules,* Eds: Ackers, G. and Johnson, M., Academic Press, in press (1994).

2

Solvent Interactions with Proteins as Revealed by X-Ray Crystallographic Studies

Edward N. Baker
Massey University, Palmerston North, New Zealand

I. INTRODUCTION

Ever since the first diffraction experiments on protein crystals [1,2], it has been recognized that the aqueous mother liquor from which they are grown is an important component of the crystals. Early studies noted changes in diffraction as the solvent content changed, until eventually the lattice collapsed as the crystal dried out. The first x-ray structural analyses revealed large solvent regions separating individual protein molecules in the crystal. The solvent was not modeled, however, because of the difficulty of reliably identifying electron density peaks with solvent molecules in the absence of suitable refinement methods. As methods have developed, however, it has become almost routine to model protein-bound solvent in high-resolution x-ray structure analyses of protein crystals, and virtually every such analysis brings with it a wealth of information on protein–solvent interactions. It is the nature and relevance of that information that is the subject of this chapter.

II. SOLVENT CONTENT OF PROTEIN CRYSTALS

The protein molecules in crystals pack quite loosely (Fig. 1), leaving large intervening spaces that are filled with solvent. Solvent content varies from a minimum of around 30% by volume for some small proteins, to a maximum as high as 80%, with an average value of about 50% [3]. Contacts between the protein molecules tend to be few and rather tenuous; for example, in actinidin, a protein of 220 amino acid residues and moderate solvent content (42%), only 20 residues make direct contacts of less than 4 Å with neighboring molecules in the crystal [4]. Additional contacts are via water bridges, which are a consistent feature of all protein crystal structures. The larger solvent regions in turn have most of the properties of bulk solvent, being in free and rapid equilibrium with the outside mother liquor.

These characteristics of protein crystals, high solvent content, and loose packing of the protein molecules have a number of important consequences:

1. The crystals tend to be soft, fragile, and easily disordered.
2. Small molecules (inhibitors, substrates, etc.) can be diffused into protein crystals via the solvent channels. Providing the active site is not blocked by a neighboring molecule, enzymes therefore tend to be active in the crystalline state, as in solution [5], and many binding studies have been carried out using protein crystals.
3. Very little of the protein structure is likely to be influenced by its crystal environment, and there is ample evidence that protein structures in the crystal are essentially the same as in solution [5,6]. Solution properties are consistently explained by crystal structures. Comparisons of the same or related proteins in different crystal environments show very high levels of similarity; for example, the homologous cysteine proteases actinidin and papain show an rms (root mean square) deviation of only 0.4 Å for 90% of main-chain atoms despite very differ-

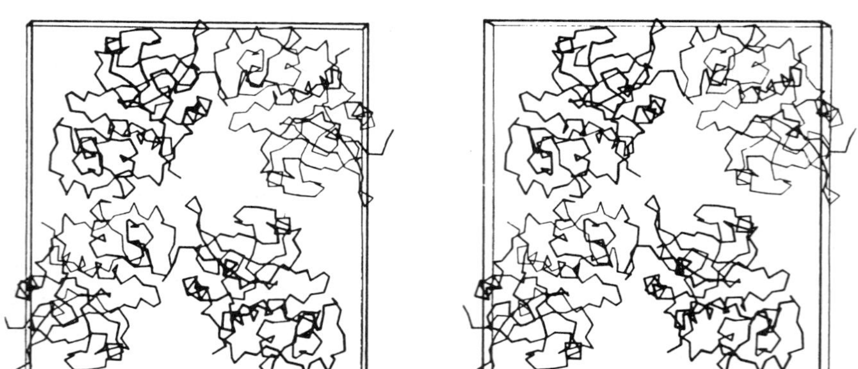

FIGURE 1 Stereo diagram showing solvent channels between the molecules in crystals of actinidin (solvent content ~42%, by volume).

ent crystallization conditions (20% ammonium sulfate, pH 6.0, and 62% methanol, pH 9.3, respectively) and different crystal packing [7]. Most recently, comparisons of structures determined in solution by NMR (nuclear magnetic resonance) spectroscopy have shown generally good correspondence with crystal structures [8–11]. The main differences arise because flexible side chains or loops can be "frozen" into one of several accessible conformations by their crystal-packing environment.

4. Of greatest relevance to protein–solvent interactions, most of the protein surface is usually free of crystal packing contacts and in contact with what is essentially bulk solution. Solvent molecules bound to these regions must experience a very similar environment to that in free solution. Only in intermolecular contact areas are significant constraints imposed by the crystal environment, and the extent of these areas depends on the solvent content of the crystal. For example, in human lysozyme crystals, with a solvent content of 37%, crystal packing contacts involve 38% of the surface, whereas for turkey egg white lysozyme (solvent content 51%), the fraction covered by crystal contacts is only 13% [13].

The major general difference between the crystal and solution is the extent to which the flexibility of protein structures is restricted. The crystal lattice does allow some flexibility of amino acid side chains, of loops, and even of whole domains [12], but generally there is a reduced mobility compared with solution [5].

III. CRYSTALLOGRAPHIC LOCATION OF SOLVENT

A. The Crystallographic Method

X-ray crystallography allows one to actually "see" the atomic and molecular structure in a crystal. The image that is seen, however, is a three-dimensional map of the distribution of electrons in the crystal, that is, an electron density map. It is the interpretation of the electron density map, in terms of atoms and groups of atoms, that constitutes the model that is finally published or deposited. It is outside the scope of this chapter to discuss crystallographic theory (for comprehensive accounts, see Refs. 6,14), but an understanding of certain features of the crystallographic method is essential to appreciate the quality and nature of the information it gives about solvent structure in protein crystals.

First, calculation of an electron density map requires two pieces of information, that is, the amplitudes and the phases, for all of the diffracted x-ray waves. The amplitudes can be measured, with an accuracy of ~5%, but the phases are very difficult to determine accurately, especially for high-resolution data. Thus, the electron density map from which an initial model of the protein structure is derived inevitably contains errors and ambiguities because of the errors in the phases. This is less of a problem for interpreting the protein part of the structure than it is for the solvent, because the protein atoms are all covalently connected in the polypeptide chain. Even though ambiguities may sometimes make chain tracing difficult, the model must still conform to a known chemical connectivity. For the solvent, however, no such restriction exists; solvent molecules are small, discrete, and not covalently linked one to another. Identification of solvent sites in such an initial map would be too unreliable to be worth attempting.

The process of crystallographic refinement improves the situation greatly, however. The current model is used to calculate amplitudes and phases. The agreement of the calculated amplitudes with the experimentally measured observed amplitudes gives a measure of the correctness of the model (expressed as the crystallographic R factor, which should be less than 0.20 for a well-refined protein structure—see Refs. 6,16). Moreover, the calculated phases can be used, either on their own or in combination with the experimentally estimated phases, to give clearer electron density maps with less "noise." As the model is refined, by least squares methods, by energy minimization, or by rebuilding from electron density maps, the phases become better and so does the quality of the maps. There are still hazards in these procedures (for example, incorrectly placed atoms in the model lead to bias in the phases, which can cause the creation of false density). There is no doubt, however, that refinement greatly enhances the reliability with which solvent peaks can be picked out from the noise.

The other major factor is the resolution of the x-ray analysis. If only the inner (low scattering angle) parts of the diffraction pattern are used, a low-resolution image (electron density map) is obtained. As the outer (higher angle) data are incorporated, the resolution is increased, giving clearer definition of structural features. Thus, at low resolution (~6 Å), helices appear as solid rods; at medium resolution (~3 Å), side chains generally have recognizable shapes and peptide carbonyl oxygens appear as "bumps" projecting from the polypeptide chain density; while at high resolution (2 Å or better), much finer detail becomes apparent [6]. The resolution possible is ultimately limited by the quality of the crystals. If all of the molecules in the crystal do not have exactly the same orientation, or groups (e.g., external side chains) have a variety of conformations, the result is a blurring of the image and a loss of resolution.

B. Identification and Refinement of Solvent Sites

As implied in the foregoing, the main difficulty in reliably identifying solvent molecules in a protein structure analysis is the need to distinguish genuine solvent peaks from the noise of an electron density map. This problem is particularly acute in the early stages of an analysis or where the resolution is limited. Most workers therefore use a conservative approach, adding solvent molecules in stages, following fairly strict criteria.

Two types of electron density map are commonly used to locate solvent. If a map is calculated with coefficients $F_o - F_c$ (where F_o is the observed structure amplitude of a scattered x-ray wave, and F_c is that calculated from the current model), a so-called difference map is obtained. Where the model is correct, no density is seen (F_o and F_c cancel out); where a feature should be included in the model but is not, a positive peak is seen (because it contributes to F_o but has not been included in the calculation of F_c); where a feature is erroneously included in the model, a negative peak is seen (because it is contributing to F_c but not F_o). Difference electron density maps are thus very sensitive to features such as solvent molecules that have not yet been included in the structural model. The other kind of map frequently used in protein structure refinement employs coefficients $2F_o - F_c$. This effectively combines a difference map (coefficients $F_o - F_c$) with a map with coefficients F_o; density is present for the whole structure (contributing to F_o) but with errors emphasized through the inclusion of the $F_o - F_c$ term. Examples of such maps are shown in Fig. 2.

Solvent molecules are seldom included in a model unless the resolution is better than 2.5 Å, and it is very important not to model solvent too early in the refinement of a protein structure. This is because if solvent molecules are erroneously placed in density that really belongs to part of the protein, not only will the solvent be wrongly placed, but the protein will tend to be "locked in" to this incorrect structure for the subsequent refinement. For this reason, the solvent is usually not added until the protein structure has been well defined, typically at a crystallographic R factor of 0.25 or lower. An example of the improvement of solvent peaks during refinement is shown in Fig. 3.

The best-defined solvent molecules are included first; this usually means those in internal sites or surface crevices and pockets. Electron density peaks greater than a certain threshold, for example, three times the rms density of a difference electron density map, are identified, either by manual scrutiny of the map or using computer programs to search it for the highest peaks (e.g., [15]). These peaks are then examined to see whether they fulfill various criteria; for example, that they are within hydrogen-bonding distance of potential hydrogen-bonding groups, in appropriate geometric orientation; they are not too close (<3.0 Å) to nonhydrogen-bonding atoms; and they are not close to any part of the protein whose conformation is in doubt [16]. Often peaks are only assigned to solvent molecules if they have been noted as persisting through several phases of the refinement. It must be stressed, however, that criteria for including solvent vary greatly from one laboratory to another.

As the refinement proceeds, and the model is further improved, more weakly bound, less well-defined solvent molecules can be added as the density becomes clearer. These are, by definition, closer to the noise level, however, and it is difficult to know where to draw the line. As a rule of thumb, a "reasonable" value for the number of solvent molecules in a high-resolution model is roughly equal to the number of residues in the protein. This should include all the strongly bound solvent, most of the first hydration shell, and a little second-shell solvent.

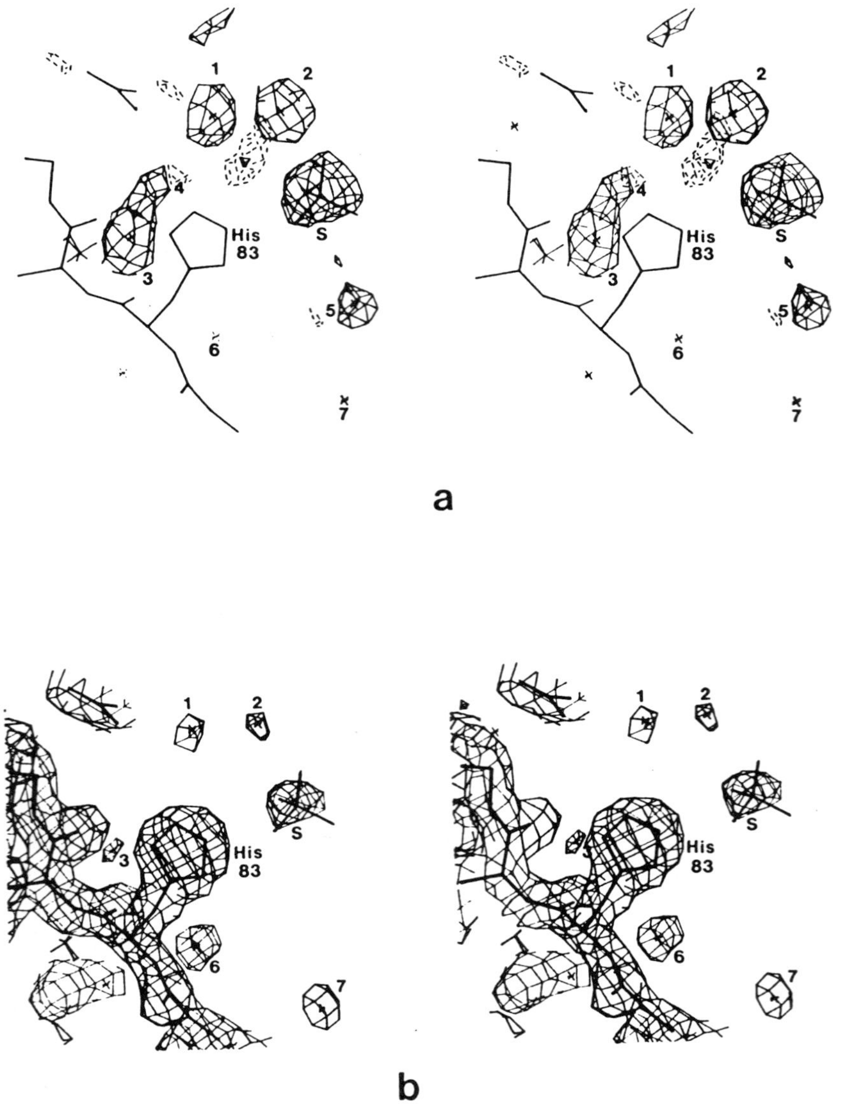

FIGURE 2 Appearance of solvent molecules in electron density maps, from the x-ray structure analysis of azurin at 1.8 Å resolution [21]. Maps shown are (a) a difference map with coefficients $F_o - F_c$ and (b) a $2F_o - F_c$ map. Peaks 1, 2, and 3 are present in both maps and are probable new, discrete solvents sites. Peaks 4 and 5 are present only in the $F_o - F_c$ map (a) and are more doubtful. The peak labeled S proved to be a sulfate ion, bound to the side chain of His 83. Peaks 6 and 7 were already correctly in the model and therefore appear in the $2F_o - F_c$ map (b) but not the $F_o - F_c$ map (a).

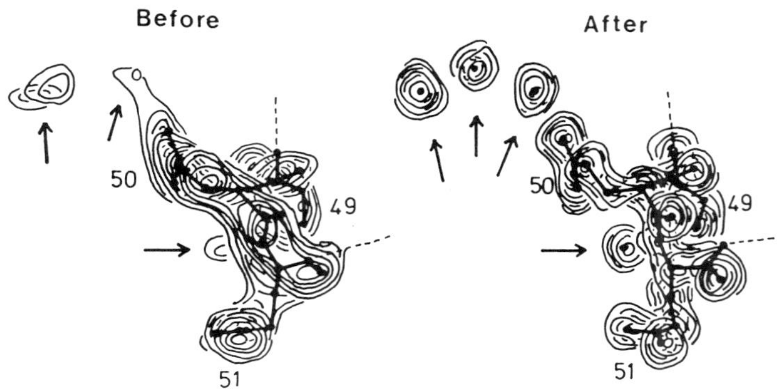

FIGURE 3 An example of the improvement of electron density as a result of crystallographic refinement. Poorly resolved electron density near the side chain of Glu 50 in actinidin suggested three possible solvent sites (arrowed) before refinement (left); these were resolved as four clear solvent molecules after refinement (right). From Baker and Dodson [16].

Two important indicators of the quality of the solvent molecules are their occupancies and *B* values. The occupancy refers to the fraction of time that a particular site is occupied in the crystal—an occupancy of 0.5 means that on average only half a solvent molecule is present at that site in the crystal. The *B* value, or temperature factor, for an atom is related to its mean amplitude of vibration, and in practice it covers both thermal vibration and positional disorder in the crystal. (A *B* value of 40 Å^2 corresponds to a mean amplitude of vibration $\bar{u}$ of ~0.7 Å, from the relation $B = 8\pi^2\bar{u}^2$.) A sharply defined atom should give a very sharp, strong electron density peak and a low *B* value, whereas a vibrating atom, or one disordered over a range of positions, would give a low-density, spread-out peak and a high *B* value. The occupancy and *B* value are correlated [17]. A weak electron density peak may mean the site is occupied only a fraction of the time (a low occupancy) or that the atom is mobile or disordered and the *B* value is high. Very high resolution data are needed to distinguish between the two possibilities. Sometimes occupancies and *B* values are varied alternately during refinement, but more often the occupancies are all set at 1.0 and only the *B* values varied; partly occupied sites then tend to assume high *B* values.

During refinement, it is common practice to periodically reexamine all the less well-defined solvent sites by omitting all solvent molecules with *B* values higher than some threshold and adding them back if the density reappears. At the end of refinement, the solvent molecules are sometimes labeled in the order of their *B* values [4] (lowest *B* = lowest serial number), or if the occupancies have also been refined, in the order of the ratio (occ)^2/*B* [18] (occ = occupancy). The best-defined solvent sites have the lowest *B* values, 10–25 Å^2, comparable with those of the protein atoms with which they are associated, and weaker, less well-defined sites have $B > 50$ Å^2.

C. Chemical Identity of Solvent Molecules

Proteins are usually crystallized either by "salting-out" from an aqueous solution or by adding small amounts of alcohols or other additives. Thus, the solvent in the crystal should also contain the corresponding inorganic ions, organic molecules, etc. Despite this, these species are not often seen in protein structure analyses. In part, this must be because they are present in much lower concentrations; even in concentrated ammonium sulfate solutions, water molecules outnumber NH_4^+ or SO_4^{2-} ions by more than 10 to 1. There are also enthalpic reasons, however, and difficulties in distinguishing species other than water when they are bound.

Inorganic ions such as sulfate, phosphate, or metal ions are highly solvated, and if they are to be bound, the energy loss of desolvation must be more than compensated by the energy of binding to the protein. Thus, ions binding to sites that offer cooperative interactions with several protein groups are most likely to be seen. Organic molecules such as alcohols may be unable to compete with water molecules for binding sites because of their greater steric bulk and lesser hydrogen bonding potential.

Bound ions are difficult to recognize from the electron density alone. The ions NH_4^+, Na^+, or Mg^{2+} give peaks of similar size to water molecules because they have a similar number of electrons. Refinement can help; a solvent molecule that is treated as water but refines to a very low B value may well be Cl^- (which has approximately twice as many electrons). A half-occupancy Cl^- ion will, however, have the same density as a full-occupancy water molecule. A well-ordered SO_4^{2-} ion may be recognized by the tetrahedral shape of its electron density peak; but since solvent peaks are usually not well defined (through disorder or mobility), less strongly bound ions may be hard to recognize.

Most often, it is the environment of a solvent peak that suggests it is an ion, for example, the association of positively charged protein groups with a bound anion or an octahedral array of ligands for a cation. Some examples are given in Sec. VI. The distance of a solvent peak from potentially interacting groups is also a consideration (for example, see below). In several cases, ion-binding sites have been identified by substituting a salt used in crystallization with an equivalent salt with "heavier" atoms, for example, NaCl by NaBr [19], or $(NH_4)_2SO_4$ by $(NH_4)_2SeO_4$ [20]. Electron density difference maps will then show peaks due to the heavier ion.

The kind of dilemma that arises is illustrated by the refinement of the copper protein azurin [21]. A strong peak in the solvent region between two protein molecules was initially refined as a water molecule. It had, however, no hydrogen-bonding neighbors within 3.5 Å, and on refinement its B value became very low compared with the rest of the structure. Given that it lay between several lysine side chains, it was most probably an anion. Its spherical shape suggested Cl^-, but the crystallization medium contained $(NH_4)_2SO_4$ and no Cl^-. It was therefore modeled as a disordered SO_4^{2-}, with only the central sulfur atom visible, and while this was satisfactory crystallographically (the *B* value then assuming a "reasonable" value, similar to the surrounding structure), the real identity of such a solvent peak clearly remains uncertain.

In most crystallographic analyses, the solvent peaks are all assigned as water molecules unless there is some convincing evidence to the contrary. This does mean that a few will probably be wrongly assigned, but in general peak sizes and shapes, and contacts with surrounding atoms, all suggest that most of the ordered solvent is water.

IV. PATTERNS OF SOLVENT STRUCTURE

A. The General Picture

Qualitative descriptions of the solvation patterns in protein crystals have been given for many different proteins analyzed by x-ray crystallography [e.g., 4,13,18, 22–25]. Some excellent general reviews have appeared [26–29] together with others on more specific aspects including water/water interactions [30], hydrogen bonding with proteins [31], and distributions of water associated with protein groups [32] and with secondary structures [33].

In most protein molecules, there are a small number of solvent molecules that occupy small cavities within the protein structure and clearly contribute to its stability. These internal solvent molecules, which are discussed more fully below, are found mostly in larger proteins (>200 amino acid residues) but do occur even in some small, single-domain proteins such as bovine pancreatic trypsin inhibitor (BPTI) (58 residues), with four internal sites [34], and tuna cytochrome c (104 residues), with three [35].

The distribution of the remaining solvent in protein crystals is nicely summed up by the description given for insulin [24]: "A fringe of peaks marked the outline of the protein framework, many very strong where the channels were narrow, others much weaker where long hydrophilic side chains extended into stretches of water. In contact with these, further peaks could be traced in an increasingly diffuse distribution" (see Fig. 4).

Most of the ordered solvent is confined to a monolayer covering the protein surface (Fig. 5). Some of the solvent in this monolayer appears to be quite strongly bound, where it occupies crevices in the surface or is bound by a number of protein groups, but the layer is not completely continuous; it becomes more diffuse over flexible or nonpolar regions. Thus, for human lysozyme [13], about 70–80% of the "available" surface area is covered with ordered solvent after allowance is made for regions made inaccessible by crystal packing and for regions of high mobility

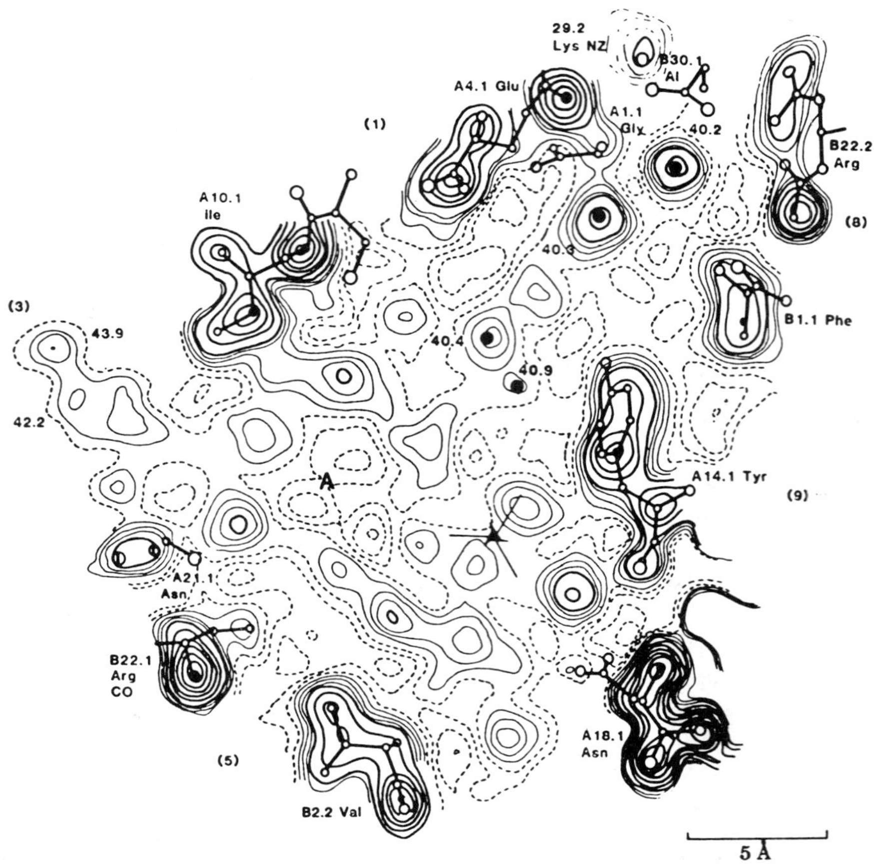

FIGURE 4 Electron density in one of solvent channels in crystals of 2-Zn insulin. Several well-defined sites can be seen close to the protein surface, e.g., 40.2 and 40.3 (top right). Elsewhere, weaker peaks, discrete or in chains, can be seen. Solid lines represent positive density, broken lines negative density. From Baker et al. [24], with permission.

to which water may be attached but not seen. For crambin, however, for which tighter crystal packing and very high resolution x-ray data has allowed the location of most of the solvent in the crystal, the monolayer is complete [25]. In most protein structure analyses, the majority of solvent molecules located belong to this monolayer, with their number approaching the number of residues in the protein.

The solvent becomes increasingly less well ordered the further it is from the protein surface. For rubredoxin, for which the solvent structure has been exten-

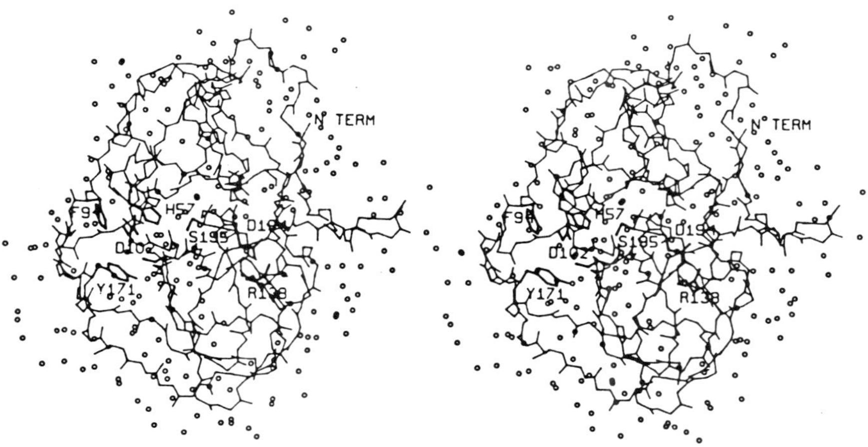

FIGURE 5 Stereo view of solvent structure associated with molecules of *Streptomyces griseus* protease A from a crystallographic analysis at 1.8 Å resolution. Solvent molecules shown as open circles. For the protein, only main-chain and catalytically important side-chain atoms are shown. The solvent molecules are mostly in a monolayer round the protein surface. From James et al. [65], with permission.

sively analyzed at very high resolution [22], solvent molecules are marked by decreasing occupancies and higher *B* values the further they are from the surface, merging into the solvent continuum beyond 5–6 Å (unless another protein molecule is within 8–10 Å). For most protein structure refinements, only a small number of second- or higher-shell solvent molecules have been included in the final model; the aspartic proteinase from *Rhizopus chinensis* would be a typical example, with 17 internal waters, 286 in the first hydration shell and 70 further solvent molecules byond that, for a protein of 324 residues [36]. The higher-shell solvent is usually in clusters, anchored by strongly bound waters on the protein surface.

Where the protein molecules are tightly packed in the crystal, as for insulin (solvent content 30% [24]) or crambin (solvent content 32% [25]), the solvent regions tend to be relatively narrow and the proportion of ordered solvent correspondingly higher. Tighter crystal packing also leads generally to better ordered crystals and the possibility of higher-resolution x-ray data; this in turn allows more complete refinement of the protein structure, corresponding reduction of "noise" levels in the electron density and the chance to see the less well-ordered solvent more clearly. Thus, protein crystals of low solvent content somewhat paradoxically are of considerable value in allowing this more mobile, diffuse solvent to be modeled. The modeling itself is difficult, however, as the electron density distribution of these solvent regions takes the form of streams, chains, or rings of peaks that are diffuse and oddly shaped, representing not individual sites but many close alternative positions ("stopping places of waters moving through the crystal" [24]; see also Fig. 4). The electron density distribution itself is probably the best indication of the character of this solvent. How it relates to the solution phase is another question, but such models could be taken as snapshots of the more mobile solvent further from the protein surface.

Finally, the solvent structure takes on a different character again in regions of intermolecular contact in the crystal. Here, solvent molecules are either excluded, where protein atoms make direct contact, or participate in bridging interactions between the protein molecules. These water bridges, involving ordered water molecules either singly or in chains or clusters (e.g., [18]), are a major contributor to the stability of protein crystals.

B. Hydration of Protein Groups

Analyses of the hydration of protein groups in a number of different proteins can be found in the analyses by Thanki et al. [32,33,37] and Baker and Hubbard [31] as well as in the descriptions of some individual protein structures [4,13,18].

A consistent observation is that there is a strong tendency for water molecules to be bound to oxygen atoms rather than nitrogen; the overall ratio of oxygen ligands to nitrogen ligands in a given structure is consistently around 2.5 to 1, on average, irrespective of the number of hydrogen bonds made to the protein by individual waters [31]. Further, although the —OH groups of Ser, Thr, and Tyr side chains can act as either donor or acceptor in hydrogen bonds, the other oxygen-containing groups (main-chain carbonyl, side-chain amide and carboxyl) are acceptors only, and as a result the water molecules tend to act as hydrogen-bond donors rather than acceptors. For example, of the water molecules making two protein hydrogen bonds in the survey by Baker and Hubbard [31], 52% bond to two oxygens (90% of which can only be acceptors), and only 5% to two nitrogens.

The preference for water molecules to be bound to protein oxygen atoms rather than nitrogen atoms, and to act as donors rather than acceptors, has been noted many times [4,13,18,22,31]. It arises from a number of factors, namely (a) the ability of all oxygen centers to be involved in more than one hydrogen bond, whereas almost all nitrogen centers can make only one, (b) the greater number of oxygen centers rather than nitrogen among the side chains of most proteins and the greater number of potential hydrogen bond acceptors rather than donors (for actinidin the ratio is 1.3 to 1), and (c) the lesser geometrical restrictions for oxygen acceptors compared with nitrogen donors [31]. Furthermore, there is evidence from hydrate crystals and small molecules that if a molecule has a potentially donatable hydrogen, then it will normally participate in a hydrogen bond via that proton [26,38]. Thus, water molecules apparently have a stronger innate tendency to act as donors, by utilizing their protons in hydrogen bonding rather their lone pairs of electrons.

A second consistent pattern is for a predominance of interactions with main-chain groups rather than side-chain; on average, about 40% with main-chain C=O groups, 15% with main-chain NH groups, and 45% with side-chain groups [31,32]. This is not because there are more main-chain sites available on the surface of a protein, as the number of polar side chains is comparable with the number of exposed C=O and NH groups, and the number of polar side-chain atoms and potential hydrogen-bonding sites is certainly greater. For example, for human lysozyme, approximately 25 main-chain NH and 40 main-chain C=O groups are solvent accessible, and almost all bind waters. This compares with 70 polar side chains, containing 126 polar atoms. Although accessibilities are not given for these, they clearly outnumber the accessible main-chain groups yet bind fewer water molecules [13].

The reason for this imbalance is probably because side chains generally have a higher mobility than main-chain atoms [39] and solvent molecules attached to the more mobile side chains will tend not to be seen in electron density maps. The solvent molecules that can be seen in crystallographic analyses are those which are best ordered and most strongly bound. This means those molecules which are bound to the less mobile parts of the protein structure and often to more than one protein group. Such sites are provided mainly by main-chain atoms and side chains whose mobility is restricted (e.g., His, Asp, Asn, and Arg, which are often hydrogen bonded to other protein atoms [31], and Tyr and Trp, which are likely to be sterically restricted); more mobile side chains such as Lys are less often involved in such sites.

The distributions of solvent sites around protein groups follow rather nicely the geometrical patterns expected from simple bonding ideas [31,32]. Interactions with NH groups are linear, and those with C=O groups show a preferred angle C–O/water of around 130°, implying interaction with a lone pair of electrons on a sp^2-hybridized oxygen [31,40]; restriction to the peptide plane is not very strong, however [31]. The ability of C=O groups to make two hydrogen bonds, using both lone pairs, has the effect that exposed C=O groups frequently bind two waters, and that C=O groups that are hydrogen bonded to other protein groups, for example, in helices, are also able to bind a water molecule by utilizing their second lone pair (see Sec. IV.E and Ref. 31).

For the polar hydrogen-bonding groups of side chains, analysis in terms of spherical polar coordinates [32] shows that Ser and Thr side chains have broad radial and angular distributions of sites, possibly due to rotation about the C_α–C_β and C_β–O_γ bonds. In contrast, the phenolic OH of Tyr side chains shows two distinct clusters of sites, at an angle C–O/water of ~120° (Fig. 6), and with a clear preference for hydrogen bonding in the plane of the phenyl ring; this presumably reflects charge delocalization of the aromatic ring. The oxygen and nitrogen centers of Asp, Glu, Asn, and Gln side chains show two clusters of sites on either side of each oxygen or nitrogen, generally in the plane of the carboxyl or amide group and

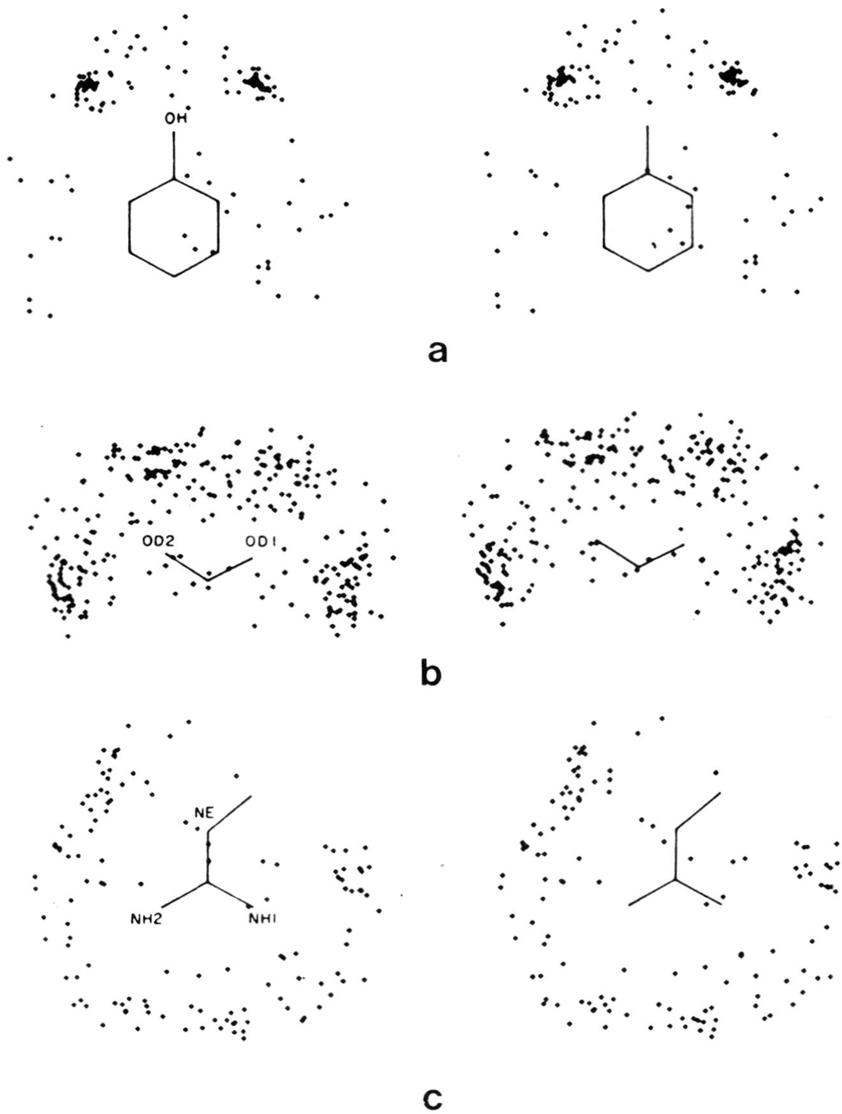

FIGURE 6 Some examples of the distribution of water molecule sites around protein side chains, from the analyses of Thanki et al. [32], with permission. Stereo views for (a) Tyr side chains, (b) the COO^- groups of Asp side chains, and (c) Arg side chains.

with angles C–O/water of ~120° (Fig. 6). Interestingly, there is little tendency for binding between the two polar atoms of these side chains. For Arg side chains, distinct clusters are found in the plane of the guanidinium group, but for Lys there is little or no geometrical preference apparent, probably because of the mobility of these side chains.

By far the most important side chains, in terms of numbers of water molecules bound, are Asp and Glu, whose COO^- groups bind on average two water molecules each [31,32]. These side chains also make a large number of hydrogen bonds with protein atoms and show a strong tendency to have their hydrogen-bonding potential fully satisfied [31]. Thus, those which have few or no hydrogen bonds with protein atoms are often highly hydrated (e.g., Fig. 7). The His side chains generally bind water molecules when not already hydrogen bonded to other protein atoms, but since they are few in number, their contribution to protein hydration overall is not great.

The hydration patterns round the smaller side chains can be influenced by their secondary structure location; preferred sites are seen for Ser and Thr, for example, which depend on secondary structure and the χ_1 angle of the side chain [37]. For a longer side chain such as Tyr, the main chain is too remote to affect hydrogen bonding preferences.

No clear patterns have been found for water interactions around nonpolar side chains [32], although it is possible that the criteria for including solvent molecules in crystallographic analyses may in some cases cause such waters to be missed. Most contacts are of the order of 3.8–4.0 Å [13,32], implying normal van der Waals contacts, with a minimum contact distance of ~3.2 Å [13]. Suggestions that hydrophobic residues would organize water molecules into five-membered ring arrays, similar to those seen in water clathrate structures [41], have not been borne out, except in occasional examples (see Sec. IV.D). What is apparently required is some "anchoring" to nearby polar groups, or constraints from crystal packing contacts, to fix the orientation of such an array before it can be seen crystallographically. The four internal water molecules in human and turkey egg white lysozymes are arranged in a semicircular ring around the methyl side chain of an Ala residue, and a cluster of three internal water molecules in penicillopepsin are adjacent to a hydrophobic surface comprising two Leu side chains [18]; in both these cases, however, the water molecules are primarily hydrogen bonded to polar groups.

C. Internal Solvent Molecules

Most globular proteins contain a small number of solvent molecules buried inside the protein structure. These molecules are well ordered and highly conserved, and they must be regarded as an integral part of the protein structure.

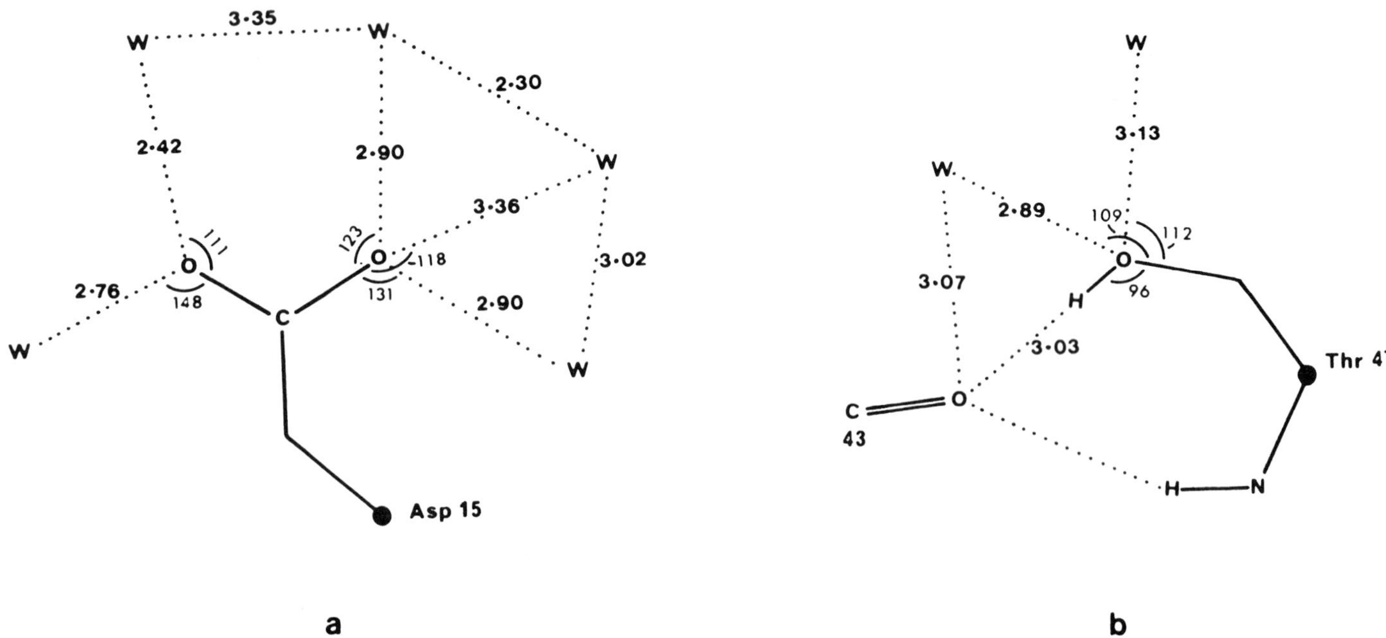

FIGURE 7 Examples of the hydration of protein side chains. (a) A highly hydrated Asp side chain in actinidin. Each carboxylate oxygen binds two water molecules, while a fifth, more distant water could be considered as a "second shell" water. (b) A Thr side chain for which water molecules take up the two available hydrogen-bonding positions. From Baker and Hubbard [31].

1. The Nature of Internal Solvent Sites

Buried solvent molecules occupy small internal cavities in the protein structure, some of them able to accommodate only a single molecule, others large enough to contain small clusters [42]. For example, the four internal solvent molecules in lysozyme are in a single four-molecule cluster [13], whereas the three in cytochrome c are each in a separate, discrete site [35]; in actinidin, the 17 internal solvent molecules are found in nine discrete sites and one eight-molecule cluster [4], while in rhizopuspepsin, there is one cluster of three, two pairs, and ten discrete sites [36].

The larger cavities tend to be found between secondary structure elements or (most often) at the interfaces between domains or protein subunits [42]. In almost every case, the buried solvent molecules are assumed, from their electron density and their hydrogen-bonding interactions, to be water, although the presence of one internal NH_4^+ ion has been proposed in actinidin [4], and there are several instances where water molecules are believed to be protonated, H_3O^+, because of their proximity to buried COO^- groups whose charge would otherwise be uncompensated [40,43]. In papain, which is crystallized from a methanol/water medium, a methanol molecule replaces one of the internal waters found in actinidin [7]. Both hydrogen–deuterium exchange [44] and NMR [45] studies show that these buried solvent molecules do exchange with the external solvent medium, presumably through a general "breathing" motion of the protein and narrow channels that link these internal cavities to the protein surface [42].

The discrete internal solvent molecules almost invariably have either three or four protein ligands to which they are hydrogen bonded [31] (see Fig. 8). The hydrogen-bond lengths tend to be close to optimal (~2.8 Å), and the geometry approximately tetrahedral. In some cases, a fifth potential ligand is within hydrogen-bonding distance, but often more distant than the other four (Fig. 8a). There is a strong tendency for both water protons to be used in hydrogen bonding in these sites; of the three-coordinate sites, almost all involve two protein acceptors and one donor [31].

Internal clusters of solvent molecules often take the form of chains in which each is hydrogen bonded to the next but is also hydrogen bonded to protein atoms. Penicillopepsin contains one such chain of three waters and another of five waters that stretches from the protein interior to a position close to the active site aspartyl residues [18]. In ribonuclease T1, a chain of five buried water molecules, lying between secondary structure elements (an α-helix and a hairpin loop), leads to a stabilizing calcium ion situated on the surface [46] (Fig. 9). In other cases, clusters are more spherical. In actinidin, the eight-water cluster in the domain interface (Fig. 10) includes one water in the center, which is bound only to other water molecules.

Solvent molecules occupying internal sites tend to be extremely well ordered, with *B* values comparable with the surrounding protein structure. For ex-

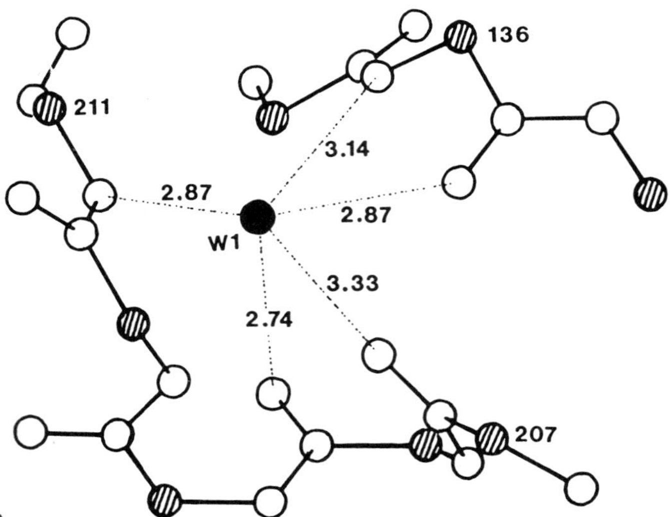

(a)

FIGURE 8 Two examples of discrete internal water sites, from the structure of actinidin [4]. In (a), the water molecule has four close protein ligands, two oxygen (C=O) and two nitrogen (NH), with a fifth (C=O) somewhat more distant. In (b), the coordination is trigonal, with two C=O groups and one NH.

ample, in actinidin, all of the internal solvent molecules have $B < 15$ Å^2, compared with B values of 10–15 Å^2 in the internal protein structure. The highest values are for several of the waters in the eight-molecule cluster, suggesting that these tend to be somewhat more mobile.

Finally, the classification of solvent molecules as internal is not always straightforward, because some bound water molecules occupy the bottoms of deep crevices or channels and are inaccessible to bulk solvent, but are nevertheless connected to other solvent molecules that are in contact with the external solution [18].

2. Conservation of Internal Solvent Sites

If buried solvent molecules are indeed an integral part of a protein structure, they would be expected to be present in free solution as well as in protein crystals, and to be conserved to the same extent as internal amino acid residues in homologous proteins. This does seem to be the case. For example, the four internal water molecules in BPTI are all found in the same positions, bound to the same ligands, in three different crystal forms of the protein [34,47,48], and NMR studies have shown their presence in solution as well [49].

An example of conservation of internal solvent in homologous proteins is given by lysozyme, where the same four internal sites, formed by the same protein ligands, are present in both human lysozyme and turkey egg white lysozyme

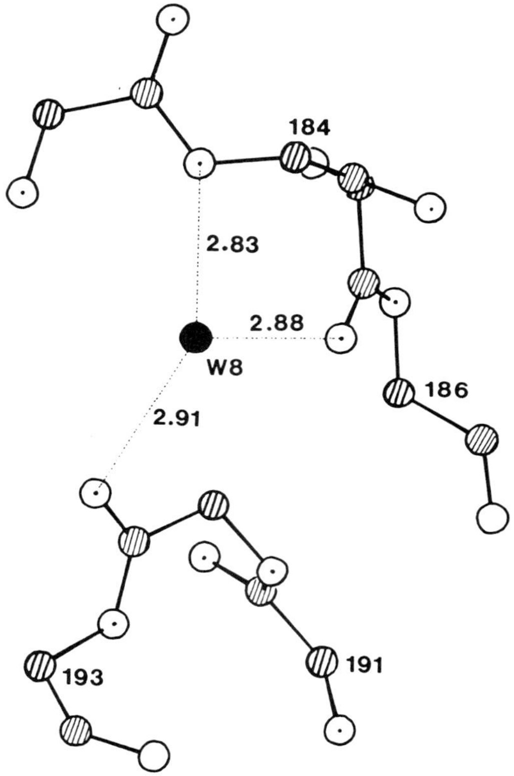

(b)

FIGURE 8 Continued

[13]. In the homologous α-lactalbumin structure, three solvent molecules are found in the same region, corresponding fairly closely, but not exactly because of the presence of a bound calcium ion at one end of this channel [50].

Comparisons of the solvent sites in actinidin and papain, which share ~48% amino acid sequence identity [7], also show a very high degree of conservation of the internal solvent molecules. Of the 17 buried solvent molecules in actinidin, 16 are also present in papain, with a very high degree of structural correspondence, that is, rms deviation in position of only 0.48 Å. One of these is in fact a methanol molecule whose oxygen atom coincides with a water position in actinidin. The only actinidin internal solvent site *not* conserved is the putative ammonium ion; a conformational change in an external loop, probably arising from sequence changes, causes a Tyr side chain to occupy this site.

An example of a discrete solvent site that is conserved in actinidin and papain is shown in Fig. 11. What is noteworthy is that both its position and hydrogen-

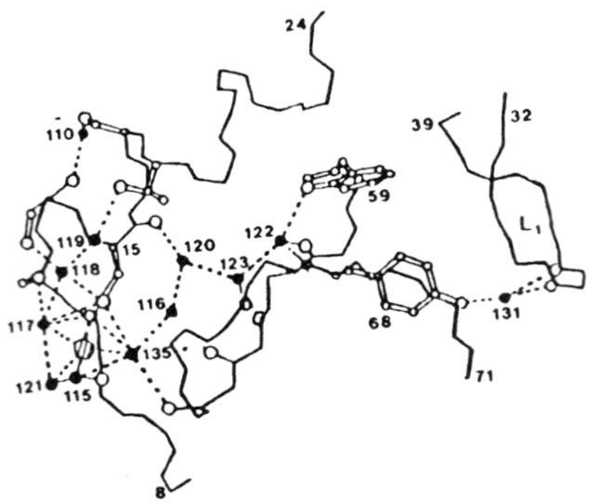

FIGURE 9 Chain of hydrogen-bonded water molecules in the structure of ribonuclease T1, from Malin et al. [46], with permission. Stereo view shows the chain of water molecules (filled circles) leading from an internal Trp residue (Trp 59) to the calcium ion. The chain of waters separates an α-helix from strand β_4 and turn t_6. Note also a water molecule linking Tyr 68 to a loop L_1.

bond geometry are essentially identical even though one of its ligands is different because of a sequence change between the two proteins. This makes the point that it is the hydrogen bonds made by internal water molecules that are of utmost importance, in the same way that other internal hydrogen bonds tend to be conserved. Another sequence change between actinidin and papain results in the movement of a buried Lys side chain. The position occupied by the Lys amino group in actinidin is then filled by a new water molecule in papain, and the hydrogen-bonding network in the internal solvent cluster is exactly preserved (Fig. 10). One may anticipate that there are many other examples in homologous proteins where substitutions of internal amino acids are associated with the appearance of water molecules to fill the place of a polar side-chain atom and maintain internal hydrogen bonding.

Conservation of internal solvent sites has been noted for many other homologous proteins. Thermitase and proteinase K, two members of the subtilisin family, have a number of internal solvent molecules in common [51], and in the trypsin family, internal water molecules are present in elastase and trypsin that match most of the 14 buried waters reported for chymotrypsin [52,53].

D. Surface Solvent Structure

The water at the protein surface is important for structural stability [26]. The folding of a globular protein results in a rather irregular surface with both crevices and projections. Some parts are highly mobile, for example, flexible loops that project from the surface and are not tied into secondary structures; other parts are much more rigid, especially when they belong to the secondary structural framework.

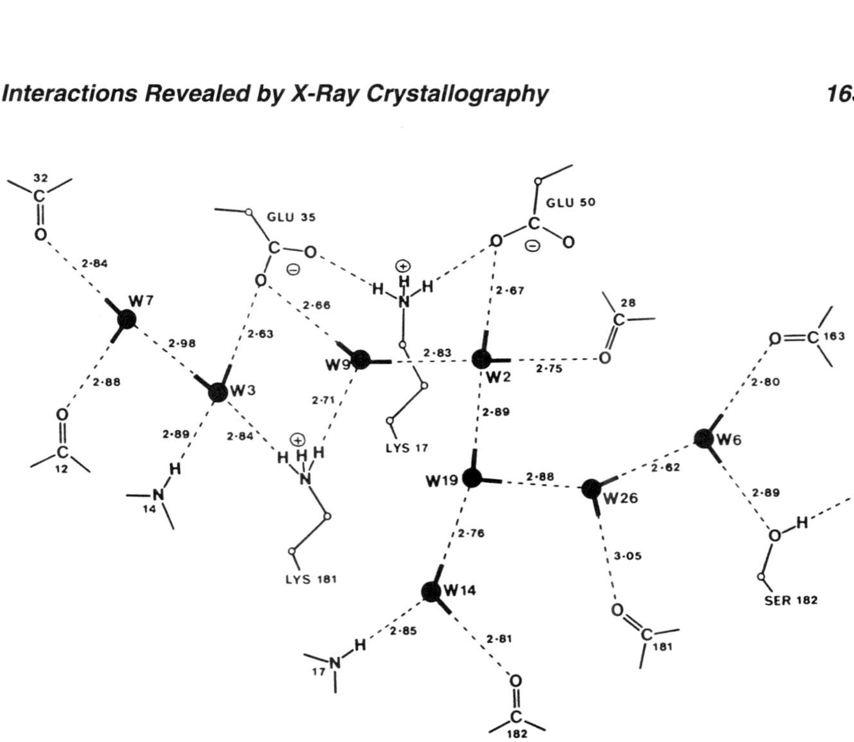

FIGURE 10 Internal cluster of water molecules occupying an interdomain cavity in the structure of actinidin [4]. All have low *B* values and three or four hydrogen bonds, and only one (W19) is not hydrogen bonded to any protein atom. Considering possible hydrogen bonds, it is also possible to place all the water protons. Note that the homologous protein papain has an identical water cluster except that Lys 17 NZ moves aside to hydrogen bond to Thr 14 OGoal (Val 14 in actinidin); an extra water molecule then replaces Lys 17 NZ, and the hydrogen bond network is exactly preserved. Note also the two water molecules, W3 and W9, bridging between two charged side chains (of Glu 35 and Lys 181), a water-mediated ion pair.

Much of the surface is polar, but parts of it are hydrophobic, and a significant proportion of the hydrophobic groups in a protein are in fact on the surface [28,54]. The combination of an irregular surface, with many projecting polar groups, both main-chain and side-chain, creates many potential binding sites for solvent molecules. An important question, however, is the extent to which these are transient or more firmly bound, and whether they make specific contributions to the structure and function of the protein.

There have been several detailed crystallographic analyses of the surface water around proteins, mostly for smaller proteins, including lysozyme [13], rubredoxin [22], crambin [25], and insulin [23,24]. More general descriptions have also been given for some larger proteins, actinidin [4], penicillopepsin [18], and others.

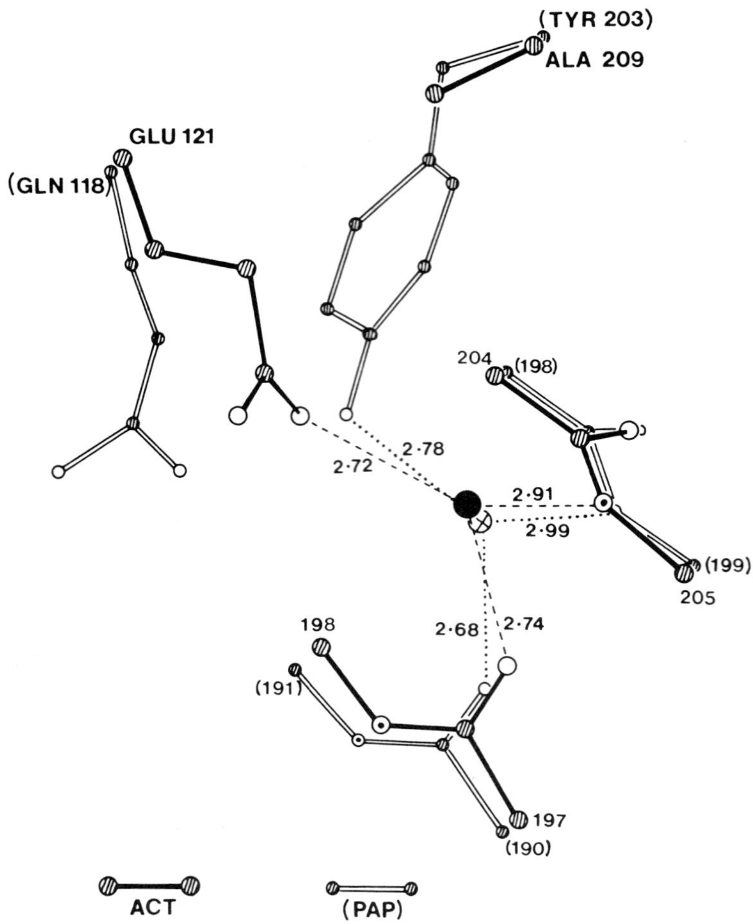

FIGURE 11 An internal water molecule and associated hydrogen bond network that is conserved in the homologous proteins actinidin and papain [7] despite sequence changes. Structure for actinidin represented by filled bonds, papain by open bonds.

A consistent observation is that there is a strong correlation between the *B* values of solvent molecules (and/or their occupancies) and the extent to which they are hydrogen bonded to protein atoms. Those which are best ordered, with low *B* values and high occupancies, are those which are closest to the surface, usually hydrogen bonded to two or more protein groups. In actinidin [4], the 54 best-ordered solvent molecules have an average of 2.2 protein ligands, the next 54 average 1.6, the next 55 average 1.1, and the rest 0.8. Similarly, in thermitase

[55], those with four hydrogen bonds to protein atoms have an average *B* value of 17.0 Å^2, those with three average 27.6 Å^2, those with two average 31.3 Å^2, those with one average 39.0 Å^2, and those with none, 42.2 Å^2. For lysozyme, the *B* values of the water molecules with two or more protein ligands are highly correlated with the *B* values of the protein atoms to which they are attached [13], implying a strong association between them.

Solvent around the protein surface clearly has a general role in hydrating polar groups, but might be expected to be rather mobile and nonspecific in its interactions if this were its only role. Well-ordered water molecules, in specific sites, can also make significant individual contributions to protein structural stability, however (see also Sec. V.B). This is because their small size and double-donor, double-acceptor hydrogen-bonding capacity makes water molecules particularly well suited for bridging between protein atoms that might not otherwise be able to interact because of steric or geometric constraints, or that might be of the wrong type to hydrogen bond directly (both acceptors or both donors). An example of water molecules in such a bridging role is shown in Fig. 12; here, a pair of water molecules bridge a surface crevice in the structure of actinidin, binding to main-chain C=O and NH groups on either side of the crevice. In the homologous protein papain, an identical arrangement exists, implying that these two water

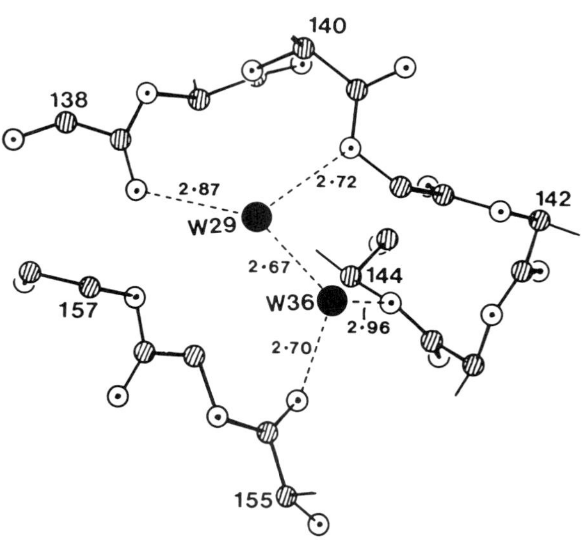

FIGURE 12 Two water molecules occupying a surface crevice in the structure of actinidin. Although exposed to the external solvent, they can be considered an integral part of the structure; solvent molecules are found in identical positions in the homologous protein papain.

molecules, although on the surface, are an important part of the protein structure. Many other bridges between polar main-chain and side-chain groups are found, involving one or more water molecules, and these undoubtedly help to stabilize the surface structure and modulate its dynamics.

There is little evidence for solvent ordering over hydrophobic groups on protein surfaces. The five-membered rings of water molecules proposed by analogy with clathrate structures [41] have rarely been seen. A rather irregular pentagonal array has been described around an exposed Val side chain in the crystal structure of 2-Zn insulin [24], but the most spectacular example is seen in crystals of the small hydrophobic protein crambin, whose structure has been analyzed at atomic resolution, 0.95 Å [25]. Here a cluster of pentagonal arrays is found covering a hydrophobic surface (Fig. 13). These arrays extend from a central array, which is confined between the hydrophobic surfaces of an intermolecular crystal contact, and there is further stabilization from hydrogen bonds between protein atoms and some water molecules of the cluster.

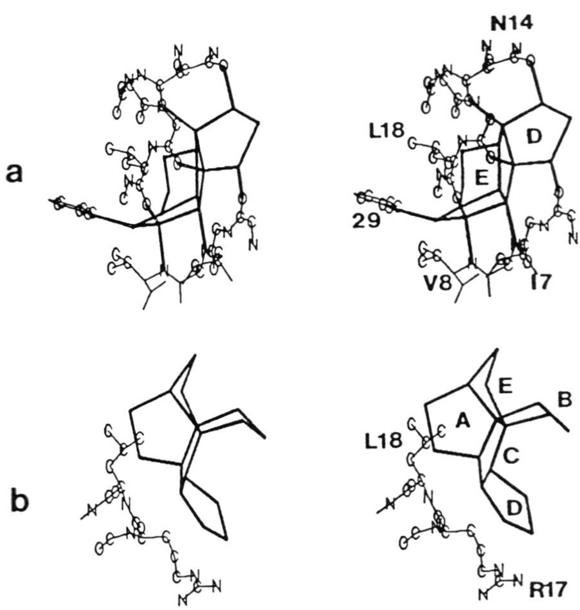

FIGURE 13 Stereo diagrams showing the pentagonal rings formed by water molecules in a largely hydrophobic crystal contact region in crystals of crambin [25]. In (a), the rings are seen to be anchored by hydrogen bonds to polar protein atoms. In (b), rings A, C, and E are seen to surround the hydrophobic side chain of Leu 18. From Teeter [25], with permission.

That arrays of this type are so rarely observed does not mean that they are not formed at protein surfaces. Transient, mobile arrays with varying orientations would not be seen by x-ray crystallography (which gives a time-averaged structure), and the few examples reported so far are either frozen out by crystal-packing contacts or are anchored by hydrogen bonding to nearby polar groups. Other ring structures are also sometimes seen spreading over small hydrophobic surfaces (e.g., Fig. 14), but for the majority of protein crystal structures, no ordered water is observed over the larger hydrophobic surfaces.

E. Association with Secondary Structures

The C=O and NH groups of the polypeptide chain are the most common groups to which ordered solvent molecules are bound (see Sec. IV.B). Many of them also are incorporated in secondary structures, and the different geometries of such assemblies suggest that characteristic solvation patterns might be seen.

Thanki et al. [33] have comprehensively analyzed the hydration of secondary structures from 24 high-resolution protein structures, while others have focused specifically on helices [31,56,57].

Several factors cause the latter to offer particularly favorable sites for solvation. First, helices are very often amphipathic with one side hydrophobic, packing against other hydrophobic structures in the protein interior, and the other side hydrophilic and exposed to solvent, with both hydrophilic side chains and main-chain C=O groups available for solvation. Second, in a helix, all the NH groups point in one direction, toward the N terminus, and all the C=O groups point in the

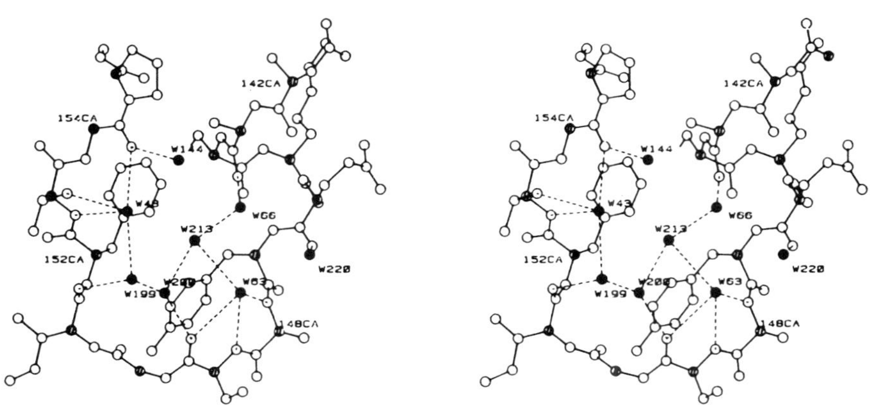

FIGURE 14 Network of eight hydrogen-bonded solvent molecules covering the exposed edges of two aromatic side chains on the surface of actinidin [4]. The network is "anchored" by hydrogen bonds to protein atoms.

opposite direction, toward the C terminus. This gives the helix polar properties, with a partial positive charge at its N terminus and a partial negative charge at its C terminus [58].

The N and C termini of helices are thus frequently highly solvated, both because of the projecting NH and C=O groups that are freely available for hydrogen bonding to solvent molecules (unless the sites are taken up by side chains) and because of the partial charge at each end, which enhances the strength of interactions. In actinidin, for example, all the free NH groups at helix N termini and 90% of the free C=O groups at helix C termini have bound solvent molecules [4], while for other proteins such as lysozyme [13] and crambin [25], the helix termini have also been noted to be highly hydrated. Analysis of the hydration patterns shows that interconnected clusters are frequently formed, linking waters and bridging neighboring "free" main-chain sites [33]. These clusters have the effect of extending the hydrogen-bonded structure of a helix. Such clusters can also link adjacent helices; for example, in proteinase K, three internal water molecules bridge between the N terminus of one helix and the C terminus of another [59]. For similar reasons, helix termini are also the most important ion-binding sites in proteins, at least for anions (see Sec. VI).

Several studies [31,56] have further shown that the C=O groups along the solvent-exposed sides of helices are frequently hydrated (e.g., Fig. 15). The helix structure apparently provides particularly favorable and specific sites for water; pockets are formed by the α- and β-carbons of residues n, $n + 3$, and $n + 4$, and the water molecule is hydrogen bonded to the C=O group of residue n [60]. The effect is enhanced by the tilting out of the carbonyl groups and by the curvature of amphipathic helices [56]. The latter effect opens up the solvent-exposed side, although whether hydration causes such curvature, as suggested, or the curvature simply arises from the packing of a helix over the hydrophobic core of a globular protein is not clear. The hydrogen bonds are short, with a favorable bond angle (C=O/water $\sim$120°), and the geometry is such that one lone pair on the peptide oxygen is directed in toward an NH group of the helix and the other out toward the water [31].

In contrast to α-helices, the hydration of β-sheets has not received much attention. Recently, however, Thanki et al. have analyzed the water structure associated with β-sheets [33] and have shown that they are, in fact, as extensively hydrated as helices. As expected, the patterns of hydration are different. The most preferred sites are at the exposed edges of β-sheets, where water molecules hydrogen bond to the "free" C=O and NH groups of the edge strand of the sheet (Figs. 16a and 16b). In doing so, they often bridge between protein atoms, forming water networks that have the effect of extending the β-sheet hydrogen bonding network by one more "strand." The next most common site is at the ends of strands, where water molecules hydrogen bond to "free" C=O and NH groups.

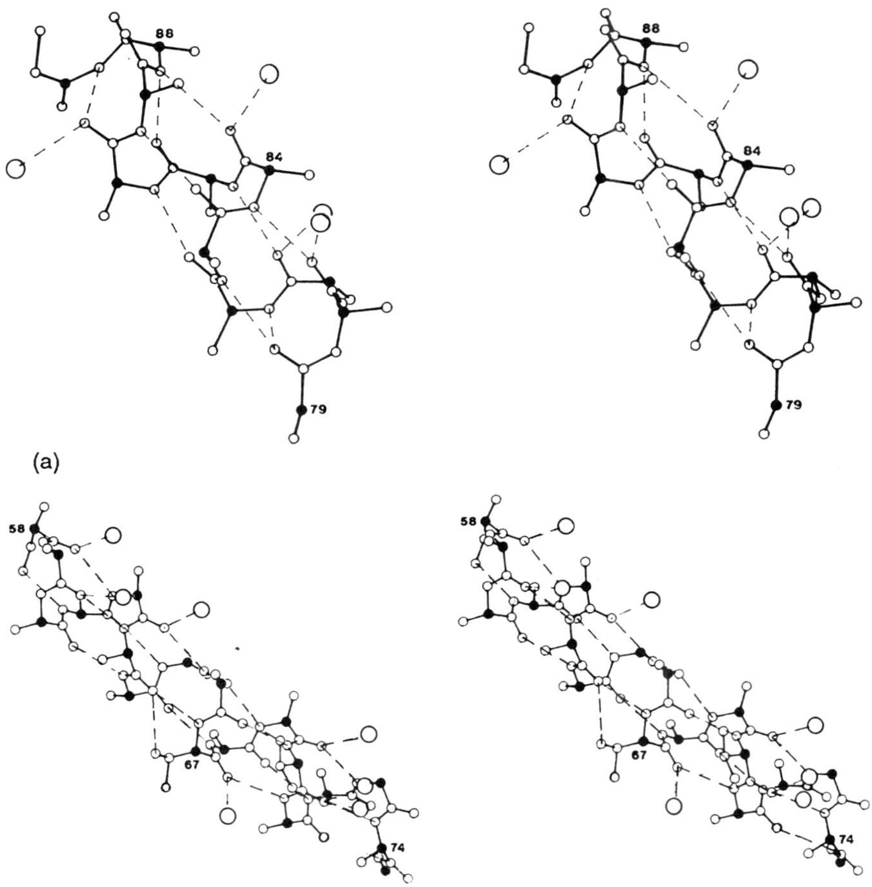

FIGURE 15 Two examples of water molecules hydrogen bonded along the sides of helices: (a) from dihydrofolate reductase, and (b) from myoglobin. For clarity, only the C_{β} atoms of side chains are included. From Baker and Hubbard [31].

Several common patterns are found (Fig. 16a), with the water molecules either bridging between strands, where they diverge, or associated with a single strand, bridging between protein atoms on the strand. An example of the former has been noted in the structure of interleukin-1β [61], and similar sites are found in other proteins. Finally, a few water molecules are found hydrogen bonded to C=O groups in the middles of β-sheets, bridging between strands; here, however, they necessarily lie above the peptide planes since the C=O groups are simultaneously involved in interstrand hydrogen bonding. There are also indications that

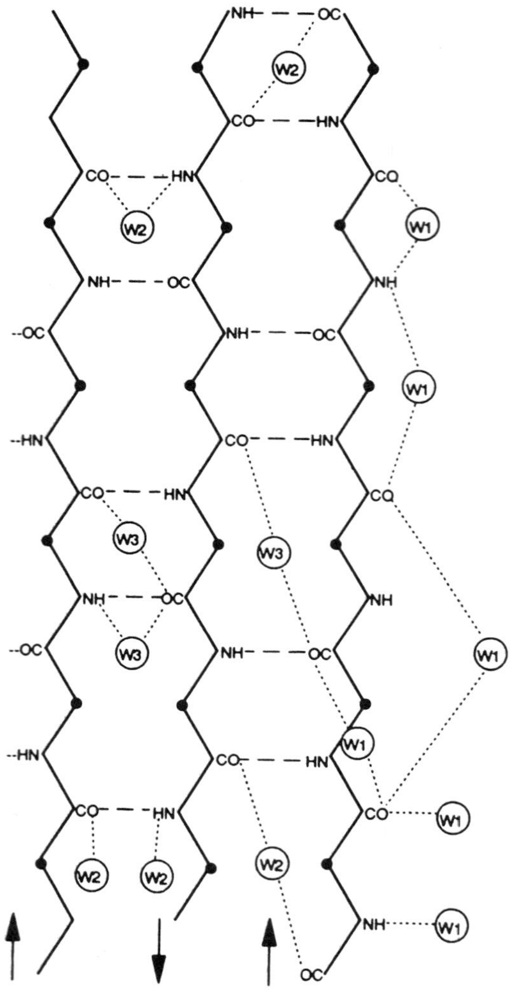

a

FIGURE 16 Hydration of β-sheets and β-turns. In (a), patterns of interaction of water with edge strands (W1), and the ends (W2) and middles (W3) of strands of an antiparallel β-sheet are shown, while in (b) an example from human Bence–Jones protein [105] is shown. From Thanki et al. [33], with permission. In (c), an example of a water-mediated β-turn is shown, from the structure of actinidin [4]. Instead of a direct 117O–120N hydrogen bond, a water molecule intervenes, hydrogen bonding to both.

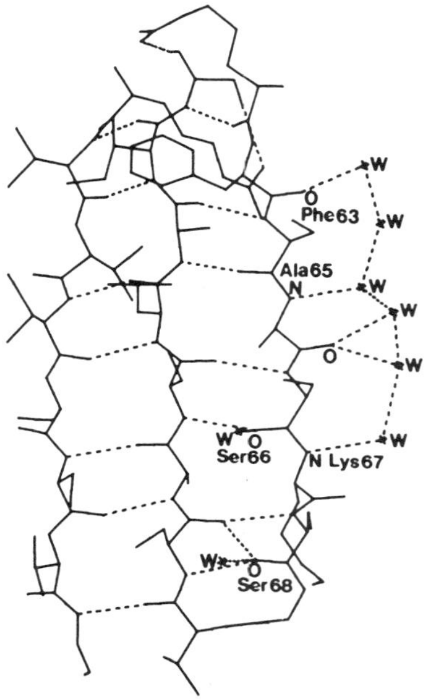

b

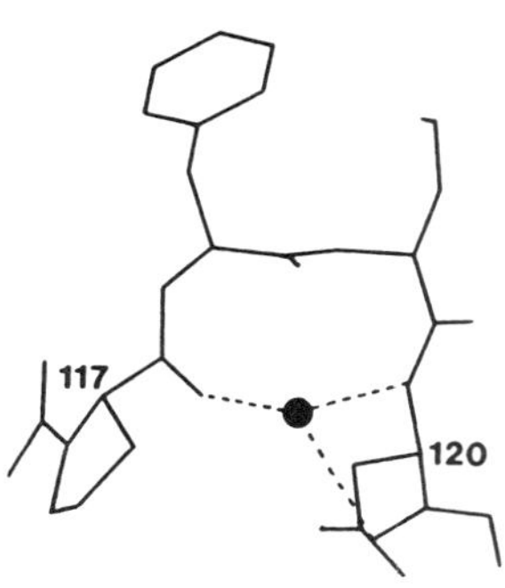

c

FIGURE 16 Continued

antiparallel β-sheets are more extensively hydrated than parallel sheets, perhaps because the latter are more often buried in the protein interior.

Reverse turns, or β-turns, tend to have a high solvent accessibility, being almost always on the protein surface. The combination of usually hydrophilic side chains and free main-chain C=O and NH groups also creates favorable binding sites, and early observations that turns tend to be highly hydrated [62] have been confirmed by systematic analyses [33]. The patterns of interaction depend on the type of turn, but water bridges of various kinds are found. If the residues constituting a turn are labeled i, $i + 1$, $i + 2$, and $i + 3$ (with the normal intraturn hydrogen bond being from $C{=}O_i$ to NH_{i+3}), there are some cases where a water molecule bridges $C{=}O_i$ and NH_{i+3}, giving a water-mediated β-turn (e.g., Fig. 16c), but more common are other bridges, for example, $C{=}O_i$ to $C{=}O_{i+3}$ or $C{=}O_{i+1}$ to $C{=}O_{i+3}$ [33]. Some of these may stabilize the β-turn structure [33,63].

The crystallographic observation of solvent associated with turns is likely to be particularly dependent on their location. Those which project from the surface or are in larger loops tend to have a higher mobility than other parts of the structure, and the solvent associated with them is then less well ordered and likely to be poorly resolved by crystallography. Those few turns that are buried are associated with sets of buried water molecules hydrogen bonded to the C=O groups [64].

F. Solvent in Active Sites

Most proteins, including enzymes, transport proteins, and immunoglobulins, are required for their biological functions to bind other molecules or ions. The binding site often takes the form of a cleft or depression in the protein surface and is thus likely to also provide a favorable environment for bound solvent molecules. Most importantly, the solvent in active sites may be of profound importance to protein function, either because it participates in binding or catalysis, or because it must be displaced by an incoming substrate.

Ordered solvent has been found in the active sites of many enzymes. The bacterial serine protease SGPA, whose active site contains 24 solvent molecules, assumed to be water, provides an excellent example [65]. The most striking feature of the solvent structure in this case is that the pattern of solvent density almost exactly matches that of the polar groups of a bound tetrapeptide product (Fig. 17). A chain of solvent molecules is ordered approximately parallel to the polypeptide chain that forms the S_1 to S_3 binding sites, and another cluster occupies a hydrophobic pocket that binds the product benzyl group. All of these solvent molecules should be displaced by substrate binding, and this is facilitated because very few of them make strong interactions with the protein; their binding is apparently dependent mainly on the steric constraints of the active site. A very similar pattern is seen in lysozyme, where the active sites of both the human and turkey egg white enzymes contain similar distributions of bound water molecules (some 12–15 in

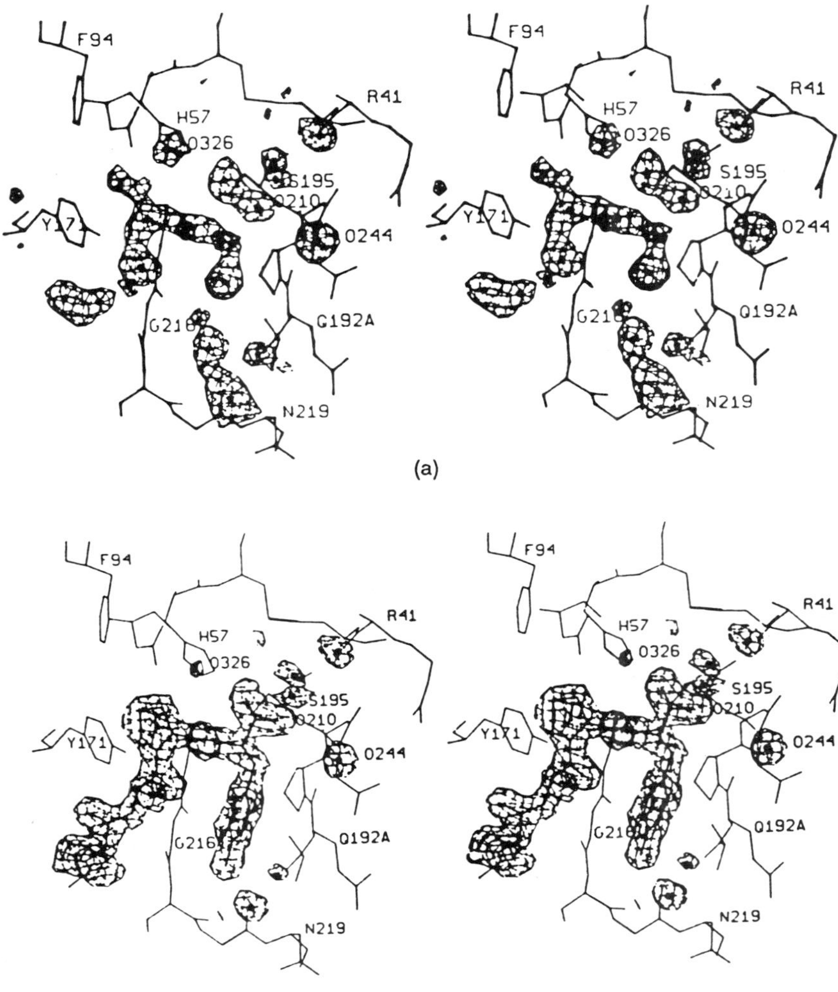

FIGURE 17 Stereo diagrams showing (a) the difference electron density distribution for the solvent structure in the active site of the serine protease SGPA, and (b) the electron density for a tetrapeptide product Ac—Pro—Ala—Pro—Tyr—OH in the active site. Note the remarkable similarity in the two distributions. From James et al. [65], with permission.

number). Comparison with hexasaccharide binding models show that these bound solvent molecules mark the positions that will be occupied by the polar groups of the bound substrate [13]. Patterns of bound solvent in active sites may then be a powerful guide to inhibitor design and modeling of enzyme–substrate interactions.

Solvent molecules in the active sites of enzymes can also help in identifying key features of enzymatic mechanisms. For example, in β-lactamase of *Staphylococcus aureus* [66], a bound water molecule is exactly in the position where it would be able to deacylate a substrate, and a second water molecule is hydrogen bonded to the two NH groups that form the "oxyanion hole," that is, it mimics the position of the carbonyl oxygen of a β-lactam substrate. Water molecules have similarly been found to occupy the "oxyanion holes" of several serine proteases [55,59,67,68]. A solvent molecule between two carboxylate groups is a conserved feature of the active sites of aspartyl proteases [69] and is believed to play a key role in catalysis (Sec. V.C).

Specificity pockets are another frequent site for bound solvent. In this case, the bound solvent may or may not be displaced by an incoming substrate; if it is not displaced, the solvent molecule may modulate the specificity and assist in binding. A good example is the specificity pocket in *Streptomyces griseus* trypsin. This pocket contains 14 solvent molecules, the outermost of which are loosely bound, with high B values and would be displaced, and the innermost of which are tightly bound and would not be displaced [70]. A Lys side chain in the P_1 position of a substrate is expected to interact with the specificity determinant, the carboxylate group of Asp 189, via bound solvent. The serine proteases in fact offer a variety of such examples that illustrate the roles bound solvent can play (Sec. V.C).

V. SIGNIFICANCE OF BOUND SOLVENT

A. Conservation of Solvent Sites

A fundamental question in considering the solvent sites identified in crystallographic analyses is the extent to which they are determined by the crystal environment or the crystallization medium. How many of them are likely to persist in free solution, and to what extent can they be regarded as a molecular property of the protein in question? Answers can be sought either from solution studies, with NMR spectroscopy as the primary structural method, or from crystallographic comparisons of the same or related proteins in different crystal environments.

1. Comparisons with NMR Studies

NMR spectroscopy is a very rapidly developing method for protein structure analysis (for recent reviews, see [71,72]), and its full potential has probably not yet been reached. Although nuclear relaxation methods have been used for some time to study the mobility of water associated with proteins, it has only been in the past few years that 2D and 3D NMR techniques have been extended to allow the detection of bound solvent molecules [45].

Several studies have now shown that the internal water molecules seen crystallographically can also be located by NMR methods. Thus, the four internal water molecules in BPTI were detected using nuclear Overhauser effect (NOE) spectroscopy [49], and similar studies on interleukin-1β located the internal water molecules in that structure as well [73], in the same positions as in the crystal structure.

The question of external water molecules is more problematical. Very few of the surface water sites determined crystallographically for BPTI, in three different crystal forms (see Sec. V.A.2), were detected by NMR [49,45], and those which could be detected were in rapid exchange with the bulk solution, having residence times on the protein of the order of 100–300 ps. This is not surprising, nor does it imply an incompatibility between x-ray and NMR results. As Otting et al. [45] point out, the two methods are sensitive to different aspects of hydration; the NMR experiments probe the residence times of water molecules near the protein surface, and hence their mobility, whereas crystallography is a time-averaged technique, which reflects the fraction of the time that a water molecule is located at a particular point in space rather than its residence time on any one visit. A high occupancy of a surface site in the crystal only implies that the site is occupied most of the time, and solvent occupying such a site would be expected to be labile and exchanging with the external solution.

2. Comparisons of Different Crystal Environments

The consistency of the solvent sites determined in crystal structure analyses of proteins has been investigated in various ways: comparison of crystallographically independent protein molecules in the same crystal form and comparison of different crystal forms of the same protein, with either different crystal packing or different mother liquor, or both.

Where the asymmetric unit (the smallest repeating unit) of a crystal contains more than one molecule, these will experience different crystal environments. For example, azurin from *Alcaligenes denitrificans* has two molecules of 129 residues each in the crystal asymmetric unit, and refinement at 1.8 Å resolution has led to the location of 281 water molecules [21]. Of these, 182 (91 pairs), representing 65% of the found water, correspond approximately between the two molecules, with 53 pairs deviating by less than 1.0 Å. Other analyses of this type show similar or slightly lower correspondences, perhaps depending on the extent to which the solvent structure has been modeled. Typical examples include *Bacillus licheniformis* β-lactamase (2 × 291 residues), with 112 pairs of water sites corresponding within 1.0 Å [71], and α-chymotrypsin (2 × 241 residues), with 49 equivalent pairs [20].

What these analyses all have in common is that it is the best-ordered, most strongly bound water molecules that are conserved. For β-lactamase, for example, the 26 water pairs that correspond within 0.25 Å have an average *B* value of 19.4 $Å^2$, compared with 28.7 $Å^2$ for those corresponding within 1.0 Å, and 37.5 $Å^2$ for all waters [74]. A similar correlation has been noted for azurin, where the conservation of solvent sites has also been related to the extent of interaction with the protein—those which are conserved average 1.5 hydrogen bonds to protein atoms, whereas those which are not average 0.7 [21].

The corollary of this conservation of well-ordered sites is that many of the other sites determined crystallographically are not conserved. In the case of azurin, the crystal packing of the two independent molecules is actually very similar, yet 35% of the solvent molecules did not correspond. Many of these are close to the noise level and may have been incorrectly assigned or have partners that are just below the threshold, but some are evidence that even very small perturbations in the environment can disturb the pattern of solvent structure.

Comparisons from structure analyses of the same protein in different crystal forms give broadly similar results. High-resolution structures have been determined for three different crystal forms of BPTI [34,47,48]. Around 60–70 solvent sites have been identified in each case (a reasonably conservative number for a protein of 58 amino acid residues), and of these 16 are conserved in all three crystal structures; these have full occupancy and very low *B* values, and make multiple hydrogen bonds to the protein [48]. The nonconservation of some sites has been rationalized in terms of different crystal contacts or different side-chain orientations [75], but a few may have been incorrect interpretations of map noise. Tetragonal and triclinic crystal forms of hen egg white lysozyme have also been compared, albeit at lower resolution than for BPTI, and a somewhat higher number of solvent sites were found to correspond, 60 altogether for a protein of 129 residues [76]. This may be because a lower fraction of the surface, and hence of the solvent, is involved in crystal contacts than for BPTI.

Several analyses have examined the effects of changed mother liquor in otherwise isomorphous crystals. Comparisons of bacteriophage T4 lysozyme at low, medium, and high ionic stengths, including independent determinations of the solvent structure in each case, showed 73 solvent molecules corresponding in all three [77]. Interestingly, there was a suggestion, from comparison of *B* values, that some of the solvent molecules originally assumed to be water in the high salt structure might actually be ions. Similar comparisons of independent refinements of the structure of ribonuclease A in crystals grown in ethanol and 2-methyl-2-propanol [78] found about 50% of solvent molecules in the same position (58, for a protein of 123 residues). This very thorough analysis showed that the strongly bound solvent occupies the same sites for mother liquors containing two different alcohols and is reliably determined by different refinements, but that misassignments of the more weakly bound solvent do occur, and that the positions of second-shell solvent should be treated with a great deal of caution. Finally, an analysis of the effects of humidity on the crystal structure of hen egg white lysozyme [63] has shown that reduction of the relative humidity caused a slight repacking of the protein molecules and some small structural changes. Associated with this were apparent changes in the first hydration shell; comparison of the solvent structures showed that even with a liberal criterion for correspondence of sites (1.8 Å) only 75 waters were found in common. Again, for the strongly bound waters, however, there is very good correspondence, with 93% of those which make three or more hydrogen bonds to the protein being conserved.

One interesting result is that in all these studies a rather similar proportion of solvent sites is conserved, about 0.4 solvent molecules per residue.

3. Conservation of Solvent Sites in Homologous Proteins

The strong tendency for internal solvent sites to be conserved in homologous proteins has already been discussed (Sec. IV.C). This reflects their important role as integral parts of the protein structure. What is less obvious is the extent to which solvent-accessible bound solvent molecules are conserved.

Several structure comparisons bear on this question. In the case of actinidin and papain, some 77 solvent sites are found within 1.5 Å between the two proteins, representing many more than just the 17 internal sites [7]. In some cases, methanol molecules in papain replace presumed water molecules in actinidin. Some of the conserved solvent molecules clearly fulfill an important structural role (e.g., Figs. 11 and 12), and although no analysis has been done, it seems likely that much of the conserved surface solvent is bound to main-chain atoms, reflecting the folding similarity of the two proteins. Comparisons of human and turkey egg white lysozyme [13] also show that a significant number of surface solvent molecules occupy common binding sites in the two proteins; a total of 37 are identified as equivalent, including the four internal water molecules. This is a somewhat similar proportion as in actinidin and papain, bearing in mind that lysozyme is a smaller protein (130 residues, compared to 218 and 212).

A feature of both comparisons is that once again it is the best-ordered solvent molecules, with lowest *B* values and generally multiple interactions with the protein, that are conserved. Moreover, for each pair, the crystallization media and crystal packing are completely different for the molecules being compared. In the case of lysozyme, the authors point out that even some of the water molecules in intermolecular sites in one crystal are still found in the other in the absence of neighboring protein molecules.

Other protein structure analyses have noted some conservation of solvent molecules, but it is important to note that given the nature of the process of identifying solvent molecules in crystallographic analyses, the proteins being compared should all be fully refined, at high resolution; otherwise, some of the well-bound solvent may not be accounted for.

B. Contributions to Stability

Two kinds of solvent molecules associated with proteins can be distinguished: those at the surface which provide a solvation shell, and those in the interior which occupy small cavities in the protein structure. (Although, as noted earlier, there is not always a clear distinction, for example, for those which occupy deep clefts in the surface.)

There are convincing reasons to believe that the internal solvent molecules and those which are firmly bound to surface clefts make an important contribution to the stability of the folded protein. First, the folding of a polypeptide chain into a globular structure, driven by the hydrophobic effect [79], inevitably results in the burial of a number of polar groups in addition to nonpolar, notably the C=O and NH groups of the polypeptide chain [80]. If these are not hydrogen bonded, for example in secondary structures, there will be a significant cost in energy when the protein folds [26,80]. Second, the imperfect packing of elements of the protein structure leaves some holes in the interior that may occasionally be quite large, at domain or subunit interfaces [42]. Mutagenesis studies [81] show that internal cavities destabilize protein structures, and most cavities, especially the larger ones, are found to be filled with solvent molecules. The few that appear not to be filled are those which are hydrophobic and either are devoid of solvent for this reason or contain solvent that is disordered (because there are no polar anchoring points) and thus not x-ray observable. Neutron diffraction studies of trypsin [82] suggest that the latter may sometimes be the case.

In general, internal solvent molecules perform a crucial role in filling small cavities in the protein interior and simultaneously satisfying the hydrogen bonding potential of buried polar groups that would otherwise not be hydrogen bonded. For example, in actinidin, 11 internal main-chain C=O groups and ten internal NH groups would lack any hydrogen-bond partners if it were not for the presence of a number of internal water molecules [4]. Similarly, in penicillopepsin, seven internal C=O groups, six internal NH groups, and two internal polar side chains would be without hydrogen bonds but for the buried water molecules associated with them [18]. It is in fact a consistent observation, from well-refined protein structures, that virtually every polar group is either explicitly bonded or is at least in contact with solvent [31].

A second function of internal solvent molecules may be a "charge spreading" role, where charged side chains are buried in the protein interior. In carboxypeptidase, the carboxyl group of the internal Glu 108 would have no compensating positive charge were it not for a bound water molecule, which may actually be protonated, H_3O^+ [40]. A similar suggestion is made for a buried carboxyl group in chymosin [43]. Sometimes, charged internal side chains do not make contact directly but are linked by water molecules; for example, in carboxypeptidase, the internal Asp 104 and Arg 59 are linked by a water molecule, and in actinidin, Glu 35 and Lys 181, both internal, are linked by two water molecules (Fig. 10). Finally, internal water molecules may help create polar microenvironments; for example, the presence of water molecules in the vicinity of the buried Asp 102 in all the serine proteases gives a locally polar environment for it, despite its internal site, and these waters may have a "charge spreading" role [26]. Similarly, the activation ion pair, involving a buried N terminus and the carboxyl of Asp 194, in members of the trypsin family, is associated with a cluster of three to five water molecules that helps dissipate the charge [70,83].

In addition to their role in stabilizing polar and charged groups, buried solvent molecules probably help to maintain the active structure of a protein by filling the larger internal cavities. In doing so, they increase the diversity of stable structures available to proteins. For example, in ribonuclease T1, a chain of five waters act as space fillers between an α-helix and a hairpin loop, and it has been speculated that without them the hairpin loop could collapse into an antiparallel β-structure and destroy the active conformation [46]. Hydrogen-bonding and van der Waals contact requirements suggest that in such a case the volume would reduce considerably if the cavity were not filled. This has clear lessons for molecular dynamics calculations, which should allow for such internal solvent. The larger clusters of internal water molecules, being found either between secondary structure elements or between domains (e.g., the eight-water cluster between the two domains of actinidin [4]), may have a further importance in giving a flexible "cushion" that could facilitate domain movements, or movements of helices, etc., which can be essential for protein function [12]. Empty cavities have been suggested as having a similar functional significance [42], although in that case there is also some cost in energy.

Direct evidence that not only internal solvent molecules contribute to the stability of the folded protein comes from mutagenesis studies on bacteriophage T4 lysozyme [84]. Mutations of Thr 157, on the surface of the molecule, destabilize the structure to varying degrees depending on the nature of the amino acid substitution. The —OH group of Thr 157 occupies a surface pocket, where it is hydrogen bonded to the main-chain NH of residue 159 and the —OH of Thr 155. Mutations that result in the loss of these hydrogen bonds are more destabilizing than those in which the hydrogen bonds are preserved. Most interestingly, the mutation Thr→Gly results in the appearance of a strongly bound water molecule at the position where the Thr—OH had been (Fig. 18), and this restores most of the stability of the native protein, a clear demonstration that solvent binding to a specific site in the folded protein structure can confer stability. Thus, we can anticipate that some solvent molecules on the protein surface, in clefts or pockets, do fill a structural and energetic role and can be regarded as integral to the protein structure just as internal solvent molecules are.

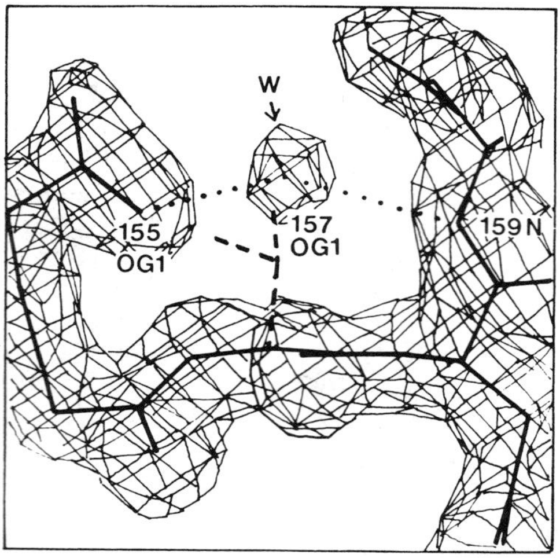

FIGURE 18 Electron density associated with the mutation Thr 157 Gly in bacteriophage T4 lysozyme. In the Gly mutant, a water molecule (w) appears in the position occupied by the Thr —OH group in the wild-type protein, thus maintaining the same hydrogen bond network. From Alber et al. [84], with permission.

C. Functional Roles of Solvent Molecules

Both the loosely associated surface water and the strongly bound water molecules in specific sites can have a role in protein function. The surface water does not freeze on cooling, and studies on hydration have shown that a certain amount is necessary to give a functional protein molecule [26–28]. Chains and clusters of solvent molecules are a distinct feature of the protein surface. These are likely to modulate the dynamics of protein groups, especially around clefts or protruberances.

The more specifically bound solvent molecules have a general importance in stabilizing the protein structure, dissipating charge, especially in the interior, and perhaps in facilitating domain and secondary structure movements (see above). Individual bound solvent molecules have also been identified with many different types of direct involvement in protein function.

In hydrolytic enzymes, water has a direct role as a nucelophile in catalysis. In some cases, for example, the cysteine proteases actinidin and papain, no ordered water molecule has been located crystallographically that might fill a direct catalytic role; in fact, the active sites contain little ordered water [4]. In other cases, however, solvent molecules are found ideally positioned for involvement in the catalytic mechanism. Two independent structure analyses of β-lactamases, both refined at high resolution [66,74], have each located a water molecule, bound to the same protein groups and accurately positioned for deacylation of

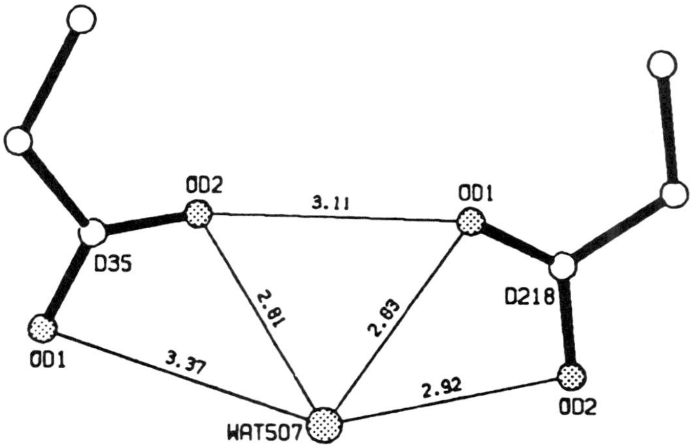

FIGURE 19 Water molecule hydrogen bonded between the two Asp side chains in the active sites of aspartyl proteases. From Suguna et al. [36], with permission.

the penicilloyl–enzyme intermediate. In the 1.35 Å structure analysis of H-*ras* p21, a water molecule is perfectly placed to be the nucleophile attacking the γ-phosphate of guanine triphosphate (GTP) [85]. Arguments based on the positions of bound solvent molecules are potentially hazardous, however. In the aspartyl proteases, a solvent molecule is found bound to the two aspartic acid residues in the catalytic site (Fig. 19). This is a consistent feature of all the aspartyl proteases so far analyzed, but its role and even its identity is a matter of debate. Pearl and Blundell [86] suggested it might be a hydroxide ion and involved in catalysis as the essential nucleophile. A more likely suggestion, given the nature of the site, is that it is a cation, H_3O^+ or NH_4^+ [69,87], but it may also be water [36]. Moreover, binding studies and modeling suggest it would actually be displaced by a bound substrate and that a second water nearby is the nucleophile [69].

As noted earlier (Sec. IV.D), the solvent found in binding pockets is not always displaced by an incoming substrate or ligand, and in these cases, it may modulate binding and help determine the specificity. For chymotrypsin, only two of the five water molecules in the specificity pocket are displaced, and it has been suggested [20] that the remaining three, because they cannot easily be displaced, help determine the specificity for smaller aromatic side chains (given that there is no other obvious determinant). Water molecules also remain in the specificity pocket in the complexes of bovine trypsin with BPTI [88] and of α-chymotrypsin with the third domain of turkey ovomucoid [89].

In the bacterial periplasmic binding proteins, the specificity for different sugars has been shown to be modulated by water molecules bound in the active site of the arabinose-binding protein [90]. Two water molecules assist the binding of L-arabinose, through favorable hydrogen bonding, but discriminate against D-fructose. In the case of D-galactose, the sugar $-CH_2OH$ replaces one water, resulting in equally tight binding as for L-arabinose. In the case of antibody–antigen complexes, it has generally been thought that water would not be involved in antigen recognition, but the recently reported structures of two antibody–antigen complexes show that this is not the case. Complexes of hen egg white lysozyme with the Fv and Fab fragments of a monoclonal antibody show that buried or partially buried water molecules mediate contacts between antibody and antigen [91]. Even more striking is an Fab-oligosaccharide complex [92] in which a bound water molecule plays a crucial role in the antibody–antigen interaction, making two hydrogen bonds to the antigen and two to the antibody (Fig. 20).

Water is also an important ligand for metal ions in metalloproteins. It is a common ligand for calcium in calcium-binding proteins (e.g., [46,59,70]). In metalloenzymes, water ligands are particularly important because they can complete the coordination requirements of the metal, in the absence of substrate, but then be relatively easily displaced by an incoming substrate—that is, the energy barrier is much lower than if all the metal ligands are protein groups. Thus, in Cu–Zn superoxide dismutase, a water molecule bound to copper marks the site that would be occupied by O_2^- [93], while in alcohol dehydrogenase, the Zn-bound water is displaced by substrates [94]. In some zinc enzymes, a Zn-bound water plays a crucial role in catalysis. In carbonic anhydrase, the net 2+ charge on the zinc reduces the *pK* of the water ligand so that at alkaline pH it is actually OH^- and acts as the essential nucleophile in catalysis; here, it apparently remains bound to zinc in the enzyme–substrate complex [95]. A similar mechanism may also apply to thermolysin and carboxypeptidase.

Yet another proposed role for bound water molecules is in electron transfer. In cytochrome c, an internal water molecule near the heme group moves ~0.9 Å during electron transfer, and this movement, with those of its hydrogen bond partners, is the largest redox state-dependent structural change observed [96]. The precise mechanism is not known, but this water molecule is clearly involved in some way. In the blue copper protein azurin, a water molecule at the center of a hydrophobic binding surface, hydrogen bonded to a His side chain (which is in turn a ligand to copper), has been suggested to be directly involved in electron transfer in a through-bond mechanism [97].

VI. BOUND IONS AND OTHER SOLVENT MOLECULES

Very few bound solvent cations have been located in crystallographic analyses; this is probably primarily because of the difficulties of identification (see Sec. III.C). These difficulties are less severe for many anions, because of their generally greater bulk and greater electron density, and a number of bound anions have been recognized.

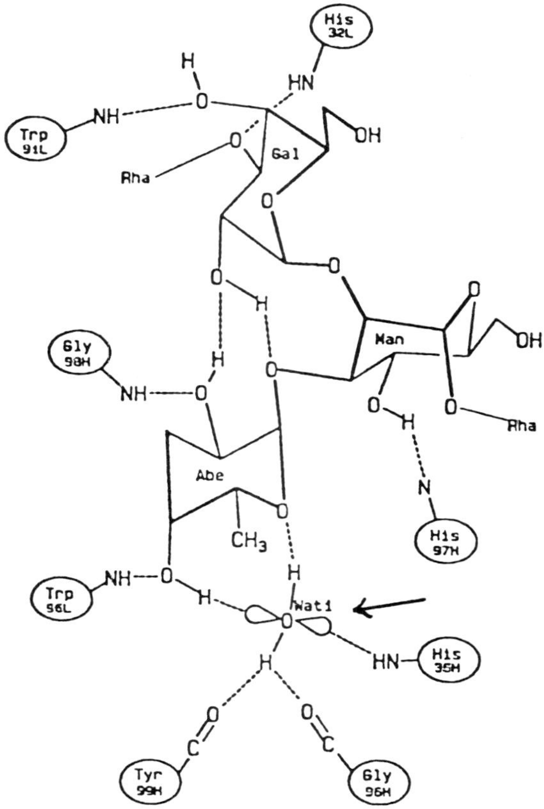

FIGURE 20 Hydrogen-bonded water molecule in the recognition site of an antibody–antigen complex. The water molecule (arrowed) makes two hydrogen bonds with the oligosaccharide antigen and two (one bifurcated) with the antibody. From Cygler et al. [92], with permission.

Helix termini, with their free NH and C=O groups and associated charge, are the most common binding sites. In cytochrome P450 cam, a cation is bound at the C terminus of a helix, coordinated to four main-chain C=O groups and two waters; its chemical identity, however, is uncertain [98]. Anions are very frequently bound to helix N termini [58] (e.g., Fig. 21). For example, in the bacterial sulfate- and phosphate-binding proteins [99,100], the anion is in each case bound between helix N termini; in lactoferrin, the binding site for carbonate is at the N terminus of a helix [101]; in phage T4 lysozyme, Cl^- ions are bound at the N termini of two of the helices [19]; and in many nucleotide-binding proteins, the nucleotide phosphate is bound to a helix N terminus [58]. A sulfate ion occupies an intermolecu-

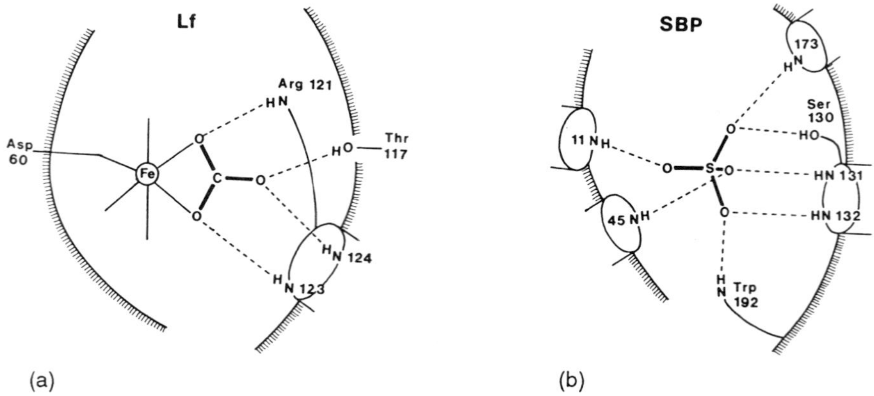

FIGURE 21 Two examples of anions bound to the N termini of α-helices. In (a), the CO_3^{2-} ion in the binding site of lactoferrin is hydrogen bonded to the N terminus of an α-helix as well as an Arg side chain and the bound Fe^{3+} ion [101]. In (b), the SO_4^{2-} ion bound by the bacterial sulfate binding protein (SBP) is bound between the N termini of four α-helices [99].

lar site in myoglobin crystals, bound between two helix N termini, one in each molecule [102]. Other anion-binding sites almost always involve positively charged side chains. In α-chymotrypsin [20], four sulfate ions are found, all adjacent to Arg or Lys side chains; all are relatively weakly bound, however, as none of the sites is fully occupied. In ribonuclease [103], a sulfate ion is attached to the two active site His residues, and in azurin, a weakly bound sulfate bridges between a His side chain and a peptide NH group [21]. Phosphate ions are found in intermolecular sites in two different crystal forms of BPTI [47,75], in each case bridging between basic groups (Arg or Lys) on two protein molecules, while a tetrahedral anion, possibly cacodylate, is bound in the active site of *B. licheniformis* β-lactamase [74]; even though the latter is bound mainly to water, it appears to be very clearly defined, perhaps because it occupies the substrate binding site.

Finally, there are a few cases where organic molecules have been located in protein crystals, generally very weakly bound, however. In papain, crystallized from 62% methanol–water solution, some methanol molecules have been identified, but they have higher *B* values than the bound water molecules and are much fewer in number, 29 compared with 195 [104]. In crystals of thioredoxin [15], seven MPD (2-methyl-2,4-pentanediol) molecules are found, most of them in surface pockets where they may stabilize loops through nonbonded interactions. The infrequency of bound organic molecules such as these may reflect not only their weaker interactions with the relatively hydrophilic protein surface, compared with water, but also the fact that the liquid composition in the crystal may not be the same as that in the crystallization medium. In crambin crystals, grown from 60% ethanol, only two ethanol molecules have been located even though the solvent structure has been almost completely modeled; apparently hydrophobic protein–protein contacts in the crystal have replaced the hydrophobic protein–protein alcohol contacts made in solution [25].

VII. CONCLUSIONS

With more than 500 protein structures determined to date, many refined at high resolution, the protein structure database [106] offers a rich resource for visualizing protein–solvent interactions. Internal solvent molecules, integral to the protein structure and stability, are clearly defined and highly conserved in related structures. The same also applies to some tightly bound solvent molecules, occupying surface pockets and crevices, and solvent molecules in active sites or binding pockets can contribute to catalysis or the specificity of protein–ligand interactions. This tightly bound solvent must clearly be taken into account in theoretical calculation of molecular dynamics and electrostatics. A monolayer of water covering the protein surface is mostly visible in crystallographic analyses, but is mobile and less well conserved. Although the time-averaged nature of models derived by x-ray crystallography is limiting in describing surface solvent structure, it is complementary to other techniques such as NMR spectroscopy and useful in giving pictures of preferred solvent sites and structural patterns. I also believe that there is much unexploited material in the crystallographic database, and comparisons of the solvent models for different proteins could be very rewarding.

ACKNOWLEDGMENTS

I am grateful to Dr. Hale Nicholson and Mrs. Heather Baker for their critical reading of this manuscript and to Dr. Bryan Anderson for help with some of the illustrations. I also gratefully acknowledge research support as an International Research Scholar of the Howard Hughes Medical Institute.

REFERENCES

1. Crowfoot, D., Riley, D., Bernal, J. D., Fankuchen, I., and Perutz, M. (1938). *Nature 141*:521–524.
2. Crowfoot, D., and Riley, D. (1939). *Nature 144*: 1011–1012.
3. Matthews, B. W. (1968). *J. Mol. Biol. 33*:491–497.
4. Baker, E., N. (1980). *J. Mol. Biol. 141*:441–484.
5. Rupley, J. A. (1969). In Structure and Stability of Biological Macromolecules (Timasheff, S. N., and Fasman, G. D., eds.), pp. 291–352, Marcel Dekker, New York.
6. Matthews, B. W. (1977). In The Proteins, 3rd Edition, Volume III (Neurath, H., and Hill, R. L., eds.), pp. 403–590, Academic Press, New York.
7. Kamphuis, I. G., Drenth, J., and Baker, E. N. (1985). *J. Mol. Biol. 182*, 317–329.

8. Wagner, G., Braun, W., Havel, T. F., Schaumann, T., Go, N., and Wuthrich, K. (1987). *J. Mol. Biol. 196*: 611–639.
9. Nilges, M., Gronenborn, A. M., Brünger, A. T., and Clore, G. M. (1988). *Protein Eng. 2*:27–38.
10. Kline, A. D., Braun, W., and Wuthrich, K. (1986). *J. Mol. Biol. 189*:377–382.
11. Pflugrath, J. W., Wiegand, G., and Huber, R. (1986). *J. Mol. Biol. 189*:383–386.
12. Bennett, W. S., and Huber, R. (1984). *CRC Crit. Rev. Biochem. 15*:291–384.
13. Blake, C. C. F., Pulford, W. C. A., and Artymiuk, P. J. (1983). *J. Mol. Biol. 167*: 693–723.
14. Blundell, T. L., and Johnson, L. N. (1976). Protein Crystallography, Academic Press, London.
15. Katti, S. K., LeMaster, D. M. and Eklund, H. (1990). *J. Mol. Biol. 212*:167–184.
16. Baker, E. N., and Dodson, E. J. (1980). *Acta Crystallogr. A36*:559–572.
17. Kundrot, C. E., and Richards, F. M. (1987). *Acta Crystallogr. B43*:544–547.
18. James, M. N. G., and Sielecki, A. R. (1983). *J. Mol. Biol. 163*:299–361.
19. Nicholson, H. H., Becktel, W. J., and Matthews, B. W. (1988). *Nature 336*:651–656.
20. Blevins, R. A., and Tulinsky, A. (1985). *J. Biol. Chem. 260*:8865–8872.
21. Baker, E. N. (1988). *J. Mol. Biol. 203*: 1071–1095.
22. Watenpaugh, K. D., Margulis, T. N., Sieker, L. C., and Jensen, L. H. (1978). *J. Mol. Biol. 122*:175–190.
23. Baker, E. N., Dodson, E. J., Dodson, G. G., Hodgkin, D. C. and Hubbard, R. E. (1987). In Crystallography in Molecular Biology (Moras, D., Drenth, J., Strandberg, B., Suck, D., and Wilson, K., eds.), pp. 179–192, Plenum Publishing Corp., New York.
24. Baker, E. N., Blundell, T. L., Cutfield, J. F., Cutfield, S. M., Dodson, E. J., Dodson, G. G., Hodgkin, D. M. C., Hubbard, R. E., Isaacs, N. W., Reynolds, C. D., Sakabe, K., Sakabe, N., and Vijayan, N. M. (1988). *Phil. Trans. Roy. Soc. B319*:369–456.
25. Teeter, M. M. (1984). *Proc. Natl. Acad. Sci. USA 81*:6014–6018.
26. Finney, J. L. (1979). In Water, a Comprehensive Treatise (Franks, F., ed.), Vol. 6, Chap. 2, pp. 47–122, Plenum Press, New York.
27. Edsall, J. T., and McKenzie, H. A. (1979). *Advan. Biophys. 10*:137–207.
28. Edsall, J. T., and McKenzie, H. A. (1983). *Advan. Biophys. 16*:53–183.
29. Saenger, W. (1987). *Ann. Rev. Biophys. Biophys. Chem. 16*:93–114.
30. Savage, H. F. J. (1986). *Water Science Rev. 2*:67–148.
31. Baker, E. N., and Hubbard, R. E. (1984). *Prog. Biophys. Molec. Biol. 44*:97–179.
32. Thanki, N., Thornton, J. M., and Goodfellow, J. M. (1988). *J. Mol. Biol. 202*: 637–657.
33. Thanki, N., Umrania, Y., Thornton, J. M., and Goodfellow, J. M. (1991). *J. Mol. Biol. 221*:669–691.
34. Deisenhofer, J., and Steigemann, W. (1975). *Acta Crystallogr. B31*:238–250.
35. Takano, T., and Dickerson, R. E. (1981). *J. Mol. Biol. 153*:79–94.
36. Suguna, K., Bott, R. R., Padlan, E. A., Subramanian, E., Sheriff, S., Cohen, G. H., and Davies, D. R. (1987). *J. Mol. Biol. 196*:877–900.
37. Thanki, N., Thornton, J. M., and Goodfellow, J. M. (1990). *Protein Eng. 3*:495–508.
38. Olovsson, I., and Jönsson, P.-G. (1976). In The Hydrogen Bond, Vol. 2 (Schuster, P., Zundel, G., and Sandorfy, C., eds.), pp. 393–456, North-Holland, Amsterdam.
39. Artymiuk, P. J., Blake, C. C. F., Grace, D. E. P., Oatley, S. J., Phillips, D. C., and Sternberg, M. J. E. (1979). *Nature 280*:563–568.

40. Rees, D. C., Lewis, M., and Lipscomb, W. N. (1983). *J. Mol. Biol. 168*:367–387.
41. Klotz, I. M. (1959). *Science 128*:815–822.
42. Rashin, A. A., Iofin, M., and Honig, B. (1986). *Biochemistry 25*:3619–3625.
43. Gilliland, G. L., Winborne, E. L., Nachman, J., and Wlodawer, A. (1990). *Proteins 8*: 82–101.
44. Kossiakoff, A. A. (1982). *Nature 296*:713–721.
45. Otting, G., Liepinsh, E., and Wüthrich, K. (1991). *Science 254*:974–980.
46. Malin, R., Zielenkiewicz, P., and Saenger, W. (1991). *J. Biol. Chem. 266*:4848–4852.
47. Wlodawer, A., Walter, J., Huber, R., and Sjölin, L. (1984). *J. Mol. Biol. 180*:301–331.
48. Wlodawer, A., Nachman, J., Gilliland, G. L., Gallagher, W., and Woodward, C. (1987). *J. Mol. Biol. 198*:469–480.
49. Otting, G., and Wüthrich, K. (1989). *J. Amer. Chem. Soc. 111*:1871–1875.
50. Acharya, K. R., Stuart, D. I., Walker, N. P. C., Lewis, M., and Phillips, D. C. (1989). *J. Mol. Biol. 208*:99–127.
51. Betzel, Ch., Teplyakov, A. V., Harutyunyan, E. H., Saenger, W., and Wilson, K. S. (1990). *Protein Eng. 3*:161–172.
52. Bode, W., and Schwager, P. (1975). *J. Mol. Biol. 98*:693–717.
53. Sawyer, L., Shotton, D. M., Campbell, J. W., Wendell, P. L., Muirhead, H., Watson, H. C., Diamond, R., and Ladner, R. C. (1978). *J. Mol. Biol. 118*:137–208.
54. Klotz, I. M. (1970). *Arch. Biochem. Biophys. 138*:704–706.
55. Teplyakov, A. V., Kuranova, I. P., Harutyunyan, E. H., Vainshtein, B. K., Frömmel, C., Höhne, W. E., and Wilson, K. S. (1990). *J. Mol. Biol. 214*:261–279.
56. Blundell, T., Barlow, D., Borkakoti, N., and Thornton, J. M. (1983). *Nature 306*: 281–283.
57. Sundaralingam, M., and Sekhurudu, Y. C. (1989). *Science 244*:1333–1337.
58. Hol, W. G. J., van Duijnen, P. T., and Berendsen, H. J. C. (1978). *Nature 273*: 443–446.
59. Betzel, Ch., Pal, G. P., and Saenger, W. (1988). *Eur. J. Biochem. 178*:155–171.
60. Bolin, J. T., Filman, D. J., Matthews, D. A., Hamlin, R. C., and Kraut, J. (1982). *J. Biol. Chem. 257*:13650–13662.
61. Finzel, B. C., Clancy, L. L., Holland, D. R., Muchmore, S. W., Watenpaugh, K. D., and Einspahr, H. M. (1989). *J. Mol. Biol. 209*:779–791.
62. Carter, C. W., Jr., Kraut, J., Freer, S. T., Xuong, N. H., Alden, R. A., and Bartsch, R. G. (1974). *J. Biol. Chem. 249*:4212–4225.
63. Kodandapani, R., Suresh, C. G., and Vijayan, M. (1990). *J. Biol. Chem. 265*: 16126–16131.
64. Rose, G. D., Young, W. B., and Gierasch, L. M. (1983). *Nature 304*:654–657.
65. James, M. N. G., Sielecki, A. R., Brayer, G. D., Delbaere, L. T. J., and Bauer, C.-A. (1980). *J. Mol. Biol. 144*:43–88.
66. Herzberg, O. (1991). *J. Mol. Biol. 217*:701–719.
67. Sielecki, A. R., Hendrickson, W. A., Broughton, C. G., Delbaere, L. T. J., Brayer, G. D., and James, M. N. G. (1979). *J. Mol. Biol. 134*:781–804.
68. Tsukada, H., and Blow, D. M. (1985). *J. Mol. Biol. 184*:703–711.
69. James, M. N. G., and Sielecki, A. R. (1987). In Biological Macromolecules and Assemblies, Vol. 3 (Jurnak, F. A., and McPherson, A., eds.), pp. 413–482, John Wiley, New York.
70. Read, R. J., and James, M. N. G. (1988). *J. Mol. Biol. 200*:523–551.

71. Bax, A. (1989). *Annu. Rev. Biochem. 58*:223–256.
72. Wright, P. E. (1989). *Trends in Biochem. Sci. 14*:255–259.
73. Clore, G. M., Bax, A., Wingfield, P. T., and Gronenborn, A. (1990). *Biochemistry 29*: 5671–5676.
74. Knox, J. R., and Moews, P. C. (1991). *J. Mol. Biol. 220*:435–455.
75. Wlodawer, A., Deisenhofer, J., and Huber, R. (1987). *J. Mol. Biol. 193*:145–156.
76. Moult, J., Yonath, A., Traub, W., Smilansky, A., Podjarny, A., Rabinovich, D., and Saya, A. (1976). *J. Mol. Biol. 100*:179–195.
77. Bell, J. A., Wilson, K. P., Zhang, X.-J., Faber, H. R., Nicholson, H., and Matthews, B. W. (1991). *Proteins 10*:10–21.
78. Wlodawer, A., Borkakoti, N., Moss, D. S., and Howlin, B. (1986). *Acta Crystallogr. B42*:379–387.
79. Kauzmann, W. (1959). *Adv. Protein Chem. 14*:1–63.
80. Chothia, C. (1976). *J. Mol. Biol. 105*:1–12.
81. Eriksson, A. E., Baase, W. A., Zhang, X.-J., Heinz, D. W., Blaber, M., Baldwin, E. P., and Matthews, B. W. (1992). *Science 255*:178–183.
82. Kossiakoff, A. A. (1987). In Biological Macromolecules and Assemblies, Vol. 3 (Jurnak, F. A., and McPherson, A., eds.), pp. 369–412, John Wiley, New York.
83. Cohen, G. H., Silverton, E. W., and Davies, D. R. (1981). *J. Mol. Biol. 148*:449–479.
84. Alber, T., Dao-pin, S., Wilson, K., Wozniak, J. A., Cook, S. P., and Matthews, B. W. (1987). *Nature 330*:41–46.
85. Pai, E. F., Krengel, U., Petsko, G. A., Goody, R. S., Kabsch, W., and Wittinghofer, A. (1990). *EMBO J. 9*:2351–2359.
86. Pearl, L., and Blundell, T. (1984). *FEBS Lett. 174*:96–101.
87. Blundell, T., Jenkins, J., Pearl, L., Sewell, T., and Pederson, V. (1985). In *Aspartic Proteinases and Their Inhibitors* (Kosta, V., ed.), pp. 151–161, Walter de Gruyter, Berlin and New York.
88. Huber, R., and Bode, W. (1978). *Acc. Chem. Res. 11*:114–122.
89. Fujinaga, M., Sielecki, A. R., Read, R. J., Ardelt, W., Laskowski, M., Jr., and James, M. N. G. (1987). *J. Mol. Biol. 195*:397–418.
90. Quiocho, F. A., Wilson, D. K., and Vyas, N. K. (1989). *Nature 340*:405–407.
91. Bhat, T. N., Bentley, G. A., Fischmann, T. O., Boulot, G., and Poljak, R. J. (1990). *Nature 347*:483–485.
92. Cygler, M., Rose, D. R., and Bundle, D. R. (1991). *Science 253*:442–445.
93. Tainer, J. A., Getzoff, E. D., Richardson, J. S., and Richardson, D. C. (1983). *Nature 306*:284–290.
94. Eklund, H., and Bränden, C.-I. (1987). In Biological Macromolecules and Assemblies, Vol. 3 (Jurnak, F. A., and McPherson, A., eds.), pp. 74–142, John Wiley, New York.
95. Eriksson, A. E., Kylsten, P. M., Jones, T. A., and Liljas, A. (1989). *Proteins 4*: 283–293.
96 Bushnell, G. W., Louie, G. V., and Brayer, G. D. (1990). *J. Mol. Biol. 214*: 585–595.
97. Nar, H., Messerschmidt, A., Huber, R., van de Kamp, M., and Canters, G. W. (1991). *J. Mol. Biol. 218*:427–447.
98. Poulos, T. L., Finzel, B. C., and Howard, A. J. (1987). *J. Mol. Biol. 195*:687–700.
99. Pflugrath, J. W., and Quiocho, F. A. (1985). *Nature 314*:257–260.

100. Luecke, H., and Quiocho, F. A. (1990). *Nature 347*:402–406.
101. Anderson, B. F., Baker, H. M., Norris, G. E., Rice, D. W., and Baker, E. N. (1989). *J. Mol. Biol. 209*:711–734.
102. Phillips, S. E. V. (1980). *J. Mol. Biol. 142*: 531–554.
103. Borkakoti, N., Moss, D. S., and Palmer, R. A. (1982). *Acta Crystallogr. B38*: 2210–2217.
104. Kamphuis, I. G., Kalk, K. H., Swarte, M. B. A., and Drenth, J. (1984). *J. Mol. Biol. 179*:233–256.
105. Furey, W., Wang, B. C., Yoo, C. S., and Sax, M. (1983). *J. Mol. Biol. 167*:661–692.
106. Protein Data Bank, Department of Chemistry, Brookhaven National Laboratory, Upton, NY 11973, USA.

3

Protein Hydration and Glass Transition Behavior

Roger B. Gregory
Kent State University, Kent, Ohio

I. INTRODUCTION

It has long been recognized that protein–water interactions play an important role in the determination and maintenance of the three-dimensional structure of proteins. Quite apart from its fundamental importance and interest, knowledge of processes occurring on hydration or dehydration of proteins is also important in biotechnological applications of proteins, such as their use as catalysts in anhydrous organic solvents, the stabilization of protein preparations for pharmaceutical use, and in food preservation. It is not surprising that protein–water interactions have been the subject of intense study and have provided very significant advances in our understanding of the involvement of water in protein stability, dynamics, and function [1–3]. Such studies can be classified into those employing protein solutions, in which the necessary variation in water activity is achieved by the use of water–cosolvent mixtures, and those employing hydrated protein powders, films, or glasses. As will become apparent in this chapter, most protein properties of interest are expressed and can be studied in the absence of a bulk water phase. Such studies also serve to emphasize the types of water of greatest importance to the determination of protein properties, namely, internal and surface water. It was once believed that drying a protein always resulted in major disruption of the native conformation to give a collection of randomly unfolded polypeptide chains—an inactive polypeptide "dust" such that studies of dry or partially hydrated proteins were of little relevance to issues of structure–function relationships. This view was challenged by the extensive systematic work of Rupley, Careri, and their coworkers on lysozyme hydration in the 1980s, which suggested that little or no conformational rearrangement occurred on dehydration [3]. Recent Fourier-transform infrared spectroscopy (FTIR) studies by Prestrelski et al. [4], however, indicate that the true situation may be more complicated than either of these extremes, the extent of dehydration-induced conformational rearrangement being quite variable and protein dependent.

Studies of protein hydration using solid samples allow the water activity to be varied over a very wide range, from the dry state to formation of the solution state, which occurs at about 0.9–1.0 *h* (g water/g protein). Although the point at which hydrated protein powders collapse to form true solutions represent concentrations that are orders of magnitude larger than those normally employed in physical studies of proteins, hydration levels in the range from 0 to 1.0 *h* prove to be highly relevant to a discussion of the role of water in protein function. Indeed, one of the most remarkable conclusions from studies of protein hydration conducted in the last 10 to 15 years is that very little water is required for the onset of functionally important protein processes. As we shall see, enzymes begin to regain their catalytic activity at hydration levels as low as 0.12–0.2 *h*, and most important changes in the thermodynamic and dynamic properties of proteins occur over the hydration range from 0 to 0.4 *h*.

The intracellular environment in which proteins function bears very little resemblance to the dilute solutions normally employed in physical studies of enzymes. The amount of intracellular water available to solvate proteins is small, about enough for two or three layers of water per protein molecule, which corresponds approximately to the hydration level at which hydrated powders collapse to form solutions. Nuclear magnetic resonance (NMR) titration of lysozyme solutions suggests that bulk water properties are only seen at hydration levels in excess of 1.4 *h*, which represents about 60% water by weight, within the range of water contents observed for some biological tissues [5]. Thus, some tissues and organisms may possess little, if any, bulk water.

Rupley and Careri [3] have recently provided an excellent and extensive review of protein hydration. We will not repeat such a monumental task here, but instead describe the events that occur as the protein is brought from the dry state to the fully hydrated state with particular emphasis on the role of water as a "plasticizer" of the protein conformation and in the reestablishment of the dynamic properties of proteins. A consideration of the results derived from hydration studies of proteins and synthetic polymers, recent work on glass transitions, and the identification of dynamically distinct structural classes in globular proteins derived from hydrogen isotope exchange results leads us to a picture of the dynamic properties of proteins in which free volume rearrangement and mobile internal water are the primary determining factors. No direct, model-independent measure of the elusive population of internal water has been made, except by x-ray crystallography, which can define only the sites with the highest water occupancies. As our understanding of the role of water in protein dynamics has developed, however, we have become increasingly convinced that the size of the population of mobile buried water and its importance to protein dynamics has been greatly underestimated.

II. PREPARATION OF SOLID STATE SAMPLES

Most work on protein hydration has involved the use of protein powders that are brought to the required hydration level by isopiestic equilibration with a solution of a salt or sulfuric acid of known water activity [6]. Sample preparation is straightforward. The protein solution is usually dialyzed against water or buffer and then lyophilized. Further drying is obtained by storage in a vacuum desiccator over phosphorus pentoxide. The lowest hydration level we have obtained in this way for lysozyme is about 0.01 *h*, which represents about 8 moles water/mole protein. Removal of the last few water molecules is extremely difficult and may require drying under high vacuum (10^{-6} torr) for several days. The dry protein powder is transferred to a sealed vessel over a solution of known water activity and is allowed to equilibrate for a period of one to a few days, depending on the physical state of the powder and the final hydration level required. Uniform hydration can also be achieved by addition of water directly to the dry protein powder. The hydration level of the sample can be determined gravimetrically or by Karl Fischer titration, although problems of protein solubility in the Karl Fischer reagent, usually a solution in pyridine, and reaction of certain protein groups with the reagent can cause problems with the latter method [7,8]. Correction for the small amount of water associated with the lyophilized protein is required for the most precise work, but this is easily determined from the weight loss of a sample of the lyophilized protein after drying in an oven at 105°C for about 24 hours.

Hydrated powder samples are not suitable for transmittance optical spectroscopy because of scattering, and instead it is possible to employ hydrated films of protein that are cast from concentrated protein solutions spread onto an appropriate optically transparent support (i.e., CaF_2 for infrared measurements). The hydration level of the films may be controlled by drying in a sealed vessel over a solution of known water activity. It is also possible to prepare large, clear "glassy" pellets by drying concentrated protein solutions in a suitable container [8].

III. ADSORPTION OF WATER VAPOR BY PROTEINS: THE SORPTION ISOTHERM

Measurement of the uptake of water by protein at constant temperature as a function of water activity or relative vapor pressure provides the adsorption isotherm, although as Kuntz and Kauzmann [1] note, the term adsorption is not particularly appropriate for what is essentially a solvation or absorption process. The adsorption isotherm is conveniently measured using the isopiestic equilibration method described in Sec. II. The sigmoidal isotherms are of the BET type [9]. The adsorption isotherm for uptake of D_2O by HEW lysozyme at 27°C obtained by Careri et al. [10] is shown in Fig. 1. At low water activities, water uptake by the protein increases rapidly. There is a distinct "knee" at a relative humidity of between 0.05 and 0.2, corresponding to a water uptake of about 0.07 *h*, beyond which the uptake of water occurs more slowly with increases in relative humidity. At relative humidities greater than about 0.9, water uptake increases rapidly again. We should note that the uptake of water by proteins at a particular relative humidity is independent of the surface area of the protein powder, in contrast to adsorption isotherms for inert gases such as nitrogen or argon, which do depend on the degree of subdivision or fineness of the powder samples [11].

One of the most significant features of the water sorption isotherms is the appearance of pronounced hysteresis below relative humidities of 0.8 to 0.9, so

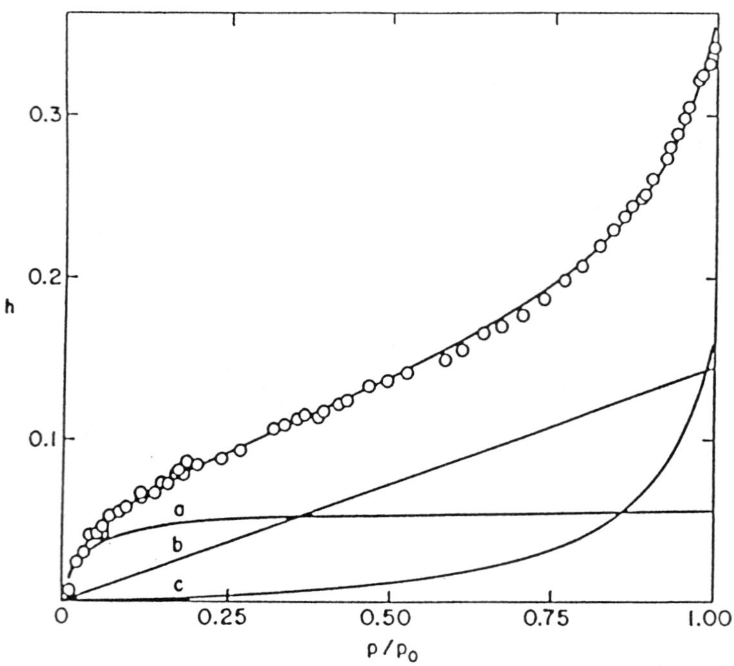

FIGURE 1 D_2O sorption isotherm for lysozyme at 27°C. The curves labeled a, b, and c represent the three classes of sorption sites defined by the D'Arcy and Watt equation (Eq. (3)). From Careri et al. [10]. Reprinted by permission of John Wiley & Sons, Inc.

that the adsorption curve lies 10–30% below the desorption curve. Adsorption isotherms for nitrogen and argon show no hysteresis [1]. Luscher-Mattli and Ruegg [12] have shown that water sorption hysteresis depends on the extent of prior dehydration of the sample. For example, adsorption–desorption cycles for α-chymotrypsin between relative humidities of 0 and 0.93 exhibit a pronounced hysteresis (Fig. 2). When samples are prepared by evaporation of a solution at a relative humidity of 0.92 (corresponding to a hydration level or water uptake of 0.3 *h*) and then cycled between relative humidities of 0.15 and 0.93, however, no hysteresis is observed within the limits of experimental error, which were reported as 0.002–0.005 g water/g protein (Fig. 2). The lowest hydration level attained was 0.065 *h*. Evidently, hysteresis is a consequence of removal of the most tightly bound water molecules.

A. Conventional Sorption Isotherms

A variety of theoretical models have been employed to describe sorption isotherms and the hysteresis effects in proteins. Much of the early work has been critically reviewed by Kuntz and Kauzmann [1], who also discuss the virtues of solution theories, such as those due to Flory [13], Hailwood and Horrobin [14,15], and D'Arcy and Watt [16], versus surface models such as Brunauer, Emmett, and Teller (BET) theory [9]. A more complex but far more powerful analysis has been developed by Cerofolini and Cerofolini [17], which accounts for the heterogeneity of water-binding sites through the use of an adsorption energy distribution. The theory is also capable of describing hysteresis phenomena.

Although protein water sorption isotherms resemble the sigmoidal BET isotherm, they are found to obey the BET equation only up to a relative humidity of about 0.6. The BET equation is given by

$$\frac{W}{W_{\mathrm{m}}} = \frac{Kx}{1 + Kx} + \frac{x}{1 - x}, \qquad x = \frac{p}{p_0} \tag{1}$$

where x is the relative vapor pressure, p is the experimental water vapor pressure, and p_0 is the water vapor pressure of pure water at the same temperature and pressure employed in the experiment; W is the hydration level or uptake of water by the protein, and K and W_{m} are constants proportional to the energy of adsorption and the number of binding sites, respectively. The BET model assumes a fixed number of independent binding sites that accommodate a monolayer of adsorbate molecules (the Langmuir term on the right-hand side of Eq. (1)) and allows subsequent layers to bind more weakly.

Other isotherms derived from surface models are equally limited in their description of water adsorption to proteins. More success in fitting the entire range of adsorption data, however, is obtained with solution models such as the Hailwood and Horrobin [14,15] equation,

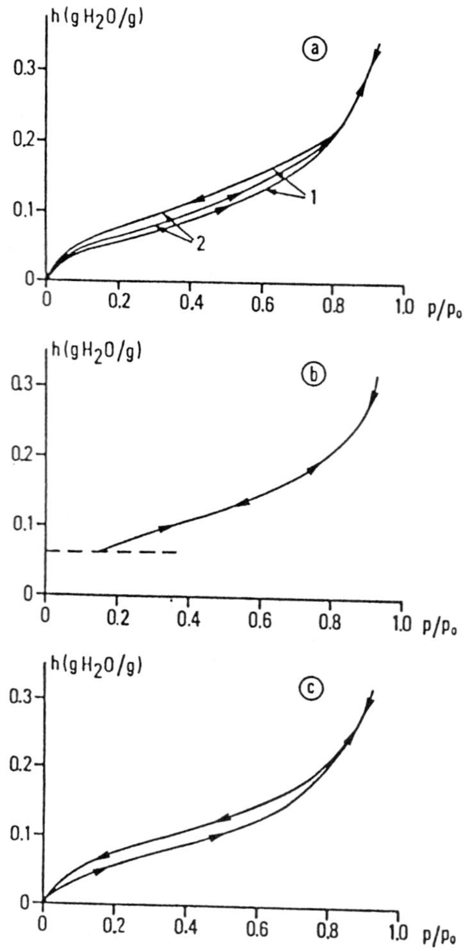

FIGURE 2 Hysteresis in sorption isotherms for α-chymotrypsin. (a) Two adsorption–desorption cycles on lyophilized α-chymotrypsin. Curves 1 represent the first cycle, curves 2 the second cycle. The relative humidities of each cycle went from 0 to 0.93 and back to 0. (b) Desorption–adsorption cycle. Relative humidities went from 0.92 to 0.15 and back to 0.92. (c) Second cycle on the sample shown in (b) over the relative humidity range from 0.92 to 0 and back to 0.92. From Luscher-Mattli and Ruegg [12]. Reprinted by permission of John Wiley & Sons, Inc.

$$\frac{W}{W_m} = \frac{Kx}{1 + Kx} + \frac{\gamma x}{1 - \gamma x} \qquad (2)$$

which assumes binding to a set of independent sites (the Langmuir term again) as well as formation of an ideal solid solution, or with the D'Arcy and Watt [16] equation,

$$W = \frac{W_m Kx}{1 + Kx} + Cx + \frac{D\gamma x}{1 - \gamma x} \qquad (3)$$

which adds a class of weak binding sites with affinity proportional to C to the Hailwood and Horrobin model. Equations (1), (2), and (3) represent two-, three-, and five-parameter models, respectively, and therefore it is not surprising that they provide successively better fits to the data. Luscher-Mattli and Ruegg [12] have employed these models to fit adsorption isotherms for a number of proteins. Kuntz and Kauzmann [1] suggest that the addition of the linear term in Eq. (3) does little to improve fits to adsorption data and is of doubtful physical significance for proteins because it corresponds to such weak binding that saturation of these sites would require an unreasonably large number of water molecules per monolayer site. Careri et al. [10] employed the D'Arcy–Watt equation in the analysis of D_2O sorption isotherms, however, and obtained coverages for the strong and weak binding sites that were in agreement with independent infrared measurements and gave values of about 100 water molecules per protein molecule for the coverage of the weak binding sites. This is not an unreasonably large number and was assigned to coverage of carbonyl groups along the polypeptide backbone. The fit of the D'Arcy and Watt equation to the sorption isotherm for lysozyme is shown in Fig. 1. Typical values of W_m for proteins obtained with Eqs. (1) and (2) vary between 0.05 and 0.1 g water/g protein. The corresponding values for K vary between 10 and 20. The parameter γ in Eq. (2) has a value of 0.8–0.9 and is related to the activity of water in the solid solution. Despite their different physical derivation and interpretation, the BET and Hailwood and Horrobin equations are very similar. In fact, the only difference is the appearance of the factor γ in the Hailwood and Horrobin equation, which thus effectively serves to adjust the degree of curvature and the size of the BET multilayer term.

Kuntz and Kauzmann [1] cite a number of reasons for preferring the solution models, among them that no phase transition is observed in the adsorption isotherm over the entire composition range, despite the change in sample appearance from solid to liquid (the lysozyme–water system forms a solution at a hydration level of about 1.0 *h*). Among the solution models, the Hailwood and Horrobin equation is the simplest that fits the data well over a broad range of relative humidities. The characterization of the "monolayer" sites by a single adsorption energy, however, is not very realistic given the variety of polar groups found at the protein surface. Nor is it reasonable to expect the protein conformation to be unaffected on binding water. Adsorption energies are significant with respect to the interaction energies among protein groups so, as Cerofolini and Cerofolini [17] note, "A kind of surface reconstruction takes place after adsorption." A theory for the adsorption of water by proteins that accounts both for the heterogeneity of the water-binding sites and the possibility that adsorption can alter the protein conformation has been developed by Cerofolini and Cerofolini [17] and is described below.

B. Site Heterogeneity and Conformational Perturbations

The heterogeneity of the protein surface is due to two distinct factors. First, the water-binding sites are chemically diverse and include a variety of charged groups and uncharged polar groups. Second, the protein samples a variety of conformational states, which gives rise to a distribution of local environments for each water-binding site. This heterogeneity is measured by the distribution of adsorption energies, q, through the probability density function (pdf) $\phi(q)$, so that $\phi(q)dq$ is the fraction of sites with adsorption energy between q and $q + dq$. The probability that a water-binding site with adsorption energy q is exposed is denoted by $P(q \mid W)$. This function accounts for the influence of hydration on the conformational state of the protein. The hydration level at a vapor pressure p is given by

$$W(p) = \int_0^\infty \theta(p, q)P(q \mid W)\phi(q)dq \tag{4}$$

where $\theta(p,q)$ is the local adsorption isotherm describing the uptake of water at sites with adsorption energy q at a vapor pressure p. Cerofolini and Cerofolini [17] employ a BET adsorption isotherm,

$$\theta_{BET}(p, q) = \frac{C(q)x}{\{[C(q) - 1]x + 1\}(1 - x)} \tag{5}$$

where $x = p/p_0$ as before and $C(q)$ is given by

$$C(q) = C_0 \exp\left(\frac{q - q_0}{kT}\right) \tag{6}$$

q_0 is the heat of vaporization of water. To simplify subsequent calculations, Cerofolini and Cerofolini [17] assume $C(q) >> 1$ and employ the following approximation for $\theta(p, q)$:

$$\theta(p, q) = \frac{p}{(1 - p/p_0)\{p + p_L \exp(-q/kT)\}} \tag{7}$$

where $p_L = C_0^{-1} p_0 \exp(q_0/kT)$. The expression for water uptake becomes

$$W(p) = \left(1 - \frac{p}{p_0}\right)^{-1} \int_0^{\infty} p\{p + p_L \exp(-q/kT)\}^{-1} P(q \mid W)\phi(q)\, dq \qquad (8)$$

Note that $p/\{p + p_L \exp(-q/kT)\}$ is simply the Langmuir isotherm $\theta_{LANG}(p,q)$.

It is necessary to specify the function $P(q \mid W)$. Cerofolini and Cerofolini [17] develop a mean field expression for $P(q \mid W)$. Sites with adsorption energy q are considered to be either exposed or masked. The probability that a site with adsorption energy q is exposed in the dry protein is assumed to be

$$P(q \mid \text{dry}) = \left[1 + \exp\left(\frac{q - q^*}{\sigma}\right)\right]^{-1} \qquad (9)$$

where q^* is the adsorption energy of a reference site with $P = 0.5$ and σ is a suitable parameter. Masking is suggested to occur via long-range electrostatic interactions between sites that are modified by the presence of water. Sites with greater adsorption energy (i.e., more polar) are therefore assumed to have a greater probability of being masked in the dry protein, while hydration increases exposure of water-binding sites. All sites are assumed to interact with a mean field generated by the presence of water, so that the energy difference between masked and exposed sites is a function of the hydration level. The probability that a water-binding site is exposed is given by

$$P(q \mid W) = \left[1 + \exp\left\{\frac{q - q^* - \alpha W(p)}{\sigma}\right\}\right]^{-1} \qquad (10)$$

where α is a constant. The presence of $W(p)$ on the right-hand side of Eq. (10) greatly complicates calculations with Eq. (8). The following simple approximation, valid when σ is very small, is therefore employed:

$$P(q \mid W) = \begin{cases} 1 & \text{for } q < q^* + \alpha W(p) \\ 0 & \text{for } q > q^* + \alpha W(p) \end{cases} \qquad (11)$$

for which the following expression for water uptake is obtained:

$$W(p) = \left(1 - \frac{p}{p_0}\right)^{-1} \int_0^{q^* + \alpha W(p)} p\left\{p + p_L \exp\left(\frac{-q}{kT}\right)\right\}^{-1} \phi(q)\, dq \qquad (12)$$

The sorption isotherm can be calculated for any given adsorption energy pdf. For example, if we suppose that the protein has a uniform distribution of adsorption energies, defined over some range $q_m < q < q_M$,

$$\phi(q) = \begin{cases} 1/\omega & \text{for } q_m < q < q_M \\ 0 & \text{otherwise} \end{cases} \qquad (13)$$

where $\omega = q_M - q_m$, then integration of Eq. (12) gives

$$W(p) = \frac{kT}{(1 - p/p_0)\omega} \ln \frac{1 + (p/p_m)\exp\{[q^* + W(p)]/kT\}}{1 + p/p_m} \tag{14}$$

An example of this sorption isotherm, with $\omega = 4kT$, $p_m/p_0 = 0.1$, $q^* = q_m + 0.5\omega$, and $\alpha = 0.25\omega$ is shown in Fig. 3.

Equation (8) is a far more realistic description of sorption isotherms for proteins than the conventional isotherms. The difficulty, of course, is to define the functions $P(q \mid W)$ and $\phi(q)$. What is of more interest is the possibility of determining these functions from experimental sorption isotherms by numerical solution of Eq. (8). Equation (8) is a Fredholm integral equation of the first kind, the solution of which is very unstable to noise in the data. Very powerful regularization methods, however, are now available for the numerical solution of Fredholm integral equations that address the problems of stability to noise and solution uniqueness. Probably the best known of these is the regularized, least squares algorithm called CONTIN, developed by Provencher [18,19], but maximum entropy methods [20,21] offer an alternative approach. Given an experimental estimate of $W(p)$ and the reasonable assumption of a Langmuir or BET model for the local sorption isotherm, these methods should yield $P(q \mid W)\phi(q)$ directly. Hysteresis phenomena complicate the analysis, but may also be amenable to analysis with integral transform methods. We outline such an analysis in the next section.

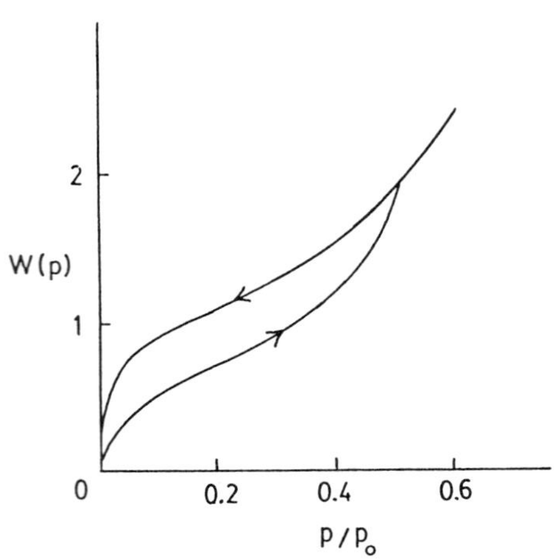

FIGURE 3 A numerical example of the adsorption and desorption isotherms given by Eq. (14). From Cerofolini and Cerofolini [17].

C. Sorption Hysteresis

We have referred, so far, to the sorption isotherm $W(p)$ without distinguishing adsorption isotherms ($W_a(p)$) from desorption isotherms ($W_d(p)$); that is, we have assumed complete reversibility. As noted earlier, however, water sorption isotherms for proteins show pronounced hysteresis. In their development, Cerofolini and Cerofolini [17] suppose that $W(p)$ can be measured under equilibrium conditions starting from the completely dry protein; that is, they employ the adsorption isotherm. This is, however, probably not possible for most proteins. For α-chymotrypsin, only sorption cycles starting at high relative humidities (p/p_0), which do not go below a relative humidity of about 0.15 (corresponding to a hydration level of about 0.07 *h*), represent equilibrium conditions. Interestingly, Luscher-Mattli and Ruegg [12] found the desorption branches of all scanning cycles to be identical. In other words, it is the desorption branch rather than the adsorption branch of sorption cycles displaying hysteresis that overlap the reversible sorption cycle. Desorption isotherms may thus represent equilibrium conditions regardless of the prior treatment of the sample, at least for α-chymotrypsin and tropocollagen.

A number of explanations of hysteresis have been suggested, including capillary condensation [22] and, most recently, the possibility that nucleation sites for condensation of water on the surface is limiting during adsorption [3,23]. There is much evidence, however, to suggest that conformational changes occur in the region of low hydration, and this strongly supports the notion that sorption hysteresis is a molecular phenomenon related to conformational change in the protein molecule [12,17,24]. As will be discussed later, the critical hydration level reported by Luscher-Mattli and Ruegg [12] corresponds to the hydration level where one begins to observe the development of internal protein mobility, suggesting that conformational flexibility and rates of conformational change are intimately linked to the hysteresis phenomena.

Bryan [24] has described several thermodynamic models of sorption hysteresis involving slow confrontational changes or intermolecular rearrangements, some of which are now given.

Model I: Slow Conformational Change Hysteresis. All protein molecules are assumed to have the same conformation at low relative humidities, but undergo a slow conformational change at some intermediate relative humidity. The conformational change is sufficiently slow that the new equilibrium position is not reached in the time required for the sorption experiment. The adsorption isotherm should be reversible below the vapor pressure range over which the conformational change takes place, while the desorption isotherm should be reversible above this range.

Model II: Conformational Hysteresis. The protein conformation changes continuously with addition or removal of water. At any particular vapor pressure, the protein is capable of adopting a number of conformations characterized by different hydration levels, $W_i(p)$ over the range $W_a(p) < W_i(p) < W_d(p)$. Each conformation corresponds to a local free energy minimum and is stable for an extended period of time. The sorption isotherm becomes reversible only when enough water has been added (p_r) to the protein to increase the rates of conformational changes sufficiently for the protein to attain a global minimum energy conformation.

Model III: Phase-Annealing Hysteresis. Relatively rapid intramolecular conformational changes can occur, but intermolecular repositioning of protein molecules relative to one another to attain the most favorable intermolecular arrangement (free energy minimum) is limited at low hydrations. Phase annealing becomes fast, and the isotherms become reversible only when enough water (p_r) is present to assist in repositioning.

Bryan [24] has described several experimental tests of these models and others involving reversibility studies around the whole hysteresis loop and long-term weight changes and vapor pressure reversal studies over part of the hysteresis loop. For example, models I and III predict long-term changes in hydration levels as the water–protein system slowly moves to its global free energy minimum. Unfortunately, no studies of long-term weight changes have been performed, and the only detailed reversibility study we are aware of is that of Luscher-Mattli and Ruegg [12].

The theory of Cerofolini and Cerofolini [17] accounts for hysteresis through the function $P(q \mid W)$, which, it will be recalled, is the probability that a site with adsorption energy q is exposed. This probability depends on the conformational state of the protein, which in turn depends on the hydration level and the previous history of the sample. Cerofolini and Cerofolini generate sorption cycles resembling those observed experimentally by adopting different functions for $P(q \mid W)$ in expressions for the adsorption and desorption branches; that is, $P(q \mid W_d)$ is assumed to equal to one (all sites are equally exposed during desorption) but $P(q \mid W_a)$ is given by Eq. (11), although it is possible that some buried water molecules may be trapped by dehydration-induced conformational changes that prevent access of these buried water molecules to the surface.

If $P(q \mid W_d) = 1$ is assumed for the desorption isotherm in Eq. (8), numerical solution of the integral in this equation using $W_d(p)$ gives $\phi(q)$ directly. Comparison with the solution $P(q \mid W_a)\phi(q)$ obtained from a similar analysis of $W_a(p)$ would then give an estimate of $P(q \mid W_a)$. This function gives the probability that a site is exposed during adsorption (relative to its exposure during desorption) and thus should indicate which sites, as indexed by their adsorption energy, are influenced by hydration of the protein.

The treatment of sorption isotherms in terms of site exposure probabilities emphasizes hydration-induced protein conformational change as the source of hysteresis. Preliminary investigations of the hysteresis loop in lysozyme do appear to support the conformational hysteresis model of Bryan [24], which is consistent with the theory of Cerofolini and Cerofolini [17] to the extent that both models propose continuous changes in protein conformation with changes in hydration. Changes in hydration, however, also influence the dynamic properties of proteins, and this would appear to be the key to understanding the irreversibility of the sorption isotherms. In solution, proteins fluctuate among a number of conformational states ("mobile defect" conformers) of approximately the same free energy, through redistribution of free volume and rearrangements in hydrogen bonding and other noncovalent interactions [25]. As water is removed, the flexibility of the protein conformation is reduced. Below a hydration level of about 0.1 h, the protein conformation becomes rigid [3]. Recent solid state NMR studies (see Sec. VI.A) indicate that the distribution of conformational states sampled by the protein is a function of the hydration level, the distribution of conformations being much broader for the dry protein than for the fully hydrated protein [26,27]. Fluctuations between these states, however, are greatly reduced or do not occur at low hydrations. Water acts as a "plasticizer" of the protein conformation, allowing conformational rearrangements to occur. In addition, water binding serves to order the protein conformation, enhancing the probabilities of certain conformational states and suppressing others, so that the width of the distribution of conformational states is greatly reduced.

As the protein is dehydrated and its conformational flexibility decreases, a protein molecule can become "trapped" in a local free energy minimum corresponding to some conformational state it happened to be sampling at the time the water molecules were removed. Activation barriers between this conformational state and neighboring states increase when certain binding sites lose water, and thus conformational transitions from this state to others are slower or not possible in the absence of water. As dehydration continues, one can imagine protein molecules becoming "trapped" in successive local free energy minima corresponding to different conformational states. The broad distribution of isotropic chemical shifts observed by solid state NMR for the dry protein would be one consequence of this accumulation of different conformational states on dehydration of the protein. In this way, a number of different conformational states, each characterized by a different hydration level, can be found at a given vapor pressure, which is a requirement of the conformational hysteresis model of Bryan [24]. At low hydrations, these conformations are not present in the equilibrium proportions characteristic of the global free energy minimum, and thus could give rise to the hysteresis observed in the adsorption isotherm.

1. Buried Water as a Source of Sorption Hysteresis

It should be noted that the protein conformation is coupled to the pattern of occupied water-binding sites, so that changes in one must modify the other. For example, tosylation is known to induce conformational changes in α-chymotrypsin that have a significant effect on the hydration of the protein [28]. Conformations sampled during adsorption that differ from those characteristic of the global free energy minimum sampled during desorption will have different patterns of occupied water-binding sites; that is, the site exposure probabilities will be different during adsorption and desorption. Evidently, the sites responsible for the hysteresis behavior are those with the highest adsorption energies (the most tightly bound water molecules). These include ionizable side chains but must also include buried water sites, since it is known from hydrogen exchange studies that internal motions recover at very low hydrations, and these motions require water to enter the protein in order to plasticize the conformation. Which type of site (ionizable side chains or buried water) is responsible for sorption hysteresis? The occupancy of buried water sites is likely to be very sensitive to changes in protein conformation, and thus one would expect these sites to be the most relevant to any conformational model of hysteresis. At a given relative humidity, buried water sites that are occupied in the desorption isotherm are not occupied in the adsorption isotherm, because the conformational rearrangements necessary for access to these sites cannot occur or at least do not occur on the time scale of the sorption measurements. Some water must enter the protein interior during the early phase of adsorption because exchange of internal peptide protons can be observed at about 0.05–0.07 h in samples hydrated by isopiestic equilibration from the lyophilized protein [29]. Exchange rates reach dilute solution values at about 0.15 h in lysozyme [29], suggesting that the full population of internal water is not achieved until hydration levels reach this value. If the source of sorption hysteresis is due to the failure to recover the full population of buried water molecules as the dry protein is hydrated up to 0.15 h, some estimate of the difference in this population for the adsorption and desorption branches of the sorption cycle can be made from the difference in the hydration levels of the two sorption branches at a given relative humidity. The data for α-chymotrypsin [12] suggest a difference of about 0.02 g water/g protein, which corresponds to about 20–25 moles of water per mole of protein. The full population of internal water must be larger than this to account for the water that enters the protein during the early phase of adsorption to begin to plasticize the protein. This appears to be a very large number of water molecules. Twenty to thirty water molecules would represent about 4–6% of the water required to fully hydrate chymotrypsin. Birktoft and Blow [30] identified 13 buried water molecules in the original crystal structure, but an analysis of this structure by Nedev et al. [31] suggested an internal capacity, based on the volume of cavities accessible to the solvent, for about 50 water molecules. Sreenivasan and Axelsen [32] identified 21 conserved, buried water sites among the serine proteases and proteins with trypsinlike substrate specificity. Sixteen of these sites were common to all the serine protease crystal structures analyzed. Not all these sites were found to be occupied by water in a particular structure. Only those water sites with high occupancies, however, can be reliable identified by x-ray crystallography. In rat trypsin, 28 buried sites were found. The pattern of water occupancy is highly dynamic, and occupancies of individual sites may be quite variable. They are coupled to the rearrangements in protein-free volume and conformation. In lysozyme, all peptide NH protons can exchange with bulk solvent from the native state via a mechanism of catalyst penetration if given sufficient time. Thus, water molecules and exchange catalysts (H^+ and OH^-) must be capable of reaching all peptide groups in the protein, albeit only very rarely in the case of the exchange-stable regions. Before discussing the plasticizing action of buried water and its effect on protein dynamics in any further detail, we turn to a description of the sorption sites and the extent of conformational change observed during hydration.

IV. IDENTIFICATION AND COVERAGE OF SORPTION SITES AND SOME CRITICAL HYDRATION LEVELS IN THE SORPTION ISOTHERM

There has been some debate as to the sites of water sorption and the extent of coverage, particularly with regard to the participation of sites along the polypeptide backbone. Even the amount of "bound" water has been difficult to define unambiguously. The physical properties of water interacting with proteins have been studied with a variety of techniques, including dielectric relaxation methods (see Chapter 4), infrared and NMR spectroscopy, Rayleigh scattering of Mossbauer radiation (RSMR) (see Chapter 5), hydrodynamic methods as well as thermodynamic measurements (see reviews by Kuntz and Kauzmann [1], Edsall and McKenzie [2], Rupley and Careri [3], Luscher-Mattli [33], and the volume of "Methods in Enzymology" edited by Hirs and Timasheff [34]). Most methods provide some estimate of the amount of water that interacts with the protein and that has structural or dynamic properties, as measured by the particular technique, that are different from those of bulk water. This water is termed either hydration water or "bound" water. Attempts by Finney and Poole [35] to rationalize the literature on this subject led them to the following conclusion: "The amount of bound water is a function of the method used to examine it and probably also of the form of the sample used."

In this section, we examine the picture of sequential hydration developed by Rupley and Careri [3]. Although the general picture that has emerged is reasonably consistent and many correlations between the results from different techniques are found, several issues remain to be fully resolved. Recent infrared results [4] cast some doubt on the interpretation of changes in the amide I band on hydration and the identification of some of the sorption sites. Finney and Poole [35] also interpret the heat capacity data somewhat differently from Rupley, which leads to a different estimate of the water necessary to cover apolar regions of the protein surface.

There are several critical hydration levels in the adsorption isotherm that mark the onset of important processes. Recovery of surface and internal mobility is obviously of particular importance, and we discuss this later in a separate section. Here, we group together three important and interesting hydration-dependent phenomena—recovery of enzymatic activity, proton percolation, and "nonfreezing water."

A. Infrared Spectroscopic Studies of Protein Hydration

Because of the sensitivity of vibrational modes to hydrogen bonding, Fourier-transform infrared spectroscopy can, in suitable cases, provide important information on the hydration process and has been used to identify the sites of water sorption. The infrared spectrum (i.e., the amide I band) is also sensitive to changes in protein conformation, however, and it can be difficult, in the absence of other evidence, to distinguish changes in protein conformation from perturbations in the environment and vibrational properties caused by removal of water. Careri et al. [10] have conducted a detailed study of the infrared spectrum of lysozyme films as a function of hydration. The spectral changes they observed were attributed to removal of water without conformational change, an assignment that may not be entirely correct in light of recent infrared [4] and NMR [26,27] spectroscopic studies (see Sec. IV). Films were cast on IR transparent windows from aqueous solutions of the protein and were exchanged with D_2O vapor in a dry box. The films were dried in a stream of dry nitrogen. Subsequent hydration of the protein films was achieved by passage of a stream of dry nitrogen and D_2O vapor over the sample. Hydration levels were monitored gravimetrically and from the integrated intensity of the 1210 cm^{-1} band, which is assigned to the OD bending mode. Difference spectra determined as a function of hydration for the amide I band at 1660 cm^{-1} and the carboxylate band at 1580 cm^{-1} established that the first D_2O molecules added to the dry protein hydrated carboxylic acid groups producing carboxylate groups, which accommodated about 40 D_2O molecules, consistent with the number of strongly bound water molecules determined from parameters of the D'Arcy–Watt isotherm. Weaker binding sites, primarily main-chain carbonyl groups, accounted for a further 100 water molecules. These results correlate well with regions identified in the sorption isotherm by heat capacity measurements [36] (see below), namely, that coverage of charged groups in lysozyme is complete at a hydration level of 0.07 h and coverage of polar groups is complete at 0.25 h. Slightly different estimates of these values have been given by Finney and Poole [35] based on their infrared studies with wedge-shaped protein films. Hydration of carboxylates appears to be complete at 0.1–0.12 h, while NH hydration is completed at about 0.26 h and carbonyl group hydration is complete at about 0.32 h.

As noted earlier, Careri et al. [10] assumed that no water-induced conformational changes occurred in their infrared studies of lysozyme hydration and attributed changes in the amide I band entirely to the removal of water. There is now evidence, however, from solid state ^{13}C NMR spectroscopic studies of lysozyme for a dehydration-induced increase in the distribution of protein conformations (static disorder) [26,27]. In addition, recent infrared spectroscopic studies [4] suggest that conformational change rather than removal of water is responsible for much of the spectral perturbations in the amide I band observed on dehydration. A number of observations suggest that such conformational changes that do occur when a protein is dehydrated are confined to the hydration range below about 0.2 *h* (see Sec. V). The largest changes in the amide I difference spectrum are associated with this hydration range. Do these changes reflect the addition of water to carbonyl groups and thus allow identification of the region in the sorption isotherm where coverage of carbonyl groups occurs, or do they reflect conformational changes, not necessarily related to the coverage of carbonyl groups? The preservation of efficient dipolar coupling in solid state ^{13}C cross-polarization magic-angle spinning (CP/MAS) NMR spectra on hydration of lysozyme suggests that hydration-induced conformational changes are relatively small in this protein [26,27]. The infrared studies described in Sec. VI, however, indicate that the extent of dehydration-induced conformational changes are highly protein dependent.

B. Heat Capacity as a Function of Hydration

The heat capacity of lysozyme has been measured as a function of hydration by Yang and Rupley [36] and by Finney and Poole [35]. Their work provides a wealth of information and defines a number of distinct regions in the sorption isotherms. The apparent specific heat capacity of lysozyme (component 2) is shown as a function of hydration in Fig. 4 and is defined by

$$\phi C_{p,2} = \frac{C_p - w_1 C_{p,1}{}^0}{w_2} \tag{15}$$

where C_p is the experimental heat capacity of the protein–water system, $C_{p,1}{}^0$ is the partial specific heat capacity of pure liquid water (component 1), *h* is the hydration of the protein in g water/g protein, and $w_1 = h/(1 + h)$ and $w_2 = 1/(1 + h)$ are the mass fractions of water and protein, respectively. Yang and Rupley [36] identified four regions that they assigned to the coverage of chemically different classes of sorption site. Region I (dilute solution to 0.38 *h*) corresponds to the addition of water to the fully hydrated protein where $\phi C_{p,2}$ is constant and equals the dilute solution value; region II (0.38–0.27 *h*) represents the condensation of water over weakly interacting nonpolar surface regions; region III (0.27–0.07 *h*) corresponds to the addition of water to main-chain carbonyls and other polar surface groups; and region IV corresponds to hydration of ionizable groups. The transition between region III and IV corresponds well with the plateau observed in the absorbance band for carboxylates (1580 cm^{-1}) and with the knee in the ad-

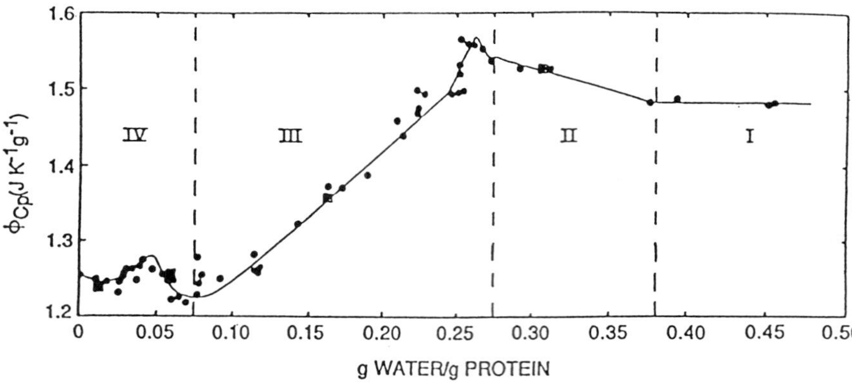

FIGURE 4 The apparent specific heat capacity of lysozyme as a function of hydration. The regions labeled I, II, III, and IV are described in the text. Reprinted with permission from Yang and Rupley [36]. Copyright (1979) American Chemical Society.

sorption isotherm. The transition between region II and III corresponds with the completion of changes in the amide I band on hydration and with the upswing in the adsorption isotherm. The peak at 0.05 *h* is attributed to a relaxation contribution to the heat capacity due to proton transfers from carboxyl groups to basic groups that accompanies the normalization of pK_a values on hydration. The peak at the transition from region III to region IV (at 0.07 *h*) and the difference in the partial specific heat capacity of the hydration water between these regions have been suggested to reflect reorganization of dispersed water molecules to form hydrogen-bonded clusters, while the peak near the transition from region II to region III (centered at 0.26 *h*) has been attributed to an order–disorder transition. These transitions at low and high coverage of water are similar to those associated with localized adsorption on heterogeneous surfaces described by Hill [37,38].

The heat capacity data of Poole shows a similar general behavior to that of Yang and Rupley, but differs in fine detail in some regions. The peak in region IV of the Yang and Rupley data appears broader in Poole's data, and its rise and fall is attributed to overlapping contributions from polar group hydration, expected to give an overall increase in heat capacity, and charged group hydration and ionization, expected to give an overall decrease in heat capacity, rather than to a reaction heat (relaxation contribution) associated with proton redistribution. There is also no evidence for the phase transitions at 0.07 and 0.26 *h* in Poole's data. There is some difference in the interpretation of events beyond 0.28 *h*. Region II is suggested to correspond to the coverage of nonpolar surface groups (i.e., groups that do not form hydrogen bonds to water) to complete the monolayer coverage of the protein surface by water [36,39]. Finney and Poole [35], however, note that such a process should be expected to give an increase in specific heat capacity rather than the decrease observed and instead suggest that the decrease in the apparent specific heat capacity of the protein is due to the formation of water bridges over apolar regions that reduce the motions of the anchoring polar groups.

Analysis of the heat capacity data of Yang and Rupley [36] suggests that hydration of the protein to give monolayer coverage is complete at 0.38 *h*, the hydration at which the apparent specific heat capacity of the protein becomes constant. Based on the calculated, exposed nonpolar group surface area, however, Finney and Poole [35] estimate the amount of water necessary for nonpolar group coverage to be 0.19 to 0.23 g water/g protein, which taken together with their estimate of polar group monolayer coverage gives an estimate of complete hydration at 0.45–0.5 *h*. Analysis of the peak area at the OD stretching frequency is in agreement with this estimate. A blue shift in the UV absorption spectrum of lysozyme, consistent with coverage of exposed tryptophan residues, is also observed at 0.45 *h* [35].

Calculations of the heat capacity for lysozyme in the dry and solution state from heat capacity data for oligomers of glycine and for amino acids in dilute solution and the crystalline state assuming group additivity are in good agreement with the experimental data [39]. This has been taken as evidence there are no special contributions to the heat capacity from soft vibrational modes peculiar to the protein. Such a conclusion would also suggest there is no hydration-induced increase in the population of soft internal modes that would contribute to the increase in heat capacity on hydration. This is a very surprising result given the sensitivity of the heat capacity to the addition of soft low-frequency modes [40] and the evidence for the role of water as a plasticizer of the protein conformation. The increase in conformational flexibility induced by addition of water beyond a critical hydration level would be expected to introduce many new low-frequency modes. In addition, there is evidence for dehydration-induced conformational changes from both infrared and solid state ^{13}C NMR data, which also might contribute to the heat capacity increase on hydration [4,26,27]. Calculations based on group additivity approaches can be subject to considerable uncertainty, at least for the folded states of proteins, and it is possible that the agreement between experimental and calculated heat capacities is simply fortuitous. We are inclined to take this view. As we discuss in Secs. VI and VII, there is now a great deal of evidence to support the view that water acts as a plasticizer of the protein conformation. At a hydration of 0.1–0.15 *h* at 298 K, the protein undergoes a glasslike dynamical transition from a rigid to a flexible state. The increase in conformational flexibility and the expansion of free volume that accompanies the transition would be expected to lead to a significant increase in the heat capacity of the protein. In the thermodynamic limit, glass transitions can be approximated as second-order transitions, characterized by a discontinuous increase in heat capacity at the transition temperature. The increase in heat capacity occurs over a broad hydration range (0.07–0.27 *h*), and so either other "chemical" terms contribute to the heat capacity change or the glass transition region must be quite broad, which simply reflects the heterogeneity of packing in the protein interior.

C. Enzyme Activity

Rupley et al. [39] have measured the catalytic activity of lysozyme in partially hydrated samples. A solution of lysozyme was mixed with a solution containing equimolar amounts of the hexasaccharide of *N*-acetylglucosamine. In order to reduce catalytic rates during sample preparation and subsequent hydration, the pH of the solution was increased. Three pH values (pH 8, 9, and 10) were examined, which gave reaction half-times of several days. The solutions were lyophilized to give a dry powder of the enzyme–substrate complex, which was then hydrated to the required hydration levels either by isopiestic equilibration or by the addition of water, hydration being completed within a day. Figure 5 shows the enzymatic activity as a function of hydration level. Activity is first observed at a hydration of about 0.20 *h* and increases quite rapidly to the solution value with hydration levels above 0.5 *h*. The rate of reaction at hydration levels where catalytic activity is first observed has a fifteenth-order dependence on water activity and thus is not limited by the availability of water as a reactant, for which an order of one would be expected. A hydration level of 0.2 *h* corresponds to about 160 moles of water per mole of protein, a hydration level where coverage of the polar groups is not yet complete. Thus solution conditions are not required for the expression of enzymatic activity; indeed, even complete coverage of the protein surface with water is unnecessary, at least for the rate-limiting rearrangement of the enzyme–substrate complex. Formation of the enzyme–substrate complex itself is likely to have a completely different hydration dependence.

A number of other enzyme activities have been studied as a function of hydration. Acylation of α-chymotrypsin by the substrate *N*-succinyl-L-phenylalanine-*p*-nitroaniline can be observed in the absence of buffer salts at about 0.12 *h* and at even lower hydrations in the presence of sodium acetate [41]. The onset of activity of urease lyophilized with ^{14}C-labeled urea occurred at a hydration level of 0.15 *h* [42]. Enzymatic activity for glucose-6-phosphate dehydrogenase, hexokinase, and fumarase was detected at 0.2 *h* and for phosphoglucose isomerase at 0.15 *h* [43].

The requirement for only small amounts of water for enzymes to express catalytic activity appears to be quite general, as has been demonstrated dramatically by the studies of enzymes in organic solvents by Klibanov, Russell, and their coworkers [44–46] (see Chapter 6). Enzymes acquire a variety of novel properties in nonaqueous solvents, including a greatly enhanced stability, altered substrate specificity, and enantiomeric selectivity, as well as the ability to catalyze new reactions. A small amount of water appears to be necessary for catalytic activity in these solvents.

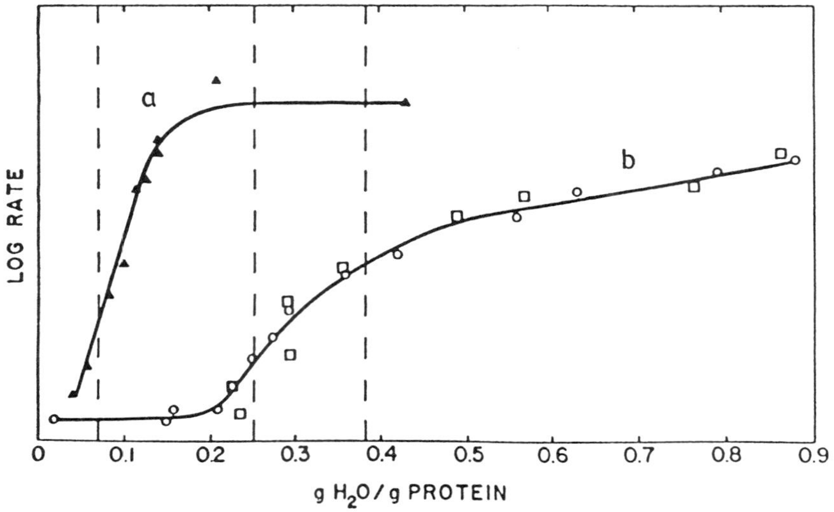

FIGURE 5 Comparison of enzyme activity, ESR spin probe rotational relaxation, and hydrogen exchange for lysozyme as a function of hydration: (a) rate of peptide hydrogen exchange, (b) enzyme activity (open squares) and the reciprocal of the rotational correlation time of the ESR spin probe TEMPONE (open circles). Reprinted from Rupley et al. [23] with permission.

D. Proton Percolation

In 1986 Careri, Giansanti, and Rupley [47] reported the hydration dependence of capacitance measurements for lysozyme over the frequency range from 10 kHz to 4 MHz at pH 3–10. They observed a sharp increase in capacitance with increases in hydration and interpreted these results in terms of percolation theory, in which the protonic conduction process was viewed as percolative proton transfer along threads of hydrogen-bonded water molecules on the protein surface. The hydration level marking the sudden increase in capacitance represents a critical hydration level, the percolation threshold, where long-range connectivity between water molecules appears [3]. The average of this critical hydration level for all samples below pH 8 was $0.152 \pm 0.016\ h$. Assuming that full hydration (i.e., completion of monolayer coverage) occurs at 0.38 h, the observed critical hydration level corresponds to a fractional coverage of 0.40, only slightly lower than the theoretical value of 0.45 for two-dimensional percolation. The critical hydration level (h_c) was found to be independent of pH over the pH range 3–8, was frequency independent, and showed no deuterium isotope effects, consistent with a percolation process, which should depend only on the fractional coverage of the protein by water. The h_c increased to 0.237 at pH 9.87, but this may reflect second-order effects such as a nonrandom distribution of ionizable groups. The h_c also increased to 0.230 in a lysozyme–oligosaccharide complex, which may reflect the disturbance of long-range connectivity over the protein surface by the oligosaccharide. It is not known if such percolation transitions have any biological significance. Careri et al. [47] note that the percolation threshold found for the lysozyme–oligosaccharide complex is close to the hydration level for the onset of enzymatic activity. Catalytic activity, however, has been observed at much lower hydration levels for other enzymes corresponding to fractional water coverages of 0.3 or less, and so the similarity of these critical hydration levels in the case of lysozyme may be coincidental. The transition is highly cooperative; a 2% change in water coverage, which is equivalent to 6 water molecules, switches the protein from a nonconducting to a conducting state. Careri et al. [47] suggest that proton percolation may be important in membrane conduction.

There are two problems preventing unambiguous comparison of theory and experiment. First, it is not at all clear to what extent second-order effects of the type noted above contribute to measured values. Second, the calculation of fractional water coverage requires knowledge of the hydration level corresponding to full hydration (i.e., completion of monolayer coverage). Careri et al. [47] employ a value of 0.38 h. The estimate of Finney and Poole [35] is somewhat higher, however, about 0.45–0.5 h. Using the latter value gives a fractional coverage at the percolation threshold of 0.3–0.34, only two-thirds of the theoretical value for two-dimensional percolation.

The need for an estimate of full hydration can be avoided by comparison of critical exponents for percolation, in particular, the critical exponent t for the conductivity σ [3]:

$$\sigma = \sigma_c + k(h - h_c)^t \qquad (16)$$

Here k is a constant dependent on the process and σ_c is the conductivity at the critical hydration h_c corresponding to the percolation threshold. The exponent t depends only on the dimensionality of the process. Analysis of the dependence of conductivity on hydration for hydrations greater than h_c gives a value for t of 1.29, in excellent agreement with the theoretical value of 1.1–1.3 for two-dimensional percolation. For comparison, the corresponding critical exponent for three-dimensional percolation is 2.0.

Our own interest in proton percolation has to do with what it might reveal about internal water. Careri et al. [47] dismiss proton percolation through the protein interior as unlikely for two reasons: (a) because the population of internal water molecules is small and does not form interconnected threads, and (b) because buried water is mainly structural and is thus likely to be tightly bound. Although the possibility of percolation through the protein interior seems unlikely, a significant population of buried water would introduce errors in estimates of fractional surface coverage. Having said this, we should note that the amounts of water that might be involved may not yield a detectable effect, particularly in the presence of the second-order effects noted above. X-ray diffraction studies of lysozyme have identified four buried water molecules [48], equivalent to 1.3% of the total hydration water or about 0.005 g water/g protein. A buried water population of 4–6% of full hydration, suggested in Sec. III, amounts to about 12 to 18 water molecules per protein molecule or 0.015–0.022 g water/g protein for lysozyme, close to the standard deviation for the critical hydration level for proton percolation. We shall pursue the arguments anyway. Two cases need to be examined:

1. The amount of mobile internal water is so small relative to surface water that it does not influence estimates of fractional surface coverage. In this case, two-dimensional proton percolation should be observed at a critical hydration level of 0.17–0.2, depending on the hydration level assumed for completion of monolayer coverage.

2. A significant fraction of hydration water enters the protein and does so early on in the hydration process as suggested by the hydrogen exchange results, but proton percolation does not occur because no paths of interconnected, buried water are formed. In this case, percolation is two-dimensional, but the observed critical hydration level, $h_{c,obs}$, will be greater than that for a protein with no buried water because the water that enters the protein contributes to the measured hydration level without contributing to surface clusters, which ultimately provide the long-range connectivity. The observed critical hydration level for two-dimensional percolation in this case would be

$$H_{c,obs} = 0.45(h_F - h_B) + h_B \tag{17}$$

where h_F represents the hydration level for full hydration and h_B is the population of buried water. For example, if we assume full hydration is 0.38 h and that the population of buried water amounts to about 0.02 h, the observed critical hydration level is 0.182 h, compared with a value of 0.171 h for a system with no buried water.

Comparison of the experimental estimates of the critical exponent for conductivity with theoretical values clearly indicates that proton percolation is two-dimensional. We would therefore expect the hydration level at the percolation threshold to be equal to or greater than 0.45, depending on the amount of water that enters the protein. In fact, the critical hydration level is similar to, or less than, that predicted for two-dimensional percolation, depending on which hydration value one uses for monolayer coverage. Only if we assume that monolayer coverage is complete at 0.38 h does one obtain a percolation threshold (0.40) consistent with determinations of the critical exponent t, and this value is still a little less than theory predicts for two-dimensional percolation (0.45). With the estimate of Finney and Poole [35] for full hydration (0.45–0.5 h), the disagreement becomes more substantial. If percolation is truly two-dimensional, as the value of the critical exponent for conductivity suggests, then the measured percolation threshold provides no evidence for a buried water population of the size discussed in Sec. III. The difference in observed critical hydration levels predicted for cases (a) and (b), however, are probably too small to allow them to be distinguished with any confidence. Only at high pH or in the enzyme–ligand complex does one observe critical hydration levels consistent with a large population of buried water. The size of this population is extremely sensitive to the estimate of hydration required for monolayer coverage. Assuming monolayer coverage is complete at 0.38 *h* yields an estimate of buried water that is unreasonably large (i.e., $h_B = 0.107\ h$, equivalent to about 85 water molecules per protein molecule). If monolayer coverage is assumed to be complete at 0.5 *h*, then the estimate of size of the buried water population is more reasonable (i.e., $h_b = 0.01$–0.02, equivalent to 7–18 water molecules).

E. Nonfreezing Water

When a protein solution is cooled to below −40°C, a fraction of the water does not freeze. The first NMR determinations of this nonfreezing fraction were made by Kuntz et al. [49,50] using high-resolution proton NMR, which shows a single Lorenzian peak, the area of which is proportional to the amount of unfrozen water. The fraction of nonfreezing water for lysozyme was found to be 0.34 g water/g protein. The fraction of nonfreezing water can also be measured by IR spectroscopy and differential scanning calorimetry (DSC). Golton obtained a value of 0.32 ± 0.02 g water/g protein for the nonfreezing fraction in lysozyme with each of these techniques [35,51]. Powder diffraction patterns for lysozyme samples at different hydrations reveal the formation of ice crystallites at hydration levels above 0.3 *h* [52]. The fraction of nonfreezing water is presumed to reflect the amount of water that is prevented from adopting an ice structure by interactions with the protein.

Kuntz [50] used the NMR method to determine the fraction of nonfreezing water in solutions of polypeptides, from which estimates of the hydration of amino acid side chains could be made. Little or no dependence of hydration on polypeptide conformation was observed. By assuming group additivity and that each amino acid is exposed to solvent, Kuntz calculated hydrations (strictly these are nonfreezing fractions) for several proteins of known composition. The calculated values are about 10% greater than the experimental values. When corrected to account for buried residues, Kuntz [50] obtained a value of 0.335 *h* for lysozyme, in excellent agreement with the experimental estimate of 0.34 *h*. In addition, the calculated (uncorrected) value for bovine serum albumin (0.445 *h*) is in good agreement with the experimentally determined nonfreezing fraction for the protein denatured in urea (0.44 *h*).

The nonfreezing fraction is often regarded as a measure of protein hydration. It is strictly an operational definition, giving the amount of water that is perturbed sufficiently by the protein to be unable to form ice. The fraction of nonfreezing water for lysozyme (0.34 *h*) is a little smaller than the estimate for full hydration derived from heat capacity measurements (0.38 *h*) and is much smaller than the estimate of Finney and Poole [35] for full hydration (0.45–0.5 *h*). Indeed, Finney and Poole suggest that the nonfreezing fraction corresponds to monolayer coverage of polar groups rather than to complete monolayer coverage of the protein surface. Fujita and Noda [53] made a similar suggestion based on an analysis of the accessible surface area occupied by polar and charged groups in chymotrypsinogen A.

F. The Effect of Hydration on Thermal Stability

The effect of hydration on protein denaturation has been determined for a number of proteins by differential scanning calorimetry (DSC) [53–56]. Hydration has a dramatic effect on protein stability. As the hydration level is decreased below 0.75 *h* in lysozyme, the denaturation temperature (T_d) increases and the enthalpy change for denaturation (ΔH_d) decreases as shown in Fig. 6. A similar increase in denaturation temperature is seen with proteins in anhydrous organic solvents [44,57,58]. Three hydration ranges have been identified from DSC studies: region I, 0–0.33 *h*; region II, 0.33–0.75 *h*; and region III, >0.75 *h*. Above 0.75 *h*, T_d and ΔH_d are almost identical to those measured in solution. Below a hydration of 0.75 *h*, T_d increases gradually until a hydration level of 0.33 *h* is reached, where it begins to increase sharply as the hydration level is reduced. ΔH_d decreases gradually between 0.75 and 0.35 *h* and then drops rapidly as hydration levels are decreased below 0.33 *h*. A similar behavior was found for chymotrypsinogen A, although the hydration ranges are a little different (region I, 0–0.41 *h*; region II, 0.41–0.82 *h*; and region III, >0.82 *h*) [53]. Fujita and Noda [53] suggest that region I and region II reflect two different hydration phases, often referred to as A- and B-shell water by analogy with ion hydration. The amount of water associated with the primary hydration phase (region I) is similar to the fraction of nonfreezing water and the amount required to complete polar group monolayer coverage. The significance of a hydration level of 0.75 to 0.81 *h*, however, which defines the upper limit of region II, is not clear. Fujita and Noda [53] suggest that this region corresponds with the hydration range over which deviations in the melting temperature and heats and entropies of fusion of water deviate from those characteristic of bulk water. But these deviations continue out to a hydration level of about 1.4 *h*, which corresponds with the amount of water with motional correlation times different from that of bulk water as determined by proton NMR relaxation [5]. Rupley and Careri [3] have offered a very appealing alternative explanation. They argue that the hydration dependence of T_d and ΔH_d should reflect contributions from the hydration of both the folded and unfolded states and assign the critical hydration level of 0.75 *h* for lysozyme to hydration of its unfolded state (the hydration level of 0.82 *h* found for chymotrypsinogen A presumably reflects the hydration of its unfolded state). These estimates would certainly be consistent with the more extensive interface with solvent expected for the unfolded state relative to the folded

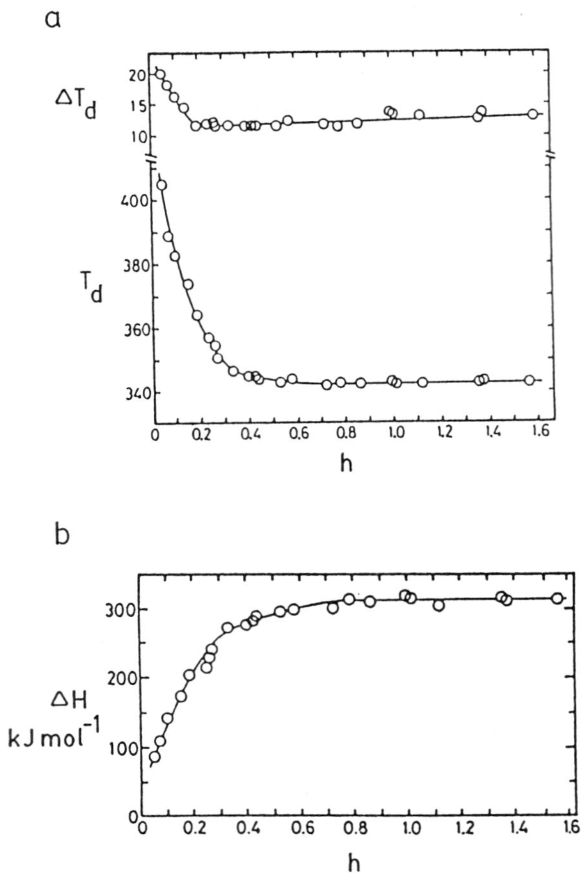

FIGURE 6 (a) The denaturation temperature (T_d) and width of the denaturation peak (ΔT_d) for lysozyme as a function of hydration. (b) The enthalpy change for denaturation of lysozyme as a function of hydration. Reprinted from Fujita and Noda [54] with permission.

state. Reducing the hydration below 0.75 *h* would destabilize the unfolded state relative to the native state and lead to an increase in T_d.

If full hydration of the unfolded state of lysozyme does indeed occur at 0.75 *h*, and the argument of Rupley and Careri [3] is a very compelling one, then something must be wrong with the calculations or interpretation of amino acid hydrations based on measurements of nonfreezing fractions in polypeptide solutions [50]. Recall that the study of the fraction of nonfreezing water in polypeptide solutions, which included polymers of amino acids with ionizable, polar, and nonpolar side chains, provided estimates of amino acid hydrations from which it was possible to calculate protein hydrations (nonfreezing fractions) from knowledge of their amino acid compositions. In lysozyme, the calculated value is 0.36 *h*. This value, supposedly for the protein with all amino acids exposed, is about half the estimate of hydration for the unfolded state suggested by Rupley and Careri [3]. The nonfreezing fraction is more consistent with completion of polar group monolayer coverage, as suggested by Finney and Poole [35].

G. Protein Surface Areas and Monolayer Coverage

The suggestion that monolayer coverage of lysozyme is complete at 0.38 *h* implies that only 300 water molecules are required to cover the surface. A hydration level of 0.45–0.5 *h* would put this figure at about 360–400 water molecules. The surface area of lysozyme is about 6000 Å^2. The area of one face of a cube equivalent in volume to the van der Waals volume of water is about 7 Å^2 while the corresponding area of a molecule of liquid water is about 10 Å^2, suggesting that the surface of lysozyme can accommodate 600–850 molecules of water [39]. Evidently, one cannot use packing arguments as a basis for preferring one estimate for monolayer coverage over another. Rupley et al. [39] note that the water arrangement in the plane perpendicular to the *c*-axis of ice I gives an area of 20 Å^2 per water molecule and point out that the large area/water molecule must imply considerable local ordering or water at the protein surface.

The area of the extended polypeptide chain of lysozyme is about 21,000 Å^2. The area/water molecule of 15–20 Å^2 found for the native state would imply that about 1000–1400 water molecules are necessary to hydrate the extended chain. The suggested hydration level of 0.75 *h* for the denatured state of lysozyme is equivalent to about 600 water molecules per protein molecule, much less than the value estimated for the extended chain. This is to be expected. Denatured states do not resemble extended chains but are more like swollen, polypeptide nets and may possess significant residual structure, albeit mostly local and transitory. Expansion of lysozyme to accommodate the additional 200–300 water molecules found in the denatured protein would suggest that the volume of the denatured state is about 1.3–1.5 times as large as the native state. Thus, the protein undergoes little expansion on denaturation, the increase in radius being perhaps 2–3 Å, consistent with estimates of Stokes radii for native and denatured proteins measured by Corbett and Roche [59].

V. HYDRATION-INDUCED CONFORMATIONAL CHANGES

It appears that no major hydration-induced conformational changes occur at hydrations above about 0.2 *h*. For example, the constancy of hydrogen exchange rates above about 0.15 *h* [29] and the detection of enzymatic activity in a number of enzymes at hydrations of 0.12–0.2 *h* [39,41–43] would suggest that the protein conformation above hydration levels of about 0.2 *h* is essentially the same as that in dilute solution. Other experimental evidence supporting this conclusion has been summarized by Rupley et al. [39] and Luscher-Mattli and Ruegg [12].

Although the appearance of hysteresis in the sorption isotherms, as well as a variety of spectroscopic evidence, suggests that conformational changes can occur at hydration levels below 0.2 *h*, the extent of the structural changes that do occur has been a matter of some debate. Kuntz and Kauzmann [1] have argued that the intermolecular voids created on dehydrating a protein would lead to large unsatisfied intermolecular forces, which would provide a large driving force for rearranging exposed protein side chains and for the refolding of the polypeptide backbone in an attempt to reduce the size of the voids. Assuming a surface tension of 50 erg cm^{-2}, the surface free energy for a spherical void of radius 5 Å would be 22.5 kcal per mole of voids. Luscher-Mattli and Ruegg [12] also conclude that pronounced conformational changes occur at hydration levels below about 0.1 *h*, based on analysis of x-ray diffraction, infrared spectra, and thermodynamic data. Analysis of spin–spin interactions between electron spin resonance (ESR) spin probes covalently linked to lysozyme [39], however suggests that little or no change in protein conformation occurs down to a hydration level of about 0.02 *h*, at least to the resolution of the line shape analysis, which was 1 Å. By contrast, laser Raman spectroscopic studies of lysozyme provide evidence for changes in the conformation of disulfide bridges, which are reordered (i.e., adopt their solution conformation) at a hydration level of 0.08 *h*, and for shifts in the position of a tryptophan side chain and the peptide backbone at hydration levels of 0.08–0.2 *h* [60].

Circular dichroism spectra between 200 and 240 nm have been reported at different hydration levels for lysozyme and ribonuclease A [61]. Comparison with spectra obtained in aqueous solution reveals differences in the relative contributions of bands at 210 and 220 nm, which may reflect changes in secondary structure at low hydration, although contributions from tryptophan side chains in this region of the spectrum are also expected and are known from Raman spectroscopic studies to undergo conformational changes on dehydration [3].

A. Solid State ^{13}C NMR Studies of Protein Hydration

Changes in the distribution of isotropic chemical shifts for lysozyme monitored by solid state ^{13}C cross-polarization magic-angle spinning (CP/MAS) NMR indicate that the distribution of conformational states becomes narrower as the protein is hydrated. These changes are observed for all peaks in the aliphatic region of the spectrum [26]. A study of ^{13}C CP/MAS NMR spectra for lysozyme as a function of hydration indicates that the change in the distribution of isotropic chemical shifts begins at a hydration level of 0.1–0.15 *h* and is largely complete at a hydration of 0.3–0.35 *h* [27]. By contrast, in bovine serum albumin (BSA), enhancements in spectral resolution on hydration are confined to the peak at 40 ppm. This begins at a hydration of about 0.2 *h* [62]. The source of this behavior is not understood, but could reflect reordering of disulfide bridges, a decrease in the linewidth of contributing resonances from lysine side chains due to increased motional averaging on hydration, or titration shifts induced by hydration. The latter two explanations, however, seem unlikely. Proton–carbon cross-relaxation times do not change very significantly on hydration, while titration shifts would be expected to occur at much lower hydrations than is observed as water adds to ionizable side chains and normalizes their pK_a values. With the exception of the changes in the peak at 40 ppm, hydration appears to have little effect on the solid state spectrum, and so the distribution of conformations sampled by fully hydrated BSA is as broad as that observed in the dry protein. The most interesting result of these studies, however, is that no significant changes are observed in either the rotating frame proton spin lattice relaxation time or cross-relaxation time on hydration. Therefore, the strength of dipolar coupling is not significantly affected by hydration of the protein. The preservation of efficient dipolar coupling suggests that the conformational changes that do occur on hydration do not lead to conformational states with any major expansion of free volume or weakened secondary interactions, which would significantly increase the reorientational freedom of protein groups [26].

B. An X-Ray Diffraction Study of a Dehydrated Protein

A recent x-ray diffraction study of dehydrated lysozyme provides the first detailed picture of the structure of a protein at low hydration levels. Protein crystals have a high solvent content (between 30 and 80% by volume) and tend to be rather soft and easily disordered. Dehydration of protein crystals usually leads to a loss of the diffraction pattern, due to intermolecular repositioning of individual protein molecules (lattice disorder). Kachalova et al. [63], however, found that triclinic crystals of hen egg white lysozyme cross-linked with glutaraldehyde retain their ability to diffract to high resolution even with a water content as low as 36 moles of water per mole of protein, the hydration level obtained at a relative humidity of 0.01. Dehydration leads to shifts in the relative positions of domains as well as numerous small displacements in the positions of individual atoms. Overall, drying leads to a contraction of the protein molecule. The molecular volume decreases from 1.93×10^4 Å^3 for the fully hydrated protein to 1.82×10^4 Å^3 for the "dry" protein, which represents an increase in the average packing density of 4–6%. The extent of displacements for main-chain atoms is comparable with the rms deviations calculated from the Debye–Waller factors for the dry protein, although there is no correlation between these parameters along the polypeptide chain. The largest dehydration-induced displacements (1.0–1.5 Å) are found at the C terminus, while values for the rest of the polypeptide backbone range from 0.2 to 0.9 Å. The average deviation for main-chain atoms is 0.6 Å, and for all atoms is 0.9 Å. The conformational changes are therefore small. The extent to which these results reflect the conformation of dehydrated or lyophilized powders of lysozyme is difficult to judge. It is possible that covalent cross-linking as well as lattice interactions prevented any large-scale conformational rearrangements on dehydration.

C. FTIR Studies of Dehydration-Induced Conformational Transitions

Prestrelski et al. [4] have measured Fourier-transform infrared spectra of a number of lyophilized proteins (basic fibroblast growth factor (basic FGF), γ-interferon, granulocyte-colony stimulating factor (G-CSF), α-lactalbumin, lysozyme, α-casein, and lactate dehydrogenase) as well as poly (L-lysine) using KBr pellets and attenuated total reflectance techniques. Comparison of the second-derivative spectra of the amide I band of the dry proteins with those obtained for proteins in aqueous solution indicate that dehydration causes significant pertubation of the amide I region of the spectrum in many cases. Although Careri et al. observed changes in the amide I band of lysozyme on dehydration, these were attributed entirely to the removal of water without conformational change [3,10]. A comparison of the amide I band of dry and hydrated poly (L-lysine), however, suggests that the predominant source of changes in the second-derivative spectrum on dehydration is related to conformational changes rather than to removal of solvent [4]. The preferred secondary structure for poly (L-lysine) in the dry state is the β-sheet. When solutions of poly (L-lysine) in unordered or α-helix conformations are dried, the amide I band reveals the dehydration-induced transition to the β-sheet structure. Apart from small frequency shifts, however, no change is observed in the amide I band when a solution of poly (L-lysine) prepared in a β-sheet conformation is dried (i.e., there is little change in the amide I band that can be attributed to solvent effects on the carbonyl stretching mode). The changes in second-derivative amide I spectra on dehydration are quite variable and protein dependent. For example, little change is observed in the spectrum of G-CSF on dehydration, while the spectrum of casein is greatly altered on dehydration, with changes characteristic of a transition from a protein with little secondary structure to one with a large fraction of β-sheet, similar to the changes observed for poly (L-lysine). In general, the resolved peaks in the second-derivative spectrum become broader on dehydration, indicative of the same type of static disorder observed by solid state ^{13}C NMR spectroscopy. In some cases, the conformational changes were irreversible. For example, lyophilized casein is essentially insoluble in water, and lactate dehydrogenase is completely inactive on rehydration.

It is well known that polyols and sugars such as sucrose stabilize proteins against denaturation in solution (see Chapters 12 and 13). A similar protective effect is observed when proteins are lyophilized in the presence of sugars [4,64]. Infrared spectra of proteins lyophilized in the presence of such additives are very similar to the spectra observed in aqueous solution, indicating that these additives preserve the solution structure of the proteins during dehydration [4]. These additives were also found to inhibit the conformational transitions observed with poly (L-lysine) on dehydration [4].

Dehydration-induced conformational changes appear to be driven by the need to compensate for the loss of hydrogen bonding to water on dehydration [4,64]. This is consistent with the transitions observed in poly (L-lysine) dehydration to a conformational state (β-sheet structure) with a lower degree of solvation and a higher degree of intermolecular hydrogen bonding relative to an α-helix or unordered structure. Additives such as sucrose interact directly with the protein, and their hydroxyl groups could serve as water substitutes, forming hydrogen bonds with protein groups that would otherwise be unsatisfied in the dry state [4].

Taken together, these results suggest that conformational changes do indeed occur as the protein is dehydrated. The extent of the changes that are induced can be quite variable and differ from protein to protein. In some proteins, little or no change in conformation occurs; in others, there is major rearrangement of the polypeptide backbone. Dehydration leads to a great reduction in conformational flexibility, and we suspect that the extent of dehydration-induced conformational changes will depend in part on where along the hydration isotherm these changes occur relative to the rigid-to-flexible transition (see also Sec. VII.E).

VI. EFFECT OF HYDRATION ON PROTEIN DYNAMICS

A variety of techniques have been employed to examine the effect of hydration on the dynamic properties of the protein and of water at the protein surface. The results of studies employing dielectric relaxation and Rayleigh scattering of Mossbauer radiation (RSMR) are described in other chapters of this book. Rupley and Careri [3] have recently provided an extensive review of the literature in this field. In many cases, the critical hydration levels found with time-averaged properties are also observed in the recovery of protein dynamic properties. In this section, we review some of these results.

A. Spectroscopic Methods

ESR spectra of the nitroxide spin probe TEMPONE, noncovalently bound to lysozyme, have been measured as a function of hydration [39]. The ratio of the amplitudes of the hyperfine lines as a function of hydration display a break at a hydration level of about 0.07–0.1 h, although this is not reflected in the correlation time. Estimates of the TEMPONE rotational correlation time, which reflect motions at the protein surface, have been determined from simulations of the ESR spectra. The behavior of the rotational correlation time with hydration is shown in

Fig. 5. At hydration levels below about 0.20–0.25 h, the correlation time is 3×10^{-9} s, but this decreases rapidly as the hydration is increased beyond this level. At the point at which the hydrated powder collapses to form a solution (1.0 h), the correlation time is about 2×10^{-10} s and continues to decrease as further water is added to the solution.

Belonogova et al. [65] have examined the influence of hydration on the mobility of ^{57}Fe in hemoglobin, myoglobin, and ferredoxin using a combination of ESR spin probes and Mossbauer spectroscopy. All proteins show a reduction in the probability of recoilless gamma-ray absorption, which reflects an increase in the amplitude of motions of the Mossbauer nuclei, at a relative humidity of about 0.4–0.6, which corresponds to a hydration level of about 0.1 h. The amplitude of vibration of ^{57}Fe was less than 0.1 Å at low hydrations but increased to about 0.3–0.4 Å at higher hydrations. The low-field component of the ESR signal was also found to shift at relative humidities greater than about 0.6, equivalent to a hydration level of about 0.15 h. This was accompanied by a decrease in the rotational correlation time from 10^{-7} to 10^{-8} s. These results suggest that recovery of intramolecular mobility in these proteins occurs at a hydration level of 0.1–0.15 h. Mossbauer and ESR spectroscopy also have been employed to examine the dynamic properties of proteins as a function of temperature, but we defer discussion of these results to a later section.

The effect of hydration on intrinsic protein fluorescence appears to be rather complex and protein dependent. Measurements of steady state fluorescence in a number of proteins as a function of hydration are consistent with an increase in protein flexibility with hydration [66]. Tryptophan fluorescence lifetimes in chymotrypsinogen A measured by Fucaloro and Forster [67] also suggest an increase in flexibility in the environment of the chromophore as hydration is increased. At low hydrations, the lifetime was about 3.0 ns, but this decreased sharply at a hydration of 0.15 h and reached a value of about 2.0 ns at 0.4 h. A global analysis of the time-resolved fluorescence of chymotrypsinogen A and α-chymotrypsin by Vermunicht et al. [68], however, found no effect of hydration on lifetimes. The discrepancy appears to be due to the different methods of analysis employed. More complex results were obtained in studies of bovine serum albumin [69]. The fluorescence lifetime increased between 0.02 and 0.1 h, was constant between 0.1 and 0.4 h, and then increased at higher hydrations, approaching the solution value of 6.5 ns. The results were interpreted in terms of water diffusion to the chromophore to form an exiplex.

Tryptophan phosphorescence lifetimes of lyophilized proteins are extremely long, extending to several seconds in some cases, consistent with a rigid chromophore environment in which dynamic quenching of phosphorescence is severely restricted [70]. The emission bands are broader that those found in solution spectra even for proteins with single tryptophans, suggesting a heterogeneous chromophore environment (static disorder), in agreement with results from solid state ^{13}C NMR [26,27]. For example, in S. nuclease that contains a single tryptophan residue, the phosphorescence emission spectrum in solution is characterized by narrow vibrational bands and the decay is exponential. In the dry state, however, the spectrum is broad and the decay is nonexponential. Hydration of proteins leads to a dramatic decrease in phosphorescence lifetimes. In liver alcohol dehydrogenase, the phosphorescence from surface tryptophans is completely quenched on hydration, so that only a well-buried tryptophan contributes to solution phosphorescence [70].

Protein–water interactions have been extensively investigated by NMR techniques using both hydrated powders and protein solutions. For reviews, see Kuntz and Kauzmann [1], Bryant [71,72], and Rupley and Careri [3]. Much of the work has been concerned with the relaxation properties of water at the protein surface using ^{1}H, ^{2}H, and ^{17}O nuclei, but several solid state ^{13}C NMR studies of hydrated powders have been reported. We have already described some of this work earlier in this chapter. The interpretation of ^{1}H NMR relaxation results has been difficult and is complicated by cross-relaxation and exchange effects. Nevertheless, it is clear that the amount of water with motional correlation times longer than that of bulk water extends to hydration levels of 1.4–1.7 *h*. Fullerton et al. [5], using an NMR titration method, identified three water fractions with different correlation times for water motions in solutions of lysozyme: "superbound" water from 0 to 0.055 *h* with $\tau_c > 10^{-6}$ s; "bound" from 0.055 to 0.25 *h* with $\tau_c = 10^{-9}$ s; and "structured" water from 0.25 to 1.4 *h* with $\tau_c = 10^{-11}$ s. Lioutas et al. [73] performed similar experiments and also found three populations of water with correlation times different from bulk water. The amount of water with motional properties different from bulk water was found to extend to large hydrations (1.7 *h*) in this case, although the water correlation times are quite different from those estimated by Fullerton et al. [5]. Studies of hydrated powders indicate that surface water undergoes rotations about the water–protein hydrogen bond with correlation times on the order of 10^{-9}–10^{-10} s. Reorientation of the water, that is, of the rotational axis, occurs more slowly, with a correlation time of about 10^{-7} s [74–78].

We have already discussed the evidence from solid state ^{13}C cross-polarization magic-angle spinning (CP/MAS) NMR for conformational changes occurring at low hydration in lysozyme (see Sec. V.A). Changes in the ^{13}C NMR spectrum are consistent with a decrease in the distribution of isotropic chemical shifts, reflecting a decrease in the distribution of conformational states sampled by the protein as it is hydrated. Recovery of the normal distribution of conformational states occurs at 0.1–0.15 *h*, very similar to the hydration levels at which conformational flexibility begins to recover as monitored by other techniques. There is little or no change in relaxation parameters, however, which suggests that the conformational changes are small and that rapid, large amplitude motions do not occur. Detailed analysis of the relaxation times is difficult because the peaks observed in the protein ^{13}C NMR spectrum represent a superposition of a large number of resonances. Proton–carbon cross-relaxation times for aliphatic carbons take values of 20–50 μs and are not much affected by hydration. A somewhat different hydration response is observed with poly (L-lysine), where significant line narrowing of δ- and ε-carbon resonances is accompanied by changes in proton–carbon cross-relaxation times from about 30–40 μs in the dry state to 80–100 μs in the hydrated state [78]. Little change in relaxation times is observed for α- or β-carbons, however. It is possible that changes in cross-relaxation times due to motional averaging do occur for some resonances in lysozyme, but these are simply overwhelmed by other resonance contributions to the aliphatic peaks of the spectrum that do not undergo any hydration-induced motions. Therefore, we cannot entirely rule out the possibility that some of the changes observed in the ^{13}C NMR spectrum on hydration may be due to the recovery of conformational flexibility at 0.1–0.15 *h*, reflecting a transition from a state of static disorder in the dry protein to a flexible state in the hydrated protein.

B. Hydrogen Isotope Exchange

Hydrogen isotope exchange studies of lysozyme as a function of hydration have been reported by Poole and Finney [79] and by Schinkel et al. [29]. Poole and Finney employed a proton NMR method. Samples of deuterated (85%) lysozyme were isopiestically equilibrated over solutions of H_2O to a known hydration for eight days. The samples were then lyophilized to remove water, dissolved in deuterated DMSO, and the NMR spectrum was measured. The residual deuterated amide expressed as a fraction of the initial deuteration was determined from the integrated amide proton resonances. Exchange that occurs below a hydration level of 0.07 *h* can be attributed to exposed peptide groups. At a hydration level between 0.07 and 0.1 *h*, the flexibility of the protein increases, allowing access of exchange catalysts to buried peptide groups. Protein internal mobility continues to increase as water is added, up to a hydration level of about 0.2 *h*. Schinkel et al. [29] determined proton exchange rates for lysozyme as a function of hydration with a tritium out-exchange technique. Samples of lyophilized, tritium-labeled lysozyme were hydrated by isopiestic equilibration over solutions of either H_2SO_4 or NaOH of appropriate water activity and were left to exchange for 24 hours. The number of hydrogens remaining unexchanged, H_{rem}, after this time was determined by a gel filtration method at pH 4.0 (to decrease exchange rates). H_{rem} decreases rapidly with increasing hydration level but eventually reaches a constant value at a hydration level of about 0.15 *h*. Further increases in hydration level therefore have no effect on the accessibility of peptide protons to the exchange catalyst, suggesting that the internal mobility of lysozyme is fully developed at this hydration level. Exchange in lysozyme powders displayed the same pH dependence as observed in solution, and the rank order of exchange was also preserved with changes in hydration, at least between 0.17 *h* and the solution state. Thus, the conformation of the protein at 0.17 *h* cannot be greatly different from that found in solution.

There has been much discussion of the mechanisms of exchange in proteins and several models of the exchange process have been proposed [80–82]. The local unfolding model developed by Englander and coworkers [83–85], based on studies of exchange in hemoglobin, assumes that exchange occurs via local unfolding of segments of the polypeptide chain that exposes peptide groups to bulk solvent where exchange takes place. Penetration models propose that water and catalyst species diffuse into the protein interior via rearrangements of free volume (mobile defects). This model, developed and explored by Rosenberg, Lumry, Woodward, Gregory, and their coworkers [25,86–90], is based primarily on exchange studies of the serine proteases, ribonuclease A, HEW lysozyme, and bovine pancreatic trypsin inhibitor. One of the most convincing arguments in favor of catalyst penetration models, derived from exchange studies of partially hydrated lysozyme powders, is that the order of the exchange reaction with respect to water is found to be about 4 for all H_{rem} values between 118 and 8, suggesting that only three water molecules, in addition to the catalyst species itself, are formally required for exchange [29]. This represents about 1% of the water required for monolayer coverage of lysozyme. This number is both small and constant for a large number of exchangeable sites, which is not easily reconciled with local unfolding models, particularly those which suggest that exchange of slower and slower hydrogens would require larger and larger "unfolding units."

C. Positron Annihilation Lifetime Spectroscopy

Positron annihilation lifetime (PAL) spectroscopy is a powerful technique for determining the size of cavities and pores in materials [91,92]. Positrons are produced in the decay of a number of neutron-deficient nuclei, such as ^{22}Na. Positrons emitted from ^{22}Na sources have up to 540 keV of kinetic energy and create a radiation track in the medium that is punctuated by spurs, created as a result of discrete deposition of energy into the medium (about 100 eV per spur), which results in ionization and radical formation in the medium. Eventually, the positron reaches thermal energies, at which point some fraction may form positronium (a bound state of a positron and an electron) by combination with an electron in the terminal spur. Positronium exists in two spin states: singlet, or para-, positronium (p-Ps), which has a lifetime of about 125 ps; and triplet, or ortho-positronium (o-Ps), which has a lifetime in vacuo of 140 ns. In condensed matter, the lifetime of o-Ps is greatly reduced through interactions with its surroundings. Positronium localizes in regions of free volume as a result of exchange repulsion from surrounding atoms. The primary mechanism of annihilation of the triplet positronium species (o-Ps) is with electrons of the host medium, so-called pick-off annihilation. The lifetime of o-Ps is therefore very sensitive to its local molecular environment and in particular to the size of the cavity in which it is localized. A number of semiempirical correlations between cavity size and annihilation rate have been developed [91–95]. Recent advances in the analysis of positron annihilation lifetime data using integral transform methods now make it possible to extract continuous distributions of annihilation rates [18,19,95,96]. These distributions may be transformed to derive free volume distributions [97].

Gregory and Chai have conducted a study of positron annihilation in lysozyme as a function of hydration at 298 K [98,99]. Positron annihilation lifetime probability density functions (pdfs) consist of three peaks, which may be assigned to the annihilation of p-Ps, free positrons, and o-Ps species. The dry protein has a mean o-Ps lifetime of about 1.6–1.7 ns, which corresponds to an average cavity volume of about 65 Å^3. The o-Ps lifetime decreases initially as water is added to the dry protein, but at a hydration of about 0.12 h (i.e., at a relative humidity of about 0.70), the lifetime begins to increase, reaching a value of about 1.8–1.9 ns, corresponding to an average cavity volume of about 90 Å^3, at 0.395 h. The fraction of positrons forming o-Ps is 0.17–0.18 in the dry protein and decreases with hydration to about 0.14. The o-Ps fraction is often assumed to reflect the population of free volume sites in materials, but other factors almost certainly influence positronium formation probabilities, such as solvation of positrons and electrons in the positron spur. As seen in Fig. 7, the changes in lifetime parameters for lysozyme are remarkably similar to those reported for the polyamides Nylon 6,6 and Nylon 6 [100] and are consistent with the existence of a glasslike dynamical transition. At low hydrations, the glass transition temperature, T_g, is higher than the measurement temperature and the protein is rigid. As the hydra-

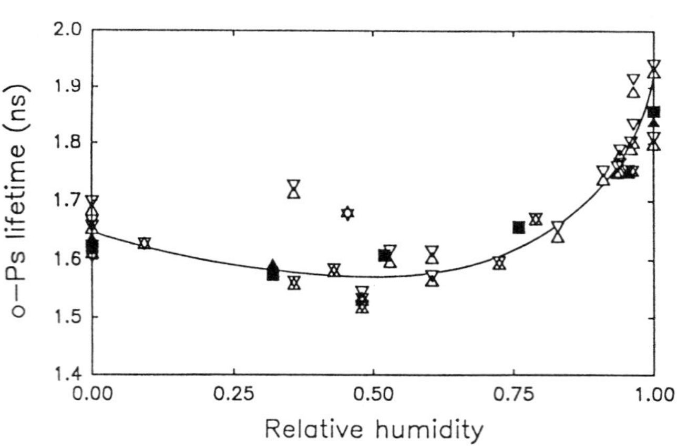

FIGURE 7 Comparison of mean o-Ps lifetime for lysozyme (open symbols), Nylon 6,6 (filled triangles), and Nylon 6 (filled squares) as a function of relative humidity. The increase in o-Ps lifetime observed above a relative humidity of 0.70–0.75 corresponds to a hydration level of 0.12–0.15 h in lysozyme. Data for lysozyme is from Gregory and Chai [99] and for the polyamides from Welander and Maurer [100].

tion level (plasticizer content) of the protein is increased, the glass transition temperature decreases. Once sufficient water has been added to the protein to cause T_g to decrease below the measurement temperature (i.e., >0.12 *h* at 298 K), the protein becomes flexible and segmental motions can occur. In flexible systems, positronium can increase the size of the cavities in which it is located as a result of exchange repulsion (Pauli exclusion) between its bound electron and the closed-shell electrons of the surrounding medium (positronium forms bubbles in liquids). This, together with an increase in free volume in the flexible state, leads to a rapid increase in o-Ps lifetime.

These studies suggest that lysozyme undergoes a glasslike dynamical transition with a transition temperature of 298 K at a hydration of about 0.12 *h* and that water acts as a plasticizer of the protein conformation. The results obtained with the other techniques described above are also consistent with transitions from a rigid to a flexible state occurring over a hydration range of 0.07–0.2 *h* at 298 K. The range of hydration levels (plasticizer content) in which transitions are observed is quite broad, but this is to be expected given the heterogeneity of protein structure and the variety of techniques employed. Indeed, Goldanskii and Krupyanskii [101] describe proteins as "heterogeneous glasses." In addition, glass transition temperatures are sensitive to the time scale of the measurement. At a given hydration level, slower measurements will give lower estimates for the glass transition temperature. At a given temperature, slower measurements will allow a glass transition to be observed at a lower plasticizer content.

VII. GLASS TRANSITIONS IN PROTEINS

The idea that water acts as a plasticizer of the protein conformation is not new; it was suggested many years ago by Bone and Pethig [102] on the basis of dielectric relaxation studies (see Chapter 4). In recent years, several studies of fully hydrated proteins and protein crystals also have indicated the existence of a glasslike dynamical transition at about 200 K in proteins. It is well known that glass transition temperatures for synthetic polymers decrease with increases in plasticizer content, yet the possibility that the transition at 200 K in fully hydrated proteins and the transition observed at 298 K at low hydrations are related has been established only recently. In this section, we describe the 200 K transition as well as work on the hydration dependence of the transition temperature. We begin by briefly describing some basic properties of glass transitions in synthetic polymers. Further details can be found in Refs. 103–105 or other standard texts on synthetic polymers.

A. Glass Transition Behavior in Polymers

At sufficiently low temperatures, all amorphous polymers enter a rigid, glassy state in which large-scale segmental motions of the polymer chain no longer occur. Those motions which do occur in the glassy state are local and are restricted to side-chain motions and small-amplitude motions of a few chain atoms. On heating, polymers become flexible in a characteristic temperature range, called the glass–rubber transition region, where atoms of the polymer chain gain sufficient thermal energy for coordinated long-range molecular motions to become possible. The glass transition region in which large changes in mechanical properties occur is usually quite broad (20–30°C), and glass transition temperatures (T_g) vary widely with chemical structure, polymer molecular weight, degree of cross-linking, and sample crystallinity. For example, the glass transition temperature of polyethylene is 140 K, while T_g values for polymers such as polyisoprene and polyisobutylene are about 200 K and those for polystyrene, poly(vinyl chloride) and poly(methyl methacrylate) are about 350–380 K. Glass transition temperatures for amorphous polyamides such as Nylon 6, Nylon 6,6, and Nylon 6,4 are about 330–350 K.

Polymers may undergo a number of transitions as various modes of motion are excited with increases in temperature. By convention, the glass transition is usually designated the α-transition and other transitions associated with side-chain rotations as well as small-amplitude chain motions are designated β-, γ-, δ-, etc. transitions in order of decreasing transition temperature. For example, polyamides show two transitions in addition to the glass transition: the β-relaxation process assigned to motions of the amide groups ($T_\beta = 190–210$ K), and the γ-relaxation process associated with motions of individual methylene groups along the main chain ($T_\gamma = 130$ K).

Many thermodynamic and mechanical properties undergo readily observable changes in the glass transition region, and so a variety of techniques are available to characterize the transition. Both dilatometric and calorimetric measurements may be used to define the glass transition temperature [103]. In addition, dilatometric methods provide information about free volume changes in the transition region. The variations in specific volume and enthalpy with temperature through the glass transition region are shown in Fig. 8a. Both show a change in

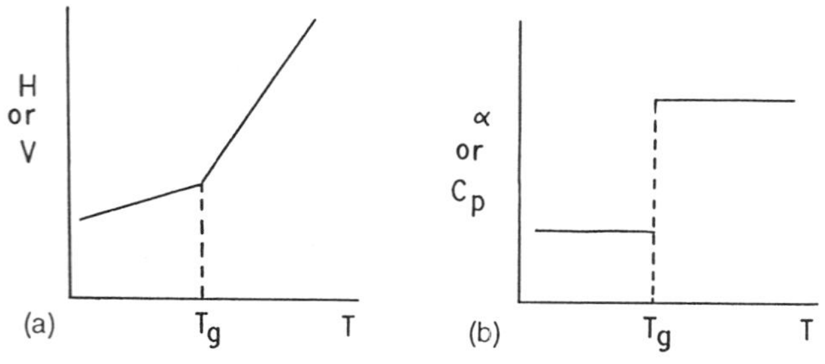

FIGURE 8 The temperature dependence of (a) the specific volume and enthalpy of an amorphous polymer through the glass transition region and (b) the corresponding changes in the thermal expansivity and heat capacity.

slope characteristic of second-order transitions. Under normal conditions of measurement, the glass transition is not a true thermodynamic transition but represents a relaxation process. In the limit of long measurement times, however, the glass transition approaches the behavior expected of a true second-order transition [106,107]. The corresponding changes in the heat capacity (C_p) and the characteristic increase in thermal expansivity (α) above the glass transition, attributed to the expansion of free volume in the flexible state, are shown in Fig. 8b. A number of static and dynamic mechanical methods are available to characterize the viscoelastic properties of polymers through the glass transition region, including the determination of Young's modulus and its component storage and loss moduli. Dielectric relaxation and broad-line proton NMR methods have also been used to characterize the transition.

The estimate of the glass transition temperature depends on the time scale of the experiment. Slower measurements give lower values for T_g. In dynamic methods, a 10-fold decrease in frequency can be expected to produce about 5–8°C decrease in the estimate of T_g, although the exact size of the effect will depend on the activation energy for the relaxation process.

The glass transition temperature is greatly influenced by the presence of small molecule "plasticizers." For example, water is a plasticizer of polyamides and causes a dramatic decrease in their glass transition temperatures, as shown in Fig. 9. At hydration levels above 0.10 g water/g polymer, T_g values for Nylon 6, Nylon 6,6 and Nylon 6,4 are about 240 K. This represents a decrease of about 80–100°C relative to T_g in the dry state [100]. In polyamides, water provides al-

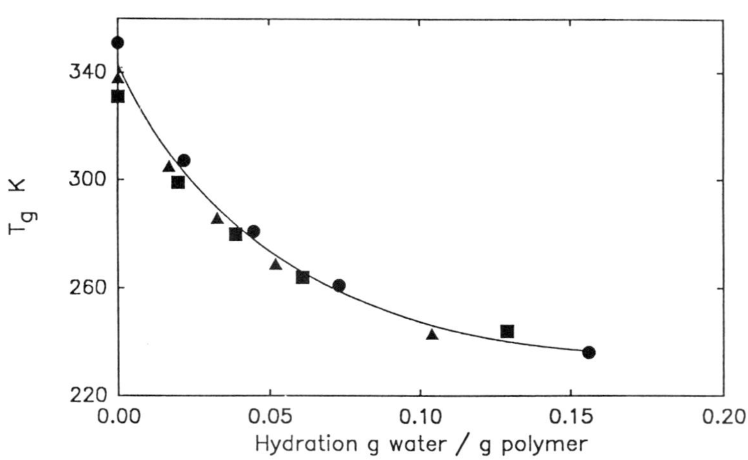

FIGURE 9 The glass transition temperature for the polyamides, Nylon 6 (squares), Nylon 6,6 (triangles), and Nylon 4,6 (circles) as a function of hydration (plasticizer content). Data from Welander and Maurer [100].

ternative mobile hydrogen bond donors and acceptors for amide groups, facilitating segmental motions of the polymer chains.

B. Free Volume in Glass Transition Theory

The concept of free volume is central to theories of the glass transition. Segmental motion of polymer chains cannot take place without adjacent regions of free volume. Obviously, such molecular motions are always accompanied by rearrangements of free volume. The thermal expansivity above the glass transition temperature is greater than that below T_g . Changes in the specific volume of an amorphous polymer are shown in Fig. 10. The specific volume of the polymer below T_g (i.e., in the glassy state) may be expressed as [104].

$$V(T) = V_o(0) + V_f(0) + \left\{\frac{\partial V_o}{\partial T}\right\} T,\ T < T_g \tag{18}$$

At T_g the specific volume is

$$V(T_g) = V_o(0) + V_f(0) + \left\{\frac{\partial V_o}{\partial T}\right\} T_g \tag{19}$$

and above T_g (i.e., in the flexible or rubbery state), it is given by

$$V(T) = V(T_g) + \left\{\frac{\partial(V_o + V_f)}{\partial T}\right\}(T - T_g),\ T > T_g \tag{20}$$

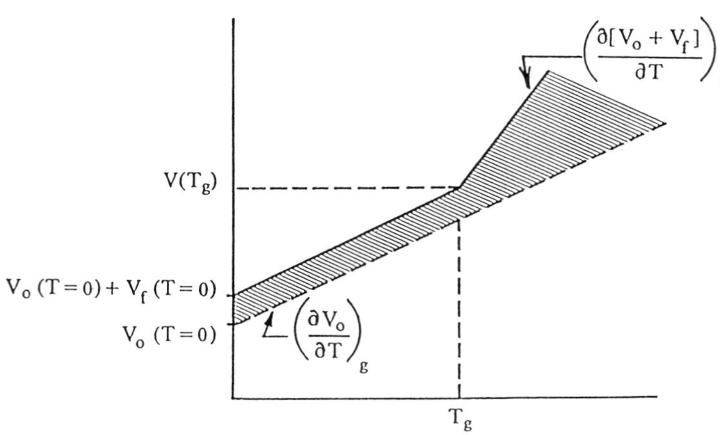

FIGURE 10 The specific volume of an amorphous polymer as a function of temperature, showing the changes in free volume in the glass transition region. The shaded region represents the free volume. See text for details. Taken from Hiemenz [104].

V_f and V_o refer to the free volume and the volume occupied by the polymer, respectively. The thermal expansivities of the rubbery and glassy states are

$$\alpha_R = \frac{\partial(V_o + V_f)/\partial T}{V(T)}$$

and

$$\alpha_G = \frac{\partial V_o/\partial T}{V(T)}$$

If expansion of the glassy state is assumed to occur at constant free volume, the free volume fraction (i.e., $f(T) = V_f(T)/V(T)$) above the glass transition temperature can be represented by [108,109]

$$f(T) = K + (\alpha_R - \alpha_G)T, T > T_g \tag{21}$$

where K is constant. Examination of volume data for a number of polymer systems by Simha and Boyer [110] suggested that the free volume fraction at T_g was a constant equal to about 11%.

Considerations of polymer melt viscosities in terms of the free volume required for molecular motion leads to the well-known Williams–Landel–Ferry (WLF) equation [111]:

$$\ln A_T = \ln \frac{\eta}{\eta_o} = \ln \frac{t}{t_o} = -\frac{B}{f_o} \frac{T - T_o}{f_o/\alpha_f + T - T_o} \tag{22}$$

which relates the viscosity, η, or relaxation time for a process, t, at a temperature T above T_g to the corresponding quantities, η_o or t_o, measured at some reference temperature T_o. Here f_o is the free volume fraction at T_o, α_f is the thermal expansivity of the free volume, and B is a constant usually assigned a value of unity, consistent with the analysis of viscosity data by Doolittle [112]. The free volume fraction at a temperature T is given by

$$f(T) = \frac{V_f(T)}{V(T)} = f_o + \alpha_f(T - T_o) \tag{23}$$

Analysis of linear amorphous polymers assuming $B = 1$ and with $T_o = T_g$ gives the following parameter estimates [103]:

$$f_o = 0.025$$

and

$$\alpha_f = 4.8 \times 10^{-4}\ K^{-1}$$

The analysis thus assigns a value to the free volume fraction at the glass transition temperature of 2.5% for all linear amorphous polymers independent of their chemical structure, which is considerably smaller than that suggested by Simha and Boyer [110]. The analysis of the glass transition by Hirai and Eyring [113,114] in terms of holes provides estimates of the free volume fraction at T_g that vary from about 4 to 12%.

C. The 200 K Transition in Fully Hydrated Proteins

There is now a considerable body of evidence that demonstrates hydrated proteins undergo a glasslike transition at 180–220 K. This includes studies of motions monitored by ESR spin labels [65,115–117], Mossbauer spectroscopy [115,117,118], phosphorescence [70,115], and neutron scattering [119] as well as by RSMR techniques [101] (see also Chapter 5). Frauenfelder and coworkers have also explored, in great detail, the hierarchy of conformational substates in myoglobin and the glasslike properties of myoglobin (see the review by Frauenfelder et al. [120]). The dynamical transition has also been reproduced in recent molecular dynamics simulations of myoglobin [121,122]. To this list must be added a remarkable x-ray crystallographic study of ribonuclease A by Tilton et al. [123] at nine temperatures over the range from 98 to 320 K that provides the temperature dependence of individual atomic Debye–Waller factors (*B* factors: $B = 8\pi^2\langle x^2\rangle$). Figure 11 shows the temperature dependence of the average mean square displacements, $\langle x^2\rangle$, for all nonhydrogen protein atoms derived from the *B* factors for ribonuclease A, together with results from inelastic neutron scattering [119] and Mossbauer spectroscopic studies [118] of myoglobin. The values derived from

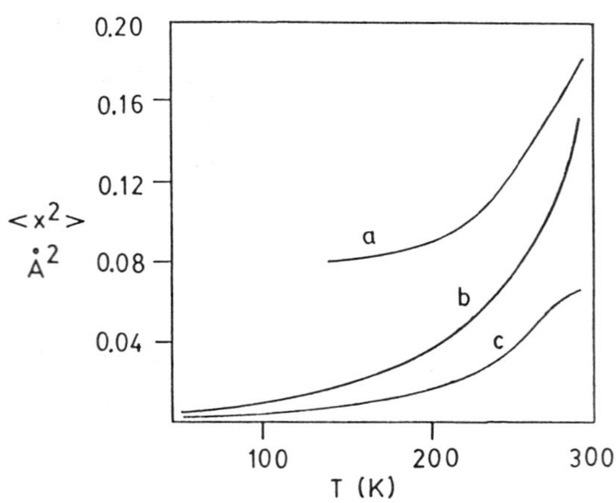

FIGURE 11 The biphasic temperature dependence of the average mean square displacements determined by (a) x-ray diffraction of ribonuclease A [123], (b) neutron scattering [119], and (c) Mossbauer spectroscopy [118] of myoglobin. Taken from Cusack [156].

B factors are somewhat higher than those measured by the other techniques, which may reflect some contribution from static lattice disorder. To a first approximation, however, static disorder is independent of temperature, and the same biphasic behavior as a function of temperature is seen with all three techniques. Below the transition region, the mean square displacements increase slowly with temperature (0.015 $Å^2$ per 100 K), but above the transition region the slope is much larger (0.081 $Å^2$ per 100 K). The behavior of the *B* factors with temperature for individual atoms is more complicated than the changes in average values would suggest, but we defer a discussion of this to a later section.

The atomic fluctuations measured by *B* factors and neutron scattering must be related to the local free volume available to atoms and groups of atoms to undergo positional fluctuations. But segmental motions are coordinated so the relationship between *B* factors and total protein free volume is complicated. Nevertheless, the biphasic temperature dependence of protein atomic mean square displacements is qualitatively consistent with the volume changes expected for a glasslike transition between 180 and 220 K. Tilton et al. [123] calculated the total protein volume using an accessibility probe to define the protein surface at each temperature studied between 98 and 320 K. There is some suggestion of a transition occurring at 220–240 K and perhaps also at 170 K, but given the considerable scatter in the data points, a linear dependence on temperature over the entire temperature range was assumed rather than the biphasic temperature dependence observed with the Debye–Waller factors. This gives a surprisingly small overall expansion of about 0.4% per 100 K, however, equivalent to a thermal expansivity of only 5.0×10^{-5} K^{-1}. Positron annihilation lifetime data show two transitions, and it is therefore possible that these are also real features of the protein volume defined from the x-ray crystallographic data.

Several studies also have revealed a broad transition in the hydration water centered at about 220 K. These include changes in the OD IR band and heat capacity of water in crystals of myoglobin [124], NMR relaxation studies of water in protein powders and crystals [72,125], RSMR studies [101], as well as ESR spectroscopic studies of spin probes [126] and ^{57}Fe-labeled ferricyanide diffused into protein crystals [127]. The transition is attributed to the melting of hydration water. The IR studies indicate that the transition temperature decreases with decreasing hydration and is not observed in spectra at low hydrations (0.22 *h*).

D. Hydration Dependence of Glass Transition Temperatures

The 200 K transition observed in proteins depends on hydration. RSMR of proteins at hydrations above 0.25 *h* show a sharp increase in the size of atomic mean square displacements ($\langle x^2 \rangle$) for the elastic scattering fraction above about 230 K, reaching values as high as 1 $Å^2$ at 300 K. By contrast, mean square displacements in dehydrated protein samples display a constant slope as a function of temperature between 100 and 300 K and only reach values of 0.05 $Å^2$ at 300 K [101].

Pissis [128,129] has recently employed thermally stimulated depolarization currents (TSDC) to monitor protein and solvent relaxations as a function of hydration. Partially hydrated protein samples are polarized in an electric field and are cooled to low temperatures, at which point the field is switched off. The sample is subsequently heated, and the depolarization current is monitored as a function of temperature. Such an experiment yields a series of depolarization bands at temperatures corresponding to the recovery of various motions in the system. Hydrated lysozyme samples display two complex depolarization bands: a "low temperature" band at 130–150 K assigned to dielectric relaxation of bound water, and a hydration-dependent "high temperature" band with band peak temperatures that decrease with increases in hydration from about 230 K at 0.13 *h* to 175 K at hydrations above 0.29 *h*. The high-temperature band has several contributions including proton percolation and the relaxation of protein polar groups. The latter contribution to the high-temperature band would appear to be closely related to the glasslike transition observed between 180 and 220 K in fully hydrated proteins.

Gregory and Chai [98,99] have recently embarked on a detailed study of the hydration and temperature dependence of positron and positronium lifetimes in lysozyme. The hydration dependence of o-Ps lifetimes suggested that lysozyme undergoes a glasslike transition with a transition temperature of 298 K at a hydration level of about 0.12 *h*. o-Ps lifetimes have also been determined as functions of temperature at 0.15 and 0.19 *h*. The results are shown in Fig. 12. The curves dis-

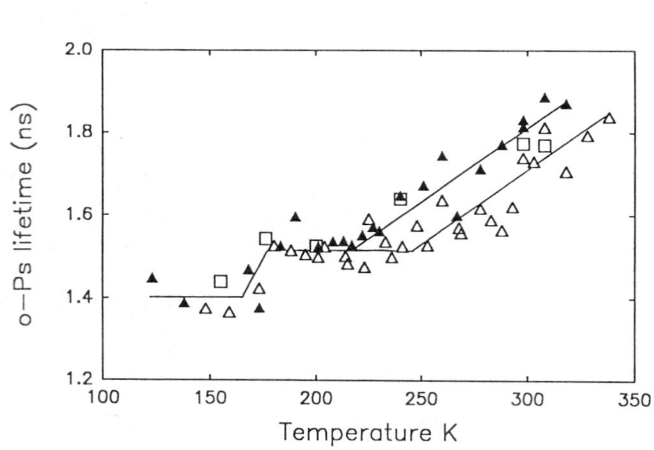

FIGURE 12 The temperature dependence of the mean o-Ps lifetimes for lysozyme at hydration levels of 0.15 *h* (open symbols) and 0.19 *h* (filled symbols). Line segments have been drawn to emphasize the transitions. The low-temperature transition occurs at a temperature of 160–170 K. The high-temperature (glass) transition occurs at about 220 K at 0.19 *h* and is shifted to about 240–250 K at 0.15 *h*. Data from Gregory and Chai [99].

play two transitions: a sharp "low-temperature" transition at 160–170 K, and a hydration-dependent "high-temperature" transition characterized by a change in cavity volume expansivity from essentially zero below the transition to about 0.6 $Å^3\ K^{-1}$ above the transition. The transition temperature at 0.15 *h* is about 240–260 K and decreases to about 220–240 K at 0.19 *h*. The hydration dependence of the transition temperature (i.e., for the "high-temperature" transition) together with those determined by other techniques are shown in Fig. 13.

There are two features we wish to emphasize: (a) the dynamical transitions observed at low hydration at 298 K and those observed in fully hydrated proteins at 180–220 K appear to be manifestations of a common hydration-dependent transition; (b) the hydration dependence, that is, the decrease in transition temperature with increasing hydration, is entirely consistent with the behavior expected for a glass transition in which water acts as a plasticizer of conformational rearrangements. The curve shown in Fig. 13 thus defines the temperature and hydration dependence of the transition between the rigid, glassy native state and the flexible native state. In the glassy state, the protein is frozen into a distribution of conformational states. Mobility is restricted to group rotations and small-amplitude vibrational and torsional motions. At sufficiently low temperatures, these motions would also be frozen. Transitions between conformational states are not possible

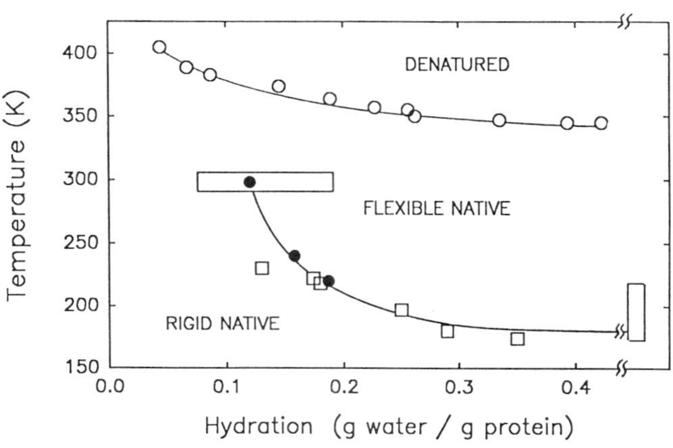

FIGURE 13 The hydration dependence of the glass transition temperature for lysozyme determined from positron annihilation lifetime spectroscopy [99] (filled circles), together with results from TSDC studies [128,129] (open squares). The glass transition regions determined from various hydration studies at 298 K and from studies of fully hydrated proteins (the 200 K transition) are indicated by bars. The hydration dependence of the denaturation temperature of lysozyme from calorimetric studies [54] is also shown (open circles). The two curves separate the rigid, glassy native state from the flexible native state (i.e., with flexible matrices and glassy knots) and the denatured state (i.e., with disrupted knots). See text for details.

because activation barriers are too high. In the dry state, it is likely that the barriers to conformational rearrangements are so high that transitions between conformational states never become probable at accessible temperatures (i.e., before chemical degradation of the protein begins). Water plasticizes proteins in much the same way it plasticizes Nylons, by providing alternative, mobile hydrogen bond donors and acceptors for peptide groups. This decreases barriers between conformational states and allows conformational transitions to occur at much lower temperatures. As is discussed in the next section, not all regions of the protein participate in this glass transition, and the heterogeneity in packing density and in the strength of interactions in those regions of the protein that are involved ensures that the glass transition region is broad.

The transition from the glassy to the flexible state is accompanied by an expansion in free volume. In the glassy state, noncovalent interactions and hydrogen bonding are maximized and the entropy of the protein is low. In principle, the configurational entropy would be close to zero in the glassy state, but dehydration-induced conformational changes and restrictions on conformational transitions as the protein approaches the glass transition cause the protein to become frozen into a broad distribution of conformational states. As the protein crosses the glass transition region into the flexible state, segmental motions and conformational rearrangements become possible and thermal expansivity is greatly increased. The enthalpy of the protein increases and noncovalent interactions and hydrogen bonding are weakened (i.e., in the flexible state "mobile defects" [25] appear). With mobile free volume now available, however, fluctuations of groups of atoms and segments of the protein into this free volume can occur with a concomitant increase in the configurational entropy of the protein, these fluctuations being catalyzed by the migration of water molecules through the protein interior.

The source of the transition observed at 160–170 K in positron lifetime studies of partially hydrated (0.15–0.19 h) lysozyme is not known. Pissis [128,129] observed a depolarization band in TSDC studies at 125–150 K, which was assigned to the dielectric relaxation of hydration water. A similar low-temperature depolarization band is observed in casein samples, which shifts to higher temperatures as the water content is decreased, but this shift with hydration is not observed in lysozyme. It is possible that the low-temperature transition observed in the positron lifetime results reflects a transition involving motions of side-chain groups.

E. Hysteresis Effects

The definition of the hydration dependence of the glass transition temperature coupled with dehydration-induced conformational changes may well provide a basis for understanding hysteresis effects in the sorption isotherms. One example will suffice to illustrate the idea. Many years ago Lumry et al. [130] reported differences in the magnetic susceptibility of samples of cytochrome c prepared by

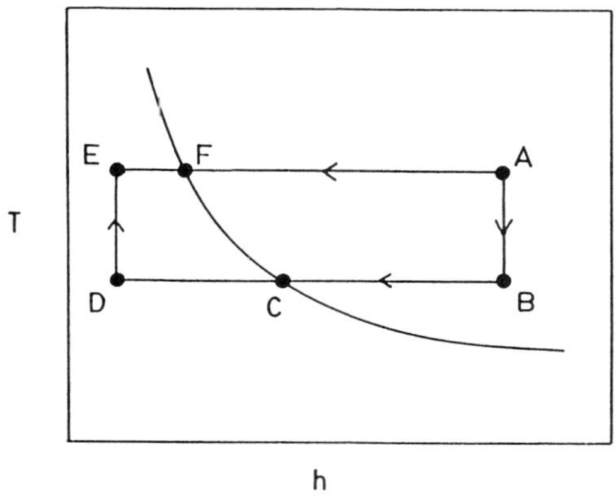

FIGURE 14 The hysteresis loop for dehydration of a protein from the fully hydrated state (point A) to the dry state (point E) by lyophilization (path A–B–C–D–E) and by drying at 27°C (path A–F–E). If continuous dehydration-induced conformational changes occur between points A and F and/or points B and C, the protein crosses from the flexible to glassy state in different average conformational states (points F and C). At point E, the two protein preparations would actually exist in different conformational states characteristic of the temperature and hydration at points F and C.

lyophilization and drying at 27°C. Figure 14 illustrates how hysteresis would result for these samples if dehydration-induced conformational changes occurred. Lyophilization follows the path A–B–C–D–E to the dry state, while drying at 27°C follows the path A–F–E. The two protein preparations cross the curve defining the hydration dependence of the glass transition temperature at different points, F and C. If continuous dehydration-induced conformational changes occurred (i.e., in the region A–F and/or B–C), the two protein preparations would be "frozen" into different average conformations in the rigid state; that is, at the temperature and hydration level corresponding to point E in Fig. 14, the two protein preparations would actually exist in conformations characteristic of points F and C.

VIII. DYNAMICALLY DISTINCT STRUCTURAL CLASSES IN GLOBULAR PROTEINS

An examination of *B* factors for almost any protein in the Brookhaven Protein Database provides evidence for their dynamic heterogeneity. The existence of dynamically distinct structural classes in proteins is also revealed by hydrogen isotope exchange studies [88,131]. Detailed discussions of this topic and its significance for protein structure, function, and stability have been given by Gregory and Lumry [88], Lumry and Gregory [132], and Lumry [133] (see also Chapter 1), and here we simply outline the important features and their connection to glass transition behavior.

A. Evidence from Hydrogen Isotope Exchange

Hydrogen isotope exchange has long been recognized as a powerful method for the study of protein conformational dynamics. The exchange rates of labile protons, including the peptide NH proton, can be monitored by a variety of techniques using deuterium or tritium labels [80–82]. The rate of the hydrogen isotope exchange reaction

$$R\text{CONH}^*R_1 + \text{HOH} \rightarrow R\text{CONH}R_1 + \text{HOH}^*$$

provides a measure of the accessibility of the exchange catalyst to the exchange site. The exchange reaction is catalyzed by H^+ and OH^- ions and for model amides and random coil polypeptides is extremely rapid (the rate constant for OH^- catalyzed exchange of poly (D,L-alanine) is about 10^8 L mol^{-1} s^{-1}). In globular proteins, however, exchange rates may be attenuated by factors of up to 10^{10} and are broadly distributed. The unique power of the method thus lies in its ability to provide dynamics information for the whole protein over a very broad time range. The last ten years have seen important advances in methodology with the development of NMR measurement of site-specific exchange rates [134–139], the availability of neutron diffraction exchange data in the form of deuterium site occupancies [140–143], and the development of numerical methods for the determination of exchange rate distribution functions of total protein exchange kinetics [18,19,88, 131]. These methods provide a far more detailed description of protein internal dynamics than had been possible previously.

The existence of dynamically distinct protein domains was first established from an analysis of exchange rate distributions [88,131]. Studies of total protein exchange kinetics give the number of hydrogens remaining unexchanged as a function of time, $H(t)$, which is usually expressed as a sum of exponentials, representing the parallel exchange of n exchange sites:

$$H(t) = \sum_{i=1}^{n} \exp(-k_i t) \tag{24}$$

When n is very large, as is the case for total protein exchange kinetics, the sum may be taken as an integral in which $H(t)$ is seen to be the Laplace transform of the exchange rate probability density function (pdf) $f(k)$:

$$H(t) = \int_0^\infty f(k) \exp(-kt)\, dk \tag{25}$$

Both expressions are notoriously difficult to analyze. With the development of the computer program CONTIN for the numerical solution of Fredholm integral equations by Stephen Provencher [18,19], however, it became possible to invert the Laplace transform and obtain reliable, physically meaningful exchange rate pdfs [88,131]. Examples of exchange rate pdfs for lysozyme are shown in Fig. 15. Each consists of three broad, overlapping peaks. The kinetics of the fastest exchanging protons (peak 3) are rather poorly defined by the tritium exchange technique. They exchange with activation energies only a few kilocalories greater than peptide protons fully exposed to solvent such as those in poly (D,L alanine) and with exchange rates that are attenuated by factors of up to 10^4 relative to those in poly (D,L alanine), behavior consistent with peptides in the protein–solvent interface with high solvent accessibilities. The exchange behaviors of protons associated with the intermediate (peak 2) and slow exchange (peak 1) peaks differ in a number of respects. The difference in the distribution of exchange rates that characterizes these two classes of protons is also reflected in the distribution of activation energies. The peak 1 protons exchange with much higher activation energies than the peak 2 protons. In addition, the activation enthalpies and entropies display an enthalpy–entropy compensation behavior characterized by two compensation temperatures with values of about 450 K for the intermediate peak and 350 K for the slow exchange peak [88]. Differences are also observed in the response of these two classes of protons to the addition of glycerol [90]. Originally, all these differences were interpreted in terms of exchange mechanisms involving folded and unfolded states of the protein. It became clear from NMR exchange studies of indole protons in lysozyme by Wedin et al. [138], however, that mechanisms involving unfolded states do not contribute to exchange at temperatures below about 55°C. This important work, together with a comparison of the activation energy pdfs for exchange of the peak 1 protons with the values expected for a thermal unfolding mechanism [88], established that exchange of protons associated with both the intermediate and slow exchange peaks must occur from folded states of the protein. As a consequence, the differences in exchange behavior for the two peaks has to be due to the existence of distinct exchange environments within the folded protein. The two dynamically distinct domains in which protons associated with the slow and intermediate exchange peaks are located were called "knots" and "matrices" [88], the term knot alluding to the fact that these exchange-stable regions must become "untied" in the cooperative, rate-limiting step leading to denaturation. Exchange behavior is consistent with an exchange mechanism involving gated catalyst diffusion to buried exchange sites through rearrangements of free volume (mobile defects). This is rate-limiting for exchange of protons associated with peak 2, but for those protons located in knots, it is supplemented by rate-limiting local distortions of the knots to expose peptide protons to catalyst and water molecules that have migrated into the protein matrix [88].

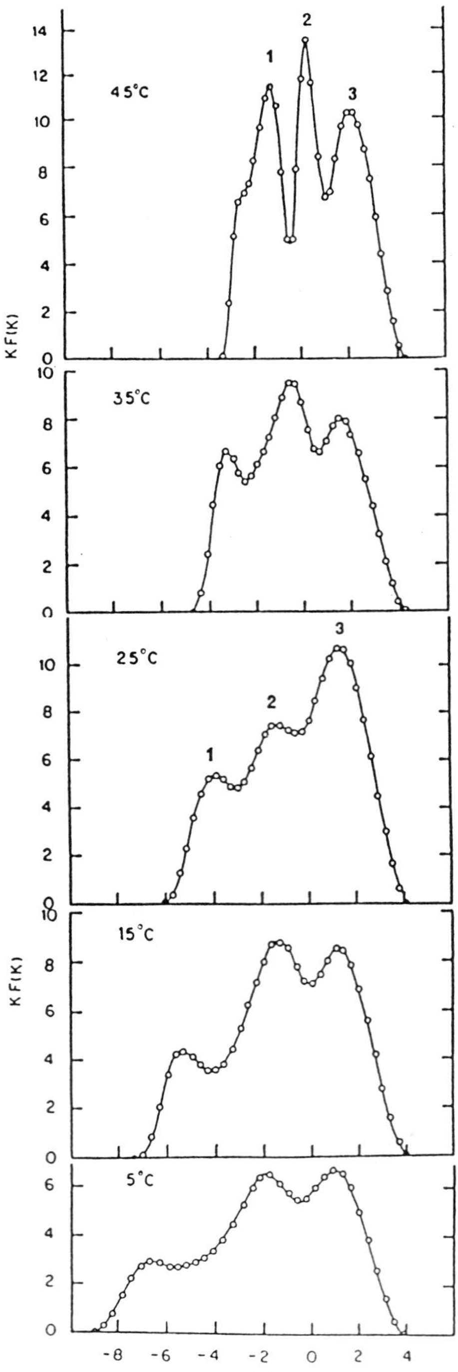

KF(K)
45°C
35°C
25°C
15°C
5°C
1
2
3
LOG K

Cooperative local unfolding of segments of the polypeptide chain to expose peptide groups to bulk solvent has long been proposed as an important mechanism of exchange in proteins [83–85]. The evidence accumulated to date, however, indicates that it is not a significant mechanism of exchange for most peptide protons, at least below 55°C. Experimental evidence favoring catalyst penetration mechanisms rather than local unfolding mechanisms includes (a) the small formal order with respect to water (three water molecules) for the exchange reaction and the constancy of this value for more than 100 exchangeable protons in lysozyme [29], (b) the low values of activation energies at least for the peak 2 protons [88], (c) inconsistencies in the influence of glycerol on the thermodynamics of exchange of lysozyme with that expected for unfolding models [145], (d) the correlation of acid and base catalyzed exchange rates with the static accessibility of carbonyl oxygens and peptide NH groups, respectively, in bovine pancreatic trypsin inhibitor (BPTI) [89], and (e) deviations in the pH dependence of exchange rates from that expected for exchange occurring in bulk solvent for slow exchanging protons in BPTI [145].

The existence of a distinct class of slow-exchanging protons was also established for ribonuclease A (see Fig. 7 in Ref. 133) and BPTI [88]. The exact number of slow-exchanging peak 1 protons is difficult to estimate given the overlap with the neighboring peak, but is approximately 16–22 in lysozyme, 10–16 in ribonuclease A, and 6–8 in BPTI. With the availability of exchange data from NMR and neutron diffraction methods, it is possible to locate these slow-exchanging protons within the protein structures. Residues with slow-exchanging protons are listed in Table 1. Several general features are immediately apparent. Residues with slow-exchanging peptide protons (knot residues) account for about 10–15% of the total residues in the protein and are dominated by nonpolar amino acid residues. Slow-exchanging protons are usually hydrogen bonded and are located in regions where helix–helix or helix–sheet packing occurs. Examination of the structures of lysozyme and ribonuclease A reveals that the knot residues are clustered together to form two well-packed regions, one on each side of the active site, which are closely associated with the essential catalytic residues (i.e., glu 35 and asp 52 in lysozyme and his 12 and his 119 in ribonuclease A) [88,146]. Each knot appears to be embedded in a more mobile protein matrix, which together may be regarded as a functional domain [132,133]. A similar picture is found on an examination of neutron diffraction data for trypsin [141]. Indeed, the pairing of functional domains each consisting of mobile protein matrix tethered to a rigid knot

---

FIGURE 15 Hydrogen exchange-rate probability density functions at five different temperatures for lysozyme determined by numerical Laplace inversion. Peaks 1, 2, and 3 correspond to the slow (knots), intermediate (matrices), and fast (surface) exchange classes, respectively. Taken from Gregory and Lumry [88]. Reprinted by permission of John Wiley & Sons, Inc.

TABLE 1 Assignment of Peptide and Side-Chain Protons Associated with the Slow Exchange Peak of Exchange Rate Probability Density Functions[a]

Lysozyme		Ribonuclease A		BPTI	
phe 3	β	gln 11	α	ile 18	β
		his 12	α	arg 20	β
leu 8	α	met 13	α	tyr 21	β
ala 9	α			phe 22	β
		val 47	β	gln 31	β
trp 28	α			phe 33	β
val 29	α	val 54	α	phe 45	β
ala 31	α	val 57	α		
ala 32	α	cys 58	α	asn 43	δ NH
lys 33	α				
		val 63	β		
phe 38	β	tyr 73	β		
asn 39	β	lys 104	β		
		ile 106	β		
asn 44	β	val 108	β		
asp 52	β	ala 109	β		
tyr 53	β	val 116	β		
gln 57	β	val 118	β		
ile 58	β				
leu 75					
leu 83					
val 92	α				
lys 96	α				
ile 98					
trp 28	indole NH				

[a]Based on a comparison with exchange rates determined by NMR [135] and amide hydrogen/deuterium occupancies determined by neutron diffraction [143,144].

appears to be a common construction principle of enzymes. The functional consequences of such a construction are described in detail by Lumry in Chapter 1.

B. The Basis of Knot Formation

Gregory and Lumry [88] proposed that the unusual exchange stability of the knot residues was due to a cooperative contraction of the knot to give a compact, rigid structure. The slow-exchanging peptides are located in hydrogen-bonded arrays in predominantly nonpolar environments. As a consequence of the low local dielectric constant, all electrostatic interactions are strengthened, hydrogen bond lengths will be shortened, and the surrounding nonpolar groups will be forced into closer contact with the array of hydrogen bonds. This restricts motions in the knots and reduces positional fluctuations of the peptide dipole clusters, which in turn reduces contributions to the local dielectric constant from orientational polarizability. The reduction in local effective dielectric constant leads to further enhancement of hydrogen bond strengths and further contraction until the repulsive limit is reached.

Comparison of hydrogen bond lengths and bond length fluctuations determined from x-ray diffraction and molecular dynamics simulations of BPTI with hydrogen exchange rates determined by NMR indicates that slow-exchanging protons are involved in hydrogen bonds that are significantly shorter than those of fast-exchanging protons, although there is no correlation between hydrogen bond length and the actual exchange rate [147]. The protons classified as "slow exchanging," however, included all those with exchange rates $<10^{-2}$ min^{-1}, giving a total of 14 protons, which thus includes a significant number of matrix protons that have short hydrogen bonds. Nevertheless, among the knot residues listed in Table 1 one does find three involved in unusually short hydrogen bonds (tyr 21, tyr 35, and phe 45 with NH–O distances of 1.71, 1.68, and 1.76 Å, respectively). The slow-exchanging (knot) peptide protons in ribonuclease A are all hydrogen bonded, and with a few exceptions these are found to be good hydrogen bonds (i.e., with NH–O distances <2.0 Å) based on an analysis of hydrogen bond lengths determined by neutron diffraction [142]. As is the case with BPTI, however, many fast-exchanging protons are also involved in hydrogen bonds of such lengths.

Quantitative correlations concerning packing and mass density characteristics of knot and matrix regions are rather sparse. This is not for lack of raw data, which, of course, are readily available from the Brookhaven Protein Database, but simple reflects that those studies of packing densities that have been performed occurred prior to the formulation of the knot–matrix construction principle and concerned other questions, and we have so far lacked the resources to conduct such a study ourselves. Approaches for determining packing densities are well described [148]. Schultz [149] has conducted an interesting study of the mass densities above and below the plane of internal β peptide groups for a number of proteins, although the need to exclude β peptides near the protein surface greatly reduced the total number of peptides that could be considered in some proteins. Nevertheless, the study provides important quantitative information about the densities in the vicinity of internal peptide groups. Densely packed regions are observed immediately interior to the catalytic residues in all the proteins studied (lysozyme, ribonuclease S, subtilisin-BPN', carboxypeptidase A, and α-chymotrypsin). In general, higher densities are observed in regions where the β sheets pack against other secondary structural elements. In those proteins for which exchange data is available or can be reliably inferred (i.e., lysozyme, ribonuclease A, and trypsin), the high-density regions correspond well with those knots involving elements of β sheets. One particularly interesting observation is that efficient packing is often sacrificed to optimize hydrogen bond geometries, and conversely, where hydrogen bonding is distorted, packing densities tend to be high [149]. Packing density and hydrogen bond strength need not always be compromised in this way, however, and it is possible that the knots represent just those regions of the protein interior where the disposition of residues allows both strong, if not always unusually strong, hydrogen bonding *and* efficient packing. Indeed, given the correlation between regions of high packing density and the location of knot residues noted above, this seems to be the case. To our knowledge, no comparable calculations of densities at α-helical peptides have been performed, but the observations of curvature in amphipathic helices [150] may be explained by the cooperative contraction process; that is, hydrogen bonds are shortened on the nonpolar face relative to hydrogen bonds on the polar face of the helix [88].

The calculation of local effective dielectric constants in protein systems is greatly complicated by the contributions from orientational polarizabilities and the effects of solvent. Nakamura et al. [151], however, have recently performed such a calculation for BPTI using a normal mode analysis to estimate dipole fluctuations. The local dielectric constant correlates closely with solvent accessibility, and contributions from orientational polarizability are found to dominate where positional fluctuations are large. As expected, the slow-exchanging knot protons are found in environments having the lowest local dielectric constant.

Examination of the amino acid composition of the knots in Table 1 indicates that the residues with slow-exchanging protons are predominantly nonpolar. Nonpolar amino acid residues (trp, phe, tyr, ala, leu, ile, val, met, and cys) account for 76% of all residues in the knots, but this is not significantly different from the amino acid composition (69%) found for the interiors of monomeric proteins [152]. Thus, the low dielectric constant for the knots cannot be due to any unusual nonpolar character relative to the rest of the protein interior. The only factors that would explain the variations in the local dielectric constant of the protein interior are the variation in orientational polarizability and solvent accessibility of the interior dipole clusters as found in the study of BPTI described above.

One feature of considerable importance to emerge from the available exchange data is that residues with slow-exchanging peptide protons are also characterized by low *B* factors for their main-chain atoms (see Chapter 1 by Lumry). Lumry distinguishes three groups of residue based on the size range of their *B* factors, which define the structural elements in which the three classes of exchange sites identified in exchange rate distributions are located. Residues with low (group I) and intermediate (group II) *B* factors are identified with the knots and matrices, respectively, while residues with the highest *B* factors (group III) are found only at the protein surface. Given the errors associated with *B* factor estimates and in defining the slow-exchanging protons, the correlation between the group I and knots residues is quite good, so that potential knot residues in proteins for which hydrogen exchange data are not available may be identified by examination of *B* factors available in the Brookhaven Protein Database. As we discuss below, the *B* factors also provide the key piece of information linking the dynamic properties measured by hydrogen exchange kinetics with glass transition behavior of proteins.

C. The Connection Between Hydrogen Exchange Properties and Glass Transition Behavior

1. Only the Matrix Regions Undergo a Glass Transition Prior to Denaturation

In their paper describing the effect of temperature on the structure and dynamics of ribonuclease A, Tilton et al. [123] present the temperature dependence of main-chain atomic *B* factors for 12 selected amino acid residues. Besides giving an idea of the range of glass transition temperatures, these data also reveal that 2 of the 12 residues do not undergo the 200 K transition and retain low *B* factor values (i.e., <7 Å^2) over the entire temperature range (98–320 K). These residues are met 13 and val 47, which have been identified as slow-exchanging (knot) residues by neutron diffraction. Ser 89 shows no clear transition but is characterized by *B* factors with a relatively strong temperature dependence, reaching *B* factor values well into the range that characterizes matrix regions at 300 K, and so the transition may be too broad to be seen. The temperature dependence of main-chain *B* factors for several residues, including met 13 and val 47, is shown in Fig. 16.

The behavior observed with met 13 and val 47 strongly suggests that only the surface and matrix regions of proteins undergo a glass transition prior to denaturation. The knots remain in a rigid state. As we have noted earlier, knots have the same amino acid composition as the remainder of the protein interior, and so their different properties cannot be due to chemical differences but instead occur because the knot residues can pack very efficiently without compromising the strength of their hydrogen bonds, while matrix regions cannot and must sacrifice one for the other to some extent. As a consequence, the knots are considerably stronger than matrices. The extent to which this appears in the free energies depends on the usual standoff between enthalpies and entropies. Knots have low enthalpies, but this is offset in the free energy by their low configurational entropies. Matrices have weaker interactions and higher enthalpies, but their free volume and configurational entropies are also higher. Lumry (see Chapter 1) has suggested that the tethering of the matrices to the knots may even force the matrices into "nonintrinsic" states, that is, states with higher free energy than they would otherwise have if their conformation was not restricted by the knots. The cooperative contraction of the knots produces packing of such efficiency that there is little free volume, certainly none sufficient to allow segmental displacements of the main chain, and thus water is unable to reach the peptide groups to plasticize the knots. Only rarely will distortions of the knots expose peptide protons for exchange. By contrast, the availability of mobile free volume ("mobile defects") in the matrices

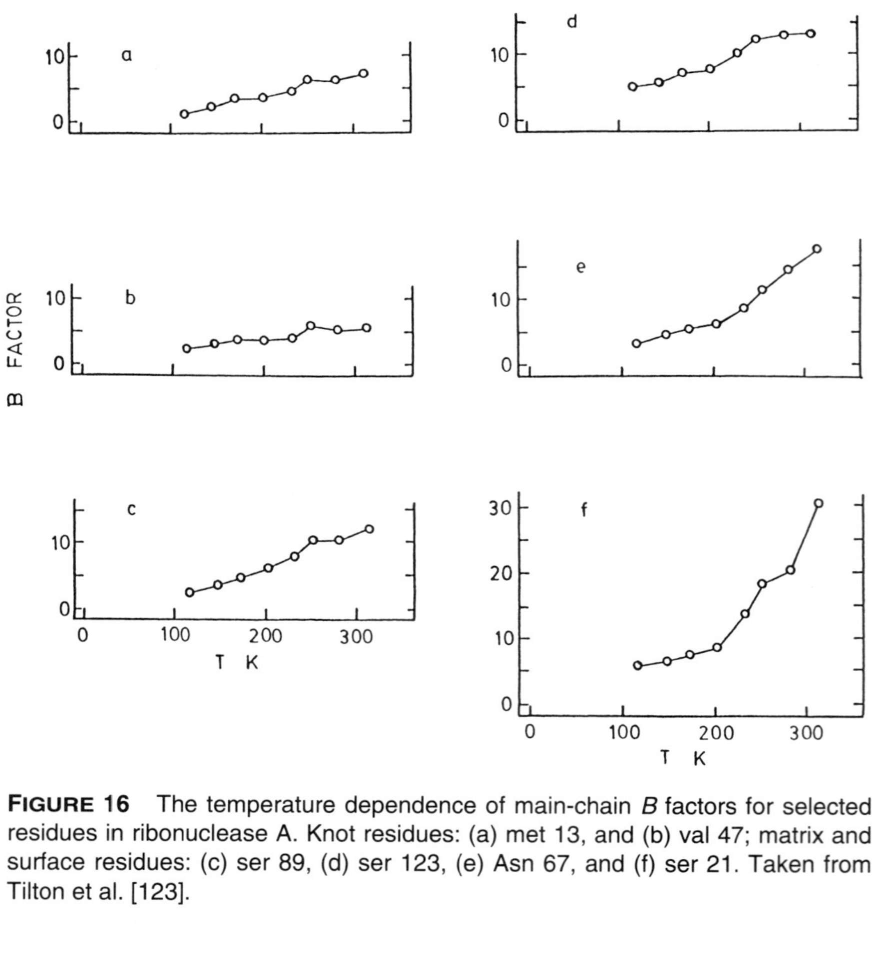

FIGURE 16 The temperature dependence of main-chain *B* factors for selected residues in ribonuclease A. Knot residues: (a) met 13, and (b) val 47; matrix and surface residues: (c) ser 89, (d) ser 123, (e) Asn 67, and (f) ser 21. Taken from Tilton et al. [123].

provides water with access to these regions of the protein interior, where it can act as a plasticizer. The different transition behavior of the knots and matrices is thus due to two main factors: the greater strength of the knots relative to the matrices, and the accessibility of plasticizing water to peptides in the matrices but not in the knots. Both factors will cause the knots to undergo a transition at much higher temperatures than the matrices.

2. Knot Disruption Occurs only on Denaturation of the Protein

There is no evidence for any temperature-dependent transitions between the 200 K glass transition and the denaturation transition that could be assigned to a glass transition in the knots. Thus, the strength and rigidity of the knots is largely maintained at all temperatures up to the denaturation temperature, and it is only on denaturation that the knots are disrupted (Fig. 13). We have suggested previously that the knots are responsible for the kinetic stability of the protein, so that cooperative disruption of the knots is what rate-limits denaturation of the protein as the knots expand and water enters and disrupts their peptide hydrogen bonding [88]. The denaturation transition, however, is not a glass transition. When the time scale of the measurement is long relative to relaxation times for segmental motion, glass transitions may be approximated as second-order transitions, characterized by smooth variations in the volume, entropy, and enthalpy with temperature but with discontinuities in thermal expansivity and heat capacity at the glass transition temperature. Protein denaturation is a weak first-order transition (weak because the size of the cooperative unit is small), and the temperature dependence of the heat capacity displays the characteristic peak at the transition temperature due to the relaxation term, $f_A f_B \Delta H_{AB}^2/kT^2$, where ΔH_{AB} is the enthalpy change for the transition between the native (A) and denatured (B) states. Both transitions involve conversion of parts of the protein from a rigid to a flexible state, so what additional factor is responsible for the first-order character of the denaturation transition if the rigid-to-flexible transition of the matrices is second order or at least can be approximated by a second-order transition in the thermodynamic limit? The answer appears to be the constraints imposed on the matrices by the knots, which force the latter to adopt conformations different from their intrinsic minimum free energy conformation (Lumry's "nonintrinsic" states). When the knots undergo a rigid-to-flexible transition, the tethered matrices are now free to adjust their conformation to minimize their free energy. Thus, rearrangement of the conformation and loss of secondary structure occurs as the protein moves from the native to the denatured state. These are distinct macrostates with different enthalpies and volumes, and the transition is first order. The glass transition of the knots leads to a flexible state that is the *transition state* for denaturation. If the matrices were at their intrinsic minimum free energy at the transition temperature, there would be no driving force to adjust the conformation, and disruption of the knots would resemble a second-order transition. The fact that denaturation is a first-order transition would appear to be evidence that matrices have a free energy that is higher than their intrinsic free energy minimum.

We have previously suggested that a significant part of the increase in heat capacity observed on denaturation might be due to the relaxation of constraints enforced on the protein by its knots [88]. The relative contributions to measured heat capacity changes in protein processes from hydrophobic hydration and increased flexibility has been a matter of some debate [88,153]. It is widely accepted that the difference in heat capacity for the native and denatured states can be attributed to the exposure of hydrophobic groups to bulk water on denaturation. This conclusion, however, is not consistent with some other data for protein denaturation. On the basis of small-molecule hydrophobes, one would expect hydrophobic hydration to produce a large decrease in system (i.e., protein + water) volume, and yet volume changes on denaturation are found to be very small [154,155]. In addition,

ΔC_p would be expected to decrease with increasing temperature if significant exposure of nonpolar groups accompanied denaturation, yet heat capacity changes for denaturation are found to be essentially temperature independent. The small expansion that apparently accompanies denaturation would also suggest that significant increase in the exposure of hydrophobic groups to water may not occur. On the basis of a comparison of temperature-induced phase transitions in proteins and lipids, Bendzko et al. [153] concluded that increases in heat capacity due to an increase in rotational freedom of polymer chains would be too small to detect calorimetrically. Significant heat capacity changes, however, are often observed in differential thermal analysis and differential scanning calorimetry of synthetic polymers through the glass transition region, which are attributed to the onset of polymer segmental motions [103].

Incidentally, there are examples of polymers consisting of regions with different interaction strengths that undergo two distinct glass transitions. These are the block copolymers, which tend to phase separate to give domains of different chemical composition, each defined by their own glass transition temperature. It is possible that a simple block copolymer could be constructed in which the drive to lower the free energy of one domain (the "master domain") through improved packing forced the other domains ("slave" domains) to adopt conformations with free energies higher than their intrinsic minimum free energy. The master domain must, by definition, be better packed and have a higher glass transition temperature than the slave domain so that the system would display a glass transition in the slave domain, followed by a glass transition in the master domain and simultaneous relaxation of constraints on the slave domain to give a first-order transition. Such a sequence of events appears to be a natural consequence of the existence of dynamically distinct polymer domains in which one domain forces others to adopt nonintrinsic states. In proteins, dynamically distinct regions are not produced by phase separation of polypeptide segments of different chemical composition, but by the cooperative contraction process made possible by the discovery and selection during evolution of segments of the polypeptide chain that can fold together without compromising good packing for strong hydrogen bonding.

The extent to which the knots and matrices interact with one another is likely to be quite variable. Matrices are tethered to the knots, and so the strength of the latter must be influenced to some extent by the properties of the matrices. For example, changes in pH that alter interactions in the matrices can influence the strength of the knots directly through conformational strain, or by changing solvent accessibility to the protein interior, which in turn can influence transitions that the knots undergo. Ligand binding, a function designated primarily to matrices, usually tightens the matrices and may reduce solvent accessibility to the protein interior. Molecules such as urea appear to function as very efficient plasticizers of proteins (more efficient than water) and enter the protein and disrupt the knots at much lower temperatures than water.

D. "Molten Globule" and Cold-Denatured States

Our discussions have focused on the glassy and flexible native state with some comments on the heat-denatured state and the nature of the transitions between these states. Two other states are known, the "molten globule" state [157] and the cold-denatured state [158]. Detailed discussion of the properties of these states is beyond the scope of the present chapter, and here we simply offer a few comments on the possible hydration dependence of the transitions between these states and the native and heat-denatured state.

Lumry (see Chapter 1) has discussed the basis for formation of the "molten globule" state in proteins with multiple functional domains in terms of the disruption of only one of the knots. In most small globular proteins, the functional domains are well coupled and the knots have similar strengths, so that they are disrupted under very similar conditions. In these systems, the protein displays the familiar all-or-none behavior in its denaturation transition, and the population of intermediates is negligible. In some proteins or under certain conditions, however, the strength of the knots may be sufficiently different that disruption of one occurs before significant disruption of the others, leading to a state, the "molten globule" state, with properties of both native and denatured states. There is every reason to expect that the hydration dependence of the transition from the flexible native state to the "molten globule" state will be similar to that observed for the all-or-none transition to the heat-denatured state.

Recent studies of cold-denaturation of phosphoglycerate kinase indicate that the cold-denatured state has a much larger molecular volume than the heat-denatured state and that it closely resembles a random coil configuration typical of polymer chains under theta conditions, that is, the point at which polymer–polymer interactions are balanced by polymer–solvent interactions [159]. This is consistent with interactions between water and nonpolar groups becoming much less unfavorable as the temperature is lowered. The amount of water associated with the cold-denatured state is therefore very large, perhaps involving several thousand water molecules for a protein the size of lysozyme. This corresponds to a hydration level in excess of 2 g water/g protein, and thus the cold-denatured state will only be accessible at hydration levels greater than those required for collapse of hydrated protein powders to true solutions.

IX. PROTEIN FOLDING

The prediction of the three-dimensional structure of a protein from its amino acid sequence (the "protein folding problem") is one of the most important challenges in molecular biology. As Lattman and Rose [160] discuss, much of the effort to date has taken a thermodynamic point of view based on Anfinsen's thermodynamic hypothesis [161]: that if the folded state is at the global free energy minimum, in principle, a determination of the factors that stabilize proteins and their contribution to the free energy of the protein system should allow the prediction of the native fold for a particular sequence.

Dill [162] has emphasized the importance of hydrophobicity over local forces in generating compact protein structures with nonpolar cores. Such a collapse of the unfolded protein in a poor theta solvent greatly reduces the conformational space available to the protein simply due to the compactness of the resulting structure. There is still a large number of possible conformations, however, and this number increases exponentially with chain length [163]. Lattice simulations of polymer chains consisting of only two residue types, polar (P) and hydrophobic (H), in which the native states are defined by lattice configurations with the maximum possible number of HH interactions, indicate that there are very few ways to configure a chain so as to form this maximum possible number of HH interactions [164]. A much larger number of sequences lead to a single native structure than lead to different native structures. For a protein consisting of 100 residues, there are 10^{130} possible sequences and 10^{120} sequences that will fold to give the same backbone conformation as a given native structure. Thus, the sequence space defining a particular fold, formed by maximizing the number of HH interactions, is very large. These considerations suggested that maximizing interactions among hydrophobic groups is the dominant factor in reducing the size of the conformation space, although it cannot be the sole factor in defining the single native conformation [162].

Consideration of the two-state behavior of denaturation and the influence of mutations on protein stability led Lattman and Rose [160] to conclude that the overall conformation of a protein is independent of its stability. As an example, they consider the introduction of several destabilizing mutations that shift the equilibrium so that the folded species represents only a small fraction of the total protein, yet this rare folded species still adopts the native fold. Lattman and Rose argue that since it is not energy per se that determines conformation, conformational specificity rather than stability might be a more fruitful focus for future studies of the protein-folding problem.

The accumulating evidence from site-directed mutagenesis experiments suggests that distributed rather than centralized control of the conformation occurs and that the conformation is likely to be overdetermined by such a distributed control [160] and thus be tolerant of amino acid substitutions. The packing of internal side chains is an obvious candidate. Protein interiors have high packing densities as a result of the high degree of complementarity between side chains. The importance of internal packing in defining conformation is also suggested because many mutations of surface residues can be tolerated with little effect on stability, while mutation of internal residues tends to be far more destabilizing. Internal residues also tend to be more highly conserved than surface residues within protein families. This type of evidence relates to the stability of the protein, however, and following the argument of Lattman and Rose [160] we are still left with the question: Does packing define the conformation? In fact, an analysis of patterns of complementarity within proteins suggests that good internal packing is not a major determinant of conformation [165]. No preferred packing interactions between pairs of side chains are found. The naturally occurring nonpolar residues appear to be able to pack together efficiently in a large number of ways. Thus, "the native fold determines packing, but packing does not determine the native fold" [165]. The elegant cassette mutagenesis experiments of Lim and Sauer [166] on λ repressor indeed reveal that there are many ways to repack the hydrophobic core and maintain the native fold in this protein. Volume, steric, and compositional constraints, however, do restrict the size of the sequence space. Not all hydrophobic core sequences can fold, but of the three constraints, hydrophobicity was found to be the most essential feature of the core residues. Volume constraints were the least important. Many different combinations of core residues were found to be accommodated in the available core volume. Volume and hydrophobicity do not account for all the restrictions on the sequence space, and Lim and Sauer concluded that additional steric constraints must operate.

Since packing does not appear to determine the protein conformation, what does and where is this stereochemical information encoded along the amino acid sequence? We suggest that the most important stereochemical information is encoded within the sequences of the knot regions. The importance of these regions in protein structure is discussed in detail by Lumry (see Chapter 1). The knots appear to be the most significant structural element in globular proteins, responsible for both the structure and the dynamic properties of the functional domains in which they are located [88]. Kim et al. [167] have come to similar conclusions on the basis of studies of destabilizing mutations in BPTI and an analysis of protection results in refolding experiments [168–173]. They propose that the "slow exchange core" (the "knot" in our terminology) of BPTI is the folding core and that the segments of structure that form the slow exchange core determine the basic fold of the protein. The slow exchange core of BPTI is suggested to consist of the NH groups of the central β-sheet residues 21–23 together with the backbone atoms of residues 20–24, 30–33, and 45 and the side-chain atoms of residues 20–23, 30, 33, 43, 45, and 51, that is, "three to eight of the slowest exchanging protons and the atoms which pack them" [167]. Kim et al. mistakenly state that the knot, as defined by the distribution of slow exchange rates, is "far larger" than their slow exchange core and includes 20–40 of the slowest-exchanging protons. This is certainly not the case in BPTI. In fact, the size of the knot originally defined in BPTI by integration of the exchange rate pdf was about 6–8 residues (residues 21–23, 31, 33, 45, and possibly residues 20, 43, or 18) [88]. Thus, the "knot" and "slow exchange core" have very similar sizes and residue compositions and are essentially one and the same structure. Perhaps the point to be made is that some residues may contribute side chains to the packing of the knots that do not have slow-exchanging peptide protons. These can be identified from their B factors but obviously would not be detected in exchange experiments unless the side chain included an exchangeable proton, as in the case of asn 43 (δ NH) in BPTI or trp 28 (indole) in lysozyme (see Table 1). As we have noted earlier, the number of knot residues will depend on the size of the protein.

Plochocka et al. [174] have developed a procedure for identifying hydrophobic microdomains in proteins of known structure. Their computational procedure selected only internal residues, as defined by their solvent accessibilities, and omitted short-range interactions ($i - j < 4$). Clusters of residues were then identified from a list of all interacting pairs (residues within 5 Å of each other). They propose that the recognition of residues forming hydrophobic microdomains is an important step in protein folding and that the basic fold is determined by interactions between residues that are distant in the sequence. About a third of the residues of the hydrophobic microdomains identified in lysozyme and ribonuclease are knot residues, as defined by their exchange rates. Each hydrophobic microdomain contains at least one knot residue. Most of the remaining knot residues are found to be adjacent to residues of the hydrophobic microdomains. Residues forming the hydrophobic microdomains are highly conserved in the amino acid sequence. Indeed, in lysozyme, they are better conserved than the knot residues themselves. The hydrophobic microdomains thus largely define the knot residues and the side chains that pack them. The latter are important for filling free volume and reducing the local dielectric constant and mobility of the knots.

The finding that destabilizing mutations in BPTI accelerate the exchange of those protons (the slow exchange core) which exchange via an unfolding mechanism but have little effect on those residues which exchange from native states [167] indicates that the slow exchange core is important for thermal stability. Gregory and Lumry [88] originally estimated a contribution to the total folding stability of proteins from the knots of about -3– -8 kcal mol^{-1}. Effects on protein stability, however, cannot be taken as evidence that the slow exchange core is the source of conformational specificity (the Lattman and Rose argument again). More pertinent to the latter question are the results from protection studies in refolding experiments for BPTI, lysozyme, and cytochrome c [168–173]. An analysis of these results reveals that the slowest-exchanging protons in the native structure are also protected early on in the refolding process [167]. Kim et al. [167] suggest that the stages of folding may actually correspond with regions of the native protein ranked in reverse order of their exchange rates, that is, initial establishment of the slow exchange cores (knots), packing of secondary structural elements that are not in the cores (matrices), followed by packing of flexible loops (surface elements). These three classes of dynamically distinct protein structure were originally identified in exchange rate pdfs for lysozyme [88].

Of particular importance, in light of the discussion by Lattman and Rose [160], is the fact that the knots appear to be responsible for the kinetic stability of proteins [88]. That is, they establish the two-state character of the denaturation transition of the functional domains in which they are located. As we have discussed, the knots remain glassy and maintain their structural integrity in the folded state. Formation of the transition state for denaturation occurs as the knots expand, become plasticized by water, and undergo a cooperative rigid-to-flexible transition. The conformation of the folded state is maintained so long as the knot regions retain their integrity; that is, knots contribute to the thermal stability, but they also *define* the conformation, and we now turn to an examination of the special characteristics of the knots that make this possible.

The knot residues are distributed throughout the amino acid sequence and in lysozyme and ribonuclease A appear as isolated residues in the sequence or as runs of two or three amino acid residues (see Table 1). The content of hydrophobic residues in the knots is not significantly different from other regions of the protein interior. They pack together very well, and on the basis of the analysis by Schultz [149], their packing is somewhat better than that which typifies the rest of the protein interior. What most distinguishes these regions from others is the synergism between good packing, strong hydrogen bonding, and the optimal interaction among peptide dipoles. Knot residues can pack very efficiently without sacrificing good hydrogen bonding (the cooperative contraction process described in Sec. VIII.B). As we have discussed, there are many ways to pack nonpolar side chains into a given volume, but the additional steric constraints that allow such packing to occur while maintaining strong hydrogen bonds and appropriate orientations and interactions among the peptide dipoles of the knots will greatly reduce the size of the sequence space. The initial collapse of the polypeptide gives a compact globular form with a nonpolar core. Its conformation, however, would not be uniquely defined. This initial compact state is sufficiently flexible for interior residues to sample a number of different interactions with each other. These rearrangements will be facilitated by the plasticizing effects of water. Formation of the compact state eliminates a large fraction of the water associated with the denatured state, but the water content of the initial compact state is still considerably larger than that of the native state. Many arrangements of internal groups will yield good side-chain packing and will bring potential hydrogen bond donors and acceptors into close proximity to initiate "trial" segments of secondary structure. Many such trials will be possible, but will not have sufficient strength to persist and be selected relative to other packing arrangements. The interaction of residues destined to form the knots and the local elimination of water, however, initiates the cooperative contraction process. The knot residues pack well together, but their arrangement in the core also allows strong hydrogen bonds to develop and optimizes the interactions between peptide dipoles. The strength of the knots is not determined just by neighboring residues in the sequence. Other residues from distant parts of the sequence contribute to fill free volume and further reduce mobility. The formation of the knots thus defines the topology of the polypeptide chain as a series of loops fixed at each end by the knot residues. The loops, destined to become matrix and surface regions, minimize their free energy as best they can, forming more extensive secondary structure and surface loops, subject to the constraints enforced on them by the knots, which may force the matrices to adopt "nonintrinsic" states, that is, the sequence of events in refolding will occur ranked in reverse order of exchange rates, as suggested by Kim et al. [167].

Hydrophobic collapse is not enough to uniquely define the native conformation. It will greatly reduce the size of the conformation space, but the compact state that results will probably resemble the transition state for unfolding rather than the native state, that is, a compact, but highly flexible state, with *no* glassy regions (knots). It is the specific interactions among the knot residues that select the native conformation from the distribution of conformations of the compact state. As noted above, Dill [162] has estimated that within the sequence space (10^{130} sequences) for a polypeptide chain of 100 amino acids there are 10^{120} sequences that will code for a particular conformation. This represents the number of sequences that will give a compact state of a particular conformation by maximizing HH interactions. Given the additional stereochemical requirements for formation of knots, however, the fraction of sequences that have appropriate residues in the correct positions to initiate the cooperative contraction process, form knots, and generate the native state is expected to be very much smaller. As we have discussed previously, formation of knots may well lead to a conformation in which hydrophobic interactions, overall, are not maximized. The matrices will be forced into "nonintrinsic" states by the knots and optimum side-chain packing may be sacrificed in some regions to improve hydrogen bonding.

Probably the most important factor in defining conformational specificity is the distribution of knot residues along the amino acid sequence. These have been defined on the basis of exchange rates, but as noted above, other residues that do not have slow-exchanging peptide protons may contribute side-chain atoms important for restricting mobility in the knots and thus contribute to the cooperative contraction process. In this regard, information from *B* factors will be invaluable in identifying these critical side chains. The knot residues impose *proximity constraints* on the amino acid sequence, which define the native fold. The knot residues, at least as defined by exchange studies, appear singly or as groups of two, three, or more neighboring residues along the sequence. In lysozyme, the knot residue groups appear as seven single residues, four doublets, and two triplets. The number of intervening (matrix and surface) residues between knot residue groups varies widely. For example, in lysozyme, the number of intervening residues varies from one to 17, and in ribonuclease A this number varies from one to 32. In lysozyme, the last knot residue is at position 98, leaving a run of 31 amino acid residues to the C-terminus. The sequence of lysozyme written to emphasize the length of runs of knot residues (K) and matrix and surface (M) residues reads

N term
2M–K–4M–3K–17M–2K–M–3K–4M–2K–4M–K–7M–2K–3M–2K–16M
–K–7M–K–8M–K–3M–K–M–K–31M C term

As we have noted, some residues listed as matrices may contribute side chains to the knots that are important in reducing knot mobility but are not revealed in exchange experiments.

Knot residues are organized into clusters of one or more knots. The segregation of knot residues into knots and the organization of knot residues within each knot is defined by the stereochemical factors that make cooperative contraction possible. These organizing factors are local, that is, constraints on efficient packing at helix–helix and helix–sheet contacts (common sites of knots residues), constraints on the residue types that can fill surfaces of secondary structures, interactions among peptide dipoles, the response of arrays of hydrogen bonds to their environment, etc. These interactions must be sufficiently specific to prevent initiation of the cooperative contraction process in nonnative associations of internal residues. A similar clustering of hydrophobic microdomains to form two hydrophobic clusters in lysozyme has been noted by Plochocka et al. [174].

The lysozyme fold is different from the ribonuclease A fold because the distribution and organization of knot residues to form knots are different in these two proteins. In lysozyme, one functional domain is defined largely by knot residue groups at the N-terminal half of the sequence, and the other domain by knot residue groups at the C-terminal half. Each of the two functional domains of ribonuclease A is defined by knot residue groups from all parts of the sequence. Imagine systematically mutating the lysozyme sequence toward the ribonuclease A sequence. One strategy might be to initially simplify the matrix sequences, say, by mutating acidic, basic, polar uncharged, and nonpolar matrix residues to a reduced generic set, say, asp, lys, ser, and ala, respectively. Would such a sequence retain the lysozyme fold? Assuming it did, one could begin systematically introducing mutations at lysozyme and ribonuclease A knot residue positions to replace lysozyme knot residues by their appropriate generic variety and introduce the ribonuclease A knot residues (because of the difference in the length of the polypeptide chains of lysozyme and ribonuclease A, five residues would need to be removed during the process). At which point would the lysozyme fold be eliminated, and at which point would the ribonuclease A fold emerge? What would emerge between these points in the process? Is it possible that, at some point in the process, a mixture of molecules, some adopting the lysozyme fold and others adopting the ribonuclease A fold, would be found, even if present as only a very small fraction of the total protein?

If the knots are the major determinants of protein conformation, how should we address the "protein folding problem"? From a thermodynamic point of view, the problem is one of recognizing and describing, in potential functions, the special properties of the knots relative to the matrices. Unfortunately, the synergism of electrostatic factors that provide the basis for knot formation (i.e., the subtleties of environmental effects on hydrogen-bond strength and geometry) are far from understood. Even if the special electrostatic properties of the knots could be so described, the size of the conformation space that would need to be searched is enormous. It may be possible to reduce the size of the conformation space by using lattice simulations on a simplified representation of the polypeptide chain in three dimensions with a suitable model for the forces that drive the chain to a compact state. We need to define, however, the correct model for formation of the compact state. The thermally denatured state, or even better, the transition state for denaturation, is likely to be a far better guide to this model than the folded state. It would then be necessary to test fit potential knot residues to see if the special characteristics of the knots emerged.

The problem looks no less daunting from a structural point of view. If, as we believe, the most important stereochemical information defining the conformation is encoded in the knot sequences, the protein folding problem could be replaced by what we might call the "knot–matrix problem" (i.e., the "*B*-factor problem" or "slow exchange problem"). That is, we must predict from the amino acid sequence which residues are located in knots and have low main-chain B factors and slow-exchanging peptide protons. An initial screening of the amino acid sequence for regions where potential knot residues are located could take advantage of the fact that knot residues and residues forming hydrophobic microdomains, which are closely associated with knot residues, tend to be more highly conserved than other residues in protein families. To be sufficiently selective, one would need to compare members of a protein family with low overall sequence homology. Of course, residues are conserved for other reasons, and so not all conserved residues would be knot residues. Nevertheless, such a screening would yield an initial set of proximity constraints for use in generating a set of potential folding topologies. Having labeled each position in the amino acid sequence as a potential knot (K) or matrix (M) residue, it is then necessary to determine how the knot residues are organized into one or more knots to define a set of folding topologies consistent with the proximity constraints imposed on the structure by the distribution of knot residues in the sequence. Knot residues cluster to form knots that define the conformation of the functional domain in which they are located. The knots appear to constitute 10–15% of the residues in a functional domain, and so selected topologies should have clusters of about this size. The development of rigorous mathematical descriptions of protein topologies [175,176] may allow determination of the different folds that could be generated subject to such proximity constraints. Additional proximity constraints could be introduced at this point to further reduce the size of the set of potential folds. For example, one might select from the set of knot-constrained topologies those folds which bring pairs of conserved cysteine residues into sufficiently close proximity to form disulfide linkages. Disulfide linkages are usually very highly conserved and would serve as very efficient filters of likely topologies. A similar strategy might be employed with potential salt linkages. Examination of protein crystallographic data may yield patterns among knot and matrix residues that might further restrict the selection of topologies. For example, it may turn out that there are preferred packing interactions among the knot residues that were not apparent in a survey of the whole protein interior [165]. In addition to proximity constraints, additional physiochemical constraints could be employed. For example, Finkelstein and Nakamura [177] have shown that "weak points" at the surfaces of antiparallel β sheets cannot be filled by their surrounding aliphatic side chains, but only by aromatic side chains from the same sheet or by residues from other parts of the protein. The known preferences for certain packing arrangements between different secondary structural elements, that is, at helix–helix and helix–sheet contacts [178] and tertiary templates, [179] represent other examples.

It remains to be seen if solution of the "protein folding problem" is facilitated in any way by its restatement as the "knot–matrix problem." Nevertheless, we believe that the identification of regions of a protein that specify its conformation (the knots or slow exchange cores) and the process responsible for their special strength (the cooperative contraction process) represent advances of some utility in the rational design and modification of proteins by site-directed mutagenesis.

X. CONCLUSIONS

Our survey of protein hydration has taken us from sorption isotherms and hysteresis to dehyration-induced conformational changes, glass transitions, and the plasticizing action of water. In the process, we have been led to a model of protein dynamic behavior in which free volume rearrangement and the plasticizing action of water play a dominant role. The chief virtue of the model of protein dynamic behavior described here is its ability to provide a unifying explanation for a number of seemingly unconnected observations, that is, hydration-induced flexibility, the 200 K dynamical transition, the dynamically distinct structural classes identified by hydrogen isotope exchange studies, and the differences in the temperature dependence of x-ray crystallographic *B* factors. We summarize the important conclusions:

1. As we stated in the introduction to this chapter, we have become increasingly convinced that the size of the population of mobile buried water and its importance to protein dynamics has been greatly underestimated. While we have no direct estimate of the size of this population, several lines of evidence suggest that it is considerably larger than the population of internal water observable by x-ray diffraction. Dry proteins are rigid and glassy but are plasticized by water, which appears to enter the protein interior early in the hydration process to provide alternative mobile hydrogen-bond donors and acceptors for

the peptide groups and thus to facilitate segmental motions and free volume expansion and rearrangement.

2. The dynamical transitions observed at low hydration at 298 K and those observed in fully hydrated proteins at 180–220 K represent a common glass transition in which water acts as a plasticizer of the protein. At low hydrations, the glass transition temperature is much higher than the measurement temperature and the protein is rigid. As the hydration level (plasticizer content) of the protein is increased, the glass transition temperature decreases rapidly. Once sufficient water has been added to lower the glass transition temperature below the measurement temperature, the protein becomes flexible. This occurs at a hydration of 0.07–0.2 *h* at 298 K; that is, the glass transition temperature in this hydration range is about 298 K. With further increases in hydration, the glass transition temperature continues to decrease, reaching a value of 180–220 K at full hydration.

3. Proteins undergo dehydration-induced conformational changes that vary in extent from protein to protein. These occur predominantly at low hydrations and may involve continuous changes in conformation, in some cases leading to a broad distribution of conformational states (static disorder) in the dry protein. The extent to which conformational changes appear in the dry protein is probably dependent on where along the desorption isotherm the changes occur relative to the glass transition. These continuous conformational changes coupled with the fact that the transition from the flexible to rigid state is both temperature and hydration dependent offer an explanation for certain hysteresis phenomena, particularly the differences in conformation that are observed when proteins are dried under different conditions.

4. Many enzymes appear to be constructed from two "functional domains," each consisting of a glassy, rigid "knot" embedded in a more mobile matrix. The knots would appear to be the most significant structural element in globular proteins and to define the conformation of the functional domains in which they are located. Apparently, there are dynamically distinct regions in globular proteins because the knot residues can pack very efficiently without compromising the strength of their hydrogen bonding. As a consequence of the low dielectric constant in the knot, the strength of all electrostatic interactions is enhanced, hydrogen bonds are shortened, and the knot region contracts. This reduces mobility and lowers the local dielectric constant, which leads to further enhancements of hydrogen-bond strength and further contraction. This cooperative contraction process continues until the repulsive limit is reached. In matrices, there is a tendency either to sacrifice optimum packing for good hydrogen bonding or to pack well at the expense of distorting hydrogen bonds from their optimum geometry. In addition, conformational constraints imposed on the matrices by the knots force matrices to adopt conformations with free energies higher than their intrinsic free energy minimum (Lumry's "nonintrinsic" states). In knots, good packing and strong hydrogen bonding are cooperatively enhanced. In matrices, one tends to be sacrificed for

the other. Enhancement of the strength of the knots by the cooperative contraction process occurs at the expense of the matrices, which further exaggerates the difference in strength of the two regions. As a consequence of this construction, only the matrices undergo the 200 K glass transition, while the knots remain glassy. Only on denaturation do the knots become flexible, but the simultaneous relaxation of the constraints enforced on the matrices by the knots now allows the matrices to adjust their conformation to minimize their free energy, so that a first-order denaturation transition rather than a second glass transition is observed.

ACKNOWLEDGMENTS

Many of the ideas developed in this chapter evolved through innumerable discussions with Dr. Rufus Lumry. He has been a constant source of inspiration and encouragement, and I am indebted to him. Special thanks are also due to Dr. Stephen Provencher for his advice over the years in the application and development of CONTIN, which has been so important to the analysis and interpretation of our hydrogen exchange and positron lifetime data. Thanks are also due to Dr. Frederick Walz for many helpful discussions of protein structure and folding, to Dr. Kyung-Jin Chai, who performed the positron annihilation lifetime studies and helped in the preparation of many of the figures, and to Dr. Stephen Prestrelski for providing his manuscript prior to publication. Part of this chapter was written while the author was on sabbatical leave at the University of Bordeaux and a special note of thanks goes to Dr. Christian Courseille and the other faculty and staff of the Department of Crystallography for their very generous hospitality. Our work has been supported by grants from Kent State University Foundation, Research Corporation, NSF (DMB-8518941) and the Army Research Office (DAAL 03-90-G-0061).

REFERENCES

1. Kuntz, I. D., and Kauzmann, W., *Adv. Prot. Chem. 28*:239–345 (1974).
2. Edsall, J. T., and McKenzie, H. A., *Adv. Biophys. 16*: 53–183 (1983).
3. Rupley, J. A., and Careri, G., *Adv. Prot. Chem. 41*: 37–172 (1991).
4. Prestrelski, S. J., Tedeschi, N., Arakawa, T., and Carpenter, J., *Biophys. J. 65*: 661–671 (1993).
5. Fullerton, G. D., Ord, V. A., and Cameron, I. L., *Biochim. Biophys. Acta 869*: 230–246 (1986).
6. Bull, H. B., and Breese, K., *Arch. Biochem. Biophys. 128*: 488–496 (1968).
7. Rao, P. B., and Bryan, W. P., *Biopolymers 17*: 291–314 (1978).
8. Poole, P. L., and Finney, J. L., in Methods in Enzymology, Ed. L. Packer, *127*, pp. 284–293. Academic Press, Orlando, Florida (1986).
9. Brunauer, S., Emmett, P. H., and Teller, E., *J. Am. Chem. Soc. 60*: 309–319 (1938).
10. Careri, G., Giansanti, A., and Gratten, E., *Biopolymers 18*: 1187–1203 (1979).

11. Benson, S. W., Ellis, D. A., and Zwanzig, R. W., *J. Am. Chem. Soc. 72*: 2102–2106 (1950).
12. Luscher-Mattli, M., and Ruegg, M., *Biopolymers, 21*: 403–418 (1982).
13. Flory, P. J., Principles of Polymer Chemistry, Cornell Univ. Press, Ithaca, New York (1953).
14. Hailwood, A. J., and Horrobin, S., *Trans. Faraday Soc. 42B*: 84–92 (1946).
15. Enderby, J. A., *Trans. Faraday Soc. 51*: 106–116 (1955).
16. D'Arcy, R. L., and Watt, I. C., *Trans. Faraday Soc. 66*: 1236–1245 (1970).
17. Cerofolini, G. F., and Cerofolini, M. J., *Colloid Interface Sci. 78*: 65–73 (1980).
18. Provencher, S. W., *Comput. Phys. Commun. 27*: 213–227 (1982).
19. Provencher, S. W., *Comput. Phys. Commun. 27*: 229–242 (1982).
20. Livesey, A. K., and Skilling, J., *Acta Cryst. A41*: 113–122 (1985).
21. Livesey, A. K., and Brochon, J. C., *Biophys. J. 52*: 693–706 (1987).
22. McLaren, A. D., and Kirwan, J. P., *J. Gen. Virol. 33, Pt. 1*: 155–157 (1976).
23. Rupley, J. A., Gratten, E., and Careri, G., *Trends Biochem. Sci. 8*: 18–22 (1983).
24. Bryan, W. P., *Biopolymers 26*: 1705–1716 (1987).
25. Lumry, R. W., and Rosenberg, A., *Colloques. Int. CNRS 246*: 53–61 (1975).
26. Kennedy, S. D., and Bryant, R. G., *Biopolymers 29*: 1801–1806 (1990).
27. Gregory, R. B., Gangoda, M., Gilpin, R. K., and Su, W., *Biopolymers 33*: 513–519 (1993).
28. Luscher, M., Schindler, P., Ruegg, M., and Rottenberg, M., *Biopolymers 18*: 1775–1791 (1979).
29. Schinkel, J. E., Downer, N. W., and Rupley, J. A., *Biochemistry 24*: 352–366 (1985).
30. Birktoft, J. J., and Blow, D. K., *J. Mol. Biol. 68*: 187–240 (1972).
31. Nedev, K. N., Volkova, R. I., Khurgin, Yu, I., and Chernavaskii, D. S., *Biofizika 19*, 983–986 (1974).
32. Sreenivasan, U., and Axelsen, P. H., Biochemistry *31*: 12785–12791 (1992).
33. Luscher-Mattli, M., in Thermodynamic Data for Biochemistry and Biotechnology, Ed. H.- J. Hinz, pp. 276–294. Springer-Verlag, Berlin (1986).
34. Hirs, C. H. W., and Timasheff, S. N., Eds, Methods in Enzymology *117*, Academic Press, Orlando, Florida (1985).
35. Finney, J. L., and Poole, P. L., *Comments Mol. Cell. Biophys. 2*, 129–151 (1984).
36. Yang, P., and Rupley, J. A., *Biochemistry 18*: 2654–2661 (1979).
37. Hill, T. L., *J. Chem. Phys. 15*: 767–777 (1947).
38. Hill, T. L., *J. Chem. Phys. 17*: 762–771 (1949).
39. Rupley, J. A., Yang, P., and Tollin, G., in Water in Polymers, Ed. S. P. Rowland, pp. 111–132. Am. Chem. Soc., Washington D.C. (1980).
40. Sturtivant, J. M., *Proc. Natl. Acad. Sci. USA 74*: 2236–2240 (1977).
41. Khurgin, Y. I., Medvedeva, N. V., and Roslyakov, V. Y., *Biofizika 22*: 1010–1014 (1977).
42. Skujins, J. J., and McLaren, A. D., *Science 158*: 1569–1570 (1967).
43. Stevens, E., and Stevens, L., *Biochem. J. 179*: 161–167.
44. Zaks, P., and Klibanov, A., *Science 224*: 1249–1251 (1984).
45. Klibanov, A., in Protein Engineering, Eds M. Inouye and R. Sarma, pp. 341–349. Academic Press, Orlando, Florida (1986).
46. Zaks, P., and Klibanov, A., *J. Biol. Chem. 263*: 8017–8021 (1984).

47. Careri, G., Giansanti, A., and Rupley, J. A., *Proc. Natl. Acad. Sci. USA 83*: 6810–6814 (1986).
48. Blake, C. C. F., Pulford, W. C. A., and Artymuik, P. J., *J. Mol. Biol. 167*: 693–723 (1983).
49. Kuntz, I. D., Brassfield, T. S., Law, G. D., Purcell, G. V., *Science 163*: 1329–1331 (1969).
50. Kuntz, I. D., *J. Am. Chem. Soc. 93*: 514–516 (1971).
51. Golton, I. C., Gellatly, B. J., and Finney, J. L., *Stud. Biophys 84*: 5 (1981).
52. Poole, P. L., Walton, A. R., and Zhang, J., *Int. J. Biol. Macromol. 9*: 245–246 (1987).
53. Fujita, Y., and Noda, Y., *Int. J. Pept. Protein Res. 18*: 12–17 (1981).
54. Fujita, Y., and Noda, Y., *Bull. Chem. Soc. Jpn. 51*: 1567–1568 (1978).
55. Fujita, Y., and Noda, Y., *Bull. Chem. Soc. Jpn. 52*: 2349–2352 (1979).
56. Fujita, Y., and Noda, Y., *Bull. Chem. Soc. Jpn. 54*: 3233–3234 (1981).
57. Wheeler, C. J., and Croteau, R., *Arch. Biochem. Biophys. 248*: 429–434 (1986).
58. Ayala, G., Gomez-Puyou, M. T., Gomez-Puyou, A. and Darszon, A., *FEBS Lett. 203*: 41–43 (1986).
59. Corbett, R., and Roche, R., *Biochemistry 23*: 1888–1894 (1984).
60. Poole, P., and Finney, J. L., *Biopolymers 22*: 255–260 (1983).
61. Chirgadze, Y. N., and Ovsepyan, A. M., *Biopolymers 11*: 2179–2186 (1972).
62. Gregory, R. B., Gangoda, M., Gilpin, R. K., and Su, W., *Biopolymers, 33*: 1871–1876 (1993).
63. Kachalova, G. S., Morozov, V. N., Morozova, T. Y., Myachin, E. T., Vagin A. A., Strokopytov, B. V., and Nekrasov, Y. V., *FEBS Lett. 284*: 91–94 (1991).
64. Carpenter, J. F., and Crowe, J. H., *Biochemistry 28*: 3916–3922 (1989).
65. Belonogova, O. V., Frolov, E. N., Krasnopol'skaya, S. A., Atanasov, B. P., Gins, V. K., Mukhin, E. N., Levina, A. A., Andreeva, A. P., Likhtenshtein, G. I., and Goldanskii, V. I., *Doklady Akad. Nauk SSSR 241*: 219–222 (1978).
66. Permyakov, E. A., and Burstein, E. A., *Stud. Biophys. 64*: 163–172 (1977).
67. Fucaloro, A. F., and Forster, L. S., *Photochem. Photobiol 41*: 91–93 (1985).
68. Vermunicht, G., Boens, N., and De Schryver, F. C., *Photochem. Photobiol. 53*: 57–63 (1991).
69. Sheats, G. F., and Forster, L. S., *Biochem. Biophys. Res. Commun. 114*: 901–906 (1983).
70. Strambini, G. B., and Gabellieri, E., *Photochem. Photobiol. 39*: 725–729 (1984).
71. Bryant, R. G., *Ann. Rev. Phys. Chem. 29*: 167–188 (1978).
72. Bryant, R. G., *Stud. Phys. Theor. Chem. 38*: 683–705 (1988).
73. Lioutas, T. S., Baianu, I. C., and Sternberg, M. P., *Arch. Biochem. Biophys. 247*: 68–75 (1986).
74. Shirley, W. M., and Bryant, R. G., *J. Am. Chem. Soc. 104*: 2910–2918 (1982).
75. Peemoeller, H., Shenoy, R. K., and Pintar, M. M., *J. Magetic Resonance 45*: 193–204 (1981).
76. Peemoeller, H., Kydon, D. W., Sharp, A. R., and Schreiner, L. J., *Can. J. Phys. 62*: 1002–1009 (1984).
77. Peemoeller, H., Yeomans, F. G., Kydon, D. W., and Sharp, A. R., *Biophys. J. 49*: 943–943 (1986).
78. Swanson, S. D., and Bryant, R. G., *Biopolymers 31*: 967–973 (1991).
79. Poole, P. L., and Finney, J. L., *Int. J. Biol. Macromol. 5*: 308–310 (1983).

80. Gregory, R. B., and Rosenberg, A., in Methods in Enzymology Eds. C. H. W. Hirs and S. N. Timasheff, *131*, pp. 448–508. Academic Press, Orlando, Florida (1986).
81. Englander, S. W., and Kallenbach, N. R., *Q. Rev. Biophys. 16*: 521–655 (1984).
82. Woodward C., Simon, I., and Tuchsen, E., *Mol. Cell. Biochem. 48*: 135–160 (1982).
83. Englander, S. W., Downer, N. W., and Teitelbaum, H., *Ann. Rev. Biochem. 41*: 903–924 (1972).
84. Englander, S. W., Calhoun, D. B., Englander, J. J., Kallenbach, N. R., Liem, R. K. H., Malin, E. L., Mandal, C., and Rogero, J. R., *Biophys. J. 32*: 577–589 (1980).
85. Englander, S. W., *Comments Mol. Cell. Biophys. 1*: 15–28 (1980).
86. Woodward, C. K., and Hilton, B. D., *Biophys. J. 32*: 561–575 (1980).
87. Ellis, L. M., Bloomfield, V. A., and Woodward, C. K., *Biochemistry 14*: 3413–3419 (1979).
88. Gregory, R. B. and Lumry, R. W., *Biopolymers 24*: 301–326 (1985).
89. Tuchsen, E., and Woodward, C., *J. Mol. Biol. 185*: 421–430 (1985).
90. Gregory, R. B., *Biopolymers 27*: 1699–1709 (1988).
91. Schrader, D. M., and Jean, Y. C., Eds. Positron and Positronium Chemistry, Studies in Physical and Theoretical Chemistry, vol. 57, Elsevier, Amsterdam (1988).
92. Jean, Y. C., *Microchemical J. 42*: 72–102 (1990).
93. Tao, S. J., *J. Chem. Phys. 56*: 5499–5510 (1972).
94. Eldrup, M., Lightbody, D., and Sherwood, J. N., *Chem. Phys. 63*: 51–58 (1981).
95. Gregory, R. B., and Zhu, Y., *Nucl. Instr. Meth. A290*: 172–182 (1990).
96. Gregory, R. B., *Nucl. Instr. Meth. A302*: 496–507 (1991).
97. Gregory, R. B., *J. Appl. Phys. 70*: 4665–4670 (1991).
98. Gregory, R. B., and Chai, K.-J., *Biochem. Soc. Trans. 21*: 478S (1993).
99. Gregory, R. B., and Chai, K.-J., in 4th International Workshop on Positron and Positronium Chemistry, Ed. I. Billard, *Journal de Physique IV 3*: 305–310 (1993).
100. Welander, M., and Maurer, F. H. J., *Materials Sci. Forum 105–110*: 1815–1818 (1992).
101. Goldanskii, V. I., and Krupyanskii, Y. F., *Q. Rev. Biophys. 22*: 39–92 (1989).
102. Bone, S., and Pethig, R., *J. Mol. Biol. 181*: 323–326 (1985).
103. Sperling, L. H., Introduction to Physical Polymer Science, Wiley-Interscience, New York (1986).
104. Hiemenz, P. C., Polymer Chemistry, Marcel Dekker, New York (1984).
105. Hedvig, P., Dielectric Spectroscopy of Polymers, Wiley, New York (1977).
106. Gibbs, J. H., and DiMarzio, E. A., *J. Chem. Phys 28*: 373–383 (1958).
107. DiMarzio, E. A., and Gibbs, J. H., *J. Polymer Sci. A1*: 1417–1428 (1963).
108. Fox, T. G., and Flory, P. J., *J. Appl. Phys. 21*: 581–591 (1950).
109. Fox, T. G., and Flory, P. J., *J. Polymer Sci. 14*: 315–319 (1954).
110. Simha, R., and Boyer, J., *J. Chem. Phys. 37*: 1003–1007 (1962).
111. Williams, M. L., Landel, R. F., and Ferry, J. D., *J. Am. Chem. Soc. 77*: 3701–3707 (1955).
112. Doolittle, A. K., *J. Appl. Phys. 22*: 1471–1475 (1951).
113. Hirai, N., and Eyring, H., *J. Appl. Phys. 29*: 810–816 (1958).
114. Hirai, N., and Eyring, H. *J. Polymer Sci. 37*: 51–70 (1959).
115. Likhenshtein, G. I., and Kotel'nikov, A. I., *Mol. Biol. (Moscow) 17*: 505–517 (1983).

116. Steinhoff, H. J., Lietenant, K., and Schlitter, J., *Z. Naturforsch., C: Biosci. 44*: 280–288 (1989).
117. Belonogova, O. V., Frolov, E. N., Illyustrov, N. V., and Likhtenshtein, G. I., *Mol. Biol. (Moscow) 13*: 567–576 (1979).
118. Parak, F., Heidemeier, J., and Neinhaus, G. U., *Hyperfine Interact. 40*: 147–158 (1988).
119. Doster, W., Cusack, S., and Petry, W., *Nature 337*: 754–756 (1989).
120. Frauenfelder, H., Parak, F., and Young, R. D., *Ann. Rev. Biophys. Biophys. Chem. 17*: 451–479 (1988).
121. Locharich, R. J., and Brooks, B. R., *J. Mol. Biol. 215*: 439–455 (1990).
122. Smith, J., Kuczera, K., and Karplus, M., *Proc. Natl. Acad. Sci. USA 87*: 1601–1605 (1990).
123. Tilton, R. F., Dewan, J. C., and Petsko, G. A., *Biochemistry 31*: 2469–2481 (1992).
124. Doster, W., Bachleitner, A., Dunau, R., Hiebl, M., and Luscher, E., *Biophys. J. 50*: 213–219 (1986).
125. Andrew, E. R., *Polymer, 26*: 190–192 (1985).
126. Ruggiero, J., Sanches, R., Tabak, M., and Nascimento, O. R., *Can. J. Chem. 64*: 366–372 (1986).
127. Parak, F., in Methods in Enzymology, Ed. L. Packer, *127*, pp. 196–206. Academic Press, Orlando, Florida (1986).
128. Pissis, P., *J. Mol. Liquids, 41*, 271–289 (1989).
129. Pissis, P., in Proton Transfer in Hydrogen Bonded Systems, Ed. T. Bountis, *NATO Adv. Study Inst. Ser. B, Physics 291*: 207–216 (1992).
130. Lumry, R. W., Solbakken, A., Sullivan, J., and Reyerson, L. H., *J. Am. Chem. Soc. 84*: 142–149 (1962).
131. Gregory, R. B., *Biopolymers, 22*: 895–909 (1983).
132. Lumry, R. W., and Gregory, R. B., in The Fluctuating Enzyme, Ed. G. R. Welch, pp. 1–190. Wiley-Interscience, New York (1986).
133. Lumry, R. W., in A Study of Enzymes, Ed. S. A. Kuby, vol. II, pp. 3–81. CRC Press, Boca Raton (1988).
134. Hilton, B. D., and Woodward, C. K., *Biochemistry, 17*: 3325–3332 (1978).
135. Richarz, R., Sehr, P., Wagner, G., and Wuthrch, K., *J. Mol. Biol. 130*: 19–30 (1979).
136. Wuthrich, K., and Wagner, G., *J. Mol. Biol. 130*: 1–18 (1979).
137. Hilton, B. D., and Woodward, C. K., *Biochemistry 18*, 5834–5841 (1979).
138. Wedin, R. E., Delepiere, M., Dobson, C. M., and Poulsen, F. M., *Biochemistry, 21*: 1098–1103 (1982).
139. Tuchsen, E., Hayes, J. M., Ramaprasad, S., Copie, V., and Woodward, C., *Biochemistry, 26*: 5163–5172 (1987).
140. Schoenborn, B. P., Ed., Neutrons in Biology, Plenum Press, New York (1984).
141. Kossiakoff, A., *Nature, 296*: 713–721 (1982).
142. Wlodawer, A., and Sjolin, L., *Proc. Natl. Acad. Sci. USA, 79*: 1418–1422 (1982).
143. Mason, S. A., Bentley, G. A., and McIntyre, G. J. in Neutrons in Biology, Ed. B. P. Schoenborn, pp. 323–334. Plenum Press, New York (1984).
144. Gregory, R. B., Rosenberg, A., Knox, D. G., and Percy, A. J., *Biopolymers 29*: 1175–1183 (1990).

145. Gregory, R. B., Crabo, L., Percy, A. J., and Rosenberg, A., *Biochemistry 22*: 910–917 (1983).
146. Gregory, R. B., and Lumry, R. W., *Biophys. J.*, *49*: 441a (1986).
147. Levitt, M., *Nature 294*, 379–380 (1981).
148. Richards, F. M., *Ann. Rev. Biophys. Bioeng.*, *6*: 151–176 (1977).
149. Schultz, D. J., Ph.D. Dissertation, Princeton University (1976).
150. Blundell, T., Barlow, D., Borkakoti, N., and Thornton, J., *Nature*, *306*: 281–283 (1983).
151. Nakamura, H., Sakamoto, T., and Wada, A., *Prot. Eng. 2*: 177–183 (1988).
152. Miller, S., Lesk, A. M., Janin, J., and Chothia, C., *J. Mol. Biol. 196*: 641–656.
153. Bendzko, P. I., Pfeil, W. A., Provalov, P. L., and Tiktopulo, E. I., *Biophys. Chem. 29*: 301–307 (1988).
154. Brandts, J., Oliveira, R. J., Westort, C., *Biochemistry 9*: 1038–1047 (1970).
155. Zipp, A., and Kauzmann, W., *Biochemistry 12*: 4217–4228 (1973).
156. Cusack, S., *Current Biol. 2*: 411–413 (1992).
157. Ptitsyn, O. B., in Protein Folding, Ed. T. E. Creighton, pp. 243–300. W. H. Freeman, New York (1992).
158. Privalov, P. L, *Crit. Rev. Biochem. Mol. Biol. 25*: 281–305 (1990).
159. Damaschun, G., Damaschun, H., Gast, K., Misselwitz, R., Muller, J. J., Pfeil, W., and Zirwer, D., *Biochemistry 32*: 7739–7746 (1993).
160. Lattman, E. E., and Rose, G. D., *Proc. Natl. Acad. Sci. USA 90*: 439–441 (1993).
161. Anfinsen, C. B., *Science 181*: 223–230 (1973).
162. Dill, K. A., *Biochemistry 29*: 7133–7155 (1990).
163. Chan, H. S., and Dill, K. A., *J. Chem. Phys. 90*: 492–509 (1989).
164. Lau, K. F., and Dill, K. A., *Proc. Natl. Acad. Sci. USA 87*: 638–642 (1990).
165. Behe, M. J., Lattman, E. E., and Rose, G. D., *Proc. Natl. Acad. Sci. USA 88*: 4195–4199 (1991).
166. Lim, W. A., and Sauer, R. T., *Nature*, *339*: 31–36 (1989).
167. Kim, K. -S., Fuchs, J., and Woodward, C. K., *Biochemistry*, *32*: 9600–9608 (1993).
168. Roder, K., and Wuthrich, K., *Prot.: Struct. Funct. Genet. 1*: 34–42 (1986).
169. Roder, K., Elove, G., and Englander, S. W., *Nature 335*: 700–704 (1988).
170. Jeng, M.-F., Englander, S. W., Elove, G., Wand, A., and Roder, H., *Biochemistry 29*: 10433–10437 (1990).
171. Miranker, A., Radford, S., Karplus, M., and Dobson, C., *Nature 349*: 633–636 (1991).
172. Radford, S., Buck, M., Topping, K., Dobson, C., and Evans, P., *Prot. Struct. Funct. Genet. 14*: 237–248 (1992).
173. Radford, S., Dobson, C., and Evans, P., *Nature 358*: 302–307 (1992).
174. Plochocka, D., Zielenkiewicz, P., and Rabczenko, A., *Prot. Eng. 2*: 115–118 (1988).
175. Nicholson, V. A., and Maggiora, G. M., *J. Math. Chem. 11*: 45–63 (1992).
176. Nicolson, V. A., and Neuzil, J. P., *J. Math. Chem. 12*: 53–58 (1993).
177. Finkelstein, A. V., and Nakamura, H., *Prot. Eng. 6*: 367–372 (1993).
178. Chothia, C., *Ann. Rev. Biochem. 53*: 537–572 (1984).
179. Ponder, J. W., and Richards, F. M., *J. Mol. Biol 193*: 775–791 (1987).

4

Dielectric Studies of Protein Hydration

Ronald Pethig
University of Wales, Bangor, Gwynedd, Great Britain

I. INTRODUCTION

As with any other experimental technique, dielectric measurements on their own cannot provide a full understanding of the dynamics associated with the interaction of water with proteins. The purpose of this chapter is to outline how dielectric measurements can add to those insights gained from the application of nuclear magnetic resonance (NMR).

Dielectric spectroscopy and NMR spectroscopy both are techniques that probe molecular orientational relaxation phenomena. The characteristic water structure induced near the surfaces of proteins arises principally from electrostatic forces associated with the water molecule possessing a large electric dipole moment, as well as through extensive hydrogen bonding of the water molecules to available proton donor and proton acceptor sites. If an electric field is imposed on such a system of protein-associated water, there will be a torque exerted on each water dipole moment so as to induce them to attempt to align along the direction of the field vector. A dielectric orientational relaxation time τ can then be defined as the time required for $1/e$ of the field-oriented water molecules to become randomly reoriented on removing the applied field. The degree of orientational polarization and the rate of reorientational relaxation will depend on how the water dipoles are influenced by local electrostatic forces and the extent to which the breaking and/or reforming of local hydrogen bonds is required to accommodate the orientational changes. Relaxations of a polar side groups and vibrations of the polypeptide backbone of the protein will also contribute to the overall polarizability of the protein–water system, and if dielectric measurements are made on protein solutions, then orientational relaxations of the protein molecule itself will also be observed.

For NMR studies, we are interested in nuclei such as the proton, deuteron, and oxygen (^{17}O) that have a spin angular momentum characterized by a spin quantum number I. On application of a magnetic field **H**, the nuclear energy levels split into $2I + 1$ states, and spin transfer from the lower to higher levels can be induced by application of a radiofrequency (RF) field whose magnetic component is perpendicular to **H**. Spins in the higher energy levels can relax to lower ones through processes induced by molecular rotation and diffusion, proton exchange, and couplings between nuclear spins and local fluctuating "lattice" fields. The so-called spin–lattice (T_1) relaxation time reflects the energy exchange between the magnetic nuclei and the lattice, and it characterizes the exponential fall to the equilibrium magnetization along the **H** field direction when the RF field is removed. Neighboring magnetic nuclei also interact with each other, and the influence of this is characterized by the spin–spin interaction time constant T_2, which represents the phase memory time of a nuclear spin state as reflected in the exponential decay of the magnetization component in directions transverse to **H**. The measured T_1 and T_2 relaxation times, together with a third one, $T_{1\rho}$ hr , relating to spin–lattice relaxation in a rotating frame, thus reflect the extent to which the rotational and translational molecular motions are coupled. In normal bulk water, there is a strong coupling between these motions, but as a result of molecular interactions, restricted motion, and the ordering of water molecules, this is not the case for protein-associated water. It is possible to formally relate the dielectric relaxation time τ with the NMR-derived relaxation times, but such relationships are sensitive to the dynamic models adopted and a discussion of this is beyond the scope of this present chapter. Some insights into this problem, for the case of water in biological systems, can be obtained from the early reviews of Packer [1] and Mathur-De Vre [2], for example.

II. DIELECTRIC THEORY AND MEASUREMENTS

The principal macroscopic parameter that characterizes the dielectric properties of a polar liquid or solution containing polar molecules is its relative permittivity ϵ, also (historically, but inaccurately) referred to as the dielectric constant. This parameter can be interpreted as a measure (on neglecting atomic and electronic polarization) of the number of molecules oriented by an external electric field of unit strength. These molecules are oriented by a torque proportional to the product of the field strength and the dipole moment of each molecule, and this is hindered by frictional forces whose extent depends on the angular frequency ω of the (sinusoidal) field and the characteristic relaxation time τ of the molecule. At low frequencies ($\omega\tau << 1$), the orienting torque overcomes the frictional forces, and the material exhibits a high relative permittivity ϵ_s, sometimes called the "static" permittivity. The electrical capacitance C of the material, when held between two plane-parallel electrodes of area A separated by a distance d, is thus defined at low frequencies (on neglecting fringe fields) by

$$C = \frac{A\epsilon_s\,\epsilon_0}{d} \tag{1}$$

where ϵ_0 is the permittivity of free space, 8.854 pF m^{-1}. At very high frequencies ($\omega\tau >> 1$), the frictional forces completely overcome the orienting torque, the relative permittivity assumes a low value ϵ_∞, and the measured capacitance drops in value accordingly. Such a fall in relative permittivity with increasing frequency is known as a dielectric dispersion, and the theory to describe this behavior was largely developed by Debye [3], who showed (to take account of the energy dissipation associated with the phase lag between the field and the orientation) that for a single relaxation time the permittivity can be written as a complex function of the form

$$\epsilon^*(\omega) = \epsilon_\infty + \frac{\epsilon_s - \epsilon_\infty}{1 + i\omega\tau} \tag{2}$$

where $i = (-1)^{1/2}$. The real and imaginary parts of the complex permittivity may also be expressed in the form

$$\epsilon^* = \epsilon' - i\epsilon'' \tag{3}$$

in which the real part ϵ', corresponding to the permittivity factor used to define capacitance as in Eq. (1), is given by

$$\epsilon'(\omega) = \epsilon_\infty + \frac{\epsilon_s - \epsilon_\infty}{1 + \omega^2\tau^2} \tag{4}$$

The imaginary component ϵ'', known as the dielectric loss factor, is given by

$$\epsilon''(\omega) = \frac{(\epsilon_s - \epsilon_\infty)\omega\tau}{1 + \omega^2\tau^2} \tag{5}$$

and takes the form of a loss peak, centered at the characteristic frequency $f_c = 1/2\pi\tau$, when plotted as a function of frequency. This loss peak is broad, with a half-height width of 1.14 decades of frequency. By defining the magnitude of a dielectric dispersion as

$$\Delta\epsilon' = \epsilon'_s - \epsilon'_\infty \tag{6}$$

we obtain, by combination with Eqs. (4) and (5)

$$\epsilon'(\omega) = \frac{\epsilon'_\infty + \Delta\epsilon'}{1 + (f/f_c)^2} \tag{7}$$

The frequency variations of ϵ' and ϵ'' that characterize the dielectric dispersion exhibited by normal bulk water are shown in Fig. 1, together with defining parameters for f_c and $\Delta\epsilon'$. At any temperature T, the magnitude of the dielectric dispersion can be related to the dipole moment m and the molecular weight M of the polar molecule by the expression

$$\Delta\epsilon' = \frac{N_A c g m^2}{2\epsilon_0 M k T} \tag{8}$$

where N_A is Avogadro's number, c is the concentration in kg m^{-3} of the polar molecule in the solvent, and g is a parameter introduced by Kirkwood [4] to account for molecular associations and correlation effects between the motions of solute and solvent molecules. Such effects are particularly important for hydrogen-bonded liquids, where the rotation of one molecule seriously disrupts the local hydrogen bonding and requires the correlated or cooperative reorientation of

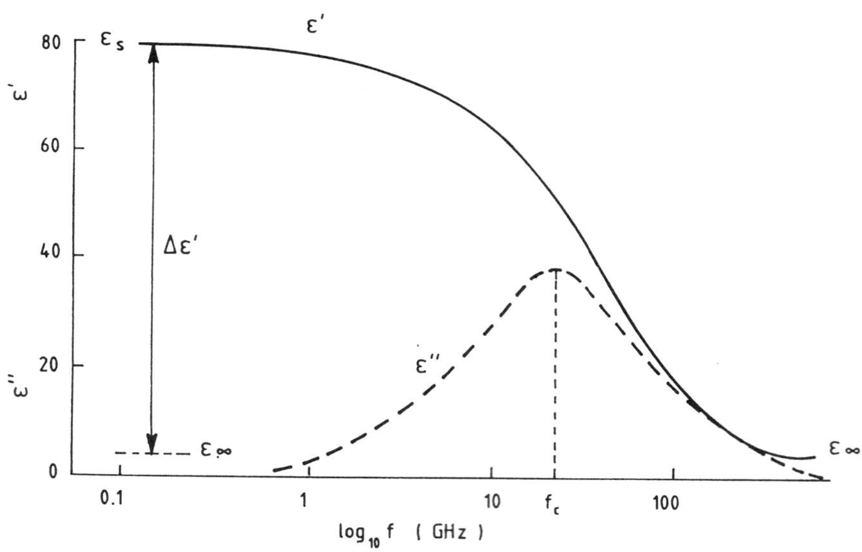

FIGURE 1 The frequency variation of the real (ϵ') and imaginary (ϵ'') parts of the permittivity as manifested in the dielectric dispersion exhibited by bulk water at 293 K.

surrounding molecules to compensate for this. In bulk water, many of the water molecules are surrounded by four others with which hydrogen bonds are formed. Depending on how its two protons are directed toward the four neighbors, the central molecule can occupy six positions (see inset of Fig. 2). On applying an electric field, the neighboring dipoles do not reorient independently, since to preserve all hydrogen bonds the molecular rotations must be cooperative. For four H-bonded water at 298 K, g has a value of about 2.8, which falls to 2.5 at 373 K [5]. A value of unity for g implies no intermolecular associations or that the net value of such associations cancels out, and this is the case for a rotating protein molecule in aqueous solution where there will be numerous solute–solvent forces acting over a wide range of directions.

A simple model, often used to describe the dielectric relaxation of a molecule with a permanent dipole moment, is to consider the dipoles to be rigid spheres whose rotation in response to an imposed electric field is opposed by frictional interaction with the surrounding viscous medium [3]. The relevant relaxation time for the orientation of such a sphere is

$$\tau = \frac{\chi}{2kT} \tag{9}$$

where χ is the molecular friction coefficient, which relates the torque exerted on the dipolar molecule by the applied electric field to the molecule's angular velocity, and k is the Boltzmann constant. If we consider the dipole to be equivalent to a rigid sphere of radius a rotating in a Newtonian hydrodynamic fluid of macroscopic viscocity η, the Stokes–Einstein relation gives $\chi = 8\pi\eta a^3$, so that the relaxation time is

$$\tau = \frac{4\pi\eta a^3}{kT} \tag{10}$$

Furthermore, if we assume that the molecular volume (and hence radius) of a globular protein is directly proportional to its molecular weight, then on substituting for the radius in Eq. (10), we have

$$\tau = \frac{3M\eta}{N_A \rho kT} \tag{11}$$

where ρ is the average density of the protein molecule. The viscocity η of water at 298 K is 8.9×10^{-4} N s m^{-2}, and as a typical protein density, we can take the value of 1.39 g cm^{-3} obtained by Kuntz [6] for carboxypeptidase. Thus, as a rough working guide, $\tau = 0.78M$ ns. On this basis, we would expect for myoglobin (M = 18 kd) a rotational relaxation time of around 14 ns, as against the experimental value of around 45 ns observed for dilute myoglobin solutions [7]. As another example, bovine serum albumin (M = 67 kd) would be expected to have a value for τ of 52 ns, compared with the experimental value [8] of 200 ns. The

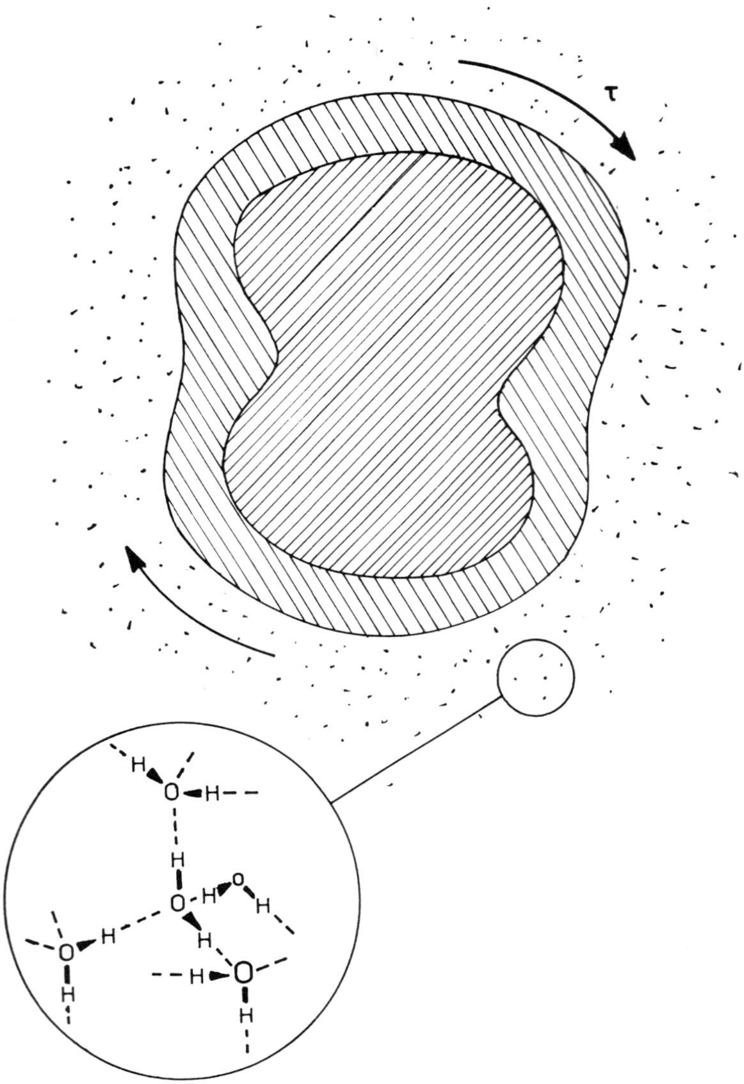

FIGURE 2 Schematic representation of a protein molecule, with a layer or two of strongly associated water, suspended in aqueous solution. The hydration layer moves with the protein molecule during orientational relaxations, and beyond this layer the water molecules progressively adopt the normal tetrahedral geometry (shown in inset) and orientational relaxation behavior of bulk water.

experimental relaxation times tend to exceed the theoretical ones expected from Eqs. (10) and (11), suggesting that the molecular weight, and hence effective volume, of the rotating entity is greater than that of the protein molecule alone. A plausible explanation is that a layer or two of water of hydration remains strongly associated with each protein molecule during its orientational relaxations, and this is shown schematically in Fig. 2.

The dielectric relaxation of pure bulk water has been extensively studied, and it has been concluded [5] that water behaves as an almost perfect Debye dielectric with a very small spread in relaxation times. The complex permittivity for water is well described by the empirical Cole–Cole equation [9]

$$\epsilon^*(\omega) = \epsilon_\infty + \frac{\epsilon_s - \epsilon_\infty}{1 + (i\omega\tau)^{1-\alpha}} \tag{12}$$

where α is a parameter that provides a qualitative indication of the spread in relaxation times. From Eqs. (2) and (12), it follows that for a single relaxation time, $\alpha = 0$. For water, α is smaller than 0.02, the parameters ϵ_s, ϵ_∞, and τ have values at 298 K of 78.4, 5.3, and 8.2×10^{-12} s ($f_c = 19.4$ GHz), respectively, and the dipole moment is 6.14×10^{-30} C m (1.84 Debye units) [5]. The relaxation time of 8.2 ps can roughly be thought of as the average lifetime of the local hydrogen bond network around each water molecule. It is highly desirable to have a detailed knowledge of the distribution of relaxation times, and so adopting an empirical approach to this problem requires explanation (for no detailed interpretation, apart from the value of zero, can be given of the values obtained for α). The spectrum of relaxation times is not directly observed in a dielectric measurement, but instead one obtains at a specific frequency ω the response of the test system to a perturbation. Because the dielectric loss peaks are so broad, such measurements will include significant contributions from any relaxation times above and below ω^{-1}, unless the separate relaxation processes have rates separated by many orders of magnitude. Explicit expressions for the spectrum of relaxation times as an intergral over all frequencies have been derived [10,11]. In practice, however, measurements of relaxation are made only at a finite number of frequencies extending over a finite range of frequency, and so use of such inversion formulas can lead to magnification of experimental errors and to derived values of relaxation times that are physically meaningless. Colonomos and Gordon [12] have described a method for analyzing dielectric data, or other measurements related to a distribution of relaxation times, but there are still difficulties owing to the possibility of being able to fit alternative dispersion functions to the data.

The addition of solutes such as inorganic salts, alcohols, amines, carboxylic acids, tetra-alkylammonium salts, and heterocyclic organic compounds to water tends to decrease ϵ_s and shift the characteristic relaxation time to either a higher or lower value, whilst the parameter α in Eq. (12) increases to a value in the range 0.1–0.2 [5]. The decrease in ϵ_s can be directly related to the replacement of water by solute molecules exhibiting a lower polarizability, while a change in the relaxation time is related to a change in entropy associated with the degrees of freedom of the water molecules and solute molecules. Thus, dielectric measurements [13] between 8 and 25 GHz have shown that on the dissolution of largely hydrophobic solutes, such as aliphatic monohydroxy alcohols and monocarboxylic acids at concentrations between 0.001 and 0.1 *M*, a slowing down of rotation motions in a monomolecular hydration layer by about a factor of two is sufficient to explain the observed increase in the macroscopic dielectric relaxation time of the bulk water. Although there might be a slight increase in the rotational freedom of the solute molecules (alcohols have rotational motions about three times slower than those of water), this is more than offset by the negative entropy of solvation associated with the formation of a hydration shell of about 20 less mobile water molecules around each solvent molecule. Water molecules freely exchange between this hydration shell and the bulk solvent, and the dielectric properties are not sensitive to the rate of this exchange.

A number of different physicochemical models have been proposed for water (see Refs. 14–16 for early reviews and 17–19 for modern analytical models). A common feature of some of these models is that a low-density quasi-lattice structure (state 1) mixes with one or more higher-density states (state 2). The presence of a hydrophobic solute favors state 1 and causes a shift toward state 1, thus lowering enthalpy and entropy. Thus, if states 1 and 2 are identified with the two exchanging dielectric species of water in the hydration shell and bulk solvent, with relaxation times τ_1 and τ_2, respectively, then for fast exchange the bulk dielectric relaxation time τ becomes the weighted average given by

$$\frac{1}{\tau} = \frac{x}{\tau_1} + \frac{1-x}{\tau_2} \tag{13}$$

where x is the mole fraction of water in state 1 (given as 0.35 at 293 K in one of the models [20]). A shift in x due to the presence of a solute now perturbs both relaxation and entropy S, such that

$$\Delta\left(\frac{1}{\tau}\right) = \frac{1/\tau_1 - 1/\tau_2}{S_2 - S_1}\,\Delta S \tag{14}$$

where ΔS is the excess entropy per mole of water. Although such a two-state model is very simplistic, and could for example be replaced by one quantifying the number of water molecules having zero to six hydrogen bonds [18], it provides a good qualitative description of the dielectric data obtained [13] for alcohols and monocarboxylic acids in water and could aid a qualitative interpretation of dielectric measurements on protein solutions.

Although this overview of dielectric theory is sufficient to serve our present purposes, it is hardly comprehensive and has not attempted to provide a full historical account of the dielectric properties of biomacromolecules or of the underlying biophysical processes involved. Such details can be found elsewhere [21–24].

III. EXPERIMENTAL RESULTS

A. Protein Solutions

The first indications that dielectric studies could provide details of protein–water association took the form of a tentative footnote in a paper by Oncley [25] describing the dielectric dispersion exhibited by aqueous solutions of carboxyhemoglobin. From the magnitude and characteristic frequency (near 2 MHz) of this dispersion, Oncley concluded that the carboxyhemoglobin molecule possessed a dipole moment of 500 D and a rotational relaxation time of 84 ns. He also noted, for frequencies above the dispersion range, that the permittivity of the protein solution was reduced below the value for bulk water by an amount exceeding that to be expected from considerations of the volume occupied by the solvated protein molecules. Oncley's quantitative consideration of this led him to conclude that 0.6 g of water was associated with 1 g of the (anhydrous) protein. In later measurements, extending over the frequency range 3 to 24 GHz, Buchanan et al. [26] refined the measurement and analysis of this effect for five different protein solutions to deduce that about 0.3 g water per g protein was carried by each protein molecule in their rotational relaxations, with around one-third of this total hydration being irrotationally bound in the sense that such water was unable to contribute to the dielectric dispersion normally exhibited by bulk water. Grant [27] combined new data for bovine serum albumin with that obtained by Buchanan et al. [26] to show that the protein-bound water exhibited a spread of rotational relaxations with characteristic frequencies extending from a few to several hundred megahertz with the overall dielectric behavior possibly being influenced by rotations of protein polar side groups. These observations by Grant were largely confirmed by the later work of Pennock and Schwan [28] and Essex et al. [29].

The most detailed characterization of the dielectric properties of a protein solution has been that provided by Grant et al. [30] for myoglobin, and their derived form of $\epsilon'(\omega)$ for myoglobin is shown in Fig. 3. As shown by Grant et al., this data can accurately be described using the sum of four Debye-type dispersions:

$$\epsilon^*(\omega) = \epsilon_\infty + \frac{\Delta_1}{1 + i\omega\tau_1} + \frac{\Delta_2}{1 + i\omega\tau_2} + \frac{\Delta_3}{1 + i\omega\tau_3} + \frac{\Delta_4}{1 + i\omega\tau_4} \tag{15}$$

From the earlier work of South and Grant [7], the Δ_1 dispersion can clearly be assigned to the rotational relaxation of the myoglobin molecule with $\tau = 74$ ns ($f_c = 2.2 \pm 0.3$ MHz). Analysis [7] of this Δ_1 dispersion also gives a dipole moment of about 150 D for the myoglobin molecule and a hydration shell (0.6 g/g) of thickness equivalent to around one-and-a-half water molecule diameters.

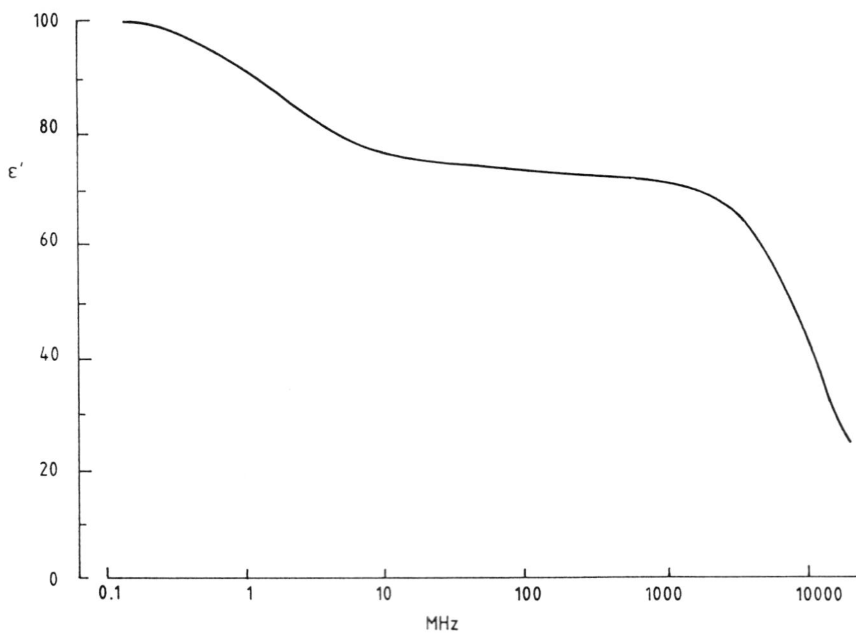

FIGURE 3 The frequency variation of $\epsilon'(\omega)$ for myoglobin in aqueous solution at 298 K. Derived from Grant et al. [30].

The concentration of myoglobin used for the data of Fig. 3 was 170 mg/ml, equivalent to a 87% volume occupation by water. Grant et al. [30] considered that, since a substantial proportion of this water will on average be many molecular diameters away from the nearest myoglobin molecule, the highest frequency dispersion Δ_4 could be assigned the relaxation time of bulk water (i.e., $\tau_4 = \tau_w = 8.29$ ps at 298 K, equivalent to $f_c = 19.2$ GHz). On "clamping" τ_4 to this value, the data of Fig. 3 is represented by Eq. (15) with Δ_2 and Δ_3 having nearly equal values (5.3 ± 1.5 and 5.4 ± 0.8, respectively), with $\tau_2 = 10.3$ ns ($f_c = 15.4 \pm 5.5$ MHz) and $\tau_3 = 40$ ps ($f_c = 4.0 \pm 0.7$ GHz). The Δ_2 dispersion can also be interpreted in terms of the so-called δ-dispersion, previously characterized for myoglobin solutions [31] with good evidence for it being associated with orientational relaxations of 0.25 g of water per g protein. Recalling that the Δ_1 dispersion indicates [7] that around 0.6 g water per g protein rotates with the protein molecule as one unit at a frequency of around 2 MHz, then the Δ_2 dispersion at 15 MHz accounts for roughly one-half of the total protein hydration. Since the Δ_3 dispersion is of the same magnitude as Δ_2, it is tempting to assign the remaining protein hydration to Δ_3. In fact, Grant et al. [30] were able to calculate, using the well-tested Maxwell–Wagner theory for heterogeneous dielectrics, that the Δ_3 dispersion can be interpreted in terms of the orientational relaxations of 0.3 g water per g protein.

Grant et al. [30] also found, as an alternative representation of the dielectric data of Fig. 3, that the Δ_3 and Δ_4 dispersions of Eq. (15) could be replaced by a single dispersion having a distribution of relaxation times as described by Eq. (12). At 298 K the corresponding average relaxation time is 8.99 ps (f_c = 17.7 ± 0.2 GHz) with the distribution parameter α having the value 0.06 ± 0.01. Resolving this dispersion for water into the two components Δ_3 and Δ_4 is thus equivalent to an interpretation in terms of a two-state model, as described by Eq. (13).

B. Solid State Studies

Rather than measure the dielectric properties of proteins in solution, an alternative approach is to make measurements on protein powders as a function of their water content. The advantages to be gained include the ability to determine dielectric phenomena associated with protein–water interactions in the absence of protein and polar side-group orientational relaxations and to investigate effects arising from the motions of ions and protons on the hydrated protein surface.

The earliest studies [32–36] of the dielectric properties of hydrated protein materials indicated that they exhibit low, frequency-independent permittivities when dry and that the absorption of water is accompanied by large frequency-dependent increases in the permittivity. The water content at which a rapid increase in permittivity occurred was termed the *critical hydration*, identified as the amount of water required to cover the protein molecule with a single layer of water [35]. For ovalbumin containing 11% water, Takashima and Schwan [36] observed a broad dielectric dispersion centered around 1 kHz that was considered to be caused by an ionic conductivity on the surface between the adsorbed water and the protein. Bayley [33] had previously observed dispersions in this frequency region for several hydrated proteins and peptides and had also suggested that they were caused by ionic effects. We shall return to this theme later.

Harvey and Hoekstra [37] were the first to observe effects that could be associated with the orientational relaxations of water bound to protein powders. They determined the dielectric properties of lysozyme between 10 MHz and 25 GHz for water contents up to 60 wt% over the temperature range 203–313 K. Two distinct dispersions were observed and assigned to two layers of adsorbed water. The dispersion associated with the first monolayer (34 wt% water) was characterized by a single relaxation time near 1 ns (f_c = 250 MHz), and this dispersion persisted to the lowest temperatures investigated, which is a behavior consistent with calorimetric and other measurements, which indicate that 0.3 to 0.5 g of water per g protein does not freeze at low temperatures [38]. The frequency range for this dispersion is at the upper range for that observed for bovine serum albumin in solution [27] and very much higher than that (15.4 MHz) determined for myoglobin [30]. A surprising result was that the relaxation time increased with increasing temperature, meaning that the activation enthalpy was negative. The activation entropy was also determined to be negative, and Harvey and Hoekstra [37] interpreted this in terms of the activation process involving the breaking of a hydrogen bond (between the protein molecule and a water molecule in the primary hydration shell) followed by the formation of better hydrogen bonds between the reorienting water molecule and its neighbors.

The second dispersion observed by Harvey and Hoekstra [37] was characterized by a distribution of relaxation times ($\alpha = 0.3$) about a mean value of 16 ps ($f_c = 9.95$ GHz) and disappeared at about 248 K, an effect taken to indicate that the water in the second layer is on average more disordered than bulk water, but that the degree of disorder is not as extreme as that in the first monolayer. In their investigations of hydrated collagen, Grigera et al. [39] observed a dispersion in the frequency range 100–800 MHz whose characteristic relaxation time increased with temperature, thus exhibiting a negative activation energy as found for lysozyme [37]. Grigera et al. also found, however, that the magnitude of the dispersion increased with increasing temperature, and they concluded that, instead of representing a dispersion due to relaxations of water bound to collagen, mobile polar side groups were instead responsible. Whether or not this conclusion should also be made concerning the results for lysozyme of Harvey and Hoekstra [37] remains an open question. The molecular details of collagen hydration were further elucidated by Grigera and Berendsen [40] by interpretation of both nuclear magnetic resonance and microwave dielectric data. For partially hydrated collagen, it was concluded that for three (unspecified) amino acids there are two firmly bound water molecules, with well-defined positions, hydrogen bonded to the macromolecular backbone. The water molecules reside in their sites for time periods ranging from 0.1 to 1 μs. The remainder of the water, comprising the major part of the total water present, is in weak interaction with the protein and rapidly exchanges with each other with rotational correlation times of less than 100 ps—in other words, not more than an order of magnitude slower than in normal bulk water. Hoeve [41] reported that dielectric measurements on hydrated collagen and elastin indicated that at room temperature the bound water molecules are rather mobile, and that their mobility decreases at lower temperatures until, not separately but collectively, they become immobilized in a glassy state in the range from 130 to 170 K.

The concept of a stepwise and sudden completion of hydration of the primary monolayer, followed by that of the secondary and subsequent ones, is of course a crude working model that ignores the thermodynamic distribution of the water molecules in the hierarchy (multilayers) of sorption sites [42]. Measurements by Kent [43] at 9.4 GHz for hemoglobin, bovine serum albumin, and actomyosin showed that at low hydrations the microwave dielectric behavior is predominantly influenced by thermally activated water in the multilayer sites. This was confirmed by Bone et al. [44] for albumin, although it was found that the primary hydration layer was more strongly bound and rotationally hindered than was concluded by Kent [43]. By comparing dielectric data obtained [45] up to 100 kHz for bovine serum albumin with that at 9.95 GHz [49], Gascoyne and Pethig [46] deduced that a dielectric dispersion existed in the megahertz range of frequencies whose magnitude could be directly related to the number of water molecules occupying the primary sorption sites of the protein. The effective dipole moment of these bound water molecules was calculated to be 0.79 D. Normal bulk water has a dipole moment of around 1.84 D, and so it was concluded that, by comparison with normal water, the water molecules bound to the primary sorption sites experience a combination of being rotationally hindered and less involved with correlated reorientations of surrounding molecules and hydrogen bonds.

C. Water as Plasticizer

It is well established [47–51] that the dynamic nature of protein structures is an essential element in mediating enzymic activity. To what extent does the presence of protein-associated water influence this?

The microwave dielectric measurements by Singh et al. [52] on myoglobin crystals provided clear evidence of the important role of adsorbed water for the dynamics of this protein. The observed temperature dependence of the dielectric relaxation time of water in the myoglobin crystals was found to be nearly identical to the temperature dependence of the conformational fluctuation rate of the protein as determined from Mossbauer linewidths. It can therefore be concluded that the protein molecule and the water around it form a strongly coupled system. Two mechanisms appear to provide a qualitative understanding of this, namely, the mechanical damping of the protein motion by the adsorbed water and a dynamic electrical coupling between the fluctuating electric dipoles of the adsorbed water and the polar side groups of the protein molecule.

Dielectric measurements on partially hydrated proteins should in principle be capable of providing information concerning not only the rotational freedom of protein-associated water, but also the vibrational freedom of the protein structure. Studies by Bone and Pethig [53] for lysozyme provided an estimate of 36 irrotationally bound water molecules (consistent with the number of multiply hydrogen-bonded water molecules later reported by Blake et al. [54] from their x-ray studies) as well as indications of the bound water acting as a plasticizer to facilitate protein flexibility. These studies were later extended [55,56] to include cytochrome c, collagen, elastin, and chymotrypsin, and some of the results obtained are outlined in Fig. 4 and Table 1.

For dielectric materials such as proteins of low hydration, the Onsager correction for the local electric field can be applied [22], so that the effective moment m of the dipoles responsible for the dielectric dispersion is given by

$$\frac{(\epsilon_s - \epsilon_\infty)(2\epsilon_s + \epsilon_\infty)}{\epsilon_s(\epsilon_\infty + 2)^2} = \frac{Ngm^2}{9\epsilon_0 kT} \tag{16}$$

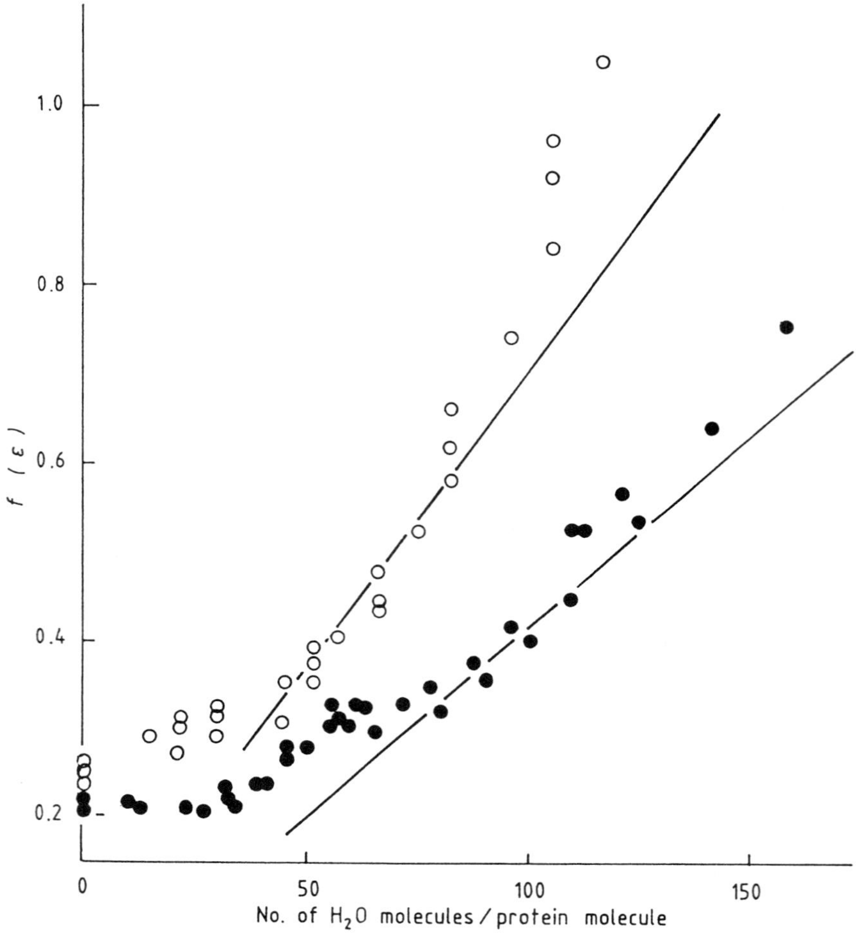

FIGURE 4 Variation of the dielectric parameter $f(\epsilon)$ of Eq. (18) as a function of the number of bound water molecules per protein molecule. ●: lysozyme; ○: cytochrome c. Derived from Bone and Pethig [53,55].

where kT is the Boltzmann energy, and ϵ_s and ϵ_∞ are the permittivity values that mark the lower and upper bounds, respectively, of the frequency range (10 MHz–1 GHz) covering the dispersions associated with the more strongly bound water molecules of number density N. This density is given by

$$N = \frac{N_A}{18} \frac{h\rho}{100} \tag{17}$$

TABLE 1 Values Derived from Dielectric and Hydration Isotherm Data for the Content of Tightly Bound Water, the Hydration Level at Which Significant Protein Structural Flexibility Begins to Occur, and the Monolayer Hydration Coverage of Primary Absorption Sites

Protein	Tightly bound water (%(w/w) ± 0.1)	Hydration for flexibility (%(w/w) ± 0.1)	Monolayer hydration (%(w/w) ± 0.1)
Chymotrypsin [56]	6.5	11.5	6.4
Collagen [55]	7.2	10.8	7.4
Cytochrome c [55]	7.9	14.5	7.5
Elastin [55]	3.3(?)	>8.5	4.2
Lysozyme [53]	4.4	12.5	7.4

Source: Refs. 53, 55, and 56.

where N_A is Avogadro's number, h is the percentage (w/w) hydration, and ρ is the specific density of the protein sample. The parameter $f(\epsilon)$, given by the left-hand side of Eq. (16),

$$f(\epsilon) = \frac{(\epsilon_s - \epsilon_\infty)(2\epsilon_s + \epsilon_\infty)}{\epsilon_s(\epsilon_\infty + 2)^2} \tag{18}$$

can be taken as a measure of the polarizability of the protein–water system, and this is plotted as a function of the number of bound water molecules N in Fig. 4 for the cases of lysozyme and cytochrome c. The straight lines shown in Fig. 4 represent the gradient expected if the protein-bound water exhibits its full dipole moment (as normal bulk water) with $m = 1.84$ D. For hydrations up to 32 water molecules per lysozyme molecule, the factor $f(\epsilon)$ is relatively insensitive to the degree of hydration. This implies that the 32 water molecules, together with the 4 that remain bound after the dehydration procedure [53], are tightly incorporated into the lysozyme structure and have relaxation times of the order or lower than that of the protein. An additional 69 water molecules appear to exhibit a polarizability typical for normal water and can therefore be considered to be bound to the protein in a way that allows them relatively unhindered rotation. It is tempting to identify these two groups of water molecules with the groups of multiply and singly hydrogen-bonded water molecules seen in the x-ray studies of Blake et al. [54]. These x-ray studies indicate that in the hydration range covered in Fig. 4, none of the bound water molecules forms multiple hydrogen bonds with other water molecules. The effect at higher hydrations shown in Fig. 4, where the rate of increase of $f(\epsilon)$ exceeds that expected for normal bulk water, is thus unlikely to be due to the onset of correlation effects between the adsorbed water molecules (i.e., g of Eqs. (8) and (16) remains close to unity). The most likely cause is the effect of increasing hydration acting as a plasticizer to increase the flexibility, and hence the polarizability, of the protein structure. It is pertinent to note that this "loosening" effect becomes quite significant for hydrations approaching 20 wt%, where for lysozyme the onset of enzyme activity takes hold [57].

Using the same logic to analyze the results for cytochrome c shown in Fig. 4, there appears to be around 40 water molecules that are tightly incorporated into the protein structure (plus the three bound to the "dry" protein [55]), and significant protein flexibility becomes apparent after the addition of another 40 water molecules that exhibit rotational freedom. Although the results for lysozyme and cytochrome c appear to follow similar trends, there is no evidence for cytochrome c exhibiting the sudden discontinuity in polarizability that occurs at a hydration of 7 wt% for lysozyme, and that appears to coincide with discontinuities in the infrared absorption and specific heat of lysozyme [57,58]. For chymotrypsin, Bone [56] identified a hydration-dependent dispersion centered around 12 MHz and a second smaller one centered around 80 MHz, which became progressively obscured by the 12 MHz dispersion as the hydration was increased above 50 water molecules per chymotrypsin molecule. Indications of a further hydration-dependent dispersion occurring above 1 GHz were also found, and this together with the 80 MHz dispersion was concluded to reflect orientational relaxations of weakly bound water. The 12 MHz dispersion was attributed to relaxations of protein polar side groups. For hydrations up to 90 water molecules per chymotrypsin, the magnitude of the polarizability parameter $f(\epsilon)$ of Eq. (18) was found to be relatively insensitive to hydration, indicating that these waters are irrotationally bound. At hydrations immediately above 90 water molecules, a discontinuity in $f(\epsilon)$ was observed prior to a steep rise with hydrations above 160 water molecules per chymotrypsin, where the bound water exhibited a dipole moment of value comparable with that of bulk water. This discontinuity could arise from the same processes responsible for the discontinuities in infrared and specific heat observed for lysozyme [57,58]. The steep rise in $f(\epsilon)$ occurring at a hydration content of 160 water molecules (i.e., 11.5 wt% water), which can be taken to indicate the onset of protein flexibility, corresponds to the hydration region where chymotrypsin begins exhibiting significant catalytic activity [59]. The (protein side-group) dispersion around 12 MHz increased rapidly for hydrations above 400 water molecules per chymotrypsin, and this was taken to reflect a general loosening of the protein structure, which Bone [56] discusses in terms of the formation of water clusters on the protein surface acting to screen electrostatic interactions between charged protein groups.

Table 1 summarizes the main interpretations of the dielectric and hydration data, and it includes values derived from the hydration isotherms [55] of the theoretical monolayer hydration capacity of the primary absorption sites. For chymotrypsin, collagen, and cytochrome c, the amount of tightly bound water, as deduced from the dielectric measurements, is close in value to the monolayer coverage of the primary absorption sites, while the onset of significant protein flexibility occurs at hydrations around one-and-a-half to twice this value. For the two enzymes (chymotrypsin and lysozyme) where comparison can be made between the dielectric and catalytic activity data, there appears to be a close match between the critical hydration contents for enzymic activity and structural flexibility.

D. Proton Conduction Effects

Reference has already been made to the observations by Bayley [33] and Takashima and Schwan [36] of a low-frequency dielectric dispersion exhibited by hydrated protein powders that was considered to arise from ionic conduction effects. More recent dielectric measurements [45,60–63] in the frequency range 10^{-4} Hz to 1 MHz for various polycrystalline proteins have revealed two dielectric dispersions, designated as the Ω- and α-dispersions, shown in Fig. 5. These dispersions move to higher frequencies as the protein hydration increases, and this is shown for bovine serum albumin [45] in Fig. 6 as the hydration dependence of the characteristic relaxation time. The Ω-dispersion has been concluded [62,64] to be associated with polarizations associated with the accumulation, or electrochemical generation, of ions at the interfaces between the electrodes and the test sample. This dispersion disappears, whereas the α-dispersion does not, if blocking electrodes composed of mica or poly(tetrafluoroethylene) are used to inhibit charge carrier injection into the test samples [64].

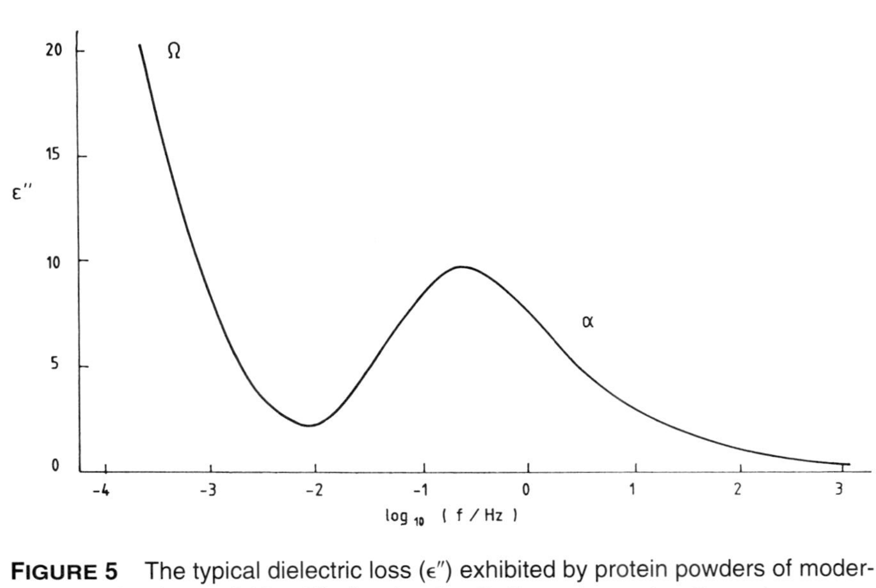

FIGURE 5 The typical dielectric loss (ϵ'') exhibited by protein powders of moderate hydration (around 10 wt%) showing the Ω-dispersion and α-dispersion.

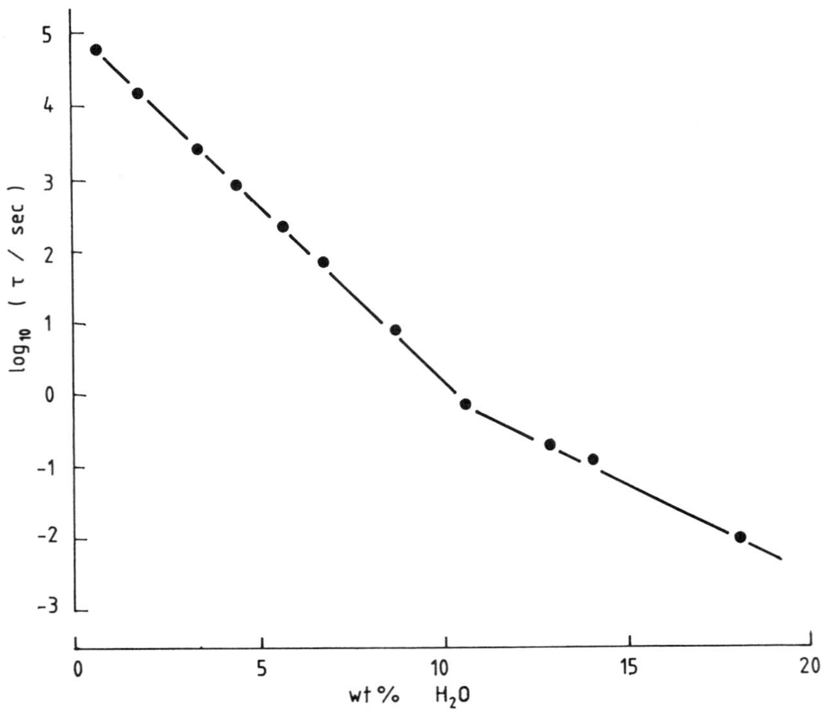

FIGURE 6 The variation of the relaxation time of the α-dispersion exhibited by bovine serum albumin, as a function of water content. Derived from Eden et al. [45].

Analyses [45,61,62] of the α-dispersion show that it is associated with charge carriers that hop between localized sites, rather than with the orientational relaxations of permanent dipoles. Furthermore, it is found that the relaxation time τ_α of the α-dispersion is related to the steady state electrical conductivity σ by

$$\tau_\alpha = \frac{\epsilon_0 \epsilon_\infty}{\sigma} \tag{19}$$

where ϵ_∞ is the permittivity at the high-frequency limit of the α-dispersion. A theoretical model has been developed [65] to show that the relationship of Eq. (19) can be understood in terms of a phonon-assisted hopping charge transport process, whereby the dielectric dispersion arises from the localized hopping of charge carriers between neighboring sites and the conductivity arises from the long-range percolation of the charges along pathways that interconnect these localized sites. Steady state conductivity and electrolysis measurements on lysozyme powders and other proteins have shown (principally through a pronounced deuterium isotope effect [64,66] and studies [67,68] using proton-injecting palladium electrodes) that for pure samples of low or moderate hydration, the conductivity is dominated by mobile protons. Since there is a direct relationship between the α-dispersion and the steady state conductivity, it follows that the α-dispersion could also be related to proton transport.

It is tempting to conclude that the mobile protons originate from the ionizable groups of the lysozyme molecule and that the sorbed water molecules form hydrogen-bond networks along which the protons can conduct. Infrared and heat capacity measurements [57,58] on lysozyme have shown that the interaction with water, even for hydrations as low as 5 wt% (i.e., 40 water molecules per lysozyme molecule), results in proton transfer from the acidic to the basic protein groups. Evidence to support the idea that the α-dispersion in lysozyme arises from proton transfer between the ionizable side groups is shown in Fig. 7, where the magnitude of the α-dispersion, as characterized by the difference between the limiting low- and high-frequency relative permittivity values ($\epsilon_s - \epsilon_\infty$), is shown as a function of the pH at which the samples were lyophilized [63]. A proton of charge q hopping between two localized sites distance s apart will exhibit an effective di-

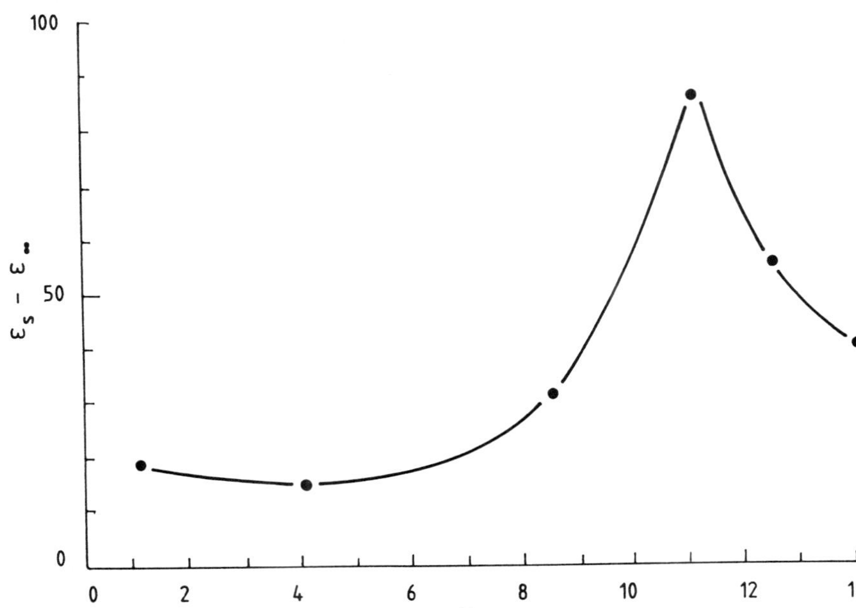

FIGURE 7 Variation of the magnitude of the α-dispersion for lysozyme of 10 wt% hydration as a function of the pH of lyophilization. Derived from Hawkes and Pethig [63].

pole moment $m = qs$. The magnitude of the dispersion will vary according to how many ionizable sites N_i exist that can accommodate a proton, how many of those sites are unoccupied by a proton, and the mean hop distance s. If for the sake of simplifying the problem the hop distance is assumed to be constant, then the dispersion magnitude can be considered to be proportional to the factor $N_i p(1 - p)$, where p is the probability that a site is occupied by a proton. Since the site must either be occupied or unoccupied by a proton, the parameter $(1 - p)$ represents the probability of a site being unoccupied. The maximum dielectric response will occur when half the sites are occupied by protons, corresponding to the factor $p(1 - p)$ having the value 0.25. Using available titration data for lysozyme, it has been shown [63] that the peak response at around pH 11 shown in Fig. 7 corresponds to the situation where proton transfer between the ionizable basic groups predominates over transfers between acidic groups. This can be understood in terms of the presence of counter-ions that become incorporated into the protein during lyophilization. Cations associated with ionized acidic groups will tend to create regions of high potential energy for protons, whereas anions associated with ionized basic groups will provide regions of low potential and attract protons.

Careri et al. [69] employed a dielectric–gravimetric technique to study the α-dispersion as a function of the hydration and pH of lysozyme powders. Four important experimental facts emerged: addition of NaCl (0 to 10 mol per mol protein) did not affect the measured characteristic relaxation time of the dispersion; the relaxation showed a deuterium isotope effect as well as a pH dependence; binding of a saccharide substrate increased the relaxation time by a factor of two. Thus, the inferred conductivity was protonic in nature and involved a mechanism most likely restricted to the surface of individual protein molecules and involving the translocation of protons between ionizable side groups of the proteins. Furthermore, the time constant of the relaxation exhibited cooperativity in its seventh-order dependence on bound protons, which developed in the hydration region (around 15 wt%) critical for the onset of the enzyme's catalytic behavior. The substantial effect of saccharide binding on the relaxation time indicated that a significant proportion of the proton flow over the protein surface is channeled through the enzyme's active site. These results were later [70] analyzed in terms of percolation theory, in which the protonic conduction process is considered as a percolative proton transfer along threads of hydrogen-bonded water molecules. A principal element of the theory is the sudden appearance of long-range connectivity and infinite clusters at the threshold hydration. The α-dispersion of the purple membrane of *Halobacterium halobium* was also investigated [71]. Results analogous to those found for lysozyme [69,70] were found, namely, an isotope effect and the onset of an "explosive" growth of the permittivity and derived conductivity above a threshold hydration level. Preferred pathways for protonic percolation appear to exist in the purple membrane, and since the percolation model is based upon stochastic behavior of a system partially populated with conducting elements, a model involving a fixed geometry, such as the proton wire concept of Nagle and Mille [72], is not the only possible basis for a mechanism of proton conduction.

IV. CONCLUDING REMARKS

The general model that has emerged from the dielectric studies is that globular proteins in solution attract to themselves from one to two layers of bound water, which represents from 0.3 to 0.7 g of protein-associated water per g dry protein. The extent of association of this water to the protein molecule is such that it can be thought of as a "shell" that contributes to the protein's effective radius of rotation, as depicted in Fig. 2. A small proportion of this hydration is either internal water that forms an integral part of the protein structure [73] or else is so strongly bound to the protein that the water molecules have orientational relaxation times equal to or longer than that of the protein molecule. Most of the hydration shell, however, consists of water that is not so strongly bound as to inhibit rotational freedom as much as this. A quite detailed picture is available for myoglobin, where it has been shown [7,30] that half of the hydration shell exhibits orientational relaxations with relaxation times on the order of 10 ns, some 1200 times longer than that exhibited by normal bulk water. This is probably the water that does not freeze at low temperatures [38] and that can also be placed under the same category as the water occupying the primary sorption sites of other proteins and determined from solid state studies [45,46,49,53,55,56] to be rotationally hindered. The other half of the hydration shell for myoglobin, presumably that part located nearest the boundary of shear of the rotating protein–water system, exhibits relaxation times centered around 40 ps [30], which is about five times longer than that of bulk water. This water can be considered to be of the same category as that identified in solid state studies for lysozyme [37], which is characterized by a broad distribution of orientational relaxation times centered around 16 ps.

The solid state microwave dielectric studies [52] for myoglobin have shown that the protein molecule and the water of hydration form a strongly coupled system, both mechanically and electrodynamically, and this is further supported by the finding [53,55,56] that the hydration appears to act as a plasticizing agent for the protein's molecular structure. Furthermore, for the cases where comparative studies have been made [53,56–58], there appears to be a close match between the critical hydration required for the onset of enzymic activity and the hydration level at which the plasticizing action lends structural flexibility to the enzyme structure.

Finally, solid state dielectric measurements, principally for lysozyme, have revealed interesting proton-transfer effects [63–71] that may provide insights into the mechanisms and possible biological roles of proton percolation in hydrogen-bond networks associated with solvated proteins and membrane-associated proteins.

REFERENCES

1. Packer, K. J., *Phil. Trans. R. Soc. London B 278*: 59–87 (1977).
2. Mathur-De Vre, R., *Prog. Biophys. Molec. Biol. 35*: 103–134 (1979).
3. Debye, P., Polar Molecules, Chemical Catalog Co., New York (1929).
4. Kirkwood, J. G., *J. Chem Phys. 7*: 911–919 (1939).
5. Hasted, J. B., Aqueous Dielectrics, Chapman and Hall, London (1973).
6. Kuntz, I. D., *J. Am. Chem. Soc. 94*: 911–919 (1972).
7. South, G. P., and E. H. Grant, *Proc. R. Soc. London A 328*: 371–387 (1972).
8. Takashima, S., *Biochim. Biophys. Acta 79*: 531–538 (1964).
9. Cole, K. S., and R. H. Cole, *J. Chem. Phys. 9*: 341–351 (1941).
10. Kirkwood, J. G., and R. M. Fuoss, *J. Chem. Phys. 9*: 329–340 (1941).
11. Davidson, D. W., and R. H. Cole, *J. Chem. Phys. 19*: 1484–1490 (1951).
12. Colonomos, P., and R. G. Gordon, *J. Chem. Phys. 71*: 1159–1166 (1979).
13. Hallenga, K., J. R. Grigera, and H. J. C. Berendsen, *J. Phys. Chem. 84*: 2381–2390 (1980).
14. Frank, H. S., in Water, a Comprehensive Treatise (F. Franks, ed.), Plenum, New York, *1*: 515–543 (1973).
15. Kell, G. S., in Water and Aqueous Solutions (R. A. Horne, ed.), Wiley-Interscience, New York, pp. 331–376 (1972).
16. Davis, C. M., and J. Jarzynski, in Water and Aqueous Solutions (R. A. Horne, ed.), Wiley-Interscience, New York, pp. 377–423 (1972).
17. Jorgensen, W. L., J. Chandrasekhar, J. D. Madura, R. W. Impey, and M. L. Klein, *J. Chem. Phys. 79*: 926–935 (1983).
18. Jorgensen, W. L., and J. D. Madura, *Mol. Phys. 56*: 1381–1392 (1985).
19. Kuwajima, S., and A. Warshel, *J. Phys. Chem. 94*: 460–466 (1990).
20. Wada, G., *Bull. Chem. Soc. Jpn. 34*: 955–962 (1961).
21. Grant, E. H., R. J. Sheppard, and G. P. South, Dielectric Behavior of Biological Molecules in Solution, Clarendon, Oxford (1978).
22. Pethig, R., Dielectric and Electronic Properties of Biological Materials, Wiley and Sons, Chichester (1979).
23. Takashima, S., Electrical Properties of Biopolymers and Membranes, Adam Hilger, Bristol (1989).
24. Pethig, R., and D. B. Kell, *Phys. Med. Biol. 32*: 933–970 (1987).
25. Oncley, J. L., *J. Am. Chem. Soc. 60*: 1115–1123 (1938).
26. Buchanan, T. J., G. H. Haggis, J. B. Hasted, and B. G. Robinson, *Proc. R. Soc. London A 213*: 379–391 (1952).
27. Grant, E. H., *J. Mol. Biol. 19*: 133–139 (1966).
28. Pennock, B. E., and H. P. Schwan, *J. Phys. Chem. 73*: 2600–2610 (1969).
29. Essex, C. G., M. S. Symonds, R. J. Sheppard, E. H. Grant, R. Lamote, F. Soetewey, M. Y. Rosseneu, and H. Peeters, *Phys. Med. Biol. 22*: 1160–1167 (1977).
30. Grant, E. H, V. E. R. McLean, N. R. V. Nightingale, R. J. Sheppard, and M. J. Chapman, *Bioelectromagnetics 7*: 151–162 (1986).
31. Grant, E. H., B. G. R. Mitton, G. P. South, and R. J. Sheppard, *Biochem. J. 139*: 375–380 (1974).

32. Fricke, H., and E. Parker, *J. Phys. Chem. 44*: 716–726 (1940).
33. Bayley, S. T., *Trans. Faraday Soc. 47*: 509–517 (1951).
34. King, G., *Trans. Faraday Soc. 43*: 601–611 (1947).
35. Rosen, D., *Trans. Faraday Soc. 59*: 2178–2191 (1963).
36. Takashima, S., and H. P. Schwan, *J. Phys. Chem. 69*: 4176–4182 (1965).
37. Harvey, S. C., and P. Hoekstra, *J. Phys. Chem. 76*: 2987–2994 (1972).
38. Kuntz, I. D., and W. Kauzmann, *Adv. Protein Chem. 28*: 239–345 (1974).
39. Grigera, J. R., F. Vericat, K. Hallenga, and H. J. C. Berendsen, *Biopolymers 18*: 35–45 (1979).
40. Grigera, J. R., and H. J. C. Berendsen, *Biopolymers 18*: 47–57 (1979).
41. Hoeve, C. A. J., in Water in Polymers (S. P. Rowland, ed.), Amer. Chem. Soc., Washington, D.C., ACS Symp. Ser. *127*: 135–146 (1980).
42. Gascoyne, P. R. C., and R. Pethig, *J. Chem. Soc., Faraday Trans. 1, 73*: 171–180 (1977).
43. Kent, M., *J. Phys. (D) 5*: 394–409 (1972).
44. Bone, S., P. R. C. Gascoyne, and R. Pethig, *J. Chem. Soc., Faraday Trans. 1, 73*: 1605–1611 (1977).
45. Eden, J., P. R. C. Gascoyne, and R. Pethig, *J. Chem. Soc., Faraday Trans. 1, 76*: 426–434 (1980).
46. Gascoyne, P. R. C., and R. Pethig, *J. Chem. Soc., Faraday Trans. 1, 77*: 1733–1735 (1981).
47. Gurd, F. R. N., and T. M. Rothgeb, *Adv. Protein Chem. 33*: 73–165 (1979).
48. Careri, G., P. Fasella, and E. Gratton, *Ann. Rev. Biophys. Bioeng. 8*: 69–97 (1979).
49. Welch, G. R., B. Somogyi, and S. Damjanovich, *Prog. Biophys. Molec. Biol. 39*: 109–146 (1982).
50. Karplus, M., and J. A. McCammon, *Ann. Rev. Biochem. 52*: 263–300 (1983) (given as vol. 53 by publishers, but printed in vol. 52).
51. Welch, G. R. (ed.), The Fluctuating Enzyme, Wiley, New York (1986).
52. Singh, G. P., F. Parak, S. Hunklinger, and K. Dransfeld, *Phys. Rev. Lett. 47*: 685–688 (1981).
53. Bone, S., and R. Pethig, *J. Mol. Biol. 157*: 571–575 (1982).
54. Blake, C. C. F., W. C. A. Pulford, and P. J. Artymiuk, *J. Mol. Biol. 167*: 693–723 (1983).
55. Bone, S., and R. Pethig, *J. Mol. Biol. 181*: 323–326 (1985).
56. Bone, S., *Biochim. Biophys. Acta 916*: 128–134 (1987).
57. Careri, G., A. Giansanti, and E. Gratton, *Biopolymers 18*: 1187–1203 (1979).
58. Careri, G., E. Gratton, P. H. Yang, and J. A. Rupley *Nature 284*: 572–573 (1980).
59. Stevens, E., and L. Stevens, *Biochem. J. 179*: 161–167 (1979).
60. Eley, D. D., N. C. Lockhart, and C. N. Richardson, *J. Chem. Soc., Faraday Trans. 1., 75*: 323–334 (1979).
61. Bone, S., J. Eden, and R. Pethig, *Int. J. Quantum Chem., Quant. Biol. Symp. 8*: 307–316 (1981).
62. Morgan, H., and R. Pethig, *Int. J. Quantum Chem., Quant. Biol. Symp. 11*: 209–216 (1984).
63. Hawkes, J. J., and R. Pethig, *Biochim. Biophys. Acta 952*: 27–36 (1988).

64. Morgan, H., and R. Pethig, *J. Chem. Soc., Faraday Trans. 1, 82*: 143–156 (1986).
65. Bone, S., J. Eden, P. R. C. Gascoyne, and R. Pethig, *J. Chem. Soc. Faraday Trans. 1, 77*: 1729–1732 (1981).
66. Behi, J., S. Bone, H. Morgan, and R. Pethig, *Int. J. Quantum Chem., Quant. Biol. Symp. 9*: 367–374 (1982).
67. Lawton, B. A., Z. H. Lu, R. Pethig, and Y. Wei, *J. Mol. Liquids 42*: 83–98 (1989).
68. Bone, S., *Biochim. Biophys. Acta 1078*: 336–338 (1991).
69. Careri, G., M. Geraci, A. Giansanti, and J. A. Rupley, *Proc. Natl. Acad. Sci. USA 82*: 5342–5346 (1985).
70. Careri, G., A. Giansanti, and J. A. Rupley, *Proc. Natl. Acad. Sci. USA 83*: 6810–6814 (1986).
71. Rupley, J. A., L. Siemankowski, G. Careri, and F. Bruni, *Proc. Natl. Acad. Sci. USA 85*: 9022–9025 (1988).
72. Nagle, J. F., and M. Mille, *J. Chem. Phys. 74*: 1367–1372 (1981).
73. Edsall, J. T., and H. A. McKenzie, *Adv. Biophys. 16*: 53–183 (1983).

5

Protein Dynamics: Hydration, Temperature, and Solvent Viscosity Effects as Revealed by Rayleigh Scattering of Mossbauer Radiation

VITALII I. GOLDANSKII AND
YURII F. KRUPYANSKII
Russian Academy of Sciences, Moscow, Russia

I. INTRODUCTION

In all contemporary models of enzyme action, the functional activity of enzymes is directly connected with the dynamic properties of proteins [1–6]. The investigation of protein dynamics is also of interest in its own right since it is well known that different and seemingly incompatible properties of proteins are often revealed by different experimental techniques. Thus, the available data from one group of experimental techniques makes one think of the protein as a solid body [7–9]. On the other hand, the data from another group of experimental methods depicts the protein as a strongly fluctuating structure [4,10,11].

Therefore, protein dynamics is currently studied by employing such powerful physical methods as nuclear magnetic resonance [12,13], neutron scattering [14], computer simulation of protein dynamics [15,16], x-ray dynamical analysis

(XRDA) [17,18], and Mossbauer absorption spectroscopy (MAS) [19–21], along with conventional research methods such as the deuterium exchange technique [11,22] and luminescence labeling [23]. It was qualitatively shown by these conventional methods that hydration degree and solvent composition strongly influence protein dynamics. It is to be noted that unlike the other experimental methods just mentioned, x-ray analysis, neutron scattering, and MAS belong to a group of methods that allow protein dynamics to be quantitatively described using the mathematics of Van Hove correlation functions [24].

In this chapter, data will be reviewed relating to the influence of hydration, temperature, and solvent composition on protein dynamics by means of yet another "correlation" technique, the Rayleigh scattering of Mossbauer radiation (RSMR).

II. BACKGROUND OF RSMR TECHNIQUE, BASIC EXPRESSIONS, AND APPROXIMATIONS

There are several surveys dealing with RSMR in which experimental and theoretical fundamentals of the method are considered in some detail [25–27].

In this chapter, we concentrate on some specific features of the application of RSMR to the study of protein–solvent systems (see also [28,29]).

Shown in Fig. 1 is a schematic of the RSMR experimental arrangement. Mossbauer radiation emitted by a ^{57}Co source mounted on a vibrator experiences Rayleigh scattering by electrons of a biopolymer B. The radiation scattered at the angle 2θ is measured by the detector D. In order to measure the elastic scattering fraction, an additional resonance adsorber is employed, which together with the

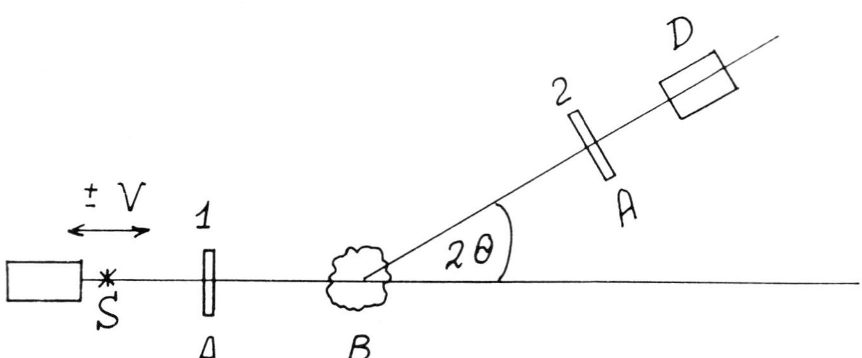

FIGURE 1 RSMR experimental setup. S is a Mossbauer source, B is a protein or other biopolymer under investigation, D is a detector, and A is a resonant absorber, which can alternatively be placed in positions 1 and 2 measure intensities $I(0)$ and $I(2\theta)$, respectively.

detector makes up the so-called Mossbauer detector. By means of the resonance absorber, first the intensity of the incident beam, $I(0)$, is measured and then the intensity of scattered radiation, $I(2\theta)$.

The elastic fraction is determined by

$$f = \frac{I_\infty(2\theta) - I_0(2\theta)}{I_\infty(0) - I_0(0)} \qquad (1)$$

where I_0 and I are measured under the resonance condition (i.e., with the source velocity $V = 0$) and far away from resonance ($V = \infty$), respectively.

In measurements of the RSMR spectral lineshape, $\eta(\nu) = [I(\nu = \infty) - I(\nu)]/I(\nu = \infty)$, the resonance absorber is to be placed only in position 2. The Mossbauer source–detector combination accomplishes the high-energy resolution of the method ($\sim 10^{-9}$ eV), which exceeds by several orders of magnitude the resolution provided by the most advanced neutron spectrometers (up to $\sim 10^{-6}$ eV), let alone that offered by XRDA (~ 1 eV). It is precisely due to this fact that comparatively slow motion with correlation times from 10^{-6} to 10^{-9} s can be detected from broadening of energy spectrum lines and motions with correlation times $\tau_c < 10^{-9}$ s from the decrease of elastic scattering fraction.

The essential advantage of this method over MAS is its versatility, for the scatterer (the biopolymer under study) need not have Mossbauer nuclei. The scope of amenable biological objects thus can be substantially widened. In addition, the possibility of changing the scattering angle (or momentum transfer) in the course of an experiment and, as a consequence, the possibility of choosing relatively small scattering angles make it possible to obtain information on motions with much larger amplitudes than with MAS.

In its first application, RSMR was employed to distinguish between elastic and inelastic Rayleigh scattering intensity for polycrystalline and single-crystal [25] scatterers.

Further application of the RSMR technique to investigation of dynamical properties of inorganic single crystals, organic and polymeric glasses, and supercooled liquids has placed RSMR among the most effective methods for studying the dynamics of atoms in a condensed phase [25–27]. Systematic investigation of biopolymers by RSMR was begun in 1980 [30].

Results on the study of protein dynamics were obtained for a relatively long period by use of the so-called incoherent version of RSMR (see [28,29] and references therein), that is, the version in which soft collimation conditions were utilized. Usually, it is necessary to fulfill two conditions for the correct use of the incoherent approximation: (a) wide divergence of the beam and (b) remoteness of the scattering angle from the main Bragg maximum [26]. For intensity reasons and to provide, in the first approach, the quantitative study of RSMR spectra, only the first condition was realized in these works. The incoherent approach may not be strictly correct in this case. The low angular resolution of the experiment, however, produced strong averaging of the interference pattern. RSMR data analyzed with the incoherent approach may therefore certainly reveal some physical features of protein dynamics.

Recently, protein dynamics were also studied by RSMR in an angular-dependent fashion with good angular resolution ("coherent" version) in order to analyze the influence of coherent effects [31,32]. Since the counting rate in these experiments is rather poor, only a few results have been obtained up to now.

Below are considered the main approximations necessary to apply formulas to data obtained from RSMR experiments.

The dependence of the intensity J of scattered γ-quanta versus momentum $\mathbf{Q}$ ($\mathbf{Q} = (4\pi \sin\theta)/\lambda$, λ = wavelength) and energy $\hbar\omega$ transfer is connected with the double differential cross section $d^2\sigma/d\Omega\, d\omega$ and in a first-order approximation is given by [33]

$$J(\mathbf{Q}, \omega) = \frac{d^2\sigma}{d\Omega\, d\omega} = \bar{\bar{r}}_0^2 p_0^2 \sum_{m,n} S_{mn}(\mathbf{Q}, \omega) \tag{2}$$

where

$$\bar{\bar{r}}_0 = \frac{e^2}{mc^2}; \qquad p_0^2 = \frac{1}{2}(1 + \cos^2 2\theta)$$

The scattering function $S_{mn}(\mathbf{Q}, \omega)$ contains all the structure and dynamic information on the scattering system and in the case of RSMR is represented by

$$S_{mn}(\mathbf{Q}, \omega) = \frac{1}{2\pi} \int dt \exp\left(\frac{-i\omega t - \Gamma|t|}{2\hbar}\right) I_{mn}(\mathbf{Q}, t) \tag{3}$$

where

$$I_{mn}(\mathbf{Q}, t) = \sum_{m,n} \phi_m \phi_n \langle \exp[-i\mathbf{Q}R_m(t)] \exp[i\mathbf{Q}R_n(0)] \rangle \tag{4}$$

is the so-called intermediate correlation function, and $\phi_i(\mathbf{Q})$ is the atomic form factor of the ith atom.

The case of hydrated proteins will be mainly discussed in this section (the crystalline case [32], which is much simpler to treat, will be touched on only briefly).

Let us now discuss only proteins and then come to the more complicated situation of water–protein systems (hydrated proteins or proteins in a solvent).

The dynamics of proteins can be considered in the following simple approach. This model introduces a division of atoms of the protein globule into two parts. The first group (fraction) of atoms (A) move collectively within segments with a mean square displacement (msd) $\langle x^2 \rangle_s$. Motional correlations between different moving segments are neglected. Another fraction of atoms (B) moves individually with $\langle x^2\rangle_i$, and motional correlations between different atoms and between atoms and segments are also neglected. In this case, the inelastic scattering intensity can be expressed as

$$J_{\text{inel}}(\mathbf{Q}) = \sum_S S_s(\mathbf{Q})\,[1 - \exp(-\mathbf{Q}^2\langle x^2\rangle_s)] + \sum_i \phi_i^2(\mathbf{Q})\,[1 - \exp(-\mathbf{Q}^2\langle x^2\rangle_i)] \qquad (5)$$

If one considers, for simplicity, that $\langle x^2\rangle_i = \langle x^2\rangle_s = \langle x^2\rangle$, it leads to the following expression:

$$\begin{aligned} J_{\text{inel}}(\mathbf{Q}) &= [1 - \exp(-\mathbf{Q}^2\langle x^2\rangle)]\left[\sum_S S_s(\mathbf{Q}) + \sum_i \phi_i^2(\mathbf{Q})\right] \\ &= [1 - \exp(-\mathbf{Q}^2\langle x^2\rangle)][AS_{\text{SEG}}(\mathbf{Q}) + BF_i^2(\mathbf{Q})] \end{aligned} \qquad (6)$$

where $A + B = 1$, $S_{\text{SEG}}(\mathbf{Q})$ represents the interference pattern from the collectively moving segment and $\mathbf{F}_i^2(\mathbf{Q}) = \Sigma_i\, \phi_i^2(\mathbf{Q})$, the sum here extending over all atoms moving individually.

Then elastic and total scattering intensity, by analogy with [32], can be written as follows:

$$\begin{aligned} J_{\text{p}}(\mathbf{Q},\infty) &= S_0^{\text{P}}(\mathbf{Q}) \exp(-\mathbf{Q}^2\langle x^2\rangle_{\text{p}}) \\ J_{\text{p}}(\mathbf{Q},0) &= S_0^{\text{P}}(\mathbf{Q}) \exp(-\mathbf{Q}^2\langle x^2\rangle_{\text{p}}) \\ &\quad + [AS_{\text{SEG}}^2(\mathbf{Q}) + BF_{ip}^2(\mathbf{Q})][1 - \exp(-\mathbf{Q}^2\langle x^2\rangle_{\text{p}})] \end{aligned} \qquad (7)$$

Here $S_0^{\text{P}}(\mathbf{Q})$ represents the interference pattern from the whole protein sample. One then obtains, for the elastic fraction,

$$f_{\text{p}}(\mathbf{Q}) = \frac{S_0^{\text{P}}(\mathbf{Q}) \exp(-\mathbf{Q}^2\langle x^2\rangle_{\text{p}})}{S_0^{\text{P}}(\mathbf{Q}) \exp(-\mathbf{Q}\langle x^2\rangle_{\text{p}}) + [AS_{\text{SEG}}^{P}(\mathbf{Q}) + BF_{ip}^2(\mathbf{Q})][1 - \exp(-\mathbf{Q}^2\langle x^2\rangle_{\text{p}})] + J_c} \qquad (8)$$

where J_c is the Compton scattering intensity. It is quite instructive to consider the limiting cases. If the motional correlations extend over the whole sample $B = 0$, $A = 1$, $S_{\text{SEG}}^{\text{P}} = S_0^{\text{P}}(\mathbf{Q})$, and

$$f_{\text{p}}(\mathbf{Q}) = \frac{S_0^{\text{P}}(\mathbf{Q}) \exp(-\mathbf{Q}^2\langle x^2\rangle_{\text{p}})}{S_0^{\text{P}}(\mathbf{Q}) + J_c} \qquad (9)$$

If all atoms within the globule move independently from each other, then $A = 0$, $B = 1$. This leads to

$$f_{\text{p}}(\mathbf{Q}) = \frac{S_0^{\text{P}}(\mathbf{Q}) \exp(-\mathbf{Q}^2\langle x^2\rangle_{\text{p}})}{S_0^{\text{P}}(\mathbf{Q}) \exp(-\mathbf{Q}^2\langle x^2\rangle_{\text{p}}) + F_{ip}^2(\mathbf{Q})[1 - \exp(-\mathbf{Q}^2\langle x^2\rangle_{\text{p}})] + J_c} \qquad (10)$$

Neglecting all interference terms for the elastic and inelastic scattering, one can come to the incoherent approximation:

$$S_0(\mathbf{Q}) = \sum_i \phi_i^2; \quad S_{SEG}(\mathbf{Q}) = \sum_i \phi_i^2$$

and

$$f_p(\mathbf{Q}) = \frac{\sum_i \phi_i^2(\mathbf{Q}) \exp(-\mathbf{Q}^2\langle x^2\rangle_p)}{\sum_i \phi_i^2(\mathbf{Q}) + J_c} = \frac{F_{ip}^2(\mathbf{Q}) \exp(-\mathbf{Q}^2\langle x^2\rangle)}{F_{ip}^2(\mathbf{Q}) + J_c} \tag{11}$$

Now consider the protein–water system. We shall use the following physical assumptions for analysis of the data: (a) For hydrated proteins (as distinct from crystalline samples), the complete absence of correlation in the location of one globule relative to another is suggested; that is, the scattering function of hydrated protein $S_p(\mathbf{Q})$ contains interference terms only from one protein globule. (For crystalline samples, the scattering function $S_p(\mathbf{Q})$ contains interference terms from all protein atoms in the crystal since the intensity is the sum from all unit cells [32].) (b) Correlation in locations of different water molecules is essential; therefore, $S_w(\mathbf{Q})$ contains interference terms from oxygen atoms of different water molecules. (c) Correlation in locations of different protein and water molecules relative to one another will be neglected (i.e., the corresponding interference terms will be considered as small).

With these assumptions, the intensity of the γ-quanta Rayleigh scattered by the protein–water system is the sum of intensities

$$J_\Sigma(\mathbf{Q}) = J_p(\mathbf{Q}) + J_w(\mathbf{Q}) \tag{12}$$

Since the hydration water is in general far more mobile than the atoms of the protein, the description demands the introduction of different msd for interprotein water, as compared with protein.

The elastic fraction for the water–protein system is equal to [26,29]

$$f_\Sigma = \frac{J_\Sigma^{el}(\mathbf{Q})}{J_\Sigma(\mathbf{Q}) + J_c} = C_p f_p + C_w f_w \tag{13}$$

In the case of the incoherent approximation, this leads to

$$f_\Sigma(\mathbf{Q}) = \frac{F_{ip}^2(\mathbf{Q}) \exp(-\mathbf{Q}^2\langle x^2\rangle_p) + F_{i}^2{}_w(Q) \exp(-\mathbf{Q}^2\langle x^2\rangle_w)}{F_{ip}^2(\mathbf{Q}) + F_{i}^2{}_w(Q) + J_c} \tag{14}$$

For a truly coherent treatment, the division of f_Σ into separate contributions is not possible due to pair correlations between protein and water. Fortunately, these correlations play a minor role at sufficiently large scattering angles and at the wide divergence of the beam employed (see assumption c). The following attempt can be made to include, at least in part, coherent effects. If Eq. (9) is assumed valid and pair correlations between protein and water are neglected, one can arrive, after averaging over scattering angle, at the expression [34]

$$
\begin{aligned}
\langle f_\Sigma(\mathbf{Q})\rangle_{\Delta Q} &= \langle C_p\rangle_{\Delta Q} f_p + \langle C_w\rangle_{\Delta Q} f_w \\
&= \frac{f_p + h\langle C_1\rangle_{\Delta Q} f_w}{1 + h\langle C_1\rangle_{\Delta Q} + \langle C_2\rangle_{\Delta Q}} \\
&\quad + \frac{\langle i_p(\mathbf{Q})\rangle_{\Delta Q} f_p + \langle i_w{}^b(\mathbf{Q})\rangle_{\Delta Q} h C_1 f_w}{1 + h\langle C_1\rangle + \langle C_2\rangle_{\Delta Q}}
\end{aligned}
\tag{15}
$$

where h is the hydration degree, ΔQ is the divergence of the beam

$$i(\mathbf{Q}) = S(\mathbf{Q}) - 1$$

$$C_1 = \frac{\phi_w{}^2 + J_c{}^w}{\phi_p{}^2 + J_c{}^P}$$

and

$$C_2 = \frac{\phi_p{}^2 i_p(\mathbf{Q}) + h\phi_w{}^2 i_w{}^b(\mathbf{Q})}{\phi_p{}^2 + J_c{}^P}$$

For the protein component, the scattering functions $S_0{}^P(\mathbf{Q})$ and $i_p(\mathbf{Q})$ can be either calculated from crystallographic coordinates or taken from experiments on dry protein. Of course, then we have to assume that no changes occur in the interatomic distances when drying or crystallizing protein. For the water component, the situation is much more difficult. It is well known that the structural and dynamical properties of interprotein (bound) water are drastically different from those for free water [35]. Strongly bound water assumes a lacy structure on the protein surface [36]. Therefore, it is natural to suggest that at one extreme the scattering function for interprotein waters, $S_w{}^b(\mathbf{Q})$, and consequently the function $i_w{}^b(\mathbf{Q})$, coincides with the scattering function for protein

$$i_w(\mathbf{Q}) = i_w{}^b(\mathbf{Q}) = i_p(\mathbf{Q}) \tag{16}$$

At the other extreme, the scattering function for interprotein water coincides with the scattering function for free (bulk) water

$$i_w(\mathbf{Q}) = i_w{}^f(\mathbf{Q}) \tag{17}$$

The numerical relationship given in Eq. (15) strongly depends on the value of the average scattering angle 2θ and the shape of the angular resolution function B, or divergence of the beam ΔQ. For the RSMR setup used in [28,29], $2\theta = 11.9°$, the angular resolution function is described by a Gaussian with a width $\sigma = 2.4°$ and the average values of $\langle i_i(\mathbf{Q})\rangle_{\Delta Q}$ are the following:

$$
\begin{aligned}
\langle i_p(\mathbf{Q})\rangle_{\Delta Q} &\simeq 0.1 \quad \text{for met-Mb and HSA} \\
\langle i_w(\mathbf{Q})\rangle_{\Delta Q} &\simeq 0.2 \quad \text{for free water}
\end{aligned}
\tag{18}
$$

According to the convolution hypothesis [38], an intermediate correlation function (see Eq. (4)), in which cooperative motions of different atoms are taken into account, may be represented by

$$I(\mathbf{Q}, t) = S(\mathbf{Q})I_s(\mathbf{Q}, t) = [1 + i(\mathbf{Q})]I_s(\mathbf{Q}, t) \tag{19}$$

The intermediate self-correlation function $I_s(\mathbf{Q}, t)$ contains information about essentially more simple individual motions of atoms and atomic groups.

It is easy to show that with this approximation, a relationship similar to (15) is valid for the spectral function that retains the nontrivial terms of η (ν):

$$g_\Sigma(\mathbf{Q}, \omega) = C_p g_p(\mathbf{Q}, \omega) + C_w g_w(\mathbf{Q}, \omega) \tag{20}$$

where

$$C_p = \frac{1 + i_p(\mathbf{Q})}{1 + hC_1 + C_2}; \qquad C_w = hC_1 \frac{1 + i_w(\mathbf{Q})}{1 + hC_1 + C_2}$$

and

$$g_{p,w}(\mathbf{Q}, \omega) = \frac{1}{2\pi}\int dt \exp\left(\frac{-i\omega t - \Gamma|t|}{2\hbar}\right) I_s^{p,w}(\mathbf{Q}, t)$$

Certainly, all assumptions made in deriving Eq. (15) are included in Eq. (20).

It is quite evident that for incoherent evaluation an expression similar to (20) is valid [29].

The use of a Gaussian approximation [37,38] for the self-correlation function $I_s(\mathbf{Q},t)$ gives the possibility to come finally to the following expression for the lineshape:

$$g_{p,w}(\mathbf{Q}, \omega) = \frac{1}{2\pi}\int dt \exp\left(-i\omega t - \frac{\Gamma|t|}{2\hbar}\right) \exp\left(-\frac{1}{2}\mathbf{Q}^2\langle[x(t) - x(0)]^2\rangle\right) \tag{21}$$

Here $\langle\ \rangle$ denotes an ensemble average and $x(t)$ is the coordinate of the atom at time t.

III. HYDRATION DEPENDENCIES OF THE ELASTIC RSMR FRACTIONS AND RSMR SPECTRA

If assumption (16) holds for interprotein water, expression (15) can be rewritten numerically as

$$\langle f_\Sigma(\mathbf{Q})\rangle_{\Delta Q} = \frac{f_p + 1.21hf_w}{1 + 1.21h} \tag{22}$$

that is, when scattering functions for interprotein water and protein coincide, the expression for the elastic fraction is identical to the expression obtained with the incoherent approximation. If, on the contrary, interprotein water behaves like free water (Eq. (17)), the elastic fraction will have the numerical form

$$\langle f_\Sigma(\mathbf{Q})\rangle_{\Delta Q} = \frac{f_{\rm P} + 0.9hf_{\rm w}}{1 + 0.9h} \tag{23}$$

In this case, the expression for the elastic fraction is slightly different from that obtained in the incoherent approximation. Within this suggestion on interprotein water properties, even wide divergence of the beam does not average the coherent phenomena.

Figure 2 shows a typical hydration curve of the elastic fraction f_Σ , for human serum albumin (HSA), obtained at room temperature [40,41].

A straightforward analysis of relationships (22) and (23) leads to the conclusion that there is no additivity of dynamical properties in the HSA–water system in the entire range of hydration studied. Indeed, if we use $f_\Sigma(h \rightarrow 0)$ for $f_{\rm P}$, $f_{\rm P} = f_\Sigma(0) = 0.8$, and assume $f_{\rm w} = 0$, as for free water, then the calculated curve for f_Σ, shown as a dotted line in Fig. 2 (see Eq. (23)), exhibits large deviations from the experimental data. For all hydration degrees studied, adding more water not only does not give the contribution to the elastic RSMR (quite in accord with $f_{\rm w} = 0$), but further loosens the protein, increasing its mobility and reducing the value

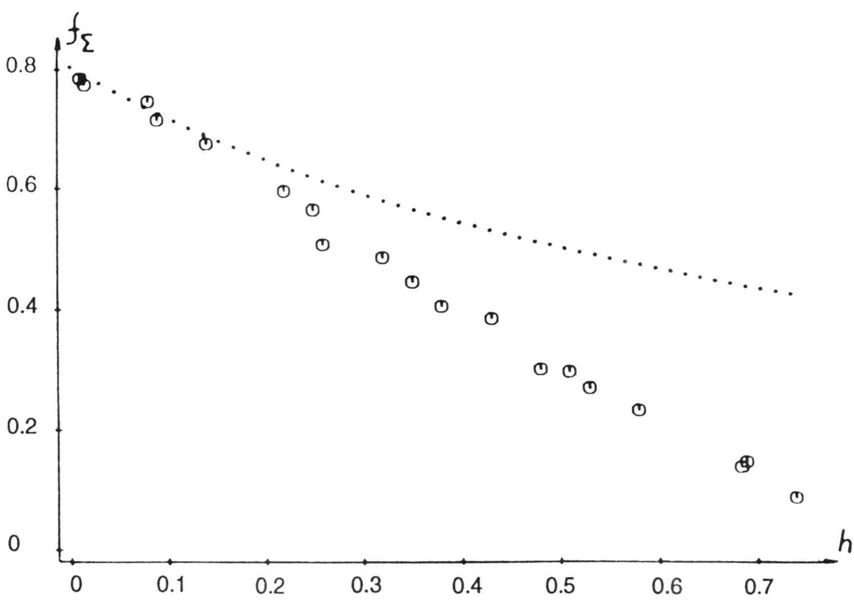

FIGURE 2 Hydration dependence of the elastic scattering fraction for an HSA sample. Dotted line is calculation in terms of Eq. (23) assuming no interaction between dry protein and free water.

of f_p. Thus, at all hydration levels investigated, $0.05 < h < 0.75$, dynamical properties of HSA are influenced by its water content. This conclusion is in no way linked to the assumptions that $f_w = 0$ or the validity of Eq. (17). Let us consider a more realistic picture of bound-water behavior. It is well known from the literature [42] that small amounts of water, up to 10% of dry protein weight, are strongly bound to the protein. According to NMR and ESR data, relaxation properties exhibited by bound water at large h, about 0.4 to 0.5, are typical of common liquids with viscosity greater than that of free water but lower than that of glycerol. Hence, at the other extreme, structural (see Eq. (16)) and dynamical properties of bound water coincide in the entire range of hydration degrees with those of the protein (see also Eq. (22)).

Shown in Fig. 3 is a family of dependencies calculated, in terms of Eqs. (22) and (23), of the elastic scattering fraction for the protein alone, ϕf_p, among which is contained the real curve representing the actual variation of the protein dynamics. The symbol ϕ denotes a calculated value rather than a measured one. The lower boundary of the cross-hatched range of ϕf_p values corresponds to the assumption that dynamical and structural properties of interprotein water are not different from those of protein for all tested values of h, and the upper one to the assumption that the water behaves as free water. The real curve of the hydration dependence of protein dynamical properties, $\phi f_p^*(h)$, shown in Fig. 3 as a heavy line, was calculated under the following assumptions: at small hydration degrees, $h \leq 0.1$, the dynamical properties of bound water are no different from those of the protein; then, at greater h, they are like those of viscous liquids, $f_w \to 0.05$ (the

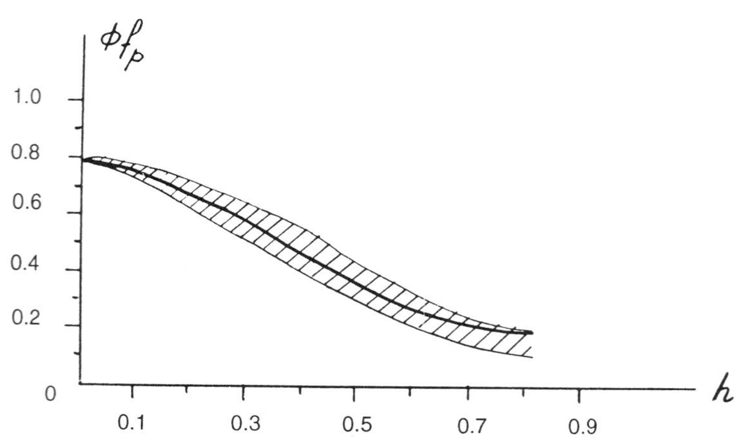

FIGURE 3 Calculated elastic fraction ϕf_p as a function of the hydration degree on the basis of Eqs. (22) and (23). Different assumptions were made concerning the behavior of the interprotein water. Details are given in the text.

magnitude of the elastic fraction for glycerol at $T = 295$ K [43]); approaching the dynamical properties of free water, $f_w = 0$, with $h \geq 0.6$ to 0.7. As seen from Fig. 3, even under these more realistic assumptions about the behavior of bound water, there still will be no additivity of dynamical characteristics in the HSA–water system in the entire hydration range studied.

The absence of additivity of the dynamical properties of water–protein systems is displayed in the dependence of the real curve, $\phi f_p^*(h)$, on h. Figure 4 shows the area under the total spectrum, S_Σ, from a metmyoglobin sample (met-Mb, $M = 17{,}800$) as a function of the hydration degree, h (in this series of measurements, spectra were taken only for a narrow velocity range, -2.3 to 2.3 mm/s). As in the case of HSA (see Fig. 2), the dotted line is calculated with Eq. (23). Shown in Fig. 5 is a family of curves $\phi S_p(h)$ calculated by using Eqs. (22) and (23) along with the realistic curve $\phi S_p^*(h)$. As one can see from Fig. 5, there is no additivity of the dynamical properties in the met-Mb–water system up to the value of $h =$ 0.6 g/g. Addition of more water above this value can be seen to have very little effect on the dynamical properties of met-Mb.

Elsewhere, we have also studied hydration dependencies for trypsin ($M =$ 23,319) [44] lysozyme ($M = 14{,}000$) [40], DNA ($M = 1$ to 3×10^6 amu) [45] and chromatophores ($M = 10^7$ amu) [46]. Hydration of all these biopolymers has been found to result in a reduction of the real elastic fraction, $\phi f_p^*(h)$, and, therefore, to

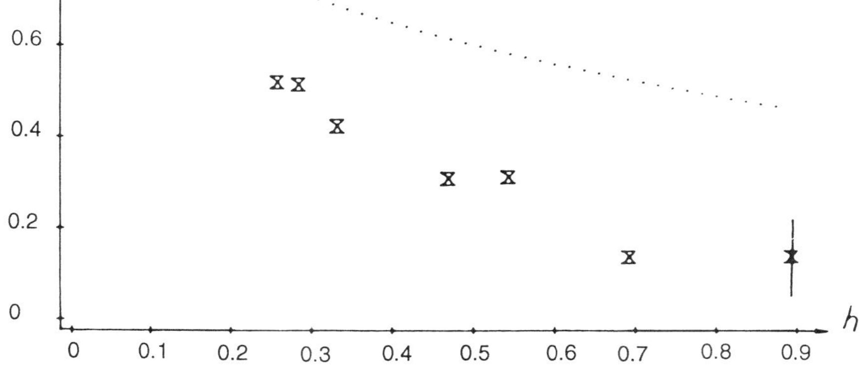

FIGURE 4 Hydration dependence of the area under the total spectrum, S_Σ, for met-Mb. Dots represent calculation under the same assumptions as in Fig. 2.

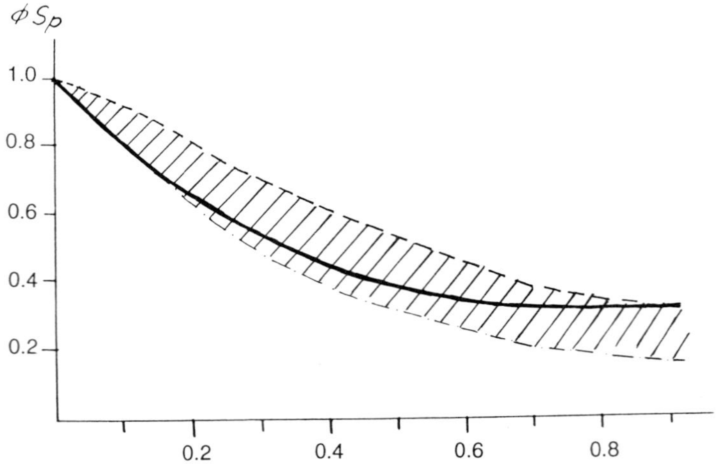

FIGURE 5 Calculated area, $\phi S_p(h)$, for met-Mb. See Fig. 3 for explanation. Heavy line is the realistic curve $\phi S_p^*(h)$.

an increase in values of $\langle x^2\rangle(h)$ for hydration degrees $0.05 < h < 0.5$ (one exception is lysozyme, where the addition of water beyond $h = 0.35$ has no effect on the dynamical properties of the protein).

Presented in Fig. 6 are RSMR energy spectra for met-Mb (for three degrees of hydration) [47], obtained over a large range of measurement velocities, $V = \pm 20$ mm/s. At large hydration degrees, $h \geq 0.6$, neither spectrum can be described by a single Lorentzian (even admitting broadening). To describe the total spectrum, a second quasi-elastic line has to be included (a "wide" component). The presented spectra are a sum of a narrow line with a width equal to the width at $h = 0$ and a "wide" line.

IV. SOLVENT COMPOSITION AND VISCOSITY DEPENDENCIES OF THE ELASTIC RSMR FRACTIONS

We have also studied the influence of viscosity, or, more precisely, solvent composition, on the dynamics of proteins and chromatophores [48,49]. A number of papers have appeared recently describing direct experimental observations of the influence of solvent viscosity on the rate of biochemical reactions [50,51] and the theoretical treatment of these data [52,53]. As a rule, it is suggested [50–53] that the solvent viscosity directly influences protein dynamics, which, in its turn, determines protein reactivity. Direct experimental evidence, however, for the influence of viscosity on protein dynamics has yet to be obtained. Therefore, we have attempted to observe directly this influence, and these attempts will be described

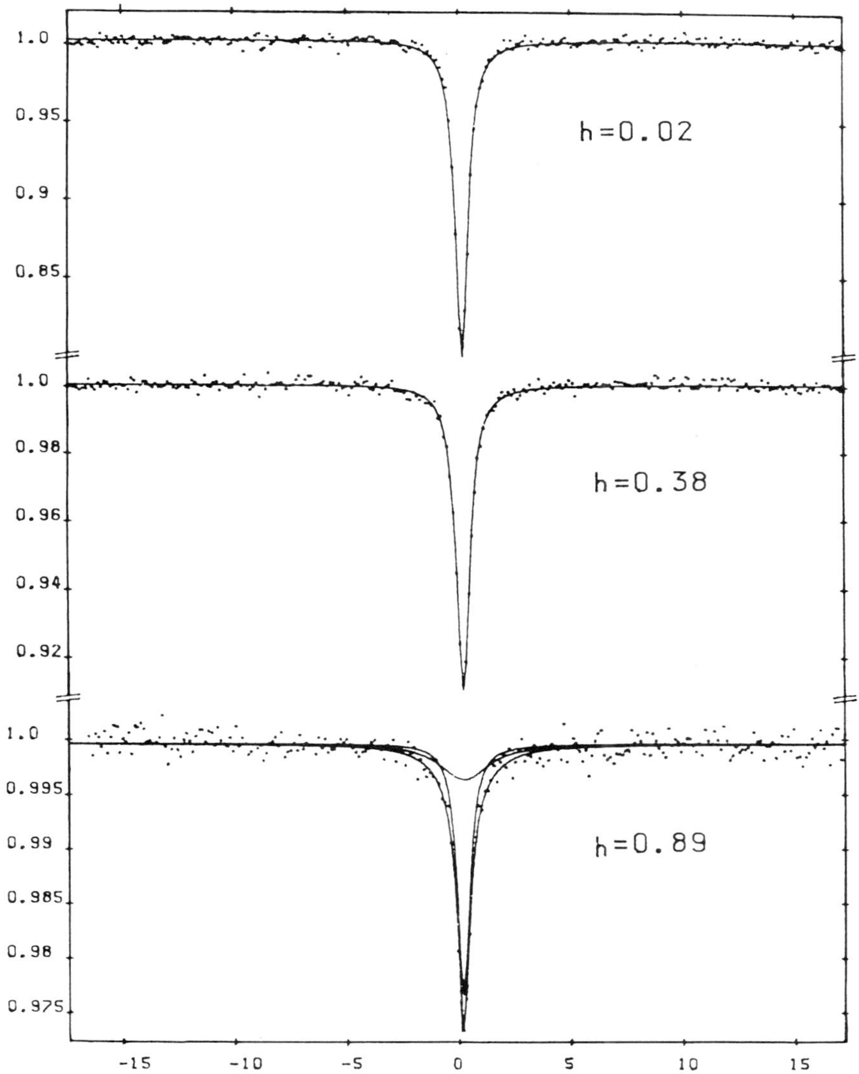

FIGURE 6 Energy spectra of met-Mb samples with different degrees of hydration.

in this section. Temperature dependencies of the elastic fraction for three samples shown in Fig. 7 were studied: I, 37% water solution of HSA; II, 37% water solution of HSA with the addition of 7.5% (by weight) of glutaric dialdehyde (GD); and III, water–glycerol solution of HSA with relative content of protein equal to 34%, water 22%, and glycerol 44%.

The addition of GD in water solution transforms sample I from the liquid to the solid state (sample II) and, consequently, leads to a very large increase in macroviscosity. Sample III (water–glycerol–protein solution) preserves fluidity and, hence, has lower visible viscosity than sample II. Nevertheless, the visible influence of water–glycerol solvent on protein dynamics is much stronger than that of GD (see Fig. 7). This fact suggests that it is the microviscosity in the vicinity of the protein surface that plays the crucial role in influencing protein dynamics.

These effects were evaluated within the framework of the incoherent approximation (see Eqs. (11) and (14)). As will be seen, more advanced treatment is impossible now because of the lack of knowledge about structural properties of water–glycerol mixtures. Anyway, as was mentioned in Sec. II, the incoherent treatment can give us a satisfactory qualitative picture.

Additional experimental temperature dependencies of the elastic fraction for water—$f_w(T)$, glycerol—$f_G(T)$, and water–glycerol solution were studied to deduce dynamical information on the protein. These curves are represented in Fig. 8.

For samples I and II, Eqs. (14) and (15) can be rewritten numerically [49] as

$$\phi f_p(T) = 3.09 f_\Sigma(T) - 2.09 f_w(T) \tag{24}$$

There exist some variants to take into account contributions of the solvent for sample III (protein in water–glycerol solution): (a) To take into account the contribu-

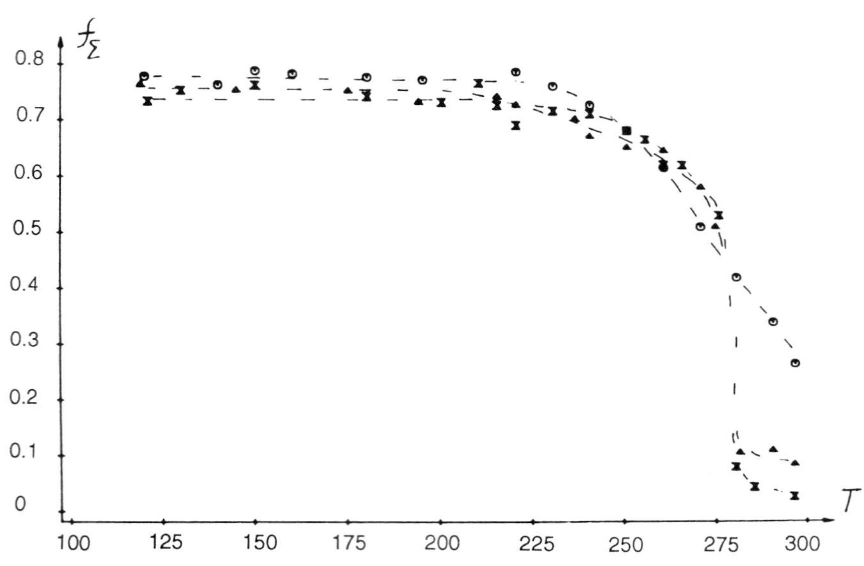

FIGURE 7 Experimental temperature dependencies of elastic fraction for 37% water solution of HSA (⧖), the same solution with the addition of 7.5% of GD (▲), and a water–glycerol solution of HSA (⊙).

tion of glycerol only, and, hence, to deduce the properties of the water–protein system,

$$\phi f_{pw}(T) = 1.77 f_{\Sigma}(T) - 0.76 f_G(T) \tag{25}$$

(b) To take into account contributions of water and glycerol separately, assuming that no interactions exist between them,

$$\phi f_p(T) = 3.17 f_{\Sigma}(T) - 0.79 f_w(T) - 1.39 f_G(T) \tag{26}$$

And (c) to take into account the contribution of the water–glycerol system,

$$\phi f_p(T) = 3.17 f_{\Sigma}(T) - 2.18 f_{wG}(T) \tag{27}$$

Calculations with the use of Eqs. (24)–(27) were done with the help of the following empirical curves, which satisfactorily describe experimental points (from the family of Fermi–Dirac curves):

$$f_{\Sigma}(T) = \left(\frac{K_1 - K_2}{1 + \exp[\alpha(T - T_0)]} + K_2\right)\left(\frac{1}{A/B(T) + 1} + \frac{\exp[-A - B(T)]}{B(T)/A + 1}\right) \tag{28}$$

where $B(T) = B_0 \exp(C/T)$ and K_1, K_2, A: and C are constants.

Figure 9 shows the dependencies of $\phi f_p(T)$ for protein in water–glycerol solution calculated on the basis of Eqs. (25)–(27). When Eqs. (26) and (27) are used,

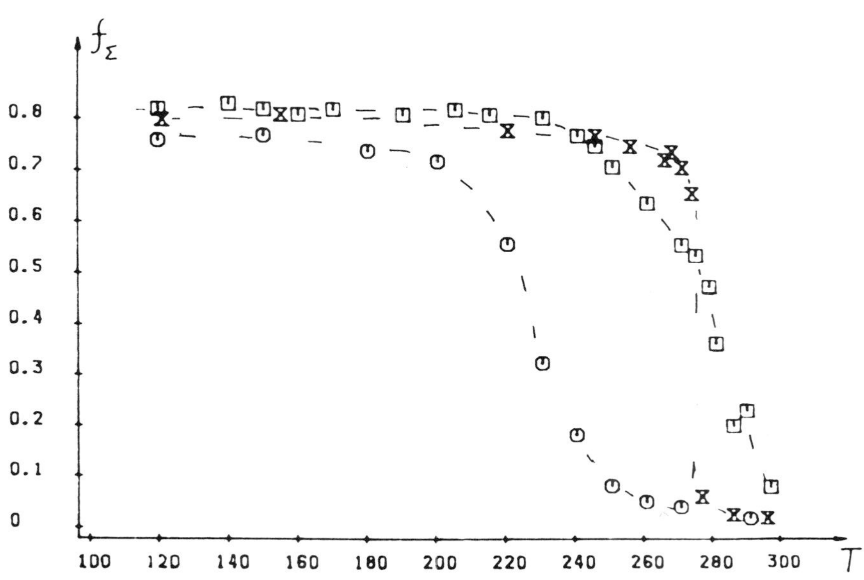

FIGURE 8 Experimental temperature dependencies of elastic fraction for water (X), glycerol (□), and water–glycerol solution (⊙).

nonmonotonic changes in $\phi f_p(T)$ are observed. Moreover, when the contribution of the water–glycerol solution (Eq. (27)) is taken into account separately, as distinct from separately taking into consideration contributions from water and glycerol (Eq. (26)), physically meaningless results are obtained in which $\phi f_p > 1$ over a wide temperature range.

The peculiarities observed above give direct evidence that dynamical properties of the real solvent, interacting with the protein surface, are considerably different from the properties of free solution. Taking into account the contribution $f_\Sigma(T)$ from glycerol only, as Fig. 9 shows, leads to the monotonic dependence of $\phi f_{wp}(T)$ on temperature for the water–protein system over the whole temperature range. This suggests that glycerol does not change remarkably its dynamical properties during interaction with the protein–water system. It is easy to evaluate from

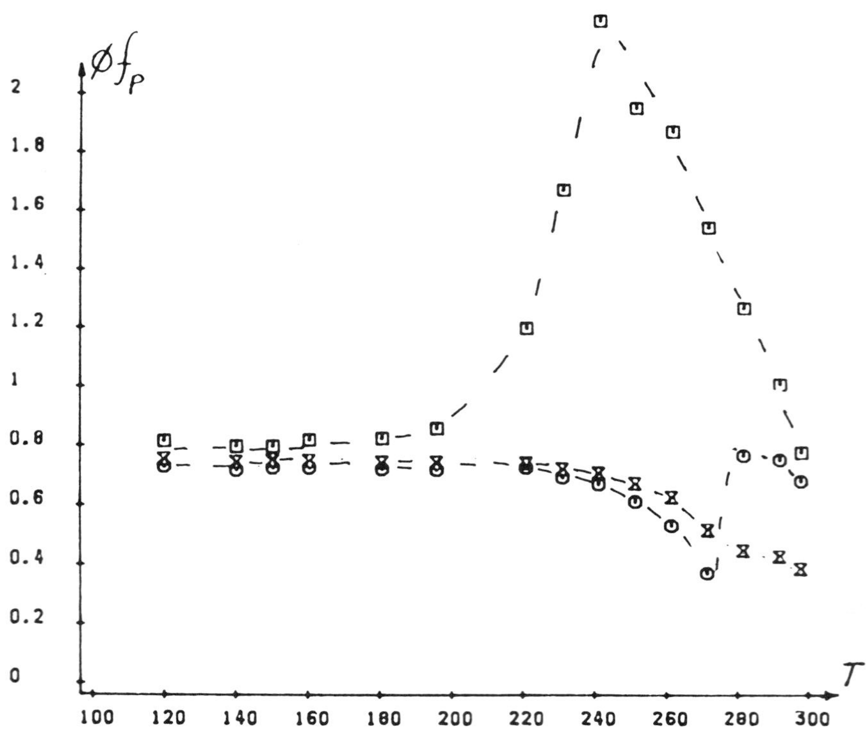

FIGURE 9 Different variants for taking into account the contribution of a solvent for HSA in water–glycerol solution: taking into account contribution from glycerol only (Eq. (25)) (⧖); taking into account contributions from water and glycerol separately (Eq. (26)) (⊡); taking into account contribution from water–glycerol system (Eq. (27)) (⊙).

Eqs. (24)–(27) that the overall msd $\langle x^2 \rangle$ for protein in water–glycerol solution at room temperature is approximately 20 times smaller than that in water. This means that the change in msd for the protein is larger than the viscosity change when glycerol substitutes part of the water in solution. Since hydration is one of the main factors that determine the intraglobular dynamics, it is possible that the drastic decrease in intraglobular dynamics can be explained by the decrease of hydration degree from $h = 1.7$ in water solution to $h = 0.65$ g H_2O/g protein in water–glycerol solution due to replacement of part of the water by glycerol. Independent study of HSA hydration, however, shows (see Sec. III) that this cannot explain entirely the decrease in ϕf_p.

It is necessary, in addition, to take into account another factor—preferential hydration that stabilizes the protein [54]. According to this model [54], glycerol is preferentially excluded from the immediately vicinity of the protein, and this thermodynamically unfavorable situation should stabilize the native structure of the protein. Recently, the exclusion of glycerol from the vicinity of proteins was independently experimentally observed by neutron scattering [55]. The additivity of the dynamical properties of glycerol and the protein–water system (see Eq. (25) and Fig. 9) is also consistent with preferential hydration.

Thus, it seems that the observed strong influence of glycerol on the intramolecular mobility of HSA is mostly connected with the effective dehydration of the protein (by substitution of part of the water by glycerol) and with a preferential hydration effect rather than with the influence of viscosity by itself.

V. TEMPERATURE DEPENDENCIES OF THE ELASTIC RSMR FRACTION AND RSMR SPECTRA

Shown in Fig. 10 are temperature dependencies of the elastic scattering f_Σ for HSA samples with different hydration degrees. A similar behavior for f_Σ (i.e., its decrease) with temperature has been observed for met-Mb samples with different hydrations [55] and for hydrated trypsin. For hydrated HSA samples [56] (Fig. 11) and for crystalline met-Mb, RSMR spectra have been studied in the temperature range 100–310 K. It has been found that at $T > 250$ K the line becomes non-Lorentzian and a second quasi-elastic (wide) component appears.

In all studied proteins with relatively high water content ($h > 0.25$)—HSA, met-Mb, and trypsin—there is, as seen from Figs. 10 and 11, a typical abrupt fall in f_Σ and $\eta_\Sigma(0)$ at $T > 230$ K, that is, a sharp increase in the effective "overall" $\langle x^2 \rangle$, becoming at room temperature and at high hydrations ($h \simeq 0.7$) as large as $\simeq 1$ Å^2. For dehydrated samples ($h \simeq 0.05$), $f_\Sigma(0)$ shows a linear decrease with a constant slope in the entire temperature interval from 100 to 300 K (with $\langle x^2 \rangle$ attaining $\simeq 0.05$ Å^2 at room temperature). Interestingly, this falloff in the elastic fraction is not accompanied by the strong linewidth broadening so typical of free unbounded diffusion. The small linewidth broadening $\Delta\Gamma$ observed in HSA

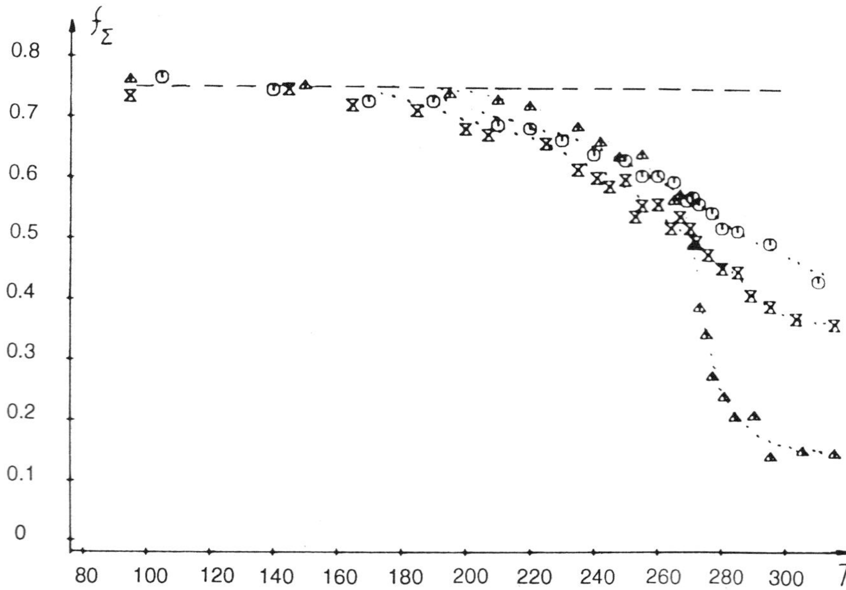

FIGURE 10 Experimental temperature dependencies of the elastic fraction for HSA with different degrees of hydration: dry protein (...), $h = 0.25$ (⊙), $h = 0.41$ (⧗), $h = 0.65$ (△).

(Fig. 11) and crystalline met-Mb samples close to room temperature is unable to account, within the scope of the above considerations, for so abrupt a drop in $f_\Sigma(T)$ and $\eta_\Sigma(0)\ (T)$. As will be clear from Sec. VIII, this behavior of the elastic scattering fraction $f_\Sigma(T)$ (or the area under the spectrum $S_\Sigma(T)$ and of the peak value of the spectral line, $\eta_\Sigma(0)\ (T)$), calls for the development of special models to account for a temperature-dependent release of conformational mobility in biopolymers.

VI. ANGULAR DEPENDENCIES OF INELASTIC RSMR INTENSITIES

Now let us turn to the discussion of spatial correlations in the dynamics of the water–protein system. In Sec. II we already mentioned that angular-dependent RSMR experiments with good angular resolution allow a determination of the sizes of moving segments. Experimental results discussed below were measured on an RSMR instrument with a beam divergence of less than 2° [31,32].

Shown in Fig. 12 is the angular dependence of the inelastic scattering intensity for crystalline met-Mb [32]. Figure 13 represents the angular dependence of the inelastic scattering intensity for hydrated ($h = 0.4$) HSA [31].

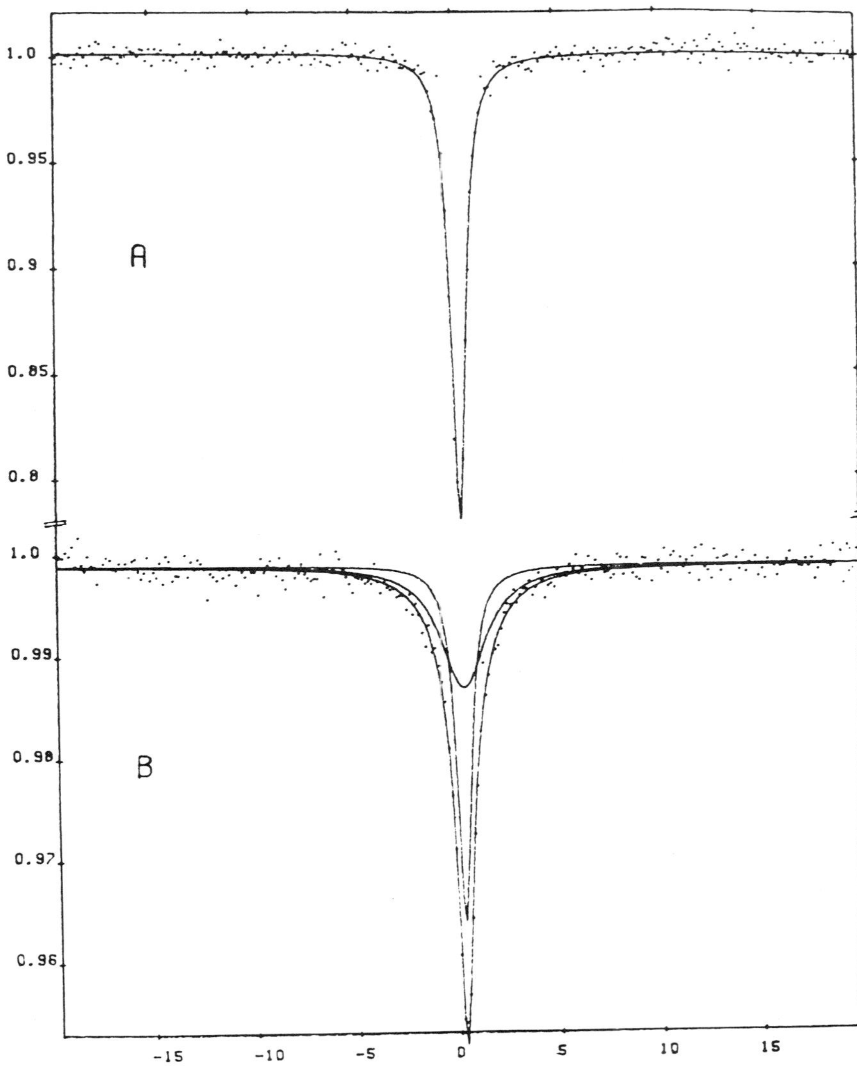

FIGURE 11 Energy spectra of HSA (with hydration $h = 0.77$) at different temperatures: (A) $T = 100$ K; (B) $T = 280$ K.

In Sec. II, a model was developed that considered the coexistence of independent and segmental motions within the protein globule, but the corresponding expressions (Eqs. (5)–(7)) do not contain the contribution from interprotein water. If we subdivide this water into two types of bound water with independent and

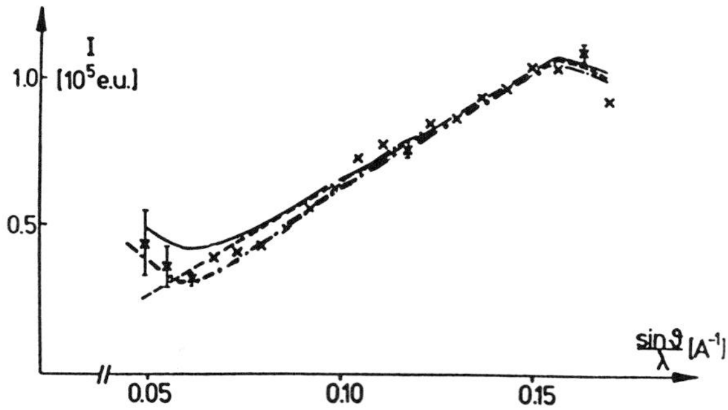

FIGURE 12 Angular dependence of inelastic scattering for met-Mb crystals. The crosses are experimental data [32]. The dashed, solid and dashed-dotted lines give the calculated inelastic intensity assuming respectively individual single atoms, α-helices, and combination of individual motions ($B = 0.3$) and spheres with $R = 6$ Å, as a moving segment ($A = 0.7$).

segmental kinds of motion and free (bulk) water with corresponding msd—$\langle x^2\rangle_{pw}{}^{i}$, $\langle x^2\rangle_{pw}{}^{s}$ and $\langle x^2\rangle_{b}$ ($\langle x^2\rangle_{b} = \infty$), then we come to the following expression for the inelastic scattering intensity in the protein–water system:

$$\begin{aligned} J_{inel}(\mathbf{Q}) = {} & [AS_{SEG}(\mathbf{Q}) + BF_{i}^{2}(\mathbf{Q})]\,[1 - \exp(-\mathbf{Q}^2\langle x^2\rangle_p)] \\ & + C_2^{s}AS_{SEG}(\mathbf{Q})\,[1 - \exp(-\mathbf{Q}^2\langle x^2\rangle_{pw}{}^{s})] \\ & + C_2^{i}BF_{iw}^{2}(\mathbf{Q})[1 - \exp(-\mathbf{Q}^2\langle x^2\rangle_{pw}{}^{i})] + C_1S_{bw} + C_3S_{ns} \end{aligned} \tag{29}$$

where S_{bw} is the well-known angular dependence for the bulk water; S_{ns} is the scattering law for the buffer molecules in crystalline Mb, which is considered the same as for bulk water; and C_1 , C_2^{i}, C_2^{s}, and C_3 are the numbers of free water molecules, bound water molecules moving individually, bound water molecules moving together with the protein segments, and buffer molecules, respectively. Information on spatial arrangements of the atoms in the protein, and correspondingly on the behavior of $S_{SEG}(\mathbf{Q})$ and S_{bw}, was taken from the x-ray analysis. Here we consider the approximation that the bound water has the same scattering law as a protein, and so Eq. (16) holds. With this additional data, the angular dependencies of inelastic scattering were analyzed. Contrary to the conclusion made in [32], application of the model described by Eq. (29) gives the best agreement between experimental data on crystalline met-Mb and calculations, when the combination of segmental and individual motions are represented in the following way: Spheres with radius $R \simeq 6$ Å should be taken as moving segments with the weight of these motions $A = 0.7$ and the weight of individual motions $B = 0.3$. In addition, $\langle x^2\rangle_p = 0.158$ Å^2; $\langle x^2\rangle_{pw}{}^{s} = 0.52$ Å^2, $\langle x^2\rangle_{pw}{}^{i} = 0.59$ Å^2; $C_1 = 225$; $C_3 = 39$; $C_2^{s} = 229$ and $C_2^{i} = 379$. The result of these calculations together with calculations when only single atom motions and motions of α-helices are considered [32] is represented in Fig. 12.

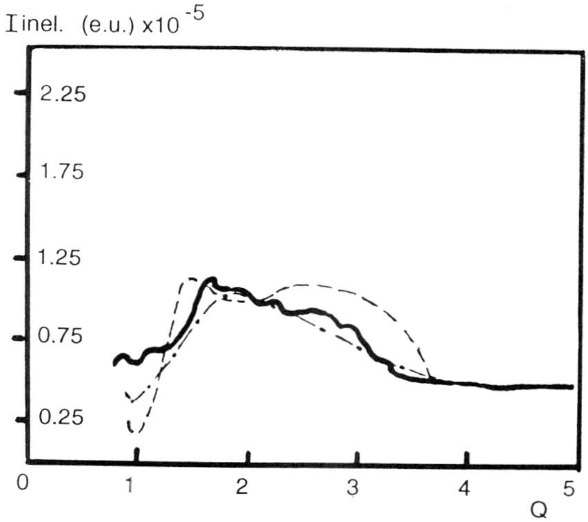

FIGURE 13 Solid line: angular dependence of inelastic scattering intensities for HSA with $h = 0.4$. Dashed line: calculated curve taking into account the motion of the protein as a whole. Dashed-dotted line: calculated curve corresponding to the motions of subdomains with $\langle x^2 \rangle = 0.19$ Å^2.

It is easy to show that another combination of segmental and individual motions, namely, α-helices as a moving segment (weight of these motions $A = 0.5$) and individual motions with weight $B = 0.5$ also gives good agreement with experiment.

Experimental results for hydrated HSA are available only for relatively large scattering angles; therefore, the calculations of inelastic intensity were done here on the basis of a simpler segmental model [32], where $B = 0$. The best coincidence between experiment and calculations was achieved assuming moving subdomains as segments (see Fig. 13) with $\langle x^2 \rangle = 0.19$ Å^2.

Unfortunately, in the model used in this section, a very crude approximation is taken from the description of the properties of interprotein (bound) water. This is because our knowledge concerning the structure and mobility of interprotein water is very poor. Further progress in the understanding of the nature of intraglobular motions is therefore closely related to progress in understanding the properties of protein-bound water.

VII. PROPERTIES OF PROTEIN-BOUND WATER

In this section we shall try to deduce some information about the properties of bound water, namely bound-water melting. Let us try to explain the drop of $f_\Sigma(T)$ and $g_\Sigma(0)$ (T) at $T > 230$ K by the melting of hydration water only (assuming that the dynamic properties of the protein remain unchanged with a temperature increase). Let us consider as an example the data for HSA (Fig. 14). The calculated temperature dependence $f_\Sigma(T)$ based on independent measurements of $f_w(T)$, for free water [48], and $f_p(T)$ for dry protein, and Eq. (22), is shown in Fig. 14 by the solid line. The use of Eq. (23) produces only small differences with the previous case.

Comparison with the experimental dependence $f_\Sigma(T)$ shows that for hydrated HSA there is a marked discrepancy between the experimental and calculated temperature dependencies above about 200 K. It manifests the nonadditivity of the RSMR-observed dynamic properties of protein and water not only for different h but also for different temperatures, confirming the above supposition of a strong interaction between protein and water. It turns out that the substitution of free water, $f_w(T)$, data in Eq. (22) leads to a strongly nonmonotonic $f_p(T)$ dependence within the range 260 K $< T <$ 290 K. Since such a nonmonotonic change in protein dynamic characteristics against temperature can hardly have any reasonable physical foundation, an attempt was made to obtain information on the hydration water properties based only on the assumption that $f_p(T)$ should be a monotonic temperature function.

Protein-bound water properties differ considerably from free water [35]; for example, the melting of bound water extends over a wide temperature range.

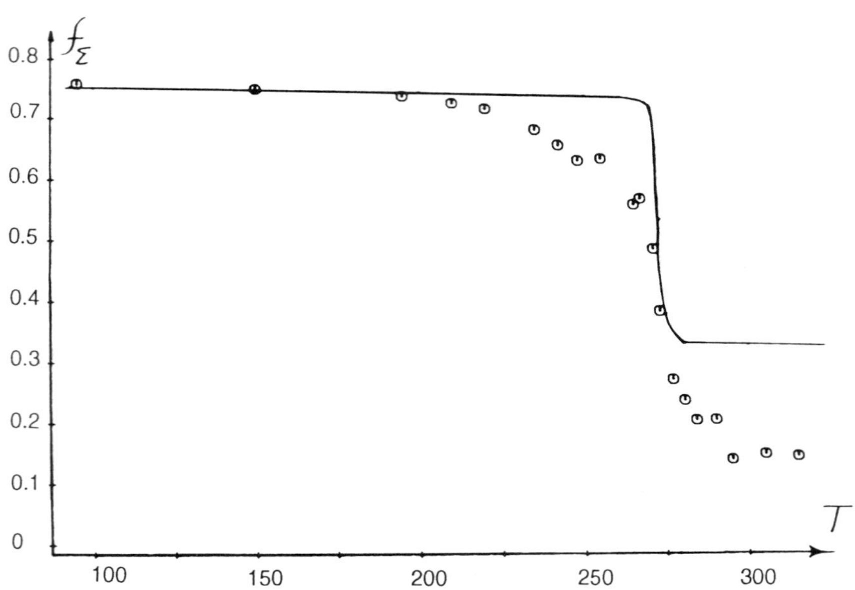

FIGURE 14 Illustration of nonadditivity of dynamical properties of protein and water in the case of HSA with $h = 0.65$; ⊙ is from experiment, and the solid line is calculated for an additive mixture of dry protein and free water.

Hence we assume that the character of the temperature dependence of $f_w(T)$ for protein-bound water is described not by a "step" at the melting point of free water (273 K) but by some function that smoothly depends on temperature, such as

$$f_w(T) = \frac{f_0}{\left[1 - \exp\left(\frac{T - T_0}{\Delta T}\right)\right]} \tag{30}$$

where f_0 is the approximate experimental value of f_w, at $T \to 0$, T_0 is the effective melting temperature ($f_w(T) = \frac{1}{2} f_0$ at $T = T_0$), and the value of ΔT characterizes the temperature range over which melting of bound water occurs; that is, its dynamic characteristics within the ΔT interval differ considerably from those of both pure ice and free water. Estimates of T_0 and ΔT parameters have been made under the requirement of the monotony of $f_p(T)$. Values of these parameters at different hydration degrees for samples of HSA and met-Mb, and for crystalline met-Mb are listed in Table 1. Since numerical values of C_p and C_w in Eq. (23) are only negligibly different from those in Eq. (22), values of T_0 and ΔT determined by Eq. (23) are also only slightly different from those in Table 1. Thus, for HSA at $h = 0.65$, T_0 does not change its value at all, and ΔT slightly decreases (within the limits of the error) from $\Delta T = 3 \pm 1$ to $\Delta T = 2.5 \pm 1$.

An increase in hydration degree is accompanied by an increase in the effective temperature of bound-water melting, which approaches 273 K, while the temperature interval of such melting gets narrower ($\Delta T \to 0$) (see Fig. 15). These results correlate well with independent calorimetric measurements of the properties of water bound to proteins [58].

More detailed information on protein dynamics, and protein-bound water properties can be deduced from the temperature dependence of line shapes of RSMR spectra (see Fig. 11). Two possible interpretations of the bound-water behavior described by Eq. (30) are examined.

The first of these corresponds to the treatment of bound water as a viscous liquid. The amplitude of a spectral line $g_w(0)$ (which is proportional to the elastic fraction f_w) for liquids can be written, according to [38], as

$$g_w(0)\,(T) \sim \frac{\Gamma}{\Gamma + \Delta\Gamma} \tag{31}$$

Here Γ is the natural linewidth, $\Delta\Gamma = 2hQ^2D$, Q is the transferred momentum, $Q = 1.5$ Å^{-1}, and D is the diffusion coefficient. Therefore, the temperature increase leads to a sharp decrease in amplitude because of exponential broadening of the line (fast rise of $\Delta\Gamma$). A combination of Eqs. (31) and (30) gives the temperature dependence of the diffusion coefficient near T_0: $D = D_0 \exp[(T - T_0)/\Delta T]$. The broadening of the line should occur in the vicinity of T_0 in this case, which, however, still was not observed within the accuracy of the experiment [57]. Besides, data on the temperature dependence of the dielectric relaxation on rate τ^{-1} for

TABLE 1 Values of Parameters of Eq. (30), Describing Properties of Protein-Bound Water

Sample	T_0 (K)	ΔT	g
HSA			
$h = 0.25$	260 ± 4	24 ± 7	4
$h = 0.41$	165 ± 3	16 ± 3	6
$h = 0.65$	271.5 ± 0.7	3 ± 1	34
Met-Mb			
$h = 0.5$	265 ± 3	15 ± 3	7
cryst.	259 ± 2	3 ± 1	31
$h = 2.5$	271 ± 0.7	2.5 ± 1	40

water in crystalline met-Mb [59] are in direct contradiction with the above suggestion. The second interpretation implies that Eq. (30) and $G_w(0)(T)$ represent the probability for bound water to be in a peculiar heterogeneous solid disordered state; that is, water melts in a stepwise fashion, in separate discrete portions (clusters), while correlation times for water motion drastically decrease down to 10^{-9}–10^{-11} s. In this case, broadening of the line should be unobservable, since melted water contributes only to inelastic scattering and unmelted bound water gives an unbroadened line. Investigation of the temperature behavior of met-Mb

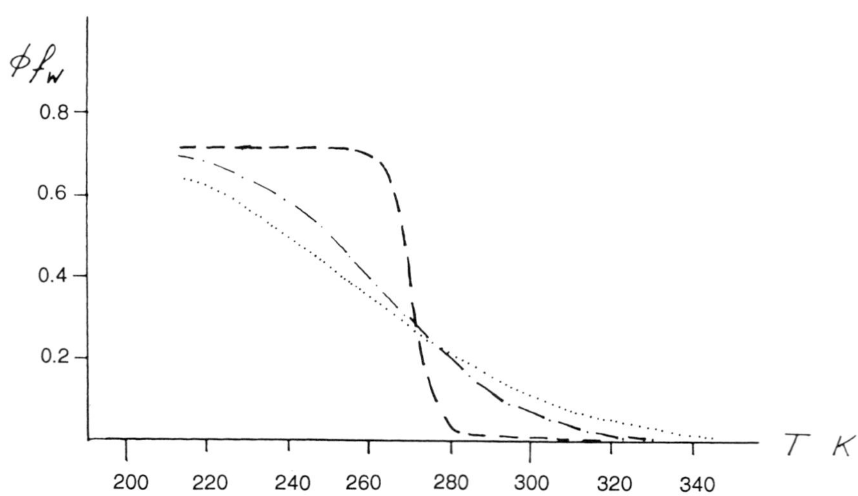

FIGURE 15 Calculation of ϕf_w for bound water (in terms of Eq. (30) and Table 1) for different hydration degrees: $h = 0.25$ (...); $h = 0.41$ (— · —); $h = 0.65$ (— — —).

spectral line shapes [47] gives additional evidence in favor of the second interpretation. The decrease in the effective melting temperature T_0 and the wide range of temperatures ΔT within which the melting occurs are also consistent with a mechanism of a cluster melting (see Table 1, specific heat data [58], and theoretical work [60], devoted to the melting of water clusters).

Let us consider a simple model for melting of bound water. We suggest that the clusters of bound water can be only in two states (phases). In the first phase, water clusters are in a solid disordered state [61], correlation times for diffusive mobility are long, and therefore these clusters have the same RSMR spectrum as a solid: $g_{w_1}(\omega)$. In the second phase, clusters of water are in a melted, liquid state, and correlation times for diffusive mobility are so short that the spectrum of this state, $g_{w_2}(\omega)$, is unobservable in RSMR. Let $\pi_{1,2}$ be a thermodynamic potential of a corresponding state per molecule, and let g the number of molecules in a cluster. The probability for a water cluster being in the first state will be described by

$$p_1 = \frac{1}{1 + \exp[-g(\pi_2 - \pi_1)/kT]} \tag{32}$$

and the probability of being in the second state by the corresponding value, $p_2 = 1 - p_1$.

In this case, the spectral function for water has the following form:

$$g_w(\omega) = p_1 g_{w_1}(\omega) + (1 - p_1)g_{w_2}(\omega) = P_1 g_{w_1}(\omega) \tag{33}$$

At temperatures close to the melting point T_0 [62],

$$\pi_2 - \pi_1 = \frac{q(T_0 - T)}{T} \tag{34}$$

where $q = T_0(S_2{}^0 - S_1{}^0)$ is the heat of melting, and $S_i{}^0$ is the entropy of the corresponding state, $i = 1,2$. It is easy to find physical meaning for the empirical function (30) and its analogue, applied above, $g_2(0)\,(T) = g_0/1 + \exp[(T - T_0)/\Delta T]$ by use of Eq. (34). Equation (30) and Eqs. (33) and (34) are equivalent if $\Delta T = kT_0^2/gq$ [63]. This means that the larger the number of molecules $g = kT_0^2/q\,\Delta T$ in a cooperative cluster of bound water, the smaller the range of melting temperatures ΔT. By substituting $q = 1.44$ kcal/mol, one can obtain values for the number of water molecules in an average cooperatively cluster, which are given in Table 1.

Thus, RSMR data give the possibility of investigating the properties of bound water in hydrated protein samples. The pattern of bound-water properties, deduced from RSMR data, is shown in Fig. 16. At low hydration levels, bound water along with protein exhibits solid state properties in the entire temperature region (up to 300 K). With increased hydration, $h > 0.1$, there appears "intermediate" water, characterized by a smooth variation of its dynamical properties with temperature. At even greater degrees of hydration, $h > 0.6$ to 0.7, interprotein water becomes very close to free water in terms of its properties. Melting of interprotein water clusters proceeds in a stepwise fashion. The dimensions of the cooperative cluster of water increase with the degree of hydration (see Table 1). The variation in the dynamical properties of bound water with hydration h shown in Fig. 16, was used in Sec. III to calculate the real curve for the variation of protein dynamical properties.

VIII. DYNAMICAL PROPERTIES OF HYDRATED PROTEINS

The parameter that depends least on the way decomposition of the observed RSMR spectrum into two components is performed is the line area. Magnitudes of these areas for both the narrow (⊙) and wide (⧗) lines of HSA spectra at T = 200–310 K are shown in Fig. 17 (the contribution from bound water is already subtracted here by a procedure similar to that described above (see Eq. (30) and Table 1). Similar dependencies for crystalline met-Mb are represented in [47]. Data displayed in Fig. 17 demonstrated a strong decrease in total line area, accompanied by small and very specific line broadening (Fig. 11). Figure 18 summarizes data on the influence of temperature and hydration on dynamical properties of proteins. Such a behavior for spectral line parameters is typical neither of Einstein or Debye

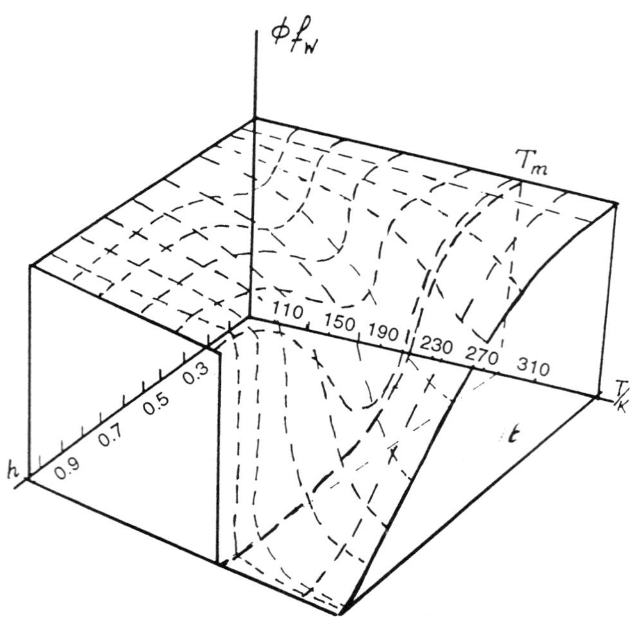

FIGURE 16 Three-dimensional plot of the dependence of the elastic scattering fraction for interprotein water, ϕf_w, versus T and h.

solids, or for liquids (where Eq. (31) is valid). Similar results were observed in Mossbauer absorption spectroscopy (MAS), which gives information on the dynamical behavior of ^{57}Fe only, usually situated in an active center of a protein. Some theoretical models have been developed to explain MAS data [64–69]. RSMR, unlike MAS, gives information on the average dynamical properties of the protein globule. An attempt will be made here to account for the real heterogeneous dynamical properties of proteins, based on the extended version [70] of the phase transition model [66] in globular proteins. Proteins are globules in their native state. Due to partial equilibrium [71], they are situated in local energy minima.

These local energy minima were first observed by Frauenfelder et al. [4], who studied the kinetics of ligand (CO,O_2) rebinding in myoglobin and x-ray dynamical analysis of this protein [17]. In the literature, they are referred to as conformation substates (CS) or quasi-degenerate conformational states. Therefore, local minima are situated or arise in a conformational potential, within which some atomic groups of proteins diffuse. As a consequence, globular proteins possess a long-range order (at distances >1 Å), but are short-range disordered—that is, they have local minima (or CS). Since protein-bound water is disordered (in crystalline samples, strongly bound water possesses only short-range disorder), it is natural that the outer (surface) shell of a protein globule has the largest degree of a short-range disorder.

We believe that the phase transition (more exactly, a series of phase transitions) takes place in a water–protein system within the temperature range $\Delta T_{tr} =$ 200–273 K. Because of a strong water–protein interaction, the phase transition in water from the solid, disordered state to a liquid state is accompanied by a phase transition in the protein globule. Since the phase transition in bound water proceeds by cluster melting and the sizes of water clusters near the protein surface are relatively small (they contain from 4 to 40 water molecules; see Table 1), it is natural to suggest that the phase transition in a protein macromolecule will proceed in the same (cluster) way.

The absence of full cooperation within the limits of the globule or stepwise character of phase transitions in a protein globule (different regions have different characteristic transition temperatures T_{tr}) is a consequence of heterogeneity in the physical properties of the globule.

Protein macromolecules and, all the more, their regions are small systems (that is, the number of links is finite inside them). Hence, phase transitions in macromolecules, unlike macroscopic bodies (where $N = \infty$), occur over some temperature region ΔT_{tr} [71].

Let us consider the protein globule as in [70] or a small part of the protein globule (the size of a cooperativity cluster, which performs such a transition, will be mentioned in more detail below).

We shall consider two globular states: a low-temperature nonfunctional and a high-temperature functional state. If a phase transition in a globule is realized by

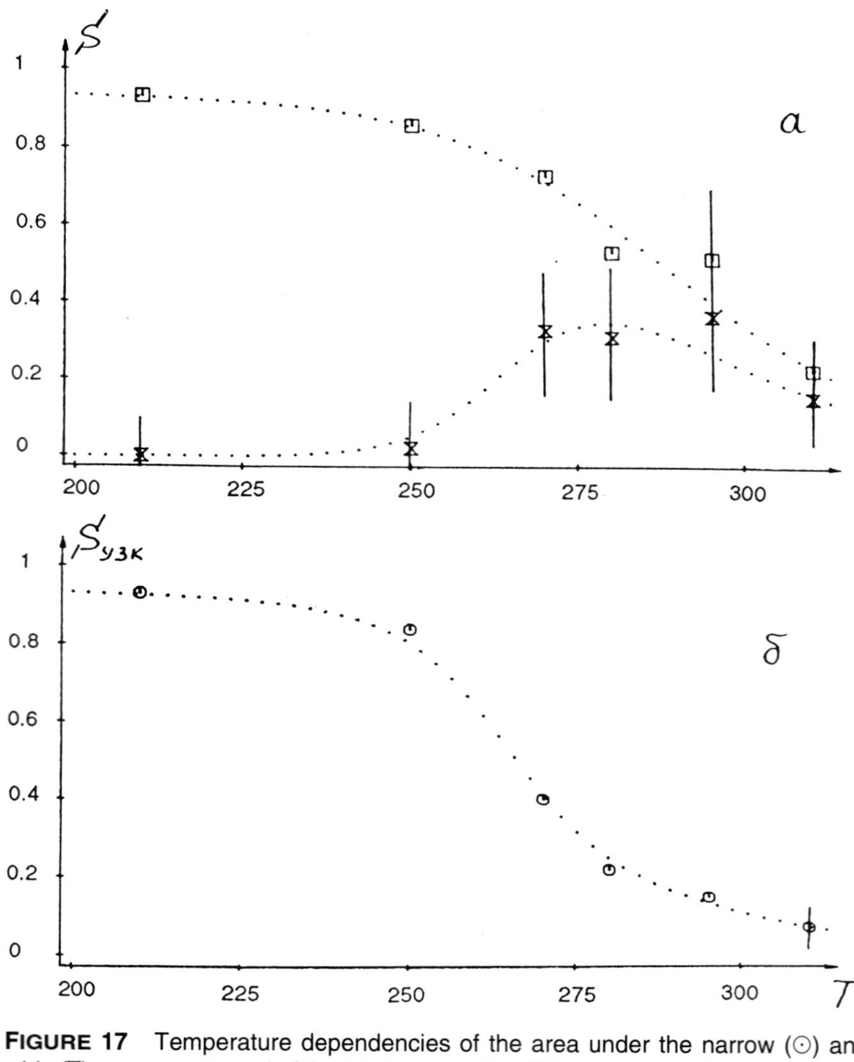

FIGURE 17 Temperature dependencies of the area under the narrow (⊙) and wide (⌛) components and of the total area under HSA spectrum (⊡) ($h = 0.77$) obtained from spectra measured at large source velocities.

stepwise transitions of (regions) clusters, it means that the globule will be in a functional state only if all globular regions are in their functional state. For globular regions, we shall also consider two states: first (at low temperatures), the "solid disordered" (on a short range) state, and second (at high temperatures), the "liquidlike" state.

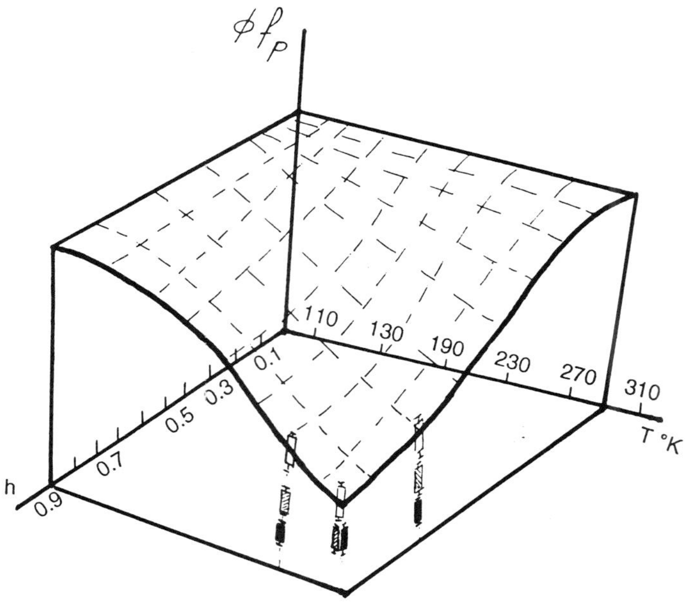

FIGURE 18 Three-dimensional plot of the elastic RSMR fraction for proteins as a function of temperature *T* and hydration *h*.

The fraction of regions in solid (S) and liquidlike (L) states will be determined by the value of their free energy $F_i = H_i - TS_i$ $(i = \mathrm{S}, \mathrm{L})$ and will be described by

$$W_s = \frac{1}{1 + \exp(\Delta F/kT)}; \qquad W_L = 1 - W_s = \frac{1}{1 + \exp(-\Delta F/kT)} \tag{35}$$

where $\Delta F = F_s - F_L = \Delta H - T\,\Delta S$.

These expressions (Eq. (35)) are analogous to those which describe phase transitions in small systems. The variation of a region from the S to L state may therefore be called a phase transition. The temperature $T_{tr} = \Delta H/\Delta S$ at which $W_s = W_L = 1/2$ is the characteristic transition temperature. The transition, as distinguished from macroscopic objects, is not discontinuous. There exists a temperature region ΔT_{tr} in which the quantities W_s and W_L are comparable and the quantity $|\Delta F| \leq kT$:

$$\Delta T_{tr} = \frac{kT_{tr}}{\Delta S} = \frac{kT_{tr}^2}{\Delta H}$$

In a protein globule, one may expect the following types of motion:

1. Solid state motions: vibrations of atoms and groups in rigid parts of the macromolecule as in any solid—with frequencies $\sim$100 cm^{-1}, root mean square displacement (rmsd) $\sqrt{\langle x^2 \rangle} \sim 0.1$–$0.2$ Å, and correlation times $\tau_c^{\mathrm{I}} \sim 10^{-12}$–$10^{-13}$ s. These motions completely explain the temperature behavior of $f(T)$ and $S(T)$ at $T < 200$ K. The msd $\langle x^2 \rangle$ at high temperature in the Debye approximation ($T \gg \theta_D$) has a linear temperature dependence:

$$\langle x^2 \rangle_{\mathrm{I}} = \frac{\beta_{\mathrm{I}} T}{Q^2}$$

2. Large-scale (conformational) individual motions of small groups of atoms of the protein globule. Their characteristic times τ_c^{II} are larger than those for solid state motions but are substantially smaller than $\tau_M \simeq 10^{-7}$ s ($\tau_c^{\mathrm{II}} \ll \tau_M$). These groups move by way of limited diffusion. (They are "liquidlike", since $\langle [\Delta x_{\mathrm{II}}(t)]^2 \rangle \sim x^2(\tau_c^{\mathrm{II}})/\tau_c^{\mathrm{II}}$ at $t \ll \tau_c^{\mathrm{II}}$ [64]. Mean square displacements of these motions ($\langle x^2 \rangle \gg 0.04$ Å^2) substantially exceed those for ordinary vibrations of atoms or atomic groups in solids ($\langle x^2 \rangle < 0.04$ Å^2) and result from the increase of short-range disorder of the above-mentioned "surface" groups of the protein globule in the course of phase transitions (let us note that the fraction of such groups is relatively large in proteins). These motions are connected with individual transitions of small atomic groups (i.e., CH_2) from side chains between local minima (conformational substates) of the conformational potential. Transitions between local minima may proceed either as thermally activated transitions or by way of fluctuatively prepared tunneling [28]. For these motions, one should expect a noticeable anharmonicity due to the further increase of "short-range disorder" with increasing temperature. In what follows, the temperature dependence of the mean square displacements in the profile of the conformational potential is approximated by the power-law dependence

$$\langle x^2 \rangle_{\mathrm{II}} = \frac{\beta_{\mathrm{II}} T^{2\nu}}{Q^2} \tag{36}$$

This corresponds to motions in a potential $V \sim x^{1/\nu}$, which is gently sloping at large x for $\nu > 1$.

3. Complex cooperative motions of large atomic groups or parts of a macromolecule. As has already been mentioned, these motions have the character of limited diffusion; that is, they are "liquidlike" at small times $t \ll \tau_c^{\mathrm{III}}$ [64]. The characteristic times of these motions are $\tau_c^{\mathrm{III}} \sim 10^{-7}$ s.

4. Motions connected with the phase transition itself. Their characteristic times are equal to $\tau_c^{\mathrm{IV}} = \tau_{tr} \gg 10^{-7}$ s. (More precisely, the transition proceeds rapidly, but the transition expectation time is large.) In the nonfunctional state, there prevail motions of the first kind. In the functional state, there prevail motions of the second and third kinds. The motions of the fourth kind proceed rarely and

do not contribute to observed quantities. In other words, during the measurement (within the time $\tau_M \simeq 10^{-7}$ s) the numbers of liquidlike and solid regions in the macromolecule do not change. It is necessary to note here that the above-mentioned four types of motions cannot be directly connected at present with FIMs in [6]. On the assumption that the transition time is much longer than τ_M, the observed energy spectrum can be written in the form

$$g(\omega) = g_s(\omega) W_s + g_L(\omega)(1 - W_s) \tag{37}$$

where $g_s(\omega)$ is a spectral function of the S state, similar to the usual spectral function for solids. In the liquidlike L state, the spectral function is determined by motions of the second and third types:

$$g_L(\omega) = \exp(-Q^2\langle x^2\rangle_{II}) \times \frac{1}{\pi}\int_0^\infty dt \exp\left(-\frac{\Gamma}{2}t - \frac{1}{2}Q^2\langle[\Delta x_{III}(t)]^2\rangle\right)\cos(\omega - \omega_0)t \tag{38}$$

where the fact that the second type of motions are much faster than motions of the third type [69] has been taken into account.

Motions of the third type will be described within the framework of a standard model for limited diffusion [69]:

$$g_s(\omega) = \exp(-Q^2\langle x^2\rangle_{III}) \frac{1}{\pi}\sum_{n=1}^{\infty}\frac{(Q^2\langle x^2\rangle)^n}{n!}\frac{(\Gamma/2 + n\alpha)}{(\Gamma/2 + n\alpha)^2 + (\omega - \omega_0)^2} \tag{39}$$

where $\alpha = 1/\tau_c$.

For type 2 motions, we believe that the transitions between local minima proceed relatively fast, $\tau_c^{II} \leq 5 \times 10^{-10}$ s; that is, the first widened component is already not observable; for this reason, this kind of motion only leads to a fall in the area with increasing temperature. Taking into account all the above types of motion, we can write the following expressions for temperature dependencies of the total area under the spectrum (S_{tot}), the area under the "narrow" component (S_n), and the area under the "wide" component (S_w):

$$\begin{aligned} S_{tot} &= \exp(-\beta_I T) W_s + \exp(-\beta_{II} T^{2\nu})(1 - W_s) \\ S_n &= \exp(-\beta_I T) W_s + \exp(-\beta_{II} T^{2\nu}) \exp(-Q^2\langle x^2\rangle_{III})(1 - W_s) \\ S_w &= \exp(-\beta_{II} T^{2\nu})[1 - \exp(-Q^2\langle x^2\rangle_{III})](1 - W_s) \end{aligned} \tag{40}$$

These expressions contain six parameters with which to describe the experimental data presented in Figs. 11 and 16 and in [47]: ΔH, ΔS, β_I, β_{II}, 2ν, $\langle x^2\rangle_{III}$. The numerical values of the parameters which best describe the experimental data are listed in Table 2.

The treatment of the data made use Eq. (16) to describe the properties of the interprotein water. The use of the assumption given in Eq. (17) does not qualitatively change the conclusions about dynamic properties of proteins. The numerical values of parameters in Table 2 change only slightly since values of C_p and C_w in Eqs. (22) and (23) are quite similar one to another. So, the use of Eq. (23) leads to a small increase (about 0.05 Å^2) in mean square displacement for high-frequency motions $\langle x^2 \rangle_{II}$.

In the described model, the changes in ΔS and ΔH are related to the cooperativity region (or cluster) of the macromolecule performing a phase transition. If the average size of a cooperativity region is comparable with the size of the macromolecule, then average values of ΔH and ΔS related to the mass unit are two orders of magnitude less than those values typical of phase transitions of ice melting. If the average size of a cooperativity region contains about 10 atomic groups, then values of ΔH and ΔS are of the same order as those in the phase transitions of the ice-melting type. We believe that the cooperativity regions have a broad size distribution, the smallest ones contain about 10 atomic groups, and largest ones are close to one-tenth or more of the protein globule. Therefore, one must expect considerable changes in dynamical properties of a globule in its functional state as compared with its nonfunctional one. A large number of individual large-scale conformational motions of the second type appear in places where the smallest cooperativity regions are situated. The existence of individual motions creates the conditions for the appearance of cooperative motions of relatively large parts of the macromolecule–solid domains (motions of the third type). These motions appear in places where the largest cooperativity regions are situated. Atomic motions inside solid domains are mainly vibrational.

Let us mention here that a similar picture of motions in proteins has emerged in Sec. VI from the study of the angular dependencies of inelastic RSMR intensities. Segmental motions are just the motions of solid domains, and individual motions are the motions of the second type. It is also necessary to stress here that coherent RSMR gives the unique possibility for separating contributions from in-

TABLE 2 Values of Parameters of Eq. (40) Yielding the Best Possible Description of the Experimental Data

Sample	β_I [K^{-1}]	β_{II}/Q^2 [K^{-1} Å^2]	2ν	$\langle x^2 \rangle_{III}{}^a$ [Å^2]	$\langle x^2 \rangle_{II}{}^a$ [Å^2]	ΔH [eV]	ΔS [eV$_K$]
HSA ($h = 0.77$)	$3.5 \pm 0.5 \times 10^{-4}$	1.3×10^{-21}	8.3	0.7 ± 0.05	0.47	0.59	2.4×10^{-3}
met-Mb crystalline	$3.5 \pm 0.5 \times 10^{-4}$	$1 \pm 0.2 \times 10^{-3}$	1.0	0.11 ± 0.02	0.13	0.86	3.2×10^{-3}

[a]Values at $T = 295$ K.

dividual and cooperative degrees of freedom, and, in addition, for determining the size (and, in principle, the shape) of moving segments. As follows from Sec. VI, the most probable moving segments (in Met-Mb) are spheres with radius $R = 6$ Å and α-helices. It is easier to associate segments or solid domains with the protein secondary structure (α-helices in the case of met-Mb).

The dynamical model of proteins described above takes into account the structural heterogeneity of a protein globule. Hierarchy in the amplitudes and correlation times of motions mentioned above is a consequence of the hierarchy in structural organization of the protein. This model has features similar to several other physical models of proteins [2,5,6].

IX. PRINCIPAL CONCLUSIONS AND OUTLOOK

The presented results of RSMR investigations of dynamical properties of hydrated (with hydration degree above 0.05) proteins and crystalline proteins at temperatures between 100 and 300 K reveal the great potential of this method for the analysis of properties of bound water and the internal dynamics of proteins. It was shown that hydration and temperature are the main factors determining the intraglobular mobility of proteins. The addition of glycerol also influences protein dynamics, but this influence is mostly connected to the effective dehydration of protein (substitution of the part of water by glycerol) and to a preferential hydration effect rather than to the influence of viscosity by itself. Consideration of available experimental data has suggested a cluster mechanism of bound-water melting (involving transition from a solid disordered state to a liquid one). The average cluster size has been shown to vary with the hydration degree and contain from 4 (at $h = 0.25$) to 40 (at $h = 2.5$) molecules of water. Theoretical analysis of the angular dependence of inelastic RSMR intensities and temperature dependencies of RSMR spectra has shown the existence of three main types of movements in protein:

1. Solid state motions, which are oscillations with amplitudes $A_{\mathrm{I}} \sim 0.1$–0.2 Å and correlation times $\tau_c^{\mathrm{I}} \sim 10^{-12}$–$10^{-13}$ s
2. Large-scale (conformational) individual motions of small atomic groups of the protein globule involving displacements A_{II} up to 0.5 Å and correlation times $\tau_c^{\mathrm{II}} \sim 10^{-9}$–$10^{-11}$ s
3. Complex cooperative motions of large atomic groups and parts of the macromolecule (i.e., segments or solids domains) proceeding by way of bounded diffusion and involving amplitudes A_{III} up to 1 Å and correlation times $\tau_c^{\mathrm{III}} \sim 10^{-7}$–$10^{-9}$ s

In proteins, there is actually the whole spectrum, $G(\tau)$, of motions of the second and third types. Just these movements are specific to proteins and are sensitive to changes in environmental parameters T, h and solvent composition. Here we should stress that all the above-mentioned results were obtained by taking into consideration coherent effects only partially. The critical analysis of experimental results [34,72] shows that all these qualitative conclusions remain unaltered. Precise quantitative results, however, cannot be extracted from these data at present. By use of an incoherent or a partially coherent approach, as in this chapter, one can obtain very important but relatively simple dynamic characteristics of proteins and water.

As was shown in Sec. VI, structural information obtained by x-ray or neutron scattering techniques are very helpful for the analysis of RSMR data. While the structures of many proteins are available, the knowledge of structural properties of interprotein water is still rather poor. Therefore, the structural properties of interprotein water should be specially studied by different physical techniques, including molecular dynamics and Monte Carlo simulations.

To deduce more delicate but much more complicated information about correlations between displacements of different atomic groups within the protein globule, about correlations in location of atoms of different protein globules, and in location of water and protein atoms, and about structural-dynamic behavior of interprotein water, it is necessary to choose those scattering angles where coherent effects are essential. Apparently, considering these coherent effects, it will be necessary to reject the separate simplified consideration of dynamic properties of strongly interacting protein and water molecules. One way of deducing this delicate information is computer modeling of the protein–water systems and comparison of model patterns with experiment.

ACKNOWLEDGMENTS

The authors would like to express their gratitude to A. M. Afanas'ev, G. Albanese, D. S. Chernavskii, A. Deriu, A. Grosberg, I. Kurinov, G. U. Nienhaus, F. Parak, I. Suzdalev, and K. Shaitan for illuminating discussions and critical remarks and F. Cavatorta, E. Gaubman, I. Sharkevich, and T. Zhuravleva for their assistance in RSMR experiments and valuable discussions.

REFERENCES

1. Blumenfeld, L. A. *Problems of Biological Physics.* Moscow, Nauka (1977) (in Russian).
2. Chernavskaya, N. M., and Chernavskii, D. S. *Tunnel Electron Transport in Photosynthesis.* Moscow, Moscow Univ. Press (1977) (in Russian).
3. Volkenstein, M. V. *Molecular Biophysics.* Moscow, Nauka (1975) (in Russian).
4. Austin, R. M., Beeson, K. W., Eisenstein, L., Frauenfelder, H., and Cunsalus, I. E. Dynamics of ligand binding to myoglobin. *Biochemistry 14*: 5355–5373 (1975).
5. Shaitan, K. V., and Rubin, A. B. Stochastic dynamics and electron-conformation interactions in proteins. *Biofizika 30*: 517–526 (1985) (in Russian).

6. Ansari, A., Berendsen, J., Bowne, S. F., Frauenfelder, H., Iben, I. E. T., Sauke, T. B., Shyamsunder, E., and Young, R. D. Protein states and protein quakes. *PNAS USA 82*: 5000 (1985).
7. Richards, M. F. Areas, volumes, packing and protein structure. *Ann. Rev. Biophys. Bioeng. 6*: 151–176 (1977).
8. Privalov, P. L. Stability of proteins. *Adv. Protein Chem. 33*: 167–241 (1979).
9. Sarvasyan, A. P., and Hemmes, P. Relaxation contributions to protein compressibility from ultrasonic data. *Biopolymers 18*: 3015–3024 (1979).
10. Lakowicz, J., and Weber, G. Quenching of protein fluorescence by oxygen. Detection of structural fluctuations in proteins on the nanosecond time scale. *Biochemistry 12*: 4171–4179 (1973).
11. Englander, S. W., and Kalenbach, N. R. Hydrogen exchange and structural dynamics of proteins and nucleic acids. *Quart. Rev. Bioph. 16*: 521–655 (1984).
12. Aksenov. S. I. Pulse NMR study of the dynamic structure of globular proteins. *Mol. Biol. 17*: 475–483 (1983) (in Russian).
13. Wuthrich, K. *NMR of proteins and nucleic acids.* New York, Wiley (1986).
14. Smith J., Cusac S., Pezzeca, V., Brooks, B., and Karplus, M. Inelastic neutron scattering analysis of low frequency motion in proteins: A normal mode study of the bovine pancreatic trypsin inhibitor. *J. Chem. Phys. 85*: 3636–3654 (1986).
15. McCammon, J. A. Protein dynamics. *Rep. Prog. Phys. 47*: 1–46 (1984).
16. Elber, R., and Karplus, M. Multiple conformational states of proteins: A molecular dynamics analysis of myoglobin. *Science 235*: 318–321 (1987).
17. Frauenfelder, H., Petsko, G. A., and Tsernoglou, D. Temperature dependent x-ray diffraction as a probe of protein structural dynamics. *Nature (London) 280*: 558–563 (1979).
18. Artymiuk, P. G., Blake, C. C. F., Grace, D. B. P., Calley, S. J., Phillips, D. C., and Sternberg, H. J. E. Crystallographic studies of dynamic properties of lysozyme. *Nature (London) 280*: 563–568 (1979).
19. Frolov, E. N., Mokrushin, A. D., Lichtenstein, G. I., Trukhtanov, V. A., and Goldanskii, V. I. An investigation of protein dynamic structure by method of gamma-resonance labels. *Doklady Ak. Nauk SSSR 212*: 165–168 (1973) (in Russian).
20. Keller, H., and Debrunner, P. Evidence for conformational and diffusional mean square displacements in frozen aqueous solution of oxymyoglobin. *Phys. Rev. Lett. 45*: 68–71 (1980).
21. Parak, F., Frolov, E. N., Mossbauer, R. L., and Goldanskii, V. I. Dynamics of metmyoglobin crystals investigated by nuclear gamma-resonance absorption. *J. Mol. Biol. 145*: 824–833 (1981).
22. Abaturov, L. V., Lebedeva, Yu. O., and Nosova, N. G. The dynamical structure of globular proteins: Conformational rigidity and fluctuational mobility. *Mol. Biol. 17*: 532–542 (1983) (in Russian).
23. Burshtein, E. A. Intrinsic protein luminescence as a tool for studying fast structural dynamics. *Mol. Biol. 17*, 455–467 (1983) (in Russian).
24. Van Hove, L. Correlations in space and time and Born approximation scattering in systems of interacting particles. *Phys. Rev. 95*: 249–262 (1954).
25. O'Connor, D. A. Crystallography using the Rayleigh scattering of gamma rays. In: *Proc. Intern. Conf. on Mossb. Spectr.,* Cracow, v. 2, 369–378 (1975).

26. Champeney, D. C. The scattering of Mossbauer radiation by condensed matter. *Rep. Prog. Phys. 42*: 1017–1054 (1979).
27. Albanese, G., and Deriu, A. High energy resolution x-ray spectroscopy. *Riv. d. Nuovo Cim. 2*: 9–41 (1979).
28. Goldanskii, V. I., Krupyanskii, Yu. F., and Fleurov, V. N. Raleigh Scattering of Mossbauer radiation data, hydration effects and glass-like dynamical model of biopolymers. *Physica Scripta 33*: 527–540 (1986).
29. Goldanskii, V. I., and Krupyanskii, Yu. F. Protein and protein-bound water dynamics studied by Raleigh scattering of Mossbauer radiation (RSMR). *Quart. Rev. Bioph. 22*: 39–92 (1989).
30. Krupyanskii, Yu. F., Parak, F., Gaubmann, E. E., Wagner, F. M., Goldanskii, V. I., Mossbauer, R. L., Suzdalev, I. P., Litterst, F. J., and Vogel, H. Investigation of the dynamics of metmyoglobin by Rayleigh Scattering of Mossbauer Radiation (RSMR). *J. de Phys. 41*: C1-489–C1-490 (1980).
31. Albanese, G., Deriu, A., Cavatorta, F., Krupyanskii, Yu. F., Suzdalev, I. P., and Goldanskii, V. I. Raleigh scattering of Mossbauer radiation in human serum albumen. *Hyperfine Inter. 29*: 1407–1409 (1986).
32. Nienhaus, G. U., Hartmann, H., Parak, F., Heinzl, J., and Huenges, E. Angular dependent Raleigh scattering of Mossbauer radiation on proteins. *Hyperfine Inter. 47*: 299–310 (1989).
33. Mossbauer, R. L. Gamma-resonance and x-ray investigations of slow motions in macromolecular systems. *Hyperfine Inter. 33*: 199–222 (1987).
34. Krupyanskii, Yu. F., Goldanskii, V. I., Kurinov, I. V., and Suzdalev, I. P. Dynamics of protein–water systems revealed by Rayleigh scattering of Mossbauer radiation (RSMR) technique. *Studia Biophysica 136*: 133–148 (1980).
35. Kuntz, I. D., and Kauzmann, W. Hydration of protein and polypeptide. *Adv. Prot. Chem. 28*: 239–345 (1974).
36. Khurgin, Yu. I. Hydration of globular proteins. *ZhVKHO im. Mendeleeva 6*: 684–690 (1976) (in Russian).
37. Vineyard, G. N. Scattering of slow neutrons by a liquid. *Phys. Rev. 110*: 999–1010 (1958).
38. Singwi, K. S., and Sjolander, A. Resonance absorption of nuclear gamma rays and the dynamics of atomic motions. *Phys. Rev. 120*: 1093–1102 (1960).
39. Krupyanskii, Yu. F., Shaitan, K. V., Gaubmann, E. E., Goldanskii, V. I., Rubin, A. B., and Suzdalev, I. P. Debye–Waller factor of Rayleigh scattering of Mossbauer radiation at substances with strong conformational motion. *Biofizika 26*: 1027–1044 (1981) (in Russian).
40. Kurinov, I. V., Krupyanskii, Yu. F., Suzdalev, I. P., and Goldanskii, V. I. The study of the influence of hydration on dynamics of some globular proteins by Rayleigh scattering of Mossbauer radiation. *Biofizika 32*: 210–214 (1987) (in Russian).
41. Kurinov, I. V., Krupyanskii, Yu. F., Suzdalev, I. P., and Goldanskii, V. I. RSMR study of hydration effects on the dynamics of some globular proteins. *Hyperfine Inter. 33*: 223–232 (1987).
42. Rupley, J. A., Gratton, E., and Careri, G. Water and globular proteins. *Trends Biochem. Sci. 8*: 18–22 (1983).

43. Krupyanskii, Yu. F., Parak, F., Hannon, D., Gaubmann, E. E., Goldanskii, V. I., Suzdalev, I. P., and Hermes, K. Determination of the mean square displacement of the atomic vibrations in myoglobin molecules by RSMR. *ZhETF 79*: 63–68 (1980) (in Russian).
44. Krupyanskii, Yu. F., Sharkevich, I. V., Khurgin, Yu. I., Suzdalev, I. P., and Goldanskii, V. I. Investigation of trypsin hydration by RSMR. *Molek. Bio. 20*: 1356–1363 (1986) (in Russian).
45. Kurinov, I. V., Krupyanskii, Yu. F., Suzdalev, I. P., and Goldanskii, V. I. Investigation of intramolecular DNA mobility by RSMR. In: *Proceedings of the Symposium Fisiko-Khimicheskiye svoistva biopolimerov v rastvore i kletkakh* (Physical and chemical properties of biopolymers in solutions and cells) Pustchino, 118–120 (1985) (in Russian).
46. Zhuravleva, T. V., Krupyanskii, Yu. F., Uspenskaya, N. Ya., Tchanorovskii, S. I., Kononeko, A. A., Suzdalev, I. P., and Goldanskii, V. I. The study of the dependence of chromatophores *Ectothiorodospira shaposhnikovii* dynamical properties on the hydration degree by RSMR. *Biol. Membrany, 4*: 495–501 (1987) (in Russian).
47. Kurinov, I. V., Rennekamp, G., Heidemeier, J., Krupyanskii, Yu. F., Parak, F., Suzdalev, I. P., and Goldanskii, V. I. Investigation of dynamics of met-Mb by means of analysis of RSMR spectra. *Khimicheskaya Fizika 7*: 3–12 (1988) (in Russian).
48. Krupyanskii, Yu. F., Bade, D., Sharkevich, I. V., Uspenskaya, N. Ya., Kononeko, A. A., Suzdalev, I. P., Parak, F., Goldanskii, V. I., Mossbauer, R. L., and Rubin, A. B. The mobility of chromatophore membranes from *Ectothiorodospira shaposhnikovii* reveals by RSMR experiments. *Eur. Bioph. J. 12*: 107–114 (1985).
49. Kurinov, I. V., Krupyanskii, Yu. F., Genkin, M. V., Davidov, R. M., Suzdalev, I. P., and Goldanskii, V. I. Study of the influence of solvent composition and viscosity on molecular dynamics of HSA by Rayleigh scattering of Mossbauer radiation. *Biofizika 33*: 407–412 (1988) (in Russian).
50. Beece, D., Eisenstein, L., Frauenfelder, H., Marden, M. C., Reinisch, L., Reynolds, A. H., Sorencen, L. B., and Yu, K. T. Solvent viscosity and protein dynamics. *Biochemistry 19*: 5147–5157 (1980).
51. Gavish, B., and Weber, M. M. Viscosity dependent structural fluctuation in enzyme catalysis. *Biochemistry 18*: 1269–1275.
52. Doster, W. Viscosity scaling and protein dynamics. *Biophys. Chem. 17*: 97–103 1983).
53. Aleksandrov, I. V., and Goldanskii, V. I. On the dependence of the rate constant of elementary chemical process in Kinetic region from the medicum viscosity. *Khimich. Fizika, 3*: 189–198 (1984) (in Russian).
54. Gekko, K., and Timasheff, S. M. Mechanism of protein stabilization by glycerol: Preferential hydration in glycerol–water mixtures. *Biochemistry 20*: 4667–4676 (1981).
55. Lehman, M. S., and Zaccai, G. Neutron small-angle scattering studies of ribonuclease in mixcol aqueous solutions and determination of preferential bound water. *Biochemistry 23*: 1939–1942 (1984).
56. Krupanskii, Yu. F., Parak, F., Goldanskii, V. I., Mossbauer, R. L., Gaubmann, E. E., Engelemann, and Suzdalev, I. P. Investigation of large intramolecular movements

within metmyoglobin by Rayleigh scattering of Mossbauer radiation. *Z. Naturfrosh. 37*: 57–62 (1982).
57. Kurinov, I. V., Krupyanskii, Yu. F., Suzdalev, I. P., and Goldanskii, V. I. The observation of quasielastic component in RSMR spectra for hydrated protein. *Doklady Ak. Nauk SSSR 290*: 738–742 (1986).
58. Mrevlishvilii, G. M. *Low-Temperature Calorimetry of Biological Macromolecules.* Metaniereba, Tbilisi (1984) (in Russian).
59. Sing, G. P., Parak, F., Hunklinger, S., and Dransfeld, K. Role of absorbed water in the dynamics of metmyoglobin. *Phys. Rev. Lett. 48*: 685–688 (1981).
60. Ping, Sheng, Cohen, R. W., and Schrieffer, J. R. Melting transition of small molecular clusters. *J. Phys. C., 14*: L565–L569 (1981).
61. Doster, W., Bachleitner, A., Dunau, R., Hiebl, H., and Luscher, E. Thermal properties of water in myoglobin crystals and solutions at sub-zero temperature. *Biophys. J. 50*: 213–219 (1986).
62. Frenkel, Ya. I. *Statistical Physics.* Moscow and Leningrad, Sov. Sc. Acad. Press (1948) (in Russian).
63. Sharkevitch, I. V., Konig, B., Krupyanskii, Yu. F., Parak, F., Suzdalev, I. P., and Goldanskii, V. I. Investigation of metmyoglobin and interprotein water dynamics by RSMR. *Khimich. Fizika 9*: 50–56 (1990).
64. Shaitan, K. V., and Rubin, A. B. Conformational mobility and theory of the Mossbauer effect in biological systems. Model of overdamped Brownian oscillator for conformational model. *Molek. Biolog. 14*: 1324–1335 (1980) (in Russian).
65. Chernavskii, D. S., Frolov, E. N., Goldanskii, V. I., Kononenko, A. A., and Rubin, A. B. Electron tunneling process and the segment mobility of macromolecules. *PNAS USA 77*: 7218–7221 (1980).
66. Knapp, E. W., Fisher, S., and Parak, F. Protein dynamics from Mossbauer spectra. The temperature dependence. *J. Phys. Chem. 86*: 5042–5047 (1982).
67. Afanas'ev, A. M., and Sedov, V. E. Mossbauer absorption spectra under continuous localized diffusion. *Phys. Stat. Sol. B 131*: 299–308 (1985).
68. Nadler, W., and Shulten, K. Theory of Mossbauer spectra of proteins fluctuating between conformational substated. *PNAS USA 81*: 5179–5723 (1984).
69. Nowik, I., Bauminger, E. R., Cohen, S. G., and Ofer, S. Spectral shapes of Mossbauer absorption and incoherent neutron scattering from harmonically bound nuclei in Brownian motion. Applications to macromolecular systems. *Phys. Rev. A 31*: 2291–2299 (1985).
70. Chernavskii, D. S., Goldanskii, V. I., Krupyanskii, Yu. F., Kurinov, I. V., and Suzdalev, I. P. *Phase Transitions in Macromolecular Constructions.* Preprint 102, Moscow, FIAN (1987).
71. Lifshitz, I. M., Grosberg, A. Yu., and Khokhlov, A. R. Volume interactions in the statistical physics of a polymer macromolecule. *Usp. Fiz. Nauk 127*: 353–390 (1979) (in Russian).
72. Krupyanskii, Yu. F., Goldanskii, V. I., Nienhaus, G. U., and Parak, F. Dynamics of water–protein systems revealed by Rayleigh scattering of Mossbauer radiation (RSMR). *Hyperfine Inter. 53*: 59–74 (1990).

6

Proteins in Essentially Nonaqueous Environments

Darrell L. Williams, Jr., Igor Rapanovich, and Alan J. Russell
University of Pittsburgh, Pittsburgh, Pennsylvania

I. INTRODUCTION

In the last 20 years, the association of enzymes with other cellular components, especially membranes, was recognized to be of utmost importance. In cells of certain types and species, roughly half of the enzymes are associated, to some extent, with membranes [1] with these interactions taking place predominantly via hydrophobic bonding. Paradoxically, this implies that studying enzymes in strictly aqueous solutions, the mainstream of biochemistry for the last century, may lead to erroneous results not particularly relevant to enzyme structure and function in vivo. We do not mean to suggest that nonaqueous enzymology (the subject of this chapter) is better, or in fact different, than its more traditional counterpart, but rather that it is reasonable to state that much can be learned about enzyme structure and function from studies of biocatalysis in organic media. Indeed, it has even been suggested (although we stress, not by these authors) that nonterrestrial "life" may be based on nonaqueous systems [2].

Presently, there are approximately 50 proteins known to work in high concentrations of organic solvents (Table 1). The enzymes included in this list generally belong to two classes: oxidoreductases (E.C. 1.) and hydrolases (E.C. 3.). It is quite natural to ask, What is the reason for such a distribution? Certainly, the first reason is commercial availability of certain enzymes and their substrates. It could be, however, that the relative simplicity of these reactions also contributes to the fact that these particular enzymes function in organic solvents. As we shall discuss later, in order to catalyze any reaction in an anhydrous environment, the enzyme must retain some flexibility [3,4]. It seems that more complex reactions, such as those catalyzed by lyases, for example, may need more flexibility of the enzyme and that such reactions would not be possible in organic solvents.

The structure of these enzymes also deserves comment. Most of them are simple, low-molecular-weight monomeric proteins without prosthetic groups, quarternary structure, or regulatory features. Notable exceptions to these rules (like lactate dehydrogenase) suggest, however, that these are not serious limitations.

One might suspect that the enzymes that function in organic solvents differ in their charge density relative to those which are inactive. The distribution of these enzymes, however, has the same characteristics as the bulk of enzymes for which this parameter has been calculated [5]. For this reason, there seems to be no criteria to predict which enzyme would work in an organic solvent and which would not. The working hypothesis of the enzymologists in this field is that any enzyme can work in organic solvents, provided that sufficiently soluble substrates and products, as well as appropriate reaction conditions, can be found [6,7].

A generally accepted principle of nonaqueous biocatalysis is that water is absolutely necessary for the catalytic functioning of the enzyme. The reason for this is that water participates either directly or indirectly in all interactions (electrostatic, hydrogen bonding, hydrophobic, van der Waals) that are necessary to form the active center of the enzyme, along with its native conformation. Therefore, without the water, the enzyme will be denatured. The actual quantity of water that an enzyme requires (termed *essential water*) is limited, however, to less than a monolayer coverage of the surface of the protein. This concept was first demonstrated in 1913 by Bourguelot and Bridel [8], who synthesized various glucosides using dry emulsin suspended in alcohol. This work revealed many features typical of more recent work. Later, Willstatter and Stoll used chlorophyllase in 92% alcoholic solution for the synthesis of chlorophyllides [9]. More than 30 years later Singer, in a review on proteins in nonaqueous solvents [10], forecast the catalytic activity of hydrolases in pure nonaqueous solvents. He supported his prediction by citing examples of trypsin and ribonuclease being active in water/cosolvent mixtures [11,12]. Only 3 to 4 years after these insightful remarks, Dastoli and coworkers [13–15] found that chymotrypsin and xanthine oxidase are active in various organic solvents. It took approximately 15 years for the next step to come. First, in 1981 Gatfield applied for the first patent in this field [16]; then,

TABLE 1 Enzymes and Cells that Have Been Studied in Organic Solvents Utilizing Various Supports

Enzymes used	E.C.
α-chymotrypsin	3.4.21.1
α-lytic protease	3.4.21.12
Alcohol dehydrogenase (equine)	1.1.1.1
Alcohol dehydrogenase (yeast)	1.1.1.1
Alkaline phosphatase	3.1.3.1
Aminoacylase	3.5.1.14
β-galactosidase	3.2.1.23
β-glucuronidase	3.2.1.31
Carboanhydrase	4.2.1.1
Carboxyesterase porcine liver	3.1.1.1
Catalase	1.11.1.6
Cholesterol oxidase	1.1.3.6
Crambin	n/a
Cytochrome c	n/a
Cytochrome oxidase	1.9.3.1
Emulsin	n/a
Enoate reductase	1.3.1.31
Esterase (pig liver)	n/a
Glucose oxidase	1.1.3.4
H^+-ATPase (mitochondrial)	3.6.1.34
Hydrogenase	1.18.99.1
Hydrogenase/lipoamide dehydrogenase	1.18.99.1/1.8.1.4
20-β-hydroxy-steroid dehydrogenase	1.1.1.53/62
Laccase	1.10.3.2
Lactate dehydrogenase	1.1.1.27
Ligninase	n/a
Lipase (many sources)	3.1.1.n
Lipoprotein lipase	3.1.1.34
Papain	3.4.22.2
Peroxidase (horseradish)	1.11.1.7
Phospholipase A2 (extracellular)	3.1.1.9
Polyphenoloxidase (mushroom)	1.10.3.1
Subtilisin	3.4.21.14
Sulfatase	3.1.6.1
Tannase	3.1.1.20
Thermolysin	3.4.24.4
Trypsin	3.4.21.4
Xanthine oxidase	1.1.3.22

TABLE 1 continued

Organism used
Aspergillus niger
A. simplex
Candida cylindracea
Chromobacterium viscosum
Corynebacterium equi
Enterobacter aerogenes
Epoxidizing cells
Flavobacterium
Kluyveromyces lactis CBS 683
Mucor miehei
Mycobacterium
Pichia pastoris
Pseudomonas
Rhizopus arrhizus
R. delemar
R. equi
R. niveus
R. nocardia
Rhodotorula minuta

Support
Alginate (calcium)
Alumina (activated, ceramic, porous)
Alumino-P-colamine complex
Amberlite
Bilayer "encagement"
Celite
Chitin
Chitosan
DEAE-cellulose
Glass (chitosan impregnated)
Glass beads
Hyflo supercel
Kiezelguhr
Polyacrylamide
Polyurethane
Polyvinylacetate
Sephadex G-50
Sepharose
Sepharose CL-6B
Silica gel
Sorsiles
TEAE-cellulose

TABLE 1 continued

Solvent
Acetone
Acetonitrile
Benzene
Benzene/heptane (1:1, water saturated)
Benzene/hexane (1:1)
Butanediol
Butanone
Butyl acetate
Carbon dioxide (supercritical)
Chloroalkanes
Chlorobenzene
Chloroform/heptane (1:1, water saturated)
Chloroform/octane/cetavlon
Cyclohexane
Cyclohexane/tetradecyl(trimethyl)ammonium bromide/water (10%)
Cyclohexanone
Dibutyl ether
Dichlorobenzene
Dichloromethylene
Di-iso-propyl ether
Dimethylformamide
Dimethylsulfoxide
Dioxane
Dodecane
Ethyl acetate
Ethyl ether
Fluoroform
Furan
Glycerol
Heptane
Hexadecane
Hexane
Iso-octane
Iso-propanol
Methylacetamide
Methyl alcohol
Methylene chloride
Methyl formate
Methyl-iso-butyl ketone
3-methyl-3-pentanol
Nitromethane
Octane
Octanol

TABLE 1 continued

Solvent
Pentanol
Petroleum ether
Polyols
Propyl acetate
Pyridine
Sec-butanol
Tert-amyl alcohol
Tetrachloromethane
Tetrahydrofuran
Thiophene
Toluene
Tributyrin
Trichlorobenzene
1,1,1-trichloromethane
Triethylamine
Xylene

All items were identified in a literature search using the keywords "enzyme" and "organic solvent." The search was performed in 1992.

the beginning of systematic research in this field was initiated by Professor A. M. Klibanov [4,17–19]. Several reviews deal extensively with this field [7,20,21]. It is interesting to note that future research in this field may be extended to studies on whole-cell activity in organic solvents [22].

II. "ANHYDROUS" AND HETEROGENEOUS SYSTEMS

The title of this section refers to the use of solid enzyme preparations in organic solvents containing little or no water. Lipase-catalyzed reactions, for example, can take place in solvents with water contents as low as 0.0015% [19]. At this concentration, lipase becomes exceedingly thermostable and its activity increases by a factor of five in comparison with its activity at 20°C. Water has an opposite influence on the activity of lipase at 20°C and at 100°C. At 20°C, water activates the enzyme by fivefold when its concentration is increased from 0.015 to 1.15%, which is consistent with our current understanding of nonaqueous enzymology [23]. At 100°C, on the other hand, there is a sharp cooperative inactivation [19], which may be the result of water-induced destabilization of the enzyme. The authors deduced from their data that although the structures of "wet" and "dry" lipases are similar, the "dry" enzyme must be more rigid, which is once again consistent with studies on the effect of water on the structure of dry proteins [3]. It should also be noted that the thermostability of lipase in nearly anhydrous organic solvents is dependent upon the nature of the solvent; that is, the half-lives changed five times going from *n*-heptanol to tributyrin. This increase in stability coincides with the rise in partition coefficient (log *P*) of the solvent, defined as the logarithm of the partition coefficient of some substance between octanol and water. Such results imply that there is a direct interaction between the organic solvent and the enzyme molecule that contributes to the structural stability of the protein. In addition, since there are areas of the enzyme surface associated directly with molecules of organic solvent, any calculations concerning protein dynamics, and/or structure-function, should include the physical characteristics of the solvent.

If we compare the enzyme in an aqueous solution, a crystal, a powdered preparation of enzyme, and a suspension of enzyme in an organic solvent, we notice that in the crystals the enzyme is hydrated approximately as much as it was in the water solution (usually >600 molecules of water/molecule of lysozyme [24]). In the case of lysozyme, full hydration, which is found at the point in which the heat capacity is the same as in dilute solution, is 0.38 $g_{water}/g_{protein}$, which corresponds to approximately 300 $moles_{water}/mole_{lysozyme}$ [25]. At this arbitrarily chosen borderline, there is barely enough water for a monolayer coverage of the protein surface. Enzymatic activity is only ~10% of that in solution. It can be concluded that a multilayer of water is necessary for some dynamic properties of the protein. As we dehydrate the protein further, there appear to be two critical points: 0.25 and 0.07 g/g, which respectively correspond roughly to 220 and 60 $molecules_{water}/molecule_{lysozyme}$. Initially, there is only enough water to cover polar groups on the protein surface (~1 H_2O per polar site). Enzyme activity is first exhibited at low levels near this point. At the second point, water is associated only with charged groups on the protein surface (~1 H_2O per charged site).

The range of protein hydration in which we deal with when working in nearly anhydrous organic solvents is as low as ~0.025 $g_{water}/g_{protein}$. Enzyme activity under these conditions, however, can still approach that of the enzyme in aqueous solutions. Klibanov found that chymotrypsin was active in octane even if the enzyme contained only 50 $molecules_{water}/molecules_{enzyme}$ [17]. It was very interesting that under these conditions the enzyme activity is dependent upon the pH of the aqueous solution from which the enzyme was prepared (in a powdered form), and that this pH dependence was very sharp. Also, the optimal activity in solvents approximated the pH optimum of the enzyme in water. It appears that the enzyme "remembers" the last pH to which it was exposed. From the point of view of hydration, it is quite possible that due to the lack of water clusters that could migrate from one polar residue to another, the degree of protonation of the charged groups is "frozen" according to its pK_a in the last buffer. Also, structural changes do not appear to take place during lyophilization [26].

If one measures the activity of different enzymes in different organic solvents with varying content of water, one can draw the following conclusion [4]:

The specific activity of a given enzyme in a particular solvent depends upon the concentration of water on the enzyme (not in the solution). In hydrophobic solvents, activity will be greater than in hydrophilic ones, since water will be preferentially partitioned onto the enzyme.

Water is not the only solvent capable of hydrogen bonding. Indeed, the essential water can be substituted to a certain extent by, for example, formamide. A solution of 3% formamide in octanol containing 1% water increases the activity of mushroom tyrosinase 35-fold [4].

Beside the effect of water, the nature of the organic solvent will also influence the enzyme activity, specificity, and stability. For example, substituting *n*-heptanol for *n*-decanol increases the half-life of lipase at 100°C from 5 to 12 hours [19]. This once again indicates we cannot neglect the direct interaction of the solvent with the protein surface. Indeed, it has been suggested that there is a direct correlation between the activity, thermostability, and operational stability of the enzyme and the hydrophobicity of the solvent [27]. The most used measure of hydrophobicity (when considering the activity of enzymes) thus far is log P [28,29]. If water solubility in organic solvents is accounted for, the correlation becomes even better [30]. There is a striking dependence of log P on the number of carbon atoms in the molecule for different homologous series. In a plot of number of carbon atoms versus log P, each homologous series is represented by a straight line, all of which are parallel. We can make the following generalization with regard to log P values: alkanes > alkenes > alkyl halides > aromatic hydrocarbons > alcohols. In this way, it is possible to deduce log P of some solvents knowing log P for homologous compounds. Another way to calculate the unknown log P is through hydrophobic fragmental constants [31]. Such calculations are of importance when attempting to determine how a solvent interacts with its environment.

Approximately five years ago, a new view of enzyme reactions [32], relevant to enzyme activity in organic solvents, was reported. Dewar proposed that each enzyme reaction takes place as it would proceed in the gas phase. This point of view is based upon the assumption that the adsorption of the substrate onto the active site can occur only if there are no molecules of solvent on the active site to interfere with adsorption. Therefore, since the active site of a protein is often hydrophobic and resembles a cleft or a hole between the domains of the protein chains, one might deduce that this cleft squeezes out the solvent. Gas reactions of ions with neutral molecules differ greatly from similar reactions in the solutions. The general implication concerning the methodology used in studying enzyme reactions was as follows: "Studies of enzymes with their active sites full of water have no relevance to their reactions with substrates" [32]. Without commenting on the validity of this statement, it is perhaps pertinent to emphasize that studies of enzymes in organic solvent are worthwhile, and they can teach us much about the process of biocatalysis.

One of the practical advantages of biocatalysis in nonaqueous media is the possibility of working with nonpolar substrates that are only slightly soluble in water. Applications involving oxidoreductases have taken particular advantage of this fact. For example, Klibanov and coworkers [33] used alcohol dehydrogenase in the preparative organic synthesis of optically active alcohols and ketones. In such reactions, it is necessary to regenerate the cofactor (nicotinamide adenine dinucleotide). This was achieved by codepositing enzyme and cofactor onto glass beads and using isopropanol as a second substrate. As a result, cofactor recycling exceeded 2 million cycles as compared with 2000 in water [34].

Another advantage of biocatalysis in nonaqueous media is the shifting of the thermodynamic equilibria to favor synthesis instead of hydrolysis (for example, in the synthesis of peptides and esters). This is probably the most popular, well-characterized, and commercially acceptable application of nonaqueous enzymology. Utilizing such enzymes as lipases, esterases, proteases, and carbohydrases, reactions including esterifications, transesterifications, lactonization, interesterifications, oximolysis, thiotransesterifications, lactonization, aminolysis, and peptide synthesis have been performed at high yields and degrees of conversion.

In these reactions, it was shown that the use of organic solvents changes the substrate specificity of enzymes, while retaining good overall regiospecificity, positional specificity, and stereoselectivity [20]. Recently, it was shown that water molecules bound to the enzyme play a determining role in affinity and specificity [35]. Some popular enzymes, such as lipases and proteases, are considered now as "bench reagents" for some of these synthetic reactions. Descriptions of these procedures, as well as many examples, may be found in the numerous reviews already mentioned.

III. "ANHYDROUS" AND HOMOGENEOUS SYSTEMS

Taking into consideration the drawbacks associated with performing kinetic measurements of insoluble enzymes in organic media, the idea of changing the solubility properties of an enzyme by chemical modification emerged. Due to its amphiphilic properties, polyethylene glycol (PEG) was especially suitable for such an application. The hydrophilic region of the molecule assures that it can be used in water to chemically modify enzymes, while its hydrophobic region enables the solubilization of the PEG–enzyme conjugate in anhydrous hydrophobic environments. Bovine serum albumin was the first protein to which PEG of molecular weight 5000 was attached [36] (for a review, see [37]). Presently, there are several enzymes that have been modified in this way. There is even an example of two modified and coupled enzyme systems working consecutively [38]. There are also reports on proteins working with proteins as substrates [39]. It is noteworthy that if the enzyme (trypsin in this case) is in suspension or is solubilized by PEG-modification, the results of, and conditions for, proteolysis are different. In the first case, reaction is dependent on the amount of water in the system (up to 1% v/v), whereas in the case of soluble trypsin, the catalyst needs an additional nucleophile, such as phenylalanine, in order to function. Proteolysis was also accompanied by a significant transpeptidation reaction for the PEG-modified enzyme.

The most commonly used method of PEG-modification is with the use of cyanuric chloride [37]. Several other methods listed make the procedure rather flexible (for instance, see [39,40]). Some of them can lead to copolymers of PEG and maleic anhydride [38].

Such enzyme conjugates are often very active in organic solvents. For example, lipase retains 80% of its original activity when measured in an emulsified aqueous system [41]. The chlorinated hydrocarbons are often the solvents of choice under these conditions (for example, 1.1.1-trichloroethane) [42]. The modified lipase is very stable in benzene solutions (after three to five months it retained ~50% of initial activity) [43]. The modified enzyme was very thermostable when dissolved in two substrates for ester exchange without the use of other solvents. The optimum temperature approached 70°C (compared with 45°C for an unmodified enzyme in aqueous systems) [44]. It was interesting that lipase was more active in ester synthesis reactions between −3 and 3°C than at elevated temperatures (at −3°C, the enzyme exhibited five times the activity when the reaction was run at 37°C) [45]. The use of PEG-modified lipase in organic solvents made it possible to measure the Michaelis constant of enzymes in organic solvents (0.1–0.3 M for lipase in a variety of solvents) [46]. It should be stressed that trace amounts of water were necessary for the reactions of ester synthesis and exchange, and that the activity was never detected in water-miscible solvents, which actually inhibited lipase already at low concentrations [47].

PEG-modification was possible with such complex proteins as catalase and peroxidase, which have prosthetic groups (heme), are glycosylated (peroxidase), and have a quaternary structure (catalase) [39]. It appears that attachment of PEG molecules to these enzymes produced certain characteristic changes in them, but did not force the dissociation of either heme or subunits. The solubility of PEG-catalase in benzene, along with its activity, increased with the degree of modification. At 55% of surface lysine modification, its solubility was >2 mg/ml and specific activity was 7.0×10^4 units/mg, which was 60% higher than in aqueous solution [48].

For practical use in industrial biotechnological applications, such PEG-modified enzymes should be made recoverable. An easy and attractive way to do this is to link them to magnetic materials such as magnetite, Fe_3O_4 [49]. This procedure was first performed with lipase, L-asparaginase, and urokinase. These magnetic enzyme conjugates are easily dispersed in aqueous solutions and organic solvents, they are stable and sufficiently active, and they could be readily recovered without loss of enzymic activity.

Recently, a new method of amphipilic modification of proteins [50] was suggested in order to make them soluble in organic solvents. The authors suggested using a lipodisaccharide possessing a reactive aldopentose function, namely, 6-O-octyl-β-D-galactopyranosyl-(1-5)-L-arabinose. By mild reductive alkylation of amino residues of the enzyme, one produces a lipo-1-deoxyglycytolated enzyme. The advantage of this method is that external lysine residues preserve their positive charges and their surface distribution remains undisturbed. The authors applied their method to chymotrypsin, which successfully catalyzed esterification of amino acids in trichloromethane, tetrahydrofuran, acetone, ethanol, acetonitrile, and ethyl acetate. With amphiphilic modification, the hydrophilic chains introduced retain water molecules (up to 3 water molecules/chain unit in the case of PEG), whereas the hydrophobic substituents can insert into the solvent. One could describe such a system as a covalent reverse micromicelle.

IV. WATER/COSOLVENT MIXTURES

As was mentioned previously, enzymes usually exhibit low activity in water-miscible organic solvents. For this reason the majority of work with enzymes in nonaqueous media is performed with rather nonpolar solvents (log $P \geq 2$). Another trend is to study enzymes in water/organic cosolvent mixtures in order to gain more insight into the reasons and the mechanism of interaction of enzyme and solvent.

There is a great deal of data in the literature concerning the inactivation of different enzymes by water-miscible solvents. In our laboratory, we have studied the effect of ethanol addition on the activity of phosphotriesterase (E.C. number not yet assigned) from *Pseudomonas diminuta*. The enzyme catalyzes the hydrolysis of the organophosphorus pesticide paraoxon (diethyl *p*-nitrophenyl phosphate). Just 30–40% of cosolvent diminishes the activity by over 90% [51]. Interestingly enough, low concentrations of alcohols can increase the activity of an enzyme. In the case of cyclodextrin glycosyltransferase (E.C. 3.2.1.19), addition of 10% of ethanol to the reaction media increased the product yield by 100% (for example, see [52]). It seems probable that this enhancement of yield is due to the increased hydrolysis of starch by the enzyme while in the presence of the alcohol.

It is known that isolated F1-ATPase hydrolyzes adenosine triphosphate (ATP) in the presence of dimethyl sulfoxide (DMSO) [53,54]. In a recent report by Sakamoto and colleagues, 35–40% (w/v) DMSO increased the rate of ATP hydrolysis and influenced the ^{31}P-NMR spectrum of ATP [53]. The authors interpreted their findings as DMSO providing a more hydrophobic environment for the asymmetric coordination of Mg^{2+} to the β-phosphoryl group of ATP and thus facilitating a more "labile" structure of ATP, which is needed for both the hydrolysis and enzymatic synthesis of ATP.

There are also reports about the stabilizing effects of lower alcohols, and other similar substances, on proteins. For example, George et al. [55] reported that

2.5–10% of methanol and ethanol stabilized beef-heart lactate dehydrogenase (E.C. 1.1.1.37) against long-term storage in the cold. For periods up to several months, in the presence of 40% DMSO, the enzyme retained complete activity. The authors speculated that this stabilization is due to the preservation of the enzyme in an "active" conformation as a result of exclusion or replacement of water molecules. In such experiments, however, it is always difficult to elucidate the role of, for example, bacterial degradation.

Solvents such as formamides, alcohols, and polyols such as glycerol (or sugars) have also been investigated with respect to their interactions with proteins. These data are mainly interpreted in terms of the influence of these solvents (and solid substances) on water activity and dielectric constant of the solution [56]. It has been shown [57] that lower primary alcohols and formamide displace the T/R conformational equilibrium of human fetal hemoglobin toward the T conformation. The data on oxygen affinity under these conditions indicate that increased electrostatic and hydrophobic protein–solvent interactions contribute to this effect.

The decrease of enzyme activity in water/cosolvent mixtures can be explained by the simultaneous occurrence of dielectric changes and competitive/noncompetitive inhibition, as proposed by Clement and Bender [58]. They predicted an exponential dependence of K_m on solvent concentration. In a recent study on trypsin [59], it was shown that a cosolvent may selectively inhibit the amidase activity of the enzyme and exert a relatively small influence on the esterase activity, thus modifying the specificity of the enzyme. For both types of reactions, there was, however, a dramatic increase in K_m with increasing concentration of an organic cosolvent (they used an unusual combination of 9:1 dimethylformamide (DMF) and dimethylsulfoxide). Since the kinetic mechanism of trypsin in different types of reactions is well known, the authors performed a detailed analysis of the reaction kinetics. They also attempted to study the influence of the cosolvents on the overall and local structure of trypsin. The decrease in k_{cat} for the esterase activity became prominent when the cosolvent concentration was $\geq$10%, which coincided with minor changes in the structure of the active site as revealed by electron paramagnetic resonance (EPR) spectroscopy.

Similar work was performed using the cysteine protease papain [60]. The authors suggest that since organic solvent molecules are more hydrophobic than water they will have a higher affinity to the active site cleft of papain, which is known to be rather hydrophobic. Thus, the increase in K_m can be attributed to competition between the substrate and cosolvent. (In this respect, it is interesting that in the case of phosphotriesterase-catalyzed hydrolysis of paraoxon, the organic cosolvents are noncompetitive inhibitors with respect to the substrate.)

High concentrations of organic solvents usually induce enzyme aggregation (e.g., 95% dioxane [61]) or shrinkage [62]. Decreasing the polarity of solvent tends to disperse the regions of protein with low polarity which results in disorganization and destabilization of the enzyme [63]. In addition, there is a decreased nonpolar association of substrate with the active site [64]. Interestingly, in anhydrous DMSO, DMF, and formamide, chymotrypsin behaves like a nonspecific nucleophilic reagent without catalytic activity [65], whereas peroxidase and catalase are active (as described above). Although the direct effect of solvent on the properties of enzymes is not the focus of this chapter, there have been reports of competitive inhibition of enzyme-catalyzed reactions by many small organic solvents (for example, methanol and ethanol). Organic solvents are also known to dissociate oligomers, cause large conformational transitions, shift equilibria between conformers, alter the helical content of proteins, and alter the dissociation of ionizable groups. It is most important to stress once more, however, that all of these studies have been performed with fully hydrated enzyme preparations and are not entirely relevant to current thought about enzyme properties in *anhydrous* media.

Another fascinating advance in recent years has been the design and selection of proteins that are more active and stable in water–solvent mixtures. Professor Arnold, at the California Institute of Technology, has pioneered this area of research and has suggested a set of design rules by which proteins can be engineered for improved properties in the presence of organic solvents. Indeed, it has been possible to produce mutant proteins that are up to 1000 times more active than wild type in 70% dimethylformamide. We feel that protein design for nonaqueous media will not only lead to more active catalysts in these unusual surroundings, but also a deeper understanding of nonaqueous enzymology as a whole.

V. CONCLUSIONS

Natural enzymes can function in a wide range of extreme environments including anhydrous organic solvents. Although it is only in the last decade that research in this field has become industrially important, evidence for the stability of proteins in nonaqueous media dates back to the last century. Current efforts (as evidenced by the publication of this book) are focusing on how the physical properties of a solvent control the specificity, activity, and stability of enzymes. Over 100 years of enzymology in aqueous media has led to a molecular understanding of how proteins function. We hope that in the years to come nonaqueous enzymology will enable us to elucidate the precise role that solvent plays in the structure/function/environment relationships of biocatalysts.

ACKNOWLEDGMENTS

This work has been funded in part by the Department of Defense, Army Research Office (Grant number DAAL03-90-M-0276), American Chemical Society (PRF Type G), Eastman Kodak, PPG, Union Carbide, and a National Science Foundation Presidential Young Investigator Award to A.J.R. (Grant number BCS 9057312).

REFERENCES

1. Butler, L. G., *Enz. Microb. Technol. 1*: 253 (1979).
2. Siegel, S. M., and Roberts, K., *Space Life Sci. 1*: 131 (1968).
3. Poole, P. L., and Finney, J. L., *Int. J. Biol. Macromol. 5*: 308 (1983).
4. Zaks, A., and Klibanov, A. M., *J. Biol. Chem. 263*: 8017 (1988).
5. Russell, A. J., unpublished data.
6. Klibanov, A. M., *Chem. Tech. 16*: 354 (1986).
7. Zaks, A., and Russell, A. J., *J. Biotech. 8*: 259 (1988).
8. Bourguelot, E., and Bridel, M., *Ann. Chim. Phys. 29*: 145 (1913)
9. Willstatter, R., and Stoll, A., *Investigations of Chlorophyll, Methods and Results*. Science Press, Princeton, New Jersey, 1928.
10. Singer, S. J., *Adv. Protein Chem. 17*: 1 (1962).
11. Heppel, L. A., and Whitfeld, P. R., *Biochem. J. 60*: 1 (1955).
12. Heppel, L. A., Whitfeld, P. R., and Makham, R., *Biochem. J. 60*: 8 (1955).
13. Dastoli, F. R., Musto, N. A., and Price, S., *Arch. Biochem. Biophys. 115*: 44 (1966).
14. Dastoli, F. R., and Price, S., *Arch. Biochem. Biophys. 118*: 163 (1967).
15. Dastoli, F. R., and Price, S., *Arch. Biochem. Biophys. 122*: 289 (1967).
16. Gatfield, I. L., *German Patent Application*, 3,108,927 (1981).
17. Zaks, A., and Klibanov, A. M., *J. Amer. Chem. Soc. 108*: 2767 (1986).
18. Zaks, A., and Klibanov, A. M., *Proc. Natl. Acad. Sci. USA 82*: 3192 (1985).
19. Zaks, A., and Klibanov, A. M., *Science 224*: 1249 (1984).
20. Dordick, J. S., *Enzym. Microb. Technol. 11*: 194 (1989).
21. Klibanov, A. M., *Acc. Chem. Res. 23*: 114 (1990).
22. Buckland, B. C., Dunnill, P., and Lilly, M. D., *Biotech. Bioeng. 17*: 815 (1975).
23. Deetz, J. S., and Rozzell, J. D., *Trends Biotech. 6*: 15 (1988).
24. Careri, G., Gratton, E., Yang, P.-H., and Rupley, J. A., *Nature 284*: 572 (1980).
25. Rupley, J. A., Gratton, E., and Careri, G., *Trends Biochem. Sci. 8*: 18 (1983).
26. Rupley, J. A., Yang, P.-H., and Tollin, G., in *Water in Polymers* Ed. S. P. Rowland, *ACS Symp. 127*, p. 111 (1980).
27. Brink, L. E. S., and Tramper, J., *Biotech. Bioeng. 27*: 1258 (1985).
28. Laane, C., Boeren, S., and Vos, K., *Trends Biotech. 3*: 251 (1985).
29. Laane, C., Tramper, J., and Lilly, M. D., Eds. *Biocatalysis in Organic Media*. Elsevier, Amsterdam, 1986.
30. Reslow, M., Adlercreutz, P., and Mattiasson, B., in *Biocatalysis in Organic Media*, Proceedings, p. 349. Elsevier, Amsterdam. 1987.
31. Rekker, R. F., *The Hydrophobic Fragmental Constants, Its Derivative and Application*. Elsevier, New York and Amsterdam, 1977.
32. Dewar, M. J. S., and Storch, D. M., *Proc. Natl. Acad. Sci., USA 82*: 2225 (1985).
33. Grunwald, J., Wirtz, B., Scollar, M. P., and Klibanov, A. M., *J. Amer. Chem. Soc. 108*: 6732 (1986).
34. Wong, C.-H., Whitesides, G. M., *J. Amer. Chem. Soc. 103*: 4890 (1981).

35. Quiocho, F., Wilson, D. K., and Vyas, N. K. *Nature 340*: 404 (1989).
36. Abuchowski A., van Es, T., Palczuk, N. C., and Davis, F. F., *J. Biol. Chem. 252*: 3578 (1977).
37. Inada, Y., Takahashi, K., Yoshimoto, T., Ajima, A., Matsushima, A., and Saito, Y., *TIBTECH 4*: 190 (1986).
38. Yoshimoto, T., Ritani, A., Ohwada, K., Takahashi, K., Kodera, Y., Matsushima, A., Saito, Y., and Inada, Y., *Biochem. Biophys. Res. Comm. 148*: 876 (1987).
39. Gaertner, H. F., and Puigserver, A. J., *Biotech. Bioeng. 36*: 601 (1990).
40. Wirth, P., Souppe, J., Tritsch, D., and Biellmann, J.-F., *Bioorg. Chem. 19*: 133 (1991).
41. Inada Y., Nishimura, T., Takahashi, K., Yoshimoto, K., Saha, R. A., and Saito, Y., *Biochem. Biophys. Res. Comm. 122*: 845 (1984).
42. Takahashi, K., Ajima, A., Yoshimoto, T., Okada, M., Matsushima, A., Tamaura, Y., and Inada, Y., *J. Org. Chem., 50*: 3414 (1985).
43. Matsushima, A., Okada, M., Takahasi, K., Yoshimoto, T., and Inada, Y., *Biochem. Int., 11*: 551 (1985).
44. Takahashi,K., Kodera, Y., Yoshimoto, T., Ajima, A., Matsushima, A., and Indada, Y., *Biochem. Biophys. Res. Comm 131*: 532 (1985).
45. Matsushima, A., Kodera, Y., Takahashi, K., Saito, Y., and Inada, Y., *Biotechnol. Lett. 8*: 73 (1986).
46. Takahashi, K., Yoshimoto, T., Ajima, A., Tamaura, Y., and Inada, Y., *Enzyme 32*: 235 (1984).
47. Takahashi, K., Hishimura, H., Yoshimoto, T., Okada, M., Ajima, A., Matsushima, A., Tamaura, Y., Saito, Y., and Inada, Y., *Biotechnol Lett. 6*: 765 (1984).
48. Takahashi, K., Ajima, A., Yoshimoto, T., and Inada, Y., *Biochem. Biophys. Res. Comm. 125*: 761 (1984).
49. Inada, Y., Takahashi, K., Yoshimoto, T., Kodera, Y., Matsushima, A., and Saito, Y., *TIBTECH 6*: 131 (1988).
50. Cabaret, D., Maillot, S., and Wakselman, M., *Tet. Lett. 31*: 2131 (1990).
51. Russell, A. J., unpublished data.
52. Lee Y.-D., and Kim, H. S., *Enz. Microb. Technol. 13*: 499 (1991).
53. Sakamoto, J., and Tonomura, Y., *J. Biol. Chem. 93*: 1601 (1983).
54. Yohda, M., Kagawa, Y., and Yoshida, M., *Biochim. Biophys. Acta 850*: 429 (1986).
55. George, H., McMahan, J., Bowler, K., and Elliott, M., *Biochim. Biophys. Acta 191*: 466 (1969).
56. Halling, P. J., *Biochim. Biophys. Acta 1040*: 225 (1990).
57. Militello, V., Vitrano, E., and Cupane, A., *Biophys. Chem. 39*: 161 (1991).
58. Clement, C. E., and Bender, M. L., *Biochem. 2*: 836 (1963).
59. Guinn, R. M., Blanch, H. W., and Clark, D. S., *Enzym. Microb. Technol. 13*: 320 (1991).
60. Fernandez, M. M., Clark, D. S., and Blanch, H. W., *Biotech. Bioeng. 37*: 967 (1991).
61. Tanizawa, K., and Bender, M. L., *J. Biol. Chem. 249*: 2130 (1974).
62. Weetall, H. H., and Vann, W. P., *Biotech./Bioeng. 18*: 105 (1976).
63. Tanford, C., *Science 200*: 1012 (1978).
64. Maurel, P., *J. Biol. Chem. 253*: 1677 (1978).
65. Klyosov, A. A., van Viet, N., and Berezin, I. V., *Eur. J. Biochem. 59*: 3 (1975).

7

Solvent Viscosity Effect on Protein Dynamics: Updating the Concepts

BENJAMIN GAVISH AND SAUL YEDGAR
The Hebrew University—Hadassah Medical School
Jerusalem, Israel

I. INTRODUCTION

It is generally accepted that the structural dynamics of proteins in solution plays an important role in achieving and regulating the biochemical functions of the proteins. The microscopic picture of the dynamics is extremely complicated and includes relaxation times that extend over a very wide range. The continuous interaction of the protein structure with the surrounding water is well known to be intimately linked with this type of structural dynamics. By what mechanisms does the solvent affect protein dynamics? A possible experimental answer is to find what property of the solvent affects protein dynamics. The Brownian theory suggests that this property is the internal friction associated with the molecular motion, which is linked to the diffusion constant and the solvent viscosity. The role of the solvent viscosity in chemical kinetics was derived by Kramers in 1940 [1]. This result, which contradicted the widely used transition state theory, was overlooked for many years. The application of Kramers's theory to protein reactions was suggested and demonstrated by Gavish in 1978 and 1979 [2,3]. Since then, many studies documented the solvent viscosity effect on biochemical reactions. Reaction rates, however, were found to display power-law dependence on viscosity ($0 > \text{power} > -1$) instead of varying linearly with 1/viscosity, as predicted by Kramers. Independently, Grote and Hynes published in 1980 a theory of chemical reactions [4], which showed that Kramers's and the transition state theories represent extreme cases of a more realistic situation. The Grote–Hynes theory necessitates the use of frequency-dependent friction, which seems to be responsible for the weaker dependence of the rate coefficients on the solvent viscosity observed also in nonbiological reactions. On the other hand, frequency dependence of the solvent viscosity is well known to be associated with the molecular dynamics of real liquids. Another aspect of the concept *viscosity* is linked with local volume fluctuations in liquids, which in densed liquids depend on the free volume available for translatory motion. In the presence of a solute, the rate of the solute structural rearrangements depends on the solvent viscosity in a power-law way, as shown theoretically and experimentally by Gegiou et al. [5].

It seems that there are common features in the above-mentioned unrelated studies that, if combined, could help us to better understand solvent effects on protein molecular dynamics. This chapter reviews systematically some related concepts, along with new ideas and suggestions for how to apply them.

In Sec. II, Brownian dynamics is critically reviewed, including its generalization to frequency-dependent friction.

In Sec. III, molecular dynamics concepts are applied for modeling chemical reactions, using energy barrier crossing, as done by the transition state, Kramers, and Grote–Hynes theories.

In Sec. IV, the experimental work concerning solvent viscosity effects on chemical and biochemical reactions is cited, and the cosolvent effect on the ultrasonic absorption in proteins is discussed at length. The possible meaning of the power-law dependence of kinetic coefficients and structural rearrangements on the solvent viscosity is discussed in Sec. V.

Conclusions are given in Sec. VI.

II. BROWNIAN DYNAMICS

The Brownian theory is an ingenious way of describing the dynamic interaction between a "particle" and its surrounding medium. Viewing this medium as a viscous continuum enables one to introduce viscosity as a friction-related property of the system. The extension of the picture to the atomic scale elicits conceptual difficulties, which are probably associated with the deviations from the "ideal" behavior observed experimentally. This section is concerned with basic concepts of Brownian dynamics: what is wrong with them if extended to atomic scale; what is the required generalization; and what properties of liquids are associated with friction.

A. Basics

1. Langevin's Equation

a. What Is Brownian Motion? Brownian motion was originally concerned with the perpetual irregular motion exhibited by small particles of colloidal size (but very large on the atomic scale) immersed in a fluid. This motion is maintained by the collisions with the molecules of the surrounding medium. Each collision can be thought of as producing a kink in the path of the particle. The assumed enormous number of such collisions gives the motion its chaotic nature (Fig. 1). The manifestation of the Brownian theory predictions led to its application to particles of atomic size, such as atoms and molecules in liquids or in complex media like solutions containing macromolecules and proteins. The dynamic interaction of a "particle" of atomic size with its surrounding "medium" requires careful considerations, however.

b. Particle–Medium Interaction The particle–medium interaction is usually expressed by three kinds of force (see Fig. 2).

An *irregular force R* is caused by the instantaneous imbalance of the interactions of an individual particle with its neighbors. The shuttling of the particle by R can be shown to increase its kinetic energy.

A *friction force* is exerted on the particle as it moves relative to its neighbors through the medium with velocity v. This force dissipates the kinetic energy of the particle. The simplest form of such friction force is $-m\gamma v$ (although other forms are possible [1]). Here m is the particle mass and γ is a constant friction coefficient having the dimension of frequency.

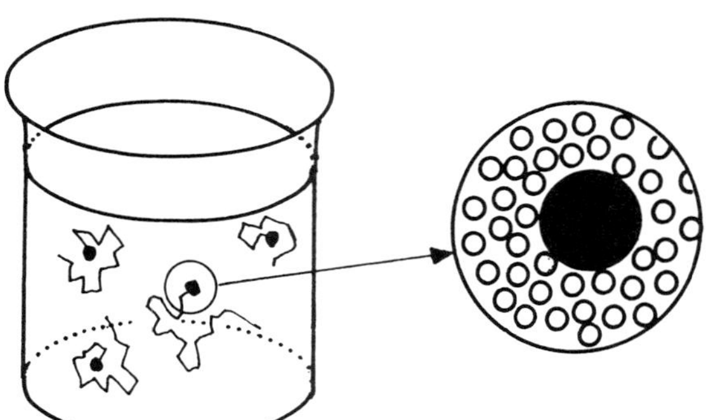

FIGURE 1 Brownian motion: This is the irregular motion displayed by small particles (filled circles) embedded in a liquid, caused by their instantaneously unbalanced interactions with the surrounding liquid molecules.

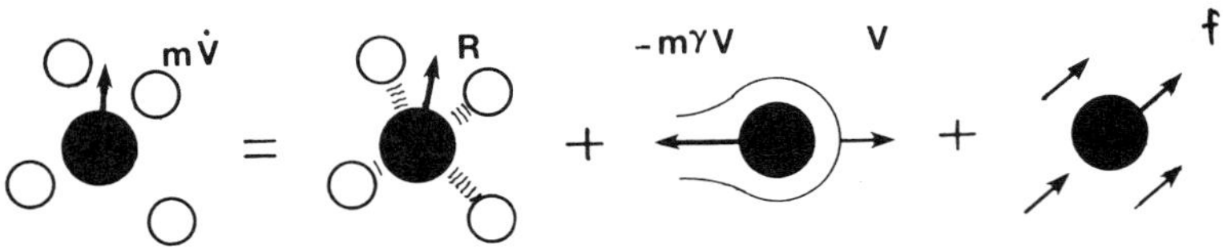

FIGURE 2 Forces involved in Brownian motion: The total force $m\dot{v}$ that acts on the particle (Fig. 1) having an instantaneous velocity v is the sum of an irregular and highly fluctuating term R, which reflects the individual interactions (stacks of lines) with the surrounding molecules (empty circles); a Stokes-like friction force $-m\gamma v$ borrowed from hydrodynamics (see stream line); and a force f exerted by an external field (arrows). This is Langevin's equation (Eq. (1)).

A *potential energy–derived force* f is contributed by an external time-independent force field, for example, an electric field, or by a mean field exerted by the surrounding particles.

c. ***Quantitative Relations*** Taking into account the three forces just mentioned, the net force mv is given by Langevin's equation (written in its one-dimensional form for simplicity):

$$m\dot{v} = -m\gamma v + R + f \tag{1}$$

As for the irregular force R, the following principal assumptions are made [6]:

1. The random force R is independent of the particle velocity v.
2. R fluctuates extremely rapidly compared with the temporal variations of v.

Since R and v are random variables, only time-averaged functions of R and v are informative. Let us denote an average over time τ of the quantity $X(\tau)$ by $\langle X(\tau)\rangle$, or just $\langle X\rangle$. The quantity $\langle X(\tau_1)X(\tau_2)\rangle$, called *the autocorrelation function of* X, can be shown to be equal to $\langle X(0)X(\tau)\rangle$ for stationary processes, where $\tau = \tau_1 - \tau_2$. This quantity equals $\langle X^2\rangle$ for $\tau = 0$ and *decays* to $\langle X\rangle^2$ for long enough time, for which $X(0)$ and $X(\tau)$ are no longer correlated (Fig. 3). Thus, the decay time τ_C of the autocorrelation function characterizes the "memory loss" X.

The following relations can be derived from Eq. (1) and are relevant for understanding the subject:

i. *Averages of velocity and random force.* In the absence of an external field, the average values of the random force ($\langle R\rangle$) and the particle velocity ($\langle v\rangle$) are zero. For a constant force f_0, Eq. (1) shows that the "drift velocity" v_d ($= \langle v\rangle$) is equal to μf_0, where $\mu = 1/m\gamma$ is the particle *mobility*.

ii. *The equipartition theorem.* Its one-dimensional form is

$$\langle v^2\rangle = \frac{k_B T}{m} \tag{2}$$

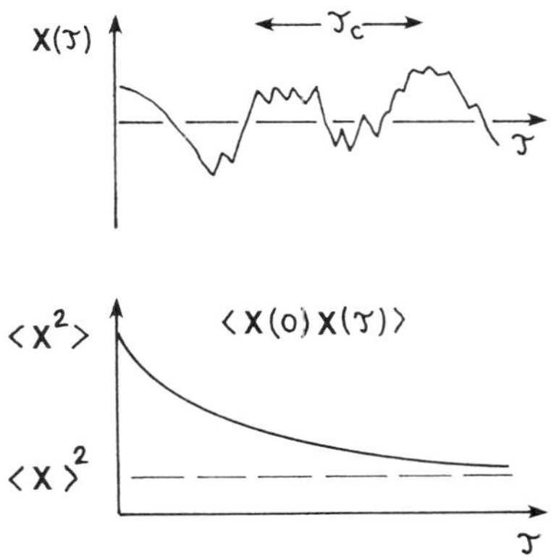

FIGURE 3 Correlation functions: This mathematical tool expresses the "memory loss" of position by a particle performing random motions (top). $X(0)$ and $X(\tau)$ are the particle positions now (0 corresponds to any time) and later at time τ, respectively. The average of $X(0)X(\tau)$ over all times, $\langle X(0)X(\tau)\rangle$, displays a decay from "full correlation" $\langle X^2\rangle$ to "no correlation" $\langle X\rangle^2$ with a characteristic time τ_C (bottom). τ_C is the correlation time.

where k_B is Boltzmann's constant and T is the absolute temperature. Thus, the gain and loss of energy due to the Brownian dynamics do not affect the average kinetic energy of the particle.

iii. *The fluctuation–dissipation theorem* [7]:

$$m\gamma = \frac{\langle R^2\rangle}{k_B T} \tag{3}$$

Equation (3) shows that the friction coefficient is related to the mean-square value of the random force fluctuations, and it reflects the fact that both collisions and friction have a common origin. A deeper insight into this relation will be given in the next section.

2. Diffusion and Viscosity

a. ***Einstein's Relation*** Let us consider a type of motion that can be described as a series of independent "jumps." The following one-dimensional expression can be shown to be valid for long enough times [6]:

$$\langle [x(\tau) - x(0)]^2\rangle = 2D\tau \tag{4}$$

where $x(0)$ and $x(\tau)$ are the locations of the moving particle at times 0 and τ, respectively, and D is the *diffusion coefficient* (the factor of 2 is replaced by 6 for a three-dimensional diffusion). This definition of D is meaningful only for τ much larger than the time required for a single "diffusion jump." When a jump involves a characteristic distance a, the time τ_D between two successive jumps is defined by

$$D = \frac{a^2}{\tau_D} \tag{5}$$

Expressions analogous to Eqs. (4) and (5) can be given for rotational diffusion, where $x(\tau)$ is replaced by the angle of rotation.

This description of diffusion fits that of a Brownian particle driven by *independent collisions*. One can relate D to the friction coefficient γ by using the Langevin's equation. This leads to Einstein's relation:

$$D = \frac{k_B T}{m\gamma} \tag{6}$$

Equation (6) suggests a simple interpretation for γ. The particle is expected to move freely during the time t_D, over the distance a. Thus, $a = v\tau_D$. The relation $v^2 = k_B T/m$ (Eq. (2)), combined with Eqs. (5) and (6) yields

$$\gamma = \frac{1}{\tau_D} \tag{7}$$

showing that γ can be interpreted as the collision rate. For intermolecular distances of the order of $a = 1$ Å and a particle of molecular weight 40 at room temperature, the thermal velocity is $v = 2.4 \times 10^4$ cm/s (Eq. (2)), and $\tau_D = a/v = 4.2 \times 10^{-13}$ s. Using Eq. (7), γ is estimated to be 2.4×10^{12}/s.

b. Stokes Law Using a hydrodynamic approach, one can identify the Brownian friction force $-m\gamma v$ with the drag force acting on a body, which moves relative to a viscous liquid. For a spherical particle of radius r moving in a homogeneous and isotropic medium, it can be shown that

$$m\gamma = c\pi r\eta \tag{8}$$

where η is the shear viscosity, and $c = 6$ when the liquid sticks to the particle and $c = 4$ when it slips. For nonspherical particles, c could take different values [8]. For compressible liquids, which display also *volume viscosity* (friction depends on the speed of compression), the shear–viscosity dependence of τ is nonlinear and can even display a saturation curve [9].

For a particle of molecular weight 40 having radius of 1 Å immersed in a liquid with viscosity 1 cp at room temperature, we find, for $c = 6$, $\gamma = 2.9 \times 10^{13}$/s, which has an order of magnitude similar to that of the γ value obtained from Eq. (7).

c. Diffusion Coefficient and Solvent Viscosity A combination of Eqs. (6) and (8) gives the Stokes–Einstein relation,

$$D = \frac{k_B T}{c \pi r \eta} \tag{9}$$

Alternatively,

$$\frac{D\eta}{T} = \text{const} \tag{10}$$

B. Generalized Approach

1. What Is Wrong?

a. Are Random Force Fluctuations "Fast"? In the above-mentioned approach, the random force exerted by the environment is assumed to vary extremely rapidly compared with the variations of the particle velocity. The assumed *independent* collisions gives the motion its "kinky" characteristics. The extension of the concept "Brownian particle" to molecular structures requires some considerations.

Can a water molecule in water be considered as a "particle" and the surrounding water molecules as "neighbors"? To answer that, one should note that all water molecules are expected to display a similar type of molecular dynamics, and their intermolecular interaction is continuous and cannot be described as series of independent "kickings" [10]. Thus, the dynamic interactions that can be described as "random force fluctuations" cannot be claimed to be "fast" in comparison with the particle velocity. The situation is more complicated for a polymer (Fig. 4). It is justified to choose as a "particle" a relatively rigid part of the molecular chain, which displayed motion *uncorrelated* with that of the rest of chain and surrounding liquid [11], taken to be the "neighbors." The random force will thus include both contributions from the liquid molecules, which could display much faster movements than that of the "particle," and from normal modes of the polymer, which contain slower movements. We may conclude that random force fluctuations cannot be assumed to be "fast" in comparison with the particle movements, especially in hydrogen-bonded liquids and macromolecules, which are of main interest for biochemical studies.

b. Too High Values of Friction Coefficient Using Stokes law (Eq. (8)) the value of γ can be calculated. For glycerol at 20°, we obtain $\gamma = 10^{19}$/s, or $\tau_D = 10^{-19}$ s. These values are too short to be interpreted as "collision frequency" or "duration of a diffusion jump" for a single molecule. In fact, experiments showed that the diffusion jumps in glycerol at room temperature occur on a

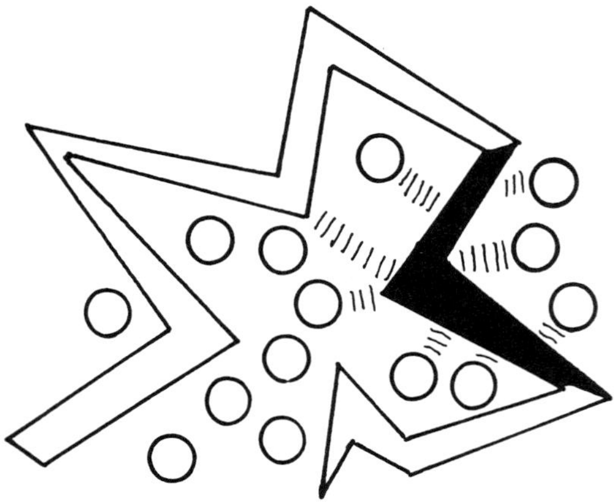

FIGURE 4 Brownian particle as part of a molecular structure: It is justified to choose as a “particle” a relatively rigid part of the molecular chain, which displays motion *uncorrelated* with that of the rest of chain and surrounding liquid molecules, taken to be the “neighbors” (interactions are marked by stacks of lines). The random force will thus include both contributions from the liquid molecules, which can display much faster movements than that of the “particle,” and from normal modes of the polymer, which contain slower movements. In this case, the assumption of “fast” random force is inapplicable.

time scale of 10^{-9} s [12,13]. Such 10 orders of magnitude differences cannot be overlooked.

A similar contradiction is reached using more general arguments. Stokes law predicts that γ increases with increasing solvent viscosity. As the solvent viscosity increases, however, Stokes law states that the faster movements of each solvent molecule are more strongly attenuated than its slower movements. Since the same solvent molecules create “the medium” for the molecule designated as “the particle,” the collision rate should decrease with increasing viscosity, contrary to the interpretation of γ as a “collision frequency.” Furthermore, in associated liquids, which display high viscosity, the strong intermolecular interactions correlate motions of neighboring molecules, which makes the assumption of “fast” random force unrealistic. These considerations suggest that for strong particle–medium interactions γ cannot be interpreted as a “collision frequency.”

c. ***Are Friction Coefficient and Solvent Viscosity Linearly Related?*** Stokes law (Eq. (8)) predicts a linear relation between the friction coefficient γ and the solvent viscosity η. Since the diffusion coefficient is a measurable quantity, the reciprocal relation between γ and the diffusion coefficient (Eq. (6)) enables us to investigate the relation between γ and η using the Stokes–Einstein relation

$D\eta/T$ = const (Eq. (10)). Stokes law was derived for a spherical particle diffusing in a *homogeneous* medium, which forms a *continuum* for practical purposes. It is not obvious, however, that this relation is applicable to a medium consisting of units of comparable or larger size than "the particle."

The relation $D\eta/T$ = const has been found to hold for a molecule or one species diffusing in a solvent of another species in many liquids [14], but not in all cases [15]. Deviations from this relation, however, have been observed for D measured with tracers at different pressures and temperatures [16,74].

A relation of the form

$$D\eta^{\alpha} = \text{const} \tag{11}$$

has been observed by Hiss and Cussler [17] using interferometry, in the diffusion of n-hexan and of anaphthalene in a series of hydrocarbon liquids. α was found to vary from 1/2 to 2/3 upon increasing the viscosity from the 1 cp range to the 5–5000 cp range. It has been concluded that α depends on the solute/solvent size ratio. Equation (11) fits data obtained by Fiorito and Meister [12] for the self-diffusion of pure glycerol, found by us to satisfy Eq. (11) with $\alpha = 0.74$ and 0.63 for 50°C and −10°C, respectively.

In studies of the viscosity effect on biochemical reactions, solvent viscosity was modified by the addition of viscous cosolvents of small molecular size (glycerol, ethylene, glycol, sucrose, glucose, and the like). In contrast, in physiological conditions, body fluid viscosity is determined by macromolecules (proteins, polysugars, and lipoproteins).

Figure 5, based on literature data, demonstrates how the Stokes–Einstein relation is violated in aqueous solutions of sucrose and glycerol (D corresponds to cosolvent). Equation 11 fits the data reasonably well with $\alpha = 0.32$ for sucrose and 0.71 for glycerol. Recently, we have systematically studied deviations from the Stokes–Einstein relation as a function of the cosolvent concentration and molecular size [20]. The results suggest that the origin of this phenomenon is the local inhomogeneity of the liquid around the diffusing species.

We may conclude that the friction coefficient does not vary linearly with the solvent viscosity; instead

$$\tau \sim \eta^{\alpha} \tag{12}$$

where $0 \leq \alpha \leq 1$. The zero limit is achieved, for example, for water molecules diffusing in an diluted aqueous solution of viscous macromolecules. The presence of the macromolecules markedly affects the bulk solvent viscosity, while it hardly affects the diffusion of water. α equals 1 when the particle size is larger than (a) the size of the solvent molecules, and (b) the average distance between the system components. When conditions (a) and (b) are fulfilled, the solution can be seen as a continuum. These considerations are of special interest for biological solutions, which contain all possible solute-to-solvent size ratios. It should be mentioned that

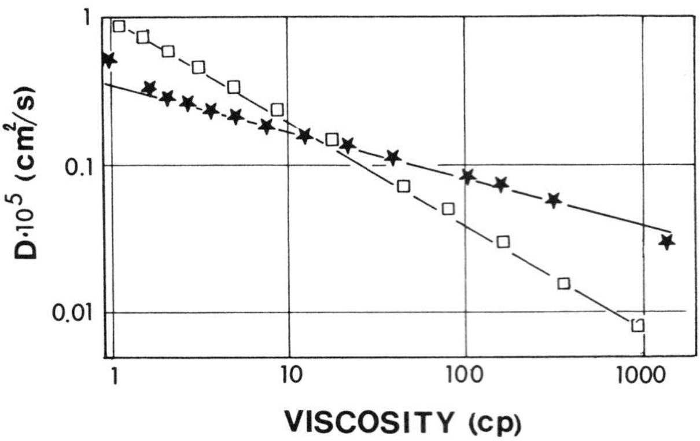

FIGURE 5 Deviations from the Stokes–Einstein relation: The diffusion coefficients of glycerol (L̇) and sucrose (⋆) in aqueous solutions at different concentrations (at 25°C) are plotted versus the solvent viscosity on a log–log scale. Linear regression gives slopes of −0.71 and −0.32, respectively, instead of the predicted −1 value (Eq. (10)). The diffusion data are from Landolt–Bornstein [18] and the viscosity values were obtained by interpolating literature data of Weast [19].

the situation in electrolyte solutions is much more complex, probably due to the ion–solvent interaction and the "dielectric friction" [21].

d. ***There Are Two Types of Solvent Viscosity*** Liquids display two types of "viscosity," depending on whether the stress applied to a liquid by an oscillating surface is parallel or perpendicular to the surface (Fig. 6). Parallel oscillations that are transmitted to the liquid are attenuated by the *shear viscosity* (η_s). Perpendicular oscillations involve both compression and shear. The compression wave causes volume and density oscillations. *Volume viscosity* (η_v) corresponds to the attenuation of this compression wave. η_v characterizes the *structural relaxation* associated with compression in the absence of volume-dependent chemical reactions [22]. In hydrogen-bonded liquids, volume and shear viscosity, measured by ultrasonic methods, were found to be comparable, probably because of the bond network, which responds similarly to both types of stress [22]. The relation between the friction coefficient and the solvent viscosities, including its frequency dependence, can be calculated theoretically from the hydrodynamics of compressible fluids [23]. This leads to nonlinear dependence of the friction coefficient on the shear viscosity [9]. We may conclude that shear viscosity contributes only part of the structural relaxation associated with the friction.

e. ***Solvent Viscosity and Friction Coefficients Are Time-Dependent*** Stokes law has been derived for *viscous* but *incompressible* fluids. Realistic liquids are *viscoelastic*, that is, displaying *viscosity* under the action of slowly vary-

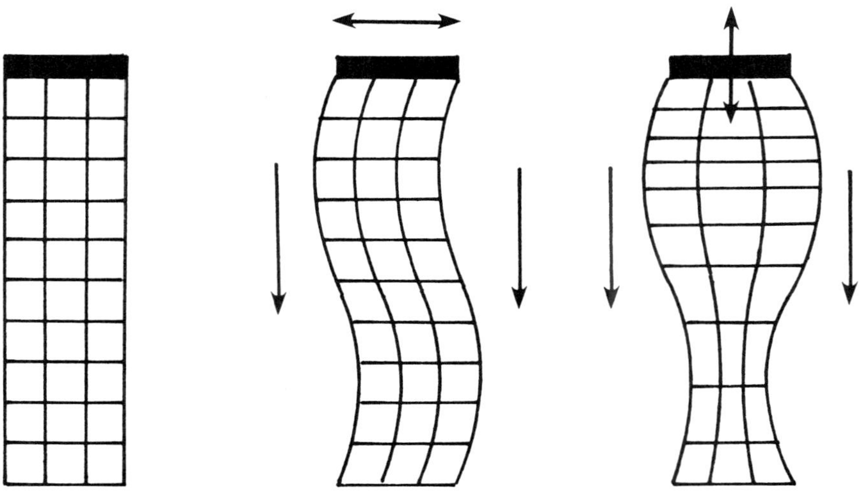

FIGURE 6 Shear and volume viscosities: A plate (bar) is found in contact with a medium (grid). On the left, the system is illustrated in its equilibrium state. When the plate vibrates parallel to its plane (middle), it excites a shear wave, which propagates in the liquid with no volume change. On the other hand, when the plate vibrates perpendicularly to its plane (right), the excited wave includes both compression and shear. The shear and volume viscosities are associated with the liquid internal friction, which attenuates the shear and compression wave, respectively.

ing stresses (e.g., the stream around a falling ball) and *elasticity* under high-frequency stresses (e.g., sound wave transmission). Thus, both viscosity and elasticity of liquids are frequency-dependent, as shown in Fig. 7. Perturbing the liquid periodically changes its behavior continuously from "liquid like" to "solid like" as the applied frequency increases. The "transition" occurs at the frequency $1/T_{\mathrm{VE}}$, where T_{VE} is the *viscoelastic relaxation time*. T_{VE} characterizes the recovery process of the medium after the application of a pressure jump. Typical values of T_{VE} for viscous liquids are 10^{-12}–10^{-9} s.

The microscopic picture of liquids viscoelasticity was first described by Frenkel [24]. He proposed that the motion of a molecule in a liquid consists of oscillations around an equilibrium position, that is, a *solid like* motion, and then a random jump into a new equilibrium position. The whole process looks like a diffusion jump characterizing a *liquid like* behavior (Fig. 8). For observation times much longer than the time between two successive jumps (T_{D}), the molecule will appear to perform Brownian motion. As a result, the response to stresses that vary on a time scale $\gg T_{\mathrm{D}}$ will be a viscous flow. On the other hand, if the stress varies on a time scale $\ll T_{\mathrm{D}}$, then the molecule will be found most of the time performing a solid

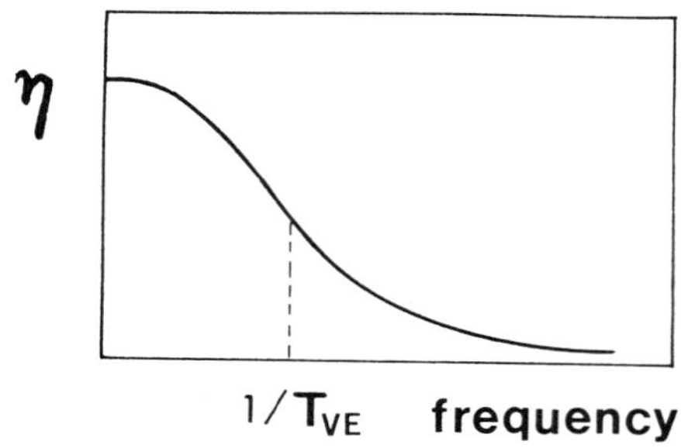

FIGURE 7 Viscoelasticity of liquids: Perturbing a liquid periodically changes its behavior continuously from "liquidlike" to "solidlike" as the applied frequency increases. The "transition" occurs at the frequency $1/T_{VE}$, where T_{VE} is the *viscoelastic relaxation time*. T_{VE} characterizes the recovery process of the medium after the application of a pressure jump.

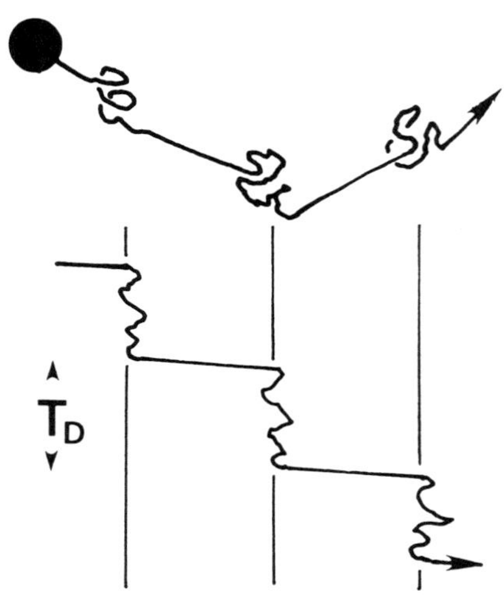

FIGURE 8 The microscopic picture of liquids viscoelasticity: It was proposed by Frenkel [24] that the motion of a molecule in a liquid consists of oscillations around an equilibrium position, that is, a solidlike motion, and then a random jump into a new equilibrium position. The figure illustrates this motion in space (top) and time (bottom). The whole process looks like a diffusion jump of duration T_D characterizing a liquidlike behavior, although most of the time could be spent at solidlike motion. T_D can be identified with the viscoelastic relaxation time T_{VE} (see Fig. 7).

like motion. This suggests that in real liquids the time between two successive diffusion jumps T_D can be identified with the viscoelastic relaxation time T_{VE}.

Viscoelasticity means, in the time domain, that if a particle starts moving at a constant speed, it takes time for the friction force to build up, as illustrated in Fig. 9 by the motion of a boat. We may conclude that the fiction coefficient in real liquids is time-dependent!

The generalization of Stokes law for a time-dependent friction $\zeta(\tau)$ is

$$-m\gamma v(t) \rightarrow -m \int_0^t \zeta(\tau) \cdot v(t-\tau)\, d\tau \tag{13}$$

For random forces, which vary much faster than the particle velocity (Fig. 10),

$$\zeta(\tau) = \gamma\delta(\tau) \tag{14}$$

The substitution of Eq. (14) in the integral of Eq. (13) reduces it to the "old" Stokes law. In other cases, the friction coefficient is "smeared" over a finite time interval (Fig. 10). At the frequency domain, a friction coefficient given by Eq. (14) has a very wide ("white") frequency spectrum. The finite response time in all other cases reflects a frequency response of a finite width. The lack of friction at higher frequencies is compatible with "viscoelasticity"; the Fourier transform of function $\zeta(\tau)$ can be identified with the frequency-dependent viscosity (friction) in liquids (Fig. 7).

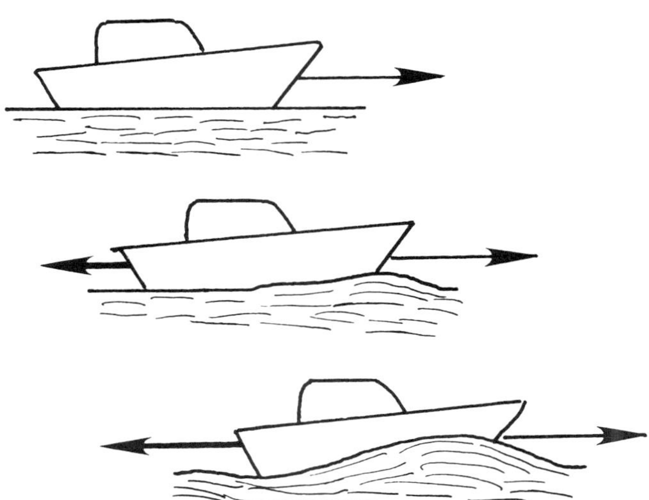

FIGURE 9 The time-dependent friction force: Illustrated here is the building up of the resistance (thick arrow) to the motion of a boat that starts moving from rest ($t = 0$) with a *constant* velocity (thin arrow).

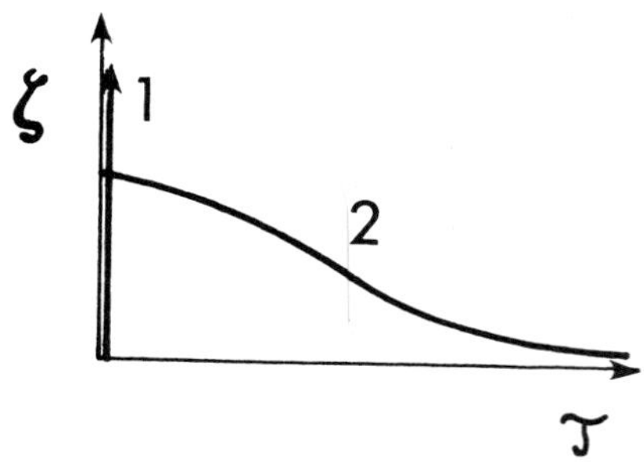

FIGURE 10 The time-dependent friction coefficient $\zeta(\tau)$: This coefficient can be shown to reflect the random-force correlation function (Eq. (16)); that is, for the classical Brownian theory, the random force has no correlation to itself for nonzero τ, which gives a delta function (curve 1, Eq. (14)). If the random force has a "memory," $\zeta(\tau)$ looks like curve 2. As a function of frequency, it reflects, in liquid, the frequency-dependent viscosity depicted in Fig. 7.

2. Generalized Langevin's Equation

Using Eq. (13), Langevin's equation (Eq. (1)) can be generalized to include time- (or frequency-) dependent friction [7]:

$$m\dot{v}(t) = -m \int_0^t \zeta(\tau) v(t-\tau)\, d\tau + R + f \tag{15}$$

The corresponding generalization of the fluctuation-dissipation theorem (Eq. (3)) is

$$\zeta(\tau) = \frac{\langle R(0)R(\tau)\rangle}{k_B T} \tag{16}$$

Equation (16) gives some insight into the nature of the time-dependent friction. It shows that the correlation time of the random force is the decay time of the friction coefficient. For example, if the dominant feature in the *random force* is the rotation of neighboring molecules, characterized by a rotational correlation time T_r, then the *friction force* acting on the "particle" at time t will include contributions from the *particle velocity* back to the time $t - T_r$.

C. Free Volume

1. Relation to Viscosity and Diffusion

When liquids reach the density characterizing the glass transition, almost no diffusion is possible. It is therefore not surprising to find that molecular motions in densed liquids are sensitive to the molecular packing, as expressed by the "free volume" V_f, that is, the excess volume per molecular over its "glass" value. In the category of densed liquids, one can find also hydrogen-bonded liquids, such as aqueous solutions of viscous cosolvents like glycerol or sucrose. Proteins are likely to fall in this category, as they show glassy behavior [25], including glass transition [26] and structural-dynamics similarity to glycerol [27,28].

With high pressures, density can be kept constant at different temperatures. It is thus not surprising to find that the main temperature dependence of both viscosity and diffusion in densed liquids can be attributed to thermal expansion. Concepts like *activation energy* for diffusion and viscosity for these liquids stem from erroneous interpretation of the Arrhenius plot (thermal dependence) of these properties, which is valid only in a narrow temperature range, while non-Arrhenius behavior is observed over a wide range of temperatures [29,30].

Under *isothermal* conditions, the phenomenological expression

$$\eta \sim \exp\left(\frac{BV_0}{V_f}\right) \tag{17}$$

where V_0 is a reference volume, was found by Doolittle [31] for hydrocarbons with B (an adjustable parameter) close to a unity. Since V_f is proportional to $T - T_g$ (T_g = glass transition temperature), Eq. (17), if expressed as a function of temperature, fits the non-Arrhenius behavior observed for densed liquids, as derived theoretically [32]. Equation (17) was corrected for the presence of activation energy E_0 by Macedo and Litovitz [33] and, later, for the temperature dependence of V_0 and the density dependence of E_0 [34].

Using the free-volume concept, Cohen and Turnbull [35] derived theoretically an expression for the diffusion constant based on calculating the probability of finding a vacancy of critical (or minimal) free volume V_0 when the *average* free volume is V_f. This expression is

$$D \sim \exp\left(\frac{-bV_0}{V_f}\right) \tag{18}$$

where $0 < b < 1$.

Equations (17) and (18) can be combined to give, under *isothermal* conditions, the relation

$$D\eta^\alpha = \text{const} \tag{19}$$

where $0 < \alpha = b/B <$ about 1, as found experimentally (see Eq. (11)).

2. Relation to Structural Rearrangements

Following the Cohen and Turnbull approach (Eq. (18)) and Doolittle expression (Eq. (17)), Gegiou et al. [5] suggested that structural rearrangements of a solute at rate k require only a fraction $p < 1$ of the critical free volume V_0 required for the translatory motion of a solvent molecule. Thus,

$$k \sim \frac{1}{\eta^p} \tag{20}$$

III. BARRIER CROSSING

A chemical reaction can be described as a process of overcoming an energy barrier from the reactant state to the product state. Behind this description stands a process of energy exchange between the reactants and the medium. This section is concerned with models that relate reaction rate to the "profile" of the energy barrier and to the molecular dynamics of the medium. The models' assumptions and limitations are discussed.

A. Basic Concepts

Figure 11 shows two examples of reactions: atom transfer (Fig. 11A) and isomerization reactions (Fig. 11B), for which the "reaction coordinates" are, respectively, distance and angle. The reaction is taken to be unidirectional—starting in the "reactant state" (*R*) and then crossing the barrier. After leaving the barrier region, the reaction is assumed to reach the "product state" (*P*) without coming back. For our

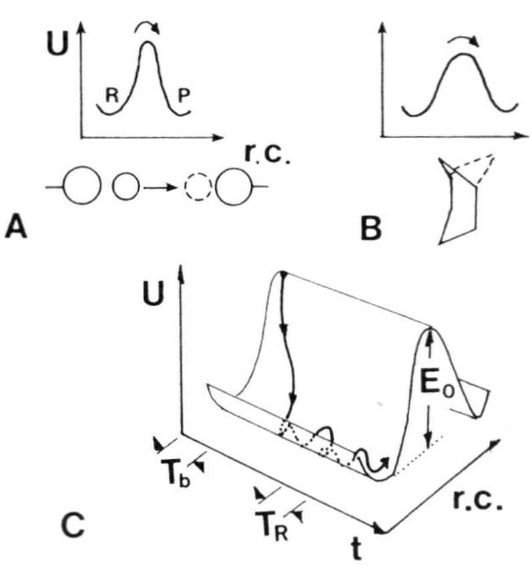

FIGURE 11 Chemical reactions and energy barriers: The process is described as a motion along a reaction coordinate (r.c.). Examples of r.c. are (A) interatomic distance for atom transfer reactions and (B) an angle for the isomerization of a carbon ring. For modeling purposes, the reaction is unidirectional, that is, it flows from a fully occupied "reactant state" (*R*) to an empty "product state" (*P*). Part (C) depicts the parameters relevant for the models, which are the height of the energy barrier E_0, the time T_b required for the system, if "placed" initially at the barrier top, to become "far from top," and the time T_R, which characterizes oscillation time around the R state. $w_b = 2\pi/T_b$ is a useful parameter, too.

purposes, the potential energy barrier can be characterized by the following parameters (Fig. 11C):

E—the barrier height.
T_b—a typical time required for falling from the barrier top to a region that can be identified as "far from top" to a region that can be identified as "far from top." $w_b = 2\pi/T_b$ is the corresponding frequency.
T_R—a typical oscillation time around the reactant state.

T_b and T_R can be calculated if the potential energy curve is known.

B. Models

1. The Transition State Theory

The transition state theory (TST) [36] assumes that the collisions of the reactants with the surrounding medium maintain an equilibrium at R and at the barrier top. In spite of that, the barrier top is crossed only once before the reaction reaches P. The reaction dynamics is illustrated in Fig. 12. The reaction rate k_{TST} is given by

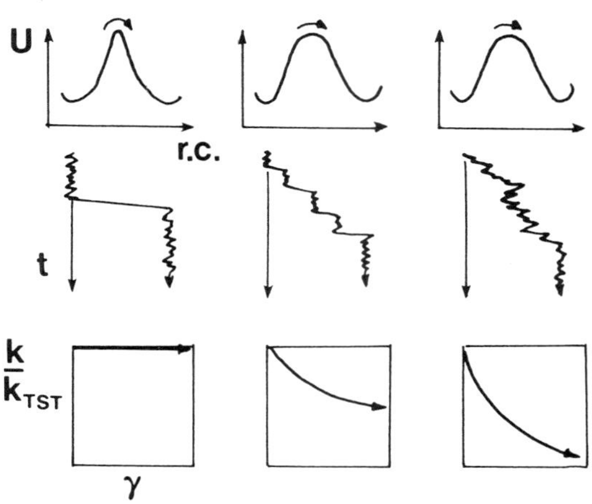

FIGURE 12 Models for chemical reactions: Depicted are the energy diagram (top), the motion along the reaction coordinate (middle), and the reaction rate k (bottom, in units of k_{TST} = the k value for the TST theory), as a function of the friction coefficient (zero-frequency component), which can be identified with the solvent viscosity. This is done for the following reaction models: transition state theory (TST) (left), Grote–Hynes's theory (middle) and Kramers's theory (right). The dynamic features of the trajectories during a transition are "undisturbed" for TST, "diffusive" for Kramers, and "mixed" for GH. The functional dependence of k/k_{TST} on γ is 1 for TST and $1/\gamma$ for Kramers. GH displays an intermediate case between these two extremes.

$$k_{\mathrm{TST}} = \frac{1}{T_{\mathrm{R}}} \exp\left(\frac{-E_0}{k_{\mathrm{B}}T}\right) \tag{21}$$

Equation (21) has been derived from equilibrium thermodynamics.

The problem with the TST is that if collisions with the environment take place at the barrier top there is no reason why the barrier could not be recrossed.

2. Kramers's Theory

Kramers [1] assumed that the barrier crossing can be described as a Brownian motion in the presence of an energy barrier. The barrier top is assumed to be parabolic as a function of the reaction coordinate. Its radius of curvature is characterized by T_{b}. The reaction rate k is given by

$$\frac{k}{k_{\mathrm{TST}}} = \left[1 + \left(\frac{\gamma}{2w_{\mathrm{b}}}\right)^2\right]^{1/2} - \frac{\gamma}{2w_{\mathrm{b}}} \tag{22}$$

where τ is a *constant friction coefficient*. The reaction dynamics is illustrated in Fig. 12.

The *low-friction limit* of Eq. (22) is obtained for $\gamma/2w_{\mathrm{b}} \ll 1$. The collisions with the medium fail to affect the reaction path and $k \to k_{\mathrm{TST}}$.

The *high-friction limit* of Eq. (22) is obtained for $\gamma/2w_{\mathrm{b}} \gg 1$, giving

$$\frac{k}{k^{\mathrm{TST}}} = \frac{w_{\mathrm{b}}}{\gamma} \tag{23}$$

This limit is also called the *diffusive limit* or the *Smoluchowski regime*. Its importance is in the explicit dependence of the reaction rate on the friction with the medium. The dependence of k/k_{TST} on γ is plotted in Fig. 12. Following Stokes law, γ is proportional to the solvent viscosity. Thus, in contrast to the TST, Kramers's theory predicts that reaction rates vary inversely with the solvent viscosity.

The conceptual problem with Kramers's approach stems from the assumption that γ is frequency-independent. This means the random force exerted by the medium on R has components of the same amplitude at any frequency ("white spectrum"). On the contrary, liquids and liquid like materials display viscoelastic and dielectric relaxation at finite frequencies (see Sec. II.B.1.). This limitation becomes a problem for the following reason: Let us suppose that the time spent in the *barrier region* while crossing is T. Thus, the random force can affect the barrier crossing process only by components at frequencies higher than $1/T$. We may conclude that if the medium relaxation time is larger than or comparable with the barrier-crossing time, then Kramers's expression is unjustified and underestimates the reaction rate, because only part of the friction force is effective.

3. Grote–Hynes's Theory

a. Introduction The GH theory [4,37] forms a conceptual link between the TST and Kramers's models by taking the friction force to be time-dependent. The GH theory shows that when the temporal behavior of the random force is much slower than T_b, the TST expression (Eq. (21)) is valid. On the other hand, when the random force fluctuates are *fast* on the T_b scale, Kramers's expression is valid (Eq. (22)). In the general case, GH theory predicts that the dependence of the reaction rate on the friction at low frequencies is *weaker* than that predicted by Kramers.

It is the "zero-frequency" component of the friction coefficient that replaces the constant friction coefficient in Stokes law, which is proportional to the "viscosity."

Following are some more details of the GH theory. Readers who are less interested in these theoretical implications are advised to focus on Fig. 12 and then to skip to subsection c.

b. Analytic Considerations The GH theory applies the generalized Langevin equation (Eq. (15)) to the problem of barrier crossing, taking into account a possible time dependence of the friction coefficient. The result is

$$\frac{k}{k_{\mathrm{TST}}} = \frac{\gamma_r}{W_b} \tag{24}$$

where γ_r is the *reactive frequency*, which reflects the decay of the system from the barrier top to the state of products. γ_r is given by the *self-consistent* relation

$$\gamma_r = \frac{w_b^2}{\gamma_r + \hat{\zeta}(\gamma_r)} \tag{25}$$

where $\hat{\zeta}(\gamma_r)$ is the Laplace transform of the time-dependent friction coefficient $\zeta(t)$ at the frequency γ_r:

$$\hat{\zeta}(\gamma_r) = \int_0^\infty dt \exp(-\gamma_r t)\, \zeta(t) \tag{26}$$

If the force fluctuations are so fast that $\zeta(t) \propto \delta(t)$, then $\hat{\zeta}(\gamma_r) = \hat{\zeta}(0)$, which could be identified with γ, the constant friction coefficient used in the ordinary Langevin equation. For that case, Eq. (25) can be solved for γ_r and gives, using Eq. (24), the Kramers result (Eq. (22).

Some insight into the results can be achieved by using the diagram depicted in Fig. 13. The X axis is identified with k/k_{TST} ($= \gamma_r/w_b$, see Eq. (24)). The Y axis describes $\hat{\zeta}(\gamma_r)/w_b = \hat{\zeta}(Xw_b)/w_b$. The intersection of this curve (plotted in four hypothetical versions, numbered $1, 2, 3, 4$) with $1/X - X$ represents a solution of Eq. (25). Two limiting cases are of special interest:

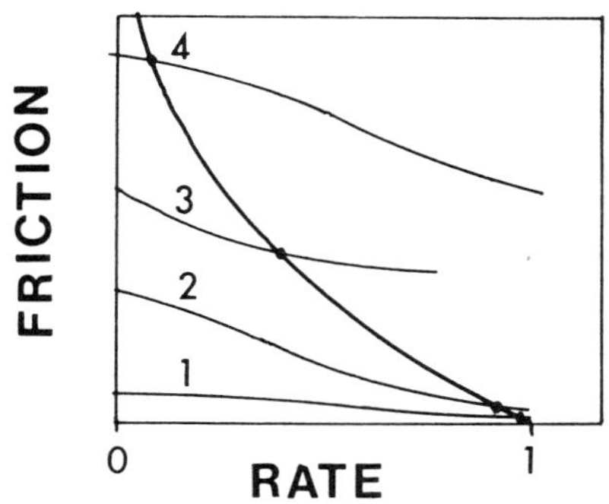

FIGURE 13 Results of the Grote–Hynes theory: The plot describes how to find the "reactive frequency" γ_r, to which the reaction rate is proportional (Eq. (24)). The "RATE" variable X equals $\gamma_r/W_b = k/k_{TST}$. $1/X - X$ is a curve (thick line) with which the "FRICTION" curve $\hat{\zeta}(\gamma_r)/W_b$ intersects (curves 1, 2, 3, 4). $\hat{\zeta}(\gamma_r)$ is the Laplace transformed $\zeta(\gamma)$. Curves 1 and 2 give the TST result, where 1 corresponds to a "sharp barrier" and 2 to "small friction." Curve 3 represents the "general" case, and curve 4 the "high-friction" case, that is, Kramers's result.

(i) Transition rate approaches its TST value (curves 1 and 2). When γ_r is very close to w_b, $\hat{\zeta}(\gamma_r) \ll w_b$ and the friction forces at the frequency w_b can hardly affect the barrier crossing process. The relevant situations are the following:

a. Friction forces are small (curve 1).
b. The friction force could be large at low enough frequencies, but its w_b component is relatively small (curve 2). This happens when the barrier is *sharp* (high w_b), as illustrated in Fig. 12 for the TST barrier shape. Thus, the time required for crossing the barrier in short in comparison with the relaxation times of the environment, and k is reduced to its TST value (Fig. 12, TST trajectory).

(ii) Only the zeroth component of the friction force is relevant. In that case, shown by curve 3 in Fig. 13, $\hat{\zeta}(\gamma_r)$ can be approximated by $\hat{\zeta}(0)$, and using Eq. (25) the expression for k/k_{TST} is reduced to the Kramers expression given by Eq. (22) with $\gamma = \hat{\zeta}(0)$ (see Fig. 12). Equation (26) shows that in order to have a case like this, $\zeta(t)$ has to decay much faster than $\exp(-\gamma_r t)$. Since the relevant range of γ_r is smaller than w_b, this means the random force fluctuates fast on time scale of T_b, resulting in diffusionlike motion along the reaction coordinate. This could be relevant to reactions that take place in *viscous* solvents, for which $\zeta(0)$ is associated with the solvent viscosity. Another interesting case is when *polar* solvents interact strongly with the reacting system. Here, the solvent dynamics dominates the motion along the reaction, and $\zeta(0)$ can be shown to be the reciprocal value of the dipole reorientation time [37,38]. The motion can described as a diffusion (see the Kramers case in Fig. 12), where the friction is of dielectric origin.

c. ***Implications*** There is an objective difficulty in testing the GH theory experimentally, since it requires the knowledge of (a) the explicit time dependence of the friction coefficient, and (b) the functional relation between the friction coefficient and the solvent viscosity. Modeling of these relation is not an easy task [39,40]. The possible specificity of the time-dependent friction with respect to the reaction type makes the problem even more difficult.

GH theory suggests that for reaction kinetics, the local friction coefficient is more relevant than viscosity. Since diffusion is linked with the friction coefficient, a frequency-dependent diffusion coefficient [7] seems to be a useful concept to bridge between molecular dynamics and reactions. Experimental tests with nonbiological systems [9,41–44] and with molecular dynamics simulations [40,41] support the point of view (not necessary associated with a specific theory, as pointed out by Chandler and Kuharski [45]) that the TST and Kramers models represent two extremes, between which fall a more realistic description of reaction kinetics, such as the GH model.

IV. VISCOSITY EFFECT

Solvent viscosity can be modified, under isothermal conditions, by adding viscous cosolvents. In that way, kinetic coefficients could be studied, as a function of the solvent viscosity, for a variety of biochemical and other reaction systems. The relation between viscosity and free volume in densed liquids, discussed in Sec. II, suggests ultrasonic absorption that probes volume-dependent structural dynamics could be a valuable tool in studying the solvent viscosity effects. These topics are discussed in relation with the observed power-law dependence of the rate of chemical reaction and structural rearrangements on the solvent viscosity.

A. Kinetic Studies

Most studies of solvent viscosity effects on biochemical processes have used viscous cosolvents for the elevation of the solvent viscosity. These mixed solvents have been shown to reduce kinetic coefficients of ligand-binding protein interactions and subsequent biochemical processes, such as peptide hydrolysis by carboxypeptidase A [2,28], binding of O_2 or CO to respiratory proteins [46–49], some steps in the bacteriorhodopsin cycle [50], kinetics of lactate dehydrogenase [51], migration of small molecules through the structure of myoglobin using viscous macromolecular cosolvents [52], ester hydrolysis by subtilisin BPN′ [53], and the activity of cell surface phospholipase A_2 [54,55], as well as membrane lipid metabolism [54,56] and cellular secretion [57]. A similar trend has been observed in the viscous cosolvent effect on hydrogen isotope exchange in lysozyme [58,59]. These observations demonstrate a deviation from the viscosity dependence predicted by Kramers's theory, $1/\eta$, and were fitted to the following phenomenological relation [46]:

$$k \sim \left(\frac{1}{\eta^p} + A\right) \cdot \exp\left(-\frac{E_0}{RT}\right) \tag{27}$$

where k is the kinetic coefficient, E_0 is the energy barrier of the reaction, and p and A are adjustable parameters. p was found to decrease as the energy barrier (or "gate") was less accessible to the solvent [46]. Except for one case, $p \leq 1$ was found to fit the data of the above-mentioned works. The only $p > 1$ case was observed in hydrogen isotope exchange in lysozyme [59] and deserves some discussion. p was found to increase from 0.6 to 1.4, proportionally to E_0, as the number of hydrogen remaining unexchanged decreased. The very slow process observed could result from a number of correlated or simultaneously occurring events, with each possibly described by Eq. (27) (with $A = 0$). If the joint probability of the process is k^n, then the observed parameters of Eq. (27) would be $p' = np$ and $E_0' = nE_0$. Here n increases as less accessible hydrogens are involved. According to this interpretation, $p > 1$ does not reflect the overcoming of a single energy barrier.

In the light of the violation of Einstein's relation (Eq. (10)), we may conclude that the solvent viscosity and the diffusion constants of water (D_W) and the cosolvent (D_C) should be tested for correlation with the kinetic coefficients as independent variables. Such an approach could give a deeper insight into the nature of the coupling between the structural fluctuations of the protein and the solution. D_W, for example, will be best correlated to k if the functionally important structural fluctuations involve mainly the exchange of water molecules with the bulk.

B. Ultrasonic Studies

The dependence of kinetic coefficients on solvent viscosity is an indirect way of probing protein dynamics. In contrast, ultrasonic absorption measurements in protein solutions can probe "internal friction" of the protein structure itself or, more precisely, volume-dependent relaxation processes. What is the significance of such processes? Since many scientists are not familiar with the application of ultrasonic methods for probing structural dynamics, it is worthwhile to provide the reader with some necessary background.

1. Basic Concepts

A propagating compression wave causes a molecular system to change its packing density periodically (Fig. 6). The energy dissipated in this process is associated, in principle, in one of the following relaxation mechanisms:

Structural relaxation. This is caused by the friction associated with the relative motion of the molecular components [22]. It should be mentioned that there is a close similarity between dielectric relaxation and structural relaxation that stems from the orientational changes that accompany volume changes in asymmetric molecules.

Thermal relaxation. This involves the energy released by shifting a chemical equilibrium in a way that follows the volume changes in accord with Le Chatelier's principle [60].

The spatial attenuation of an ultrasonic wave reflects a relaxation process, and it is quantified by a dimensionless number μ called *ultrasonic absorption per wavelength.* μ measures the fraction of ultrasonic energy dissipated per cycle (or within a spatial distance of one wavelength). Since a relaxation process has a characteristic time, or a number of characteristic times, μ depends on the applied frequency and reaches a peak at the "relaxation frequency" = 1/("characteristic time"). Hereafter, we shall refer by UA to ultrasonic absorption, whether regular, or excess, or per wavelength.

For a solution measured at a number of solute concentrations, one can obtain the *excess absorption* $\delta\mu$; of the solute itself over that of the pure solvent.

2. Ultrasonic Absorption of Proteins in Solutions

For associated liquids like glycerol that are currently used as viscous cosolvents, most of the UA reflects structural relaxation and is proportional to the solvent viscosity [22]. There is a similarity between known features of protein dynamics and glasses, or associated liquids, as pointed out in Sec. II.C.1. This similarity makes the UA measurements in protein solution, regarding the solvent viscosity effects on the internal friction of proteins, of particular interest.

Proteins in solutions reveal large UA around the neutral pH region in the range 1–10 MHz. Its origin is still puzzling. The relevant observations are the following:

1. It is associated with the tertiary structure of the proteins, since it disappears upon denaturing [61].
2. The relaxation is of structural origin. $\delta\mu$ is proportional to the protein compressibility and is associated with volume fluctuations [62].
3. Neither proton transfer [63] nor helix-coil transitions [64] are likely to be the origin of that excess absorption.

On the other hand, the following evidence suggests that hydration dominates the UA of proteins and of some macromolecules.

4. UA of some macromolecules is likely to have a contribution from a coupling between thermal motion of the polymer segments and water molecules that changes the strength of their interaction with the polymer as a result of its motion [65].
5. Starting with dried macromolecules, UA increases with increasing the water content [65].
6. Measurements of dielectric relaxation in protein films demonstrated that an increase of hydration to a range in which the biological function of the protein was restored is associated with a relaxation process in the

1–10 MHz region [66]. As stated above, dielectric relaxation probes molecular dynamics features that are closely related to the structural relaxation measured by UA [22].

3. Cosolvent Effect on Ultrasonic Absorption

Following the prediction that solvent viscosity modulates the internal viscosity (or friction) of a protein [67], the UA of bovine serum albumin (BSA) was measured in aqueous solutions of glycerol, sucrose, and ethylene glycol at the frequency range of 3–4 MHz [68]. A power law of the form

$$\delta\mu \sim \eta^{\varepsilon}$$

with $\varepsilon = 0.47 \pm 0.07$ was found for *all* tested cosolvents (including also a 1 *M* NaCl solution), which is compatible with the above prediction. On the other hand, $\delta\mu$ data for different cosolvents failed to fall on the same curve if plotted versus the cosolvent weight fraction, the activity, or the dielectric constant of the solution. We may conclude that a *dynamic* variable (viscosity), and not *thermodynamic* ones, dominate the observed cosolvent effect on the UA of BSA.

A similar conclusion was drawn from measuring cosolvent effects on the exchange kinetics of the fast-exchanging Trp-63 indole proton of lysozyme [58], on hydrogen exchange kinetics and fluorescence quenching at indole side chains [69], and on hydrogen isotope exchange rates in lysozyme [59]. In these works, the relevance of a number of thermodynamic variables was investigated.

It should be stressed that the validity of the foregoing conclusions is limited, in principle, to *nonequilibrium* properties of proteins. Cosolvents have pronounced effects on the *equilibrium* properties of proteins, amongst them average conformation, stability, hydrophobicity, and hydration. Preferential hydration is a fruitful concept in that respect [70]. The relevance of a dynamic or thermodynamic variable for the description of a nonequilibrium or equilibrium phenomenon for a change in the system that affects both is illustrated schematically in Fig. 14.

What could be the origin of the observed solvent viscosity effect of the UA of proteins (at least BSA)?

Since protein hydration seems to contribute significantly to the protein UA, should one ask what kind of water is involved in that effect?

Both water tightly bound to charged residues by electrostriction and water loosely bound to the protein, which is accessible by salt solution, were found not to contribute significantly to the effect in BSA [68].

This elimination and observations (1)–(6) above favor the special water species that acts as "lubricant" or "plasticizer" and provides the protein with the conformational mobility vital for its function. More specifically, this "lubrication" is attributed to the ability of certain water molecules to form intermittent hydrogen bonds between residues belonging to different but adjacent parts of the

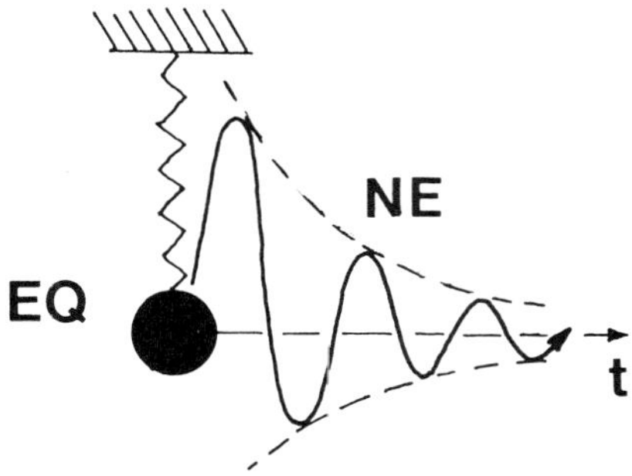

FIGURE 14 Cosolvent effects on equilibrium and nonequilibrium properties of proteins: The cosolvent effects on the viscoelastic properties of proteins, which can affect at least ultrasonic measurements, can be illustrated schematically in the following way: The cosolvent can change the protein rigidity by thermodynamic-related phenomena, for example, preferential hydration. In addition, cosolvents can affect the internal viscosity of the proteins. A selected part of a protein is represented by a mass connected to a spring, and its local rigidity and internal viscosity are represented by the spring constant and the medium viscosity, respectively. The position of the mass corresponds to the protein density. The presence of cosolvent can cause a density shift (equilibrium change, EQ) and a change in the attenuation of the mass oscillations (nonequilibrium change, NE). What effect dominates depends on what property is probed. Free-volume theories and experiments suggest that a change of packing density in densed liquids is expected to affect attenuation-related phenomena much more than rigidity-related ones.

polypeptide chain [71] (Fig. 15). "Lubricant" water prevents intrachain interactions between residues, which can stabilize a specific conformation. The important role of this water species in catalysis under high cosolvent concentrations was studied and discussed by Zak and Klibanov [72].

How could "lubricant" water contribute to the observed cosolvent effect on the UA of proteins?

The following mechanism was suggested [68] (Fig. 16): It is well known from the Brownian theory that the effect of viscosity on the thermal fluctuations of a molecular chain is to shift its spectrum of motion to lower frequencies (the total kinetic energy is conserved). Since segmental motion of a polymer chain in the MHz chain is considered as "slow," we expect the elevation of the solvent viscosity to intensify the chain motions and volume fluctuations, to which the "lubricant" water is coupled (Fig. 16). Since this water species is a source of volume-dependent friction, the final result will be a higher UA of structural origin.

This suggestion does not give any quantitative indication for the observed power law. It, however, rationalizes a modulation of internal protein friction by the solvent viscosity.

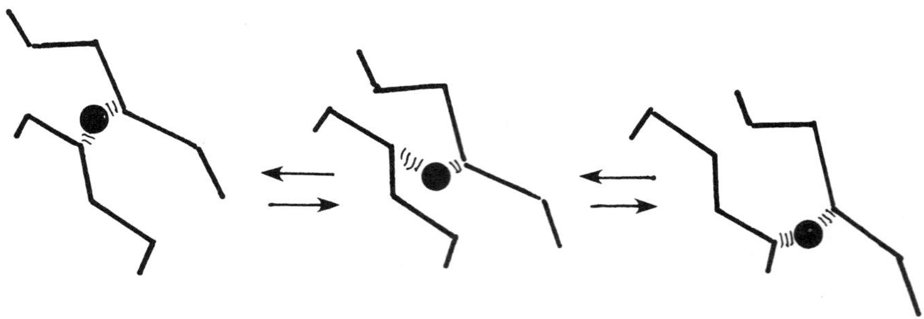

FIGURE 15 "Lubricant" water: Following Chirgadze and Ovsepyan [71], this "lubrication" is attributed to the ability of certain water molecules to form intermittent hydrogen bonds between residues that belong to different but adjacent parts of the polypeptide chain, thus preventing intrachain interactions between residues, which can stabilize a specific conformation. "Lubricant" water plays an important role in protein function by maintaining the conformational mobility [72].

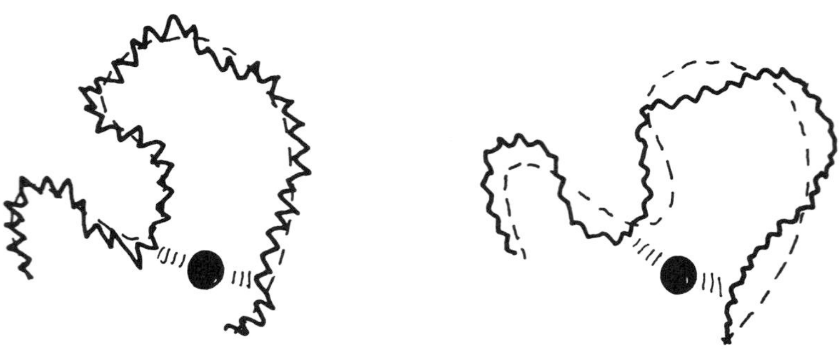

FIGURE 16 A suggested mechanism for the solvent viscosity effect on the ultrasonic absorption: The elevation of solvent viscosity shifts the spectrum of the polypeptide chain movements to lower frequencies. The resulting increase in the amplitude of low-frequency segmental motion intensifies the dissipation caused by the "lubricant" water dynamics, which is coupled to the chain motions. The figure sketches hypothetical motions of a molecular chain with dominant high-frequency motion (left, low viscosity) and dominant low-frequency motion (right, high viscosity). A "lubricant" water molecule is shown as well. The equilibrium conformation is marked by a dashed line.

V. WHY A POWER LAW?

Although a number of explanations have been suggested for the cosolvent effect on proteins [67,73], one should remember that he same law has been observed in simple nonbiological reaction systems [9,41–44]. Therefore, the search for its origin should start on a more basic level. A simple explanation of this law was given by Gegiou et al. [5] on the basis of the free-volume concept (see Sec. II.B.3). They have shown the following:

The rate of structural rearrangements k in a solute surrounded by solvent molecules is $k \sim 1/\eta^p$ (Eq. (20)), where $0 < p < 1$ is the fraction of the critical (minimal) free volume required by a solvent molecule for translatory motions.

This description is compatible with the concepts suggested by Grote–Hynes theory (Sec. III.B.3): If the solute follows accurately the structural rearrangements in the solvent, that is, maximum "effective collisions," the solute is expected to display a diffusionlike motion. This corresponds to the $k = \sim 1/\eta$ prediction of Kramers, and $p = 1$ here. On the other hand, if the solvent is able to rearrange without any critical volume, it means that there are no solvent–solute "collisions" during that rearrangement process, and thus no viscosity dependence of k is expected ($p = 0$). GH theory predicts an intermediate behavior between these two extremes, as shown in Fig. 12, but not quantification in terms of solvent viscosity was possible. Here, the observed intermediate behavior $1/\eta^p$ is a straightforward result. These two approaches are not easily comparable, however, since the basic assumption systems are completely different.

VI. CONCLUSIONS

1. Accumulating evidence suggests that solvent viscosity affects the kinetics of proteins and simple reaction systems and also structural rearrangements that are associated with the solute–solvent packing dynamics.
2. Viscous cosolvents affect the ultrasonic absorption of proteins, probably reflecting "lubricant" water, which is functionally important. This interesting possibility deserves further study.
3. The diffusion constant of water and the cosolvent should be tested as independent correlates in addition to the solvent viscosity.
4. The cosolvent effect on the equilibrium properties of proteins should be evaluated in addition to the nonequilibrium properties, in order to be able to find which properties dominate the observations.
5. A theoretical linkage between the Grote–Hynes theory of chemical reactions and the Gegiou–Muszkat–Fischer approach to structural rearrangements could contribute to a better understanding of the effects of solvent viscosity on reactions systems.

REFERENCES

1. Kramers, H. A. (1940). Brownian motion in a field of force and the diffusion model of chemical reactions. *Physica (Utrecht)* *7*: 284–304.
2. Gavish, B. (1978). The role of geometry and elastic strains in dynamic states of proteins. *Biophys. Struc. Mech.* *4*: 37–52.

3. Gavish, B., and Werber, M. M. (1979). Viscosity-dependent structural fluctuations in enzyme catalysis. *Biochemistry 18*: 1269–1275.
4. Grote, R. F., and Hynes, J. T. (1980). The stable picture of chemical reactions. II. Rate constants for condensed and gas phase reaction models. *J. Chem. Phys. 73*: 2715–2732.
5. Gegiou, D., Muszkat, K. A., and Fischer, E. (1968). Temperature dependence of photoisomerization. VI. The viscosity effect. *J. Am. Chem. Soc. 90*: 12–18.
6. Chandrasekhar, S. (1943). Stochastic problems in physics and astronomy. *Rev. Mod. Phys. 15*: 1–89.
7. Kubo, R. (1969). The fluctuation–dissipation theorem, in *Progress in Physics: Many-Body Problems*, W. E. Parry, R. E. Turner, D. terHaar, J. S. Rowlinson, G. V. Chester, N. M. Hugenholtz, and R. Kubo (Eds.). W. A. Benjamin, New York, pp. 255–283.
8. Landau, L. D., and Lifshitz, E. M. (1963). *Fluid Mechanics*. Pergamon, Oxford, pp. 63–71.
9. Bagchi, B., and Oxtoby D. W. (1983). The effect of frequency dependent friction on isomerization dynamics in solution. *J. Chem. Phys. 78*: 2735–2741.
10. Rahman, A., and Stillinger, F. H. (1971). Molecular dynamics study of liquid water. *J. Chem. Phys. 55*: 3336–3359.
11. Imry, Y., and Gavish, B. (1974). Correlation functions and structure factors for a lattice in a viscous medium. *J. Chem. Phys. 61*: 1554–1558.
12. Fiorito, R. B., and Meister, R., (1972). Pressure and temperature studies of NMR translational relaxation in hydrogen bonded liquids. *J. Chem. Phys. 56*: 4605–4619.
13. Barnett, L. J., and Hardon, J. F. (1972), Self-diffusion in viscous liquids: Pulse NMR measurements. *J. Chem. Phys. 57*: 1293–1297.
14. Sitaraman, R., Ibrahim, S. H., and Kuloor, N. R. A. (1963). Generalized equation for diffusion in liquids. *J. Chem. Eng. Data 8*: 198–200.
15. Defries, T., and Jonas, J. (1977). Molecular motions in compressed liquid heavy water at low temperatures. *J. Chem. Phys. 66*: 5393–5399.
16. Ertl, H., Ghai, R. K., and Dullien, F. A. L. (1974). Liquid diffusion of nonelectrolytes. *AIChE J. 20*: 1–20 (1974).
17. Hiss, T. G., and Cussler, E. L. (1973). Diffusion in high viscosity liquids. *AIChE J. 19*: 698–703.
18. Landolt-Bornstein, (1969). In *Transportphanomene, Kinetik Homogene Gasgleichgewichte*, H. Borchers, H. Hausen, K.-H. Helluege, K. L. Schäfer, and E. Schmidt (Eds.). Springer-Verlag, Vol. 5/II, pp. 566–697.
19. Weast, R. C. (Ed.) (1979) *CRC Handbook of Chemistry and Physics*, 60th ed. Boca Raton, Florida.
20. Barstein, G., Almagor, A., Yedgar, S., and Gavish, B. (unpublished results).
21. Smedley S. I. (1980). *The Interpretation of Ionic Conductivity in Liquids*. Plenum, New York, Chap. 2, pp. 11–74.
22. Litovitz, T. A., and Davis, C. M. (1965). Structural and shear relaxation in liquids. In Physical Acoustics, Vol IIA. W. P. Mason (ed.). Academic Press, New York, pp. 282–349.

23. Sette, D. (1961). Dispersion and absorption of sound waves in liquids and mixtures of liquids, in *Encyclopedia of Physics: Acoustics I*, S. Flugge (Ed.). Springer-Verlag, Berlin, pp. 276–360.
24. Frenkel, J. (1946). *Kinetic Theory of Liquids*. Dover, New York, Chap. 4.
25. Iben, I. E. T., Braunstein, D., Frauenfelder, H., Hong, H. K., Johnson, J. B., Luck, S., Ormos, P., Schulte, A., Steinbach, P. J., and Xie, A. H. (1989). Glassy behavior of a protein. *Phys. Rev. Let. 62*: 1916–1919.
26. Doster, W., Bachleitner, A., Dunau, R., Hiebl, M., and Luscher E. (1986). Thermal properties of water in myoglobin crystals and solutions at subzero temperatures. *Biophys. J. 50*: 213–219.
27. Parak, F., Fischer, M., and Nienhaus, G. U. (1989). The similarity in the dynamics of myoglobin and glycerol as seen from Mossbauer spectroscopy on ^{57}Fe. *J. Molec. Liq. 42*: 145–153.
28. Gavish, B. (1986). Molecular dynamics and the transient strain model of enzyme catalysis, in *The Fluctuating Enzyme*, G. R. Welch (Ed.). John Wiley, New York, pp. 263–339.
29. Jonas, J. (1975). Nuclear magnetic resonance at high pressures. *Am. Rev. Phys. Chem. 26*: 167–190.
30. Stephan, K., and Lucus, K. (1979). *Viscosity of Densed Fluids*. Plenum, New York, pp. 3–24.
31. Doolittle, A. K. (1951). Studies in Newtonian flow. II. The dependence of the viscosity of liquids on free-space. *J. Appl. Phys. 22*: 1471–1475.
32. Bueche, F. (1959). Mobility of molecules in liquids near the glass temperature. *J. Chem. Phys. 30*: 748–752.
33. Macedo, P. B., and Litovitz, T. A. (1965). On the relative role of free volume and activation energy in the viscosity of liquids. *J. Chem. Phys. 42*: 245–256.
34. Gubbins, K. E., and Tham, M. J. (1969). Free volume theory for viscosity of simple monopolar liquids. *AIChE J. 15*: 264–271.
35. Cohen, M. H., and Turnbull, D. (1959). Molecular transport in liquids and glasses. *J. Chem. Phys. 31*: 1164–1169.
36. Glasstone, K., Laidler, J., and Eyring, H. (1941). *The Theory of Rate Processes*. McGraw-Hill, New York.
37. Hynes, J. T. (1986). Chemical reaction rates and solvent friction. *J. Stat. Phys. 42*: 149–168.
38. Hynes, J. T. (1986). Outer-sphere electron-transfer reactions and frequency-dependent friction. *J. Phys. Chem. 90*: 3701–3706.
39. Velsko, S. P., and Fleming, G. R. (1982). Photochemical isomerization in solution. Photophysics of diphenyl butadiene. *J. Chem. Phys. 76*: 3553–3562.
40. Hynes, J. T., Carter, E. Y., Ciccotti, G., Kim, H. J., Zichi, D. A., Ferrario, M., and Kapral, R. (1990). Environmental dynamics and electron transfer reactions, in *Perspectives in Photosynthesis*, J. Jortner and B. Pullman (Eds.). Kluwer Academic, The Netherlands, pp. 133–148.
41. Bullock, A. T., and Cameron, T. G. G. (1974). Electron spin resonance studies of spin-labelled polymers. *J. Chem. Phys. Farad. Trans. II 70*: 1202–1221.
42. Rothenberger, G., Negus, D. K., and Hochstrasser, R. M. (1983). Solvent influence on photoisomerization dynamics. *J. Chem. Phys. 79*: 5360–5367.

43. Flom, S. R., Brearly, A. M., Kahlow, M. A., Nagarajan, V., and Barbara, P. F. (1985). Double well isomerization in solution: A new comparison of theory and experiment. *J. Chem. Phys. 83*: 1993–1995.
44. Rosenthal, S. J., Xie, X., Du, Mei, and Fleming, G. R. (1991). Fentosecond solvation dynamics in acetonitrile: Observation of the inertial contribution to the solvent response. *J. Chem. Phys. 95*: 4715–4718.
45. Chandler, D., and Kuharski, R. A. (1988). Two simulation studies of chemical dynamics in liquids. *Farad. Discuss. Chem. Soc. 85*: 329–340.
46. Beece, D., Eisenstein, H., Frauenfelder, H., Good, D., Marden, M. C., Reinisch, L., Sorensen, L. B., and Yue, K. T. (1980). Solvent viscosity and protein dynamics. *Biochemistry 19*: 5147–5157.
47. McKinnie, R. E., and Olson, J. S. (1981). Effect of solvent composition and viscosity on the rates of CO binding to heme proteins. *J. Biol. Chem. 256*: 8923–8932.
48. Sawicki, C. A., and Khaleque, M. A. (1983). Laser photolysis study of conformational change rate for hemoglobin in viscous solutions. *Biophys. J. 44*: 191–199.
49. Lavalette, D., and Tetreau, C. (1988). Viscosity-dependent energy barrier and equilibrium conformational fluctuations in oxygen recombination with hemerythrin. *Eur. J. Biochem. 177*: 97–108.
50. Beece, D., Bowne, S. F., Czege, J., Eisenstein, L., Frauenfelder, H., Good, D., Marden, M. C., Marque, J., Ormos, P., Reinisch, L., and Yue, K. T. (1981). The effect of viscosity on the photocycle of bacteriorhodopsin. *Photochem. Photobiol. 33*: 517–522.
51. Demchenko, A. P., Rusyn, O. I., and Saburova, E. A. (1989). Kinetics of the lactate dehydrogenase reaction in high-viscosity media. *Biochim. Biophys. Acta 998*: 196–203.
52. Feitelson, J., and Yedgar, S. (1991). The effect of the solvent viscosity on the migration of small molecules through the structure of myoglobin. *Biorheology 28*: 99–105.
53. Ng, K., and Rosenberg, A. (1991). Possible coupling of chemical to structural dynamics in subtilisin BNP′ catalyzed hydrolysis. *Biophys. Chem. 39*: 57–68.
54. Yedgar, S., Reisfeld, N., Halle, D., and Yuli, I. (1987). Medium viscosity regulates the activity of membrane-bound and soluble phospholipase A_2. *Biochemistry 26*: 3395–3401.
55. Hovav, E., Halle, D., and Yedgar, S. (1987). Viscous macromolecules inhibit erythrocyte hemolysis induced by snake venom phospholipase A_2. *Biorheology 24*: 377–384.
56. Yedgar, S., Reisfeld, N., and Sela, B. A. (1986). Regulation of cell ganglioside composition by extracellular fluid viscosity. *Lipids 21*: 629–633.
57. Yedgar, S., Weinstein, D., Patsch, W., Schonfeld, G., Casanada, F., and Steinberg, D. (1982). Viscosity of culture medium as a regulator of synthesis and secretion of very low density lipoproteins by cultured hepatocytes. *J. Biol. Chem. 257*: 2188–2192.
58. Somogyi, B., Norman, J. A., Zempel, L., and Rosenberg, A. (1988). Viscosity and transient solvent accessibility of Trp-63 in the native conformation of lysozyme. *J. Biophys. Chem. 32*: 1–33.
59. Gregory, R. G., Rosenberg, A., Knox, D., and Percy, A. J. (1990). The thermodynamics of hydrogen isotope exchange in lysozyme. The influence of glycerol. *Biopolymers 29*: 1175–1183.

60. Lamb, J. (1965). Thermal relaxation in liquids, in *Physical Acoustics*, W. P. Mason (Ed.). Academic Press, New York, Vol. II/A, pp. 203–280.
61. O'Brien, W. D., and Dunn, F. (1972). Ultrasonic absorption mechanisms in aqueous solutions of bovine hemoglobin. *J. Phys. Chem. 76*: 528–533.
62. Cho, C. K., Leung, W. P., Mok, H. Y., and Choy, C. L. (1985). Ultrasonic absorption in myoglobin and other globular proteins. *Biochim. Biophys. Acta 830*: 36–44.
63. Slutsky, L. J., and Madsen, L. (1977). Acoustic absorption in solutions of small peptides and proteins, in *IEEE Ultrasonic Symposium (Proceedings)*, J. de Klerk and B. R. McAvoy (Eds.), pp. 148–152.
64. Barksdale, A. D., and Stuehr, J. E. (1972). Kinetics of the helix-coil transition in aqueous poly(L-glutamic acid). *J. Acous. Soc. Am. 94*: 3334–3338.
65. Kato, S., Suzuki, T., Nomura, H., and Miyahara, Y. (1980). Ultrasonic relaxation in aqueous solutions of dextran. *Macromolecules 13*: 889–892.
66. Careri, G., Geraci, M., Giansanti, A., and Rupley, J. A. (1985). Protonic conductivity of hydrated lysozyme powders at megahertz frequencies. *Proc. Natl. Acad. Sci. USA 82*: 5342–5346.
67. Gavish, B. (1980). Position-dependent viscosity effects on rate coefficients. *Phys. Rev. Lett. 44*: 1160–1163.
68. Almagor, A., Yedgar, S., and Gavish, B. (1992). Viscous cosolvent effect on the ultrasonic absorption of bovine serum albumin. *Biophys. J. 61*: 480–486.
69. Rosenberg, A., Ng, K., and Punyiczki, M. (1989). Activity and viscosity effects on the structural dynamics of globular proteins in mixed solvent systems. *J. Molec. Liq. 42*: 31–43.
70. Gekko, K., and Timasheff, S. N. (1981). Mechanism of protein stabilization by glycerol: Preferential hydration in glycerol–water mixtures. *Biochemistry 20*: 4667–4676.
71. Chirgadze, Y. N., and Ovsepyan, A. M. (1972). Hydration mobility in peptide structures. *Biopolymers 11*: 2179–2186.
72. Zak, A., and Klibanov, A. M. (1988). The effect of water on enzyme action in organic media. *J. Biol. Chem. 263*: 8017–8021.
73. Doster, W. (1983). Viscous scaling and protein dynamics. *Biophys. Chem. 17*: 97–103.
74. Easteal, A. J. (1990). Tracer diffusion in aqueous sucrose and urea solutions. *Can. J. Chem. 68*: 1611–1615.

8

Effect of Solvent on Protein Internal Dynamics: The Kinetics of Ligand Binding to Myoglobin

Wolfgang Doster, Thomas Kleinert, Frank Post, and Marcus Settles
Technische Universität München, Munich, Germany

I. INTRODUCTION

The contribution of water to the stability, flexibility, and function of proteins has been a matter of controversy for many years. Recent investigations of protein denaturation seem to show that the stability of the native protein structure is entirely due to intramolecular hydrogen and van der Waals bonding, while the ordering action of the nonpolar groups on water exerts only a destabilizing effect [1,2]. This implies that removal of water should enhance the stability of the native state. Small proteins usually can be lyophilized without showing a significant loss of secondary and tertiary structure. Once dry, the stability range is increased to temperatures far above the denaturation temperature observed in solution [3]. As was demonstrated for lysozyme [4,5], the subtle changes in structure in response to drying concern mainly the network of hydrogen bonds established by the polar groups. The removal of hydrogen bond donors and acceptors enhances intramolecular hydrogen bonding and the packing density, which can result in a loss of structural flexibility. The protein–water interaction modifies the properties known for bulk water. For instance, bound water does not crystallize but forms an amorphous solid at low temperatures [6,7]. Bulk water consists of patches of molecules forming an open network structure. Each molecule is bonded to four neighbors, which tend to be tetrahedrally disposed as in low-density ices. Size fluctuations of these patches explain the unusual properties of bulk water [8,9]. Recent work suggests that charged and polar groups on protein and DNA surfaces act as patch breakers, thus preventing crystallization [10,11].

Hydration can affect protein function in various ways: The high mobility of bound water with respect to internal residues may result in a liquidlike surface, facilitating the exchange of substrate molecules. This property is lost at low hydration. Dehydration of polar groups will change the network of hydrogen bonds, which in turn may affect the activity. Dynamic aspects of protein–solvent interactions contribute to enzyme catalysis. In particular, the entry and release of substrate molecules and changes in hydration upon binding to the active site, large-scale conformational changes, and charge fluctuations on protein surfaces [12] depend on the mobility of the solvent. It was even proposed that fluctuations in the solvent structure may trigger enzyme catalysis [13]. To investigate the effect of solvent on protein flexibility, one usually modifies the solvent either by changing the degree of hydration or the chemical composition. Experiments performed as a function of the degree of hydration reveal how much water is required to establish structural flexibility and protein function. An important goal is to correlate the onset of biological activity to changes in the dynamic parameters and the hydration of charged and polar groups [4,5]. On the theoretical side, molecular dynamic simulations of proteins were performed in vacuo and with solvent [14,15]. Experiments probing the hydration dependence of structural and dynamic properties may contribute to the search for an effective force-field describing protein–solvent interactions [16]. A second approach is based on variable solvent experiments. One goal is to monitor protein motions in response to changes of the solvent viscosity [17–21,34]. This method allows one to investigate solvent damping of intramolecular processes in proteins. In both cases, one encounters the problem of differentiating between purely dynamic and structure-mediated effects. Changes in hydration and solvent composition may induce structural shifts, which in turn modify the mobility of the molecule. Instead of solvent perturbation, one may alternatively perturb the active site and monitor the response of the solvent. In this way one can establish direct correlation between function and solvent relaxation.

To gain insights into dynamic and structure-mediated effects on protein function, we discuss a well-known reaction, the binding of carbon monoxide (CO) to myoglobin. This system has been investigated in solution by flash photolysis within a wide range of temperatures, time scales, and solvent conditions [18,22]. Its behavior in solution is now well characterized [23], and a variety of structural markers are available [24,34]. Three major CO conformers, which correspond to different binding angles, were identified according to their C–O stretching frequencies [25–27]. Their relative population is sensitive to structural changes [34] and depends on the degree of hydration [28]. The access of ligand to the heme pocket is controlled by density fluctuations of the polypeptide chain. Myoglobin is thus a model system to study the relation of motion to function in proteins. Because of the partitioning of side chains forming a polar surface and a hydrophobic core, myoglobin can be considered as the prototype of the oil drop model of proteins. Removal of water is thus expected to strongly modify the exchange of ligands at the protein–solvent interface.

II. THE FLASH PHOTOLYSIS EXPERIMENT

One uses the optical transitions of the iron-porphyrin ring to photodissociate the ligand from the iron. The recombination is recorded via changes in the absorption spectrum as a function of time. Figure 1 shows the absorption of spectra of deoxy and carboxy myoglobin. In closed-shell porphyrins, there is a very intense band near 400 nm, called the Soret or B-band, and much weaker bands between 500 and 600 nm, called Q-bands. These are due to $\pi \rightarrow \pi^*$ transitions of the porphyrin π electron ring. Photolysis is achieved through a short pulse (8 ns) of a frequency-doubled Nd : YAG laser at 532 nm. The pulse energy amounts to 120 mJ. The monitor wavelength was typically 436 nm. Our apparatus can record changes in the extinction coefficient with a resolution of 10^{-4} between 50 ns and 1 ks [29]. A weak charge transfer band III at 760 nm can be observed in deoxymyoglobin. Band III depends sensitively on the position of the iron with respect to the heme plane and was used to monitor structural changes [23]. Below we discuss the

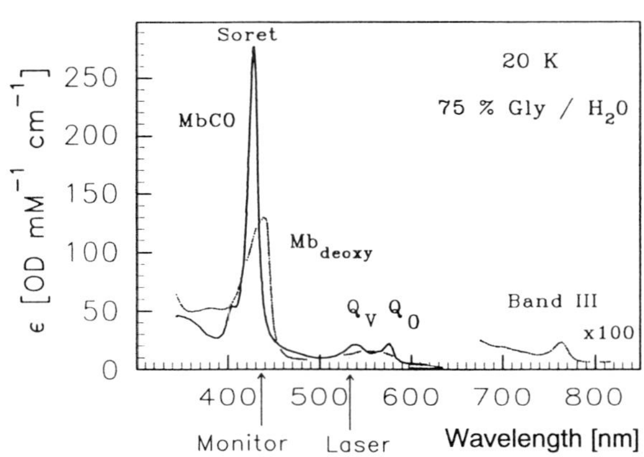

FIGURE 1 Absorption spectrum of deoxy and CO myoglobin. The deoxy state was generated by illumination with white light at a sample temperature of 20 K.

kinetics of CO binding to horse myoglobin (Sigma) in 75% and 90% (v/v) glycerol–water solvents and hydrated myoglobin films. The films were prepared by drying an aqueous protein solution on a glass plate under constant humidity. The water content was determined using a dry weight method.

III. THE KINETICS OF CO BINDING TO MYOGLOBIN

Figure 2a shows the survival fraction $N(t)$ of the dissociated state, Mb + CO, versus time on a double logarithmic scale. At 280 K in a solvent mixture of 75% glycerol–water, nearly all ligands escape to the solvent, resulting in pseudo-first-order recombination, the rate of which is proportional to the CO concentration. This process is denoted by $S + L \rightarrow A$, where A is the bound state. Also, a fast ki-

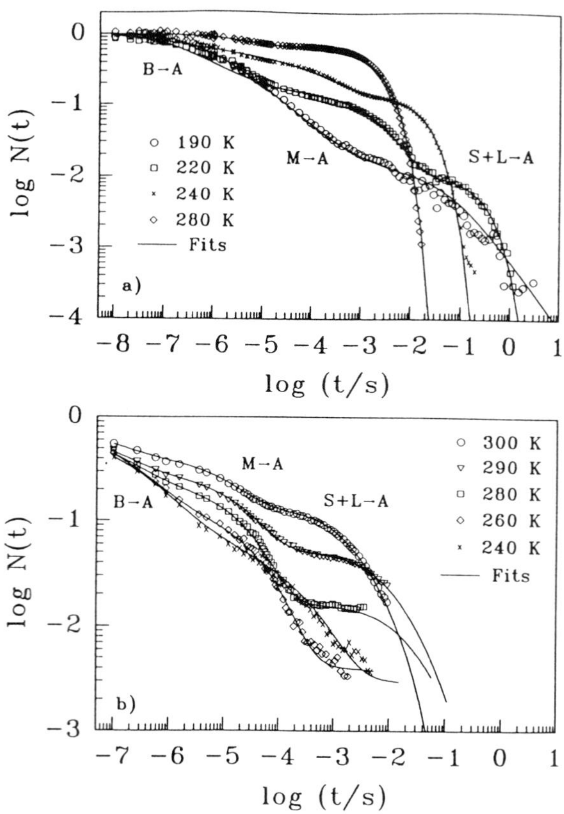

FIGURE 2 Survival fraction of photolyzed MbCO monitored at 436 nm: (a) solvent 75% glycerol/water; (b) hydrated film, $h = 0.3$ g/g.

netic phase is observed, which does not depend on the ligand concentration in the solvent, implying internal or geminate recombination. With decrease in temperature (220 K), the internal process splits into two components, denoted by $M \rightarrow A$ and $B \rightarrow A$. At 190 K the escape to the solvent becomes negligible and only internal recombination takes place. Below 160 K (Fig. 3) only the fastest component, $B \rightarrow A$, survives, with a profound change in the kinetic shape: More than 10 exponentials are required to fit the data. This suggests a nearly continuous rate distribution. Assuming the Arrhenius law, one derives a temperature-independent distribution of activation energies $g(H)$ [22,29]. Thus,

$$k(H) = \mathrm{A} \exp - \frac{H}{RT} \tag{1}$$

$$N(t) = \int g(H) \exp[-k(H)t]\, dH$$

The temperature independence of $g(H)$ and multiple flash photolysis experiments prove that structural motion is frozen in at low temperatures. Thus, each protein molecule assumes a particular conformational substate specified by an activation energy H. Since we observe three kinetic components, the analysis of the data discussed below is based on a sequential four-state model [30]:

$$A \leftrightarrow B \leftrightarrow M \leftrightarrow S + L \tag{2}$$

At short times after photolysis, the structure is rigid, confining the CO molecule to positions within the heme pocket B. Only direct recombination $B \rightarrow A$ occurs. As structural relaxation proceeds, a new kinetic state M with enhanced activation energy becomes available [31]. Evidence for a structural change comes

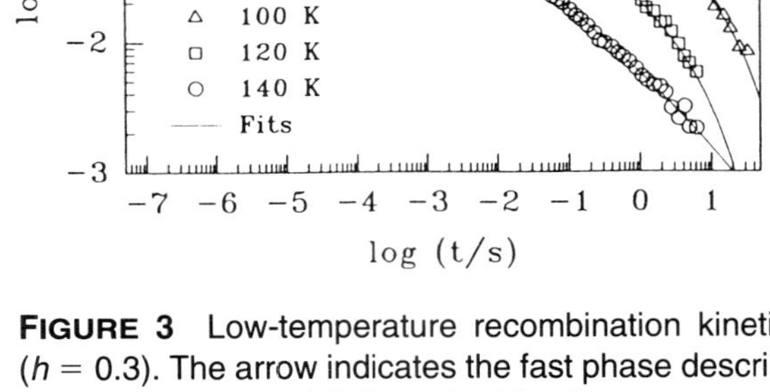

FIGURE 3 Low-temperature recombination kinetics of a hydrated MbCO film (h = 0.3). The arrow indicates the fast phase described in the text; the lines correspond to fits to Eq. (1). See also Fig. 8.

from the observation of a time-dependent shift of band III [23,32]. From M, the ligand may either rebind or escape to the solvent ($S + L$).

In the following, we discuss how the internal processes, specified by $g(H)$ and the structural relaxation rate k_{BM}, and the ligand release rate k_{MS} depend on solvent conditions.

IV. THE SURFACE BARRIER

In aqueous solutions, nearly all dissociated ligands escape to the solvent at physiological temperatures, $N_S \approx 0.96$ (Fig. 2a). the barrier at the surface is therefore small compared with the internal barriers. In partially hydrated films, however, N_S is much reduced (Fig. 2b).

Figure 4 shows the CO binding kinetics of a myoglobin film initially hydrated to 0.25 g/g that is exposed to a gentle vacuum. Both the amplitude, N_S, and the recombination rate of $S + L \rightarrow A$ decrease in similar proportion with decreasing hydration. Note, however, that the kinetic shape of the internal process $M \rightarrow A$ does not change. This result indicates that the surface barrier increases drastically at low hydration, while internal recombination is much less affected.

Figure 5 shows the fraction of ligands that escape to the solvent, N_S, versus degree of hydration. N_S decreases strongly when the degree of hydration is reduced below the nonfreezable water content, $h \approx 0.4$ [6]. From temperature-dependent experiments, we estimate a barrier height of at least 140 kJ/mol at $h = 0.2$. A similar, but less drastic, reduction in N_S occurs in glycerol–water solvents when the glycerol content is increased. To determine whether this effect results from structural or dynamic changes at the surface we investigate the viscosity dependence of N_S and the escape rate k_{MS}:

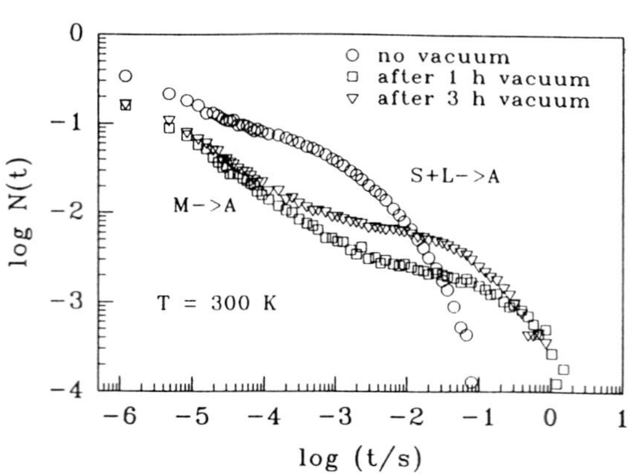

FIGURE 4 Recombination kinetics of a hydrated MbCO film ($h = 0.25$) that is exposed to a gentle vacuum (10^{-1} bar).

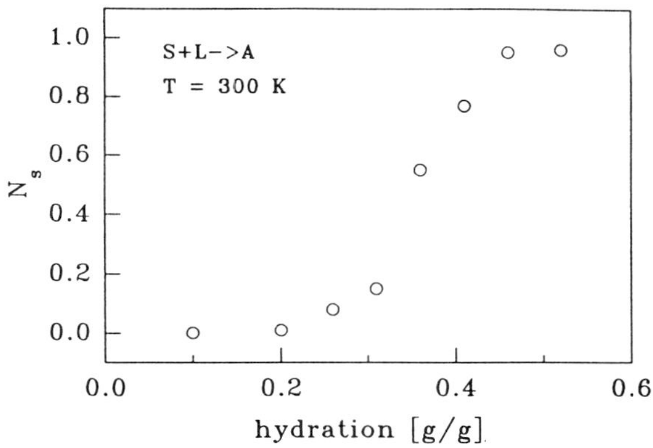

FIGURE 5 Fraction of CO molecules that escape to the solvent versus degree of hydration.

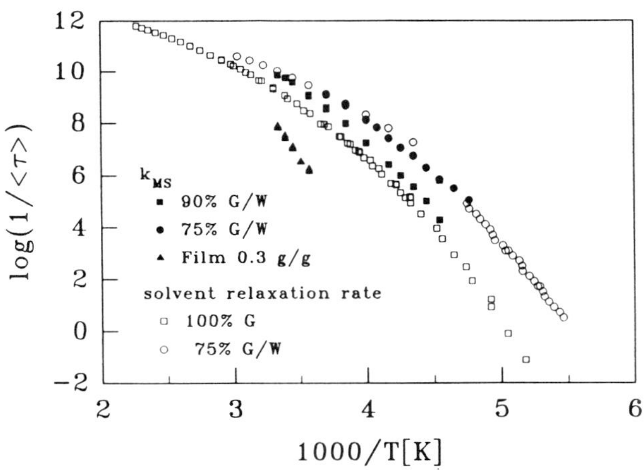

FIGURE 6 Arrhenius plot of solvent relaxation rates (open symbols) and k_{MS} (filled symbols).

$$N_S = \frac{k_{MS}}{k_{MA} + k_{MS}} \tag{3}$$

Figure 6 displays the structural relaxation rates (proportional to $1/\eta$) of glycerol and a 75% glycerol–water mixture [12]. The figure also shows the escape rates k_{MS} of a hydrated film ($h = 0.3$), 90% and 75% glycerol–water solutions. For comparison, the protein escape rates were multiplied by a common factor (3.0×10^4).

The protein and solvent relaxation rates show approximately the same temperature dependence: $k_{MS} \propto 1/\eta$.

This result provides evidence that the solvent viscosity controls the release of ligands and points to a dynamic origin of the surface barrier. The escape rate may be approximated by Kramers's law:

$$k_{MS} = \frac{A}{\eta} \exp \frac{-H_{MS}}{RT} \tag{4}$$

where H_{MS} is small compared with the activation enthalpy of the solvent viscosity η. A similar viscosity-dependent regime was observed for the rate of a conformational change in myoglobin (sperm whale) induced by photolysis at ambient temperatures [21].

The escape rates in the hydrated film (Fig. 6) are about 100 times smaller than in 90% glycerol. The hydration shell ($h = 0.3$) is therefore apparently more viscous than glycerol.

V. THE INTERNAL BARRIERS

The solvent effects on the internal transitions $B \rightarrow A$ and $B \rightarrow M$ are small compared with what is found for the entry-release mechanism (Fig. 4). Figure 7 illustrates this statement for structural relaxation rates k_{BM}. The figure also shows the solvent relaxation rates (75% glycerol–water) for comparison. The temperature dependence and, thus, the activation energies are entirely different. Furthermore, below 200 K the solvent dynamics is slower than the protein conformational change. The protein internal transitions are therefore decoupled from the relaxation on the surface.

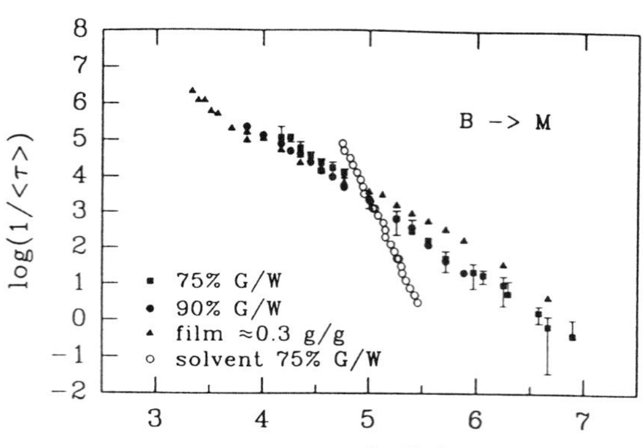

FIGURE 7 Arrhenius plot of k_{BM} for various solvents (filled symbols) and of solvent relaxation rate for 75% glycerol (○).

There are, however, noticeable variations in the kinetics depending on the solvent composition. These differences become apparent after the following data analysis: We invert Eq. (1), using a model-independent best fit to the data. This yields the activation enthalpy distribution $g(H)$, describing the decay of the intermediates B, M, and $S + L$ as shown in Fig. 8 [29]: The B decay is approximately independent of the solvent. But the position, shape, and magnitude of the M peak vary with solvent composition. The M-decay kinetics in the hydrated film is much more stretched compared with the glycerol solutions, giving rise to a broad shoulder in $g(H)$ instead of a peak. Increasing the glycerol concentration enhances the activation enthalpy of $M \to A$.

VI. CONCLUSION

The flash photolysis experiments discussed here demonstrate a strong correlation between ligand transfer at the protein–solvent interface and structural relaxation of the solvent. One can imagine a fluctuating gate whose open time varies in proportion to the solvent viscosity. By contrast, the internal recombination events seem to be entirely decoupled from the dynamics of the solvent. This clear-cut separation of external and internal processes may be a consequence of the side-chain distribution forming a polar surface and a hydrophobic core around the heme. The heme, however, is not entirely shielded from the solvent: The distribution of CO-bound states as monitored by the CO-stretching vibration in sperm whale myoglobin depends on the solvent composition and degree of hydration [28,34]. We thus tentatively assign the solvent composition–dependent kinetic effects to changes in the heme–ligand geometry.

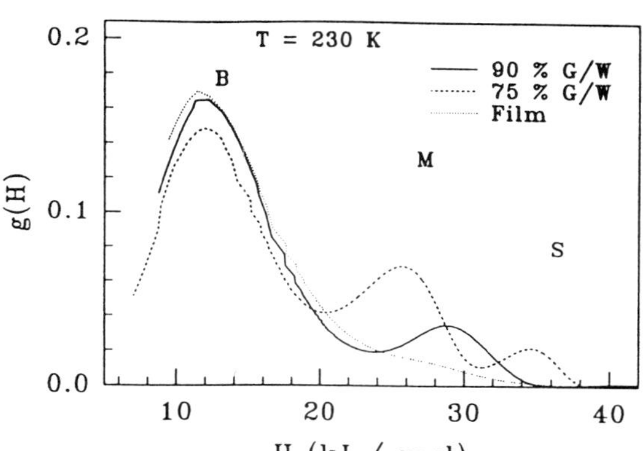

FIGURE 8 Activation enthalpy distributions of $g(H)$ of CO myoglobin at 230 K obtained by numerical inversion of kinetic data according to Eq. (1) [29].

It was proposed that the pressure jump induced relaxation of the *A* state distribution depends on the solvent viscosity [33]. A detailed comparison of protein and solvent relaxation [12] reveals a coincidence of the initial time dependence only. The corresponding relaxation time distributions are quite different, however. This result may imply that the protein responds to changes of the average solvent structure. And this response cannot be faster than the structural relaxation time of the solvent. In such large perturbation experiments, the distinction between structure-mediated and dynamic solvent effects becomes meaningless.

REFERENCES

1. Privalov, P. L. (1990). *Biochem. Mol. Biol. 25*: 281–305.
2. Privalov, P. L. (1989). Ann. *Rev. Biophys. Chem. 18*: 47–61.
3. Fujita, Y., and Noda, Y., *Bull. Chem. Soc. Jpn. 51*: 1567–1568, 1978.
4. Rupley, J. A., Gratton, E., and Careri, G. *Trends Biochem. Sci. 8*:18–23, 1983.
5. Finney, J. L., and Poole, P. L. (1984). *Biopol. 23*: 1647–1655.
6. Doster, W., Bachleitner, T., Dunau, R., Hiebl, M., Lüscher, E. (1986). *Biophys. J. 50*: 213–221.
7. Sartor, G., Hallbrucker, A., Hofer, K., and Mayer, E. (1992). *J. Phys. Chem 96*: 5733.
8. Angell, C. A. (1982). In *Water: A Comprehensive Treatise*, ed. Franks (Plenum, New York), Vol. 7, 1–81.
9. Geiger, A., and Stanley, E. H. (1982). *Phys. Rev. Lett. 49*: 1749–1752.
10. Hiebl, M., and Maksymiw, R. (1991). *Biopol. 31*: 161–167.
11. Tao, N. J., Lindsay, S. M., and Rupprecht, A. (1989). *Biopol. 28*: 1019–1030.
12. Settles, M., Post, F., Doster, W., Schulte, A., and Müller, D. (1992). *Biophys. Chem. 43*: 107–116.
13. Careri, G. (1974). In *Quantum Statistical Mechanics in the Natural Sciences* (Plenum, New York).
14. Smith, J., Kuczera, M., and Karplus, M. (1990). *PNAS* USA *87*: 1601–1605.
15. Loncharich, R. J., and Brooks, B. R. (1990). *J. Mol. Biol. 215*: 439–452.
16. Ben-Naim, A. (1989). *J. Chem. Phys. 90*: 7412–7425.
17. Gavish, B., and Werber, M. M. (1979). *Biochem. 18*: 1269.
18. Beece, D., Eisenstein, L., Frauenfelder, H., Good, D., Marden, M. C., Reinisch, L., Reynolds, A. H., Sorensen, L. B., and Yue, K. T. (1980). *Biochem. 19*: 5147–5157.
19. Demchenko, A. P., Rusyn, O. I., and Saburova, E. A. (1989). *Biochim Biophys. Acta 998*: 196.
20. Settles, M., Doster W., Kremer F., Post F., and Schirmacher, W. (1992). *Phil. Mag. 65*: 861–866.
21. Ansari, A., Jones, M., Henry, E. R., Hofrichter, J., and Eaton, W. A. (1992). *Science 256*: 1796–1798.
22. Austin, R. H., Beeson, K. W., Eisenstein, L., Frauenfelder, H., and Gunsalus, I. C. (1975). *Biochem. 14*: 5355–5373.
23. Steinbach, P. J., Ansari, A., Berendzen, J., Braunstein, D., Chu, K., Cowen, B. R., Ehrenstein, D., Frauenfelder, H., Johnson, J. B., Lamb, D. C., Luck, S., Mourant,

J. R., Nienhaus, G. U., Ormos, P., Philipp, R., Xie, A., and Young, R. D. *Biochemistry 30*: 3988–4001, 1991.
24. Morikis, D., Champion, P. M., Springer, B. A., and Sligar, S. G. (1989). *Biochem. 28*: 4791–4800.
25. Alben, J. O., Beece, D., Bowne, S. F., Doster, W., Eisenstein, L., Frauenfelder, H., Good, D., McDonald, Marden, M. C., Moh, P. P., Reinisch, L., Reynolds, A. H., Shyamsunder, E., and Yue, K. T (1982). *PNAS, USA 79*: 3744–3748.
26. Moore, J. N., Hansen, P. A., and Hochstrasser, R. M. (1988). *PNAS, USA 85*: 5062–5066.
27. Ormos, P., Braunstein, D., Frauenfelder, H., Hong, M. K., Lin, S. L., Sauke, T. B., and Young, R. (1988). *PNAS, USA*, 8492–8496.
28. Brown, W. F., Sutcliffe, J. W., and Pulsinelli, P. D. (1983). *Biochem. 22*: 2914–2923.
29. Post, F., Doster, W., and Settles, M. (1993). *Biophys. J. 64*: 1833–1842.
30. Doster, W., Beece, D., Bowne, S. F., Di Iorio, E., Eisenstein, L., Frauenfelder, H., Reinisch, L., Shyamsunder, E., Winterhalter, K. H., and Yue, K. T. (1982). *Biochem. 21*: 4831–4839.
31. Agmon, N., and Hopfield, J. J. (1983). *J. Chem. Phys. 79*: 2042–2053.
32. Dunn, R. C., and Simon, J. O. (1991). *Biophys. J. 60*: 884–889.
33. Iben, I., Braunstein, D., Doster, W., Frauenfelder, H., Hong, M., Johnson, J., Huck, S., Ormos, P., Schulte, A., Steinbach, P., Xie, A., and Young, R. (1989). *Phys. Rev. Lett. 62*: 1916–1919.
34. Ansari, A., Berendzen, J., Braunstein, D., Cowen, B. R., Frauenfelder, H., Hong, M. K., Iben, I. E. T., Johnson, J. B., Ormos, P., Sauke, T. B., Scholl, R., Schulte, A., Steinbach, P. J., Vittitow, J., and Young, R. D. *Biophys. Chem.*, *26*: 337–355, 1987.

9

Solvent Effects on Protein Stability and Protein Association

ARIEH BEN-NAIM
The Hebrew University, Jerusalem, Israel

I. INTRODUCTION: A HISTORIC PERSPECTIVE

Biochemical processes involving proteins and nucleic acids have been the focus of intensive research by chemists, physicists, biochemists, and biologists. These macromolecules are known to form highly specific structures that are essential for their function in biological systems. Examples are the folding of proteins to a well-defined three-dimensional (3-D) structure, the association of proteins to form multisubunit macrostructures such as hemoglobin and various regulatory enzymes, and the binding of proteins to DNA—a crucial step in the control and regulation of the transcription of genetic information.

All of these processes are highly specific in the sense that the end product must have a structure defined within very narrow limits. It is also recognized that these specific structures are selected from an enormously large number of possible outcomes.

What mechanism do biomolecules use to reach the right target, in a relatively short time?

The simplest answer would be that they do so by trial and error, that is, the molecules wander randomly in their configurational space until they reach the point of maximum stability and remain there for a sufficiently long time to perform their biochemical function. Rough estimates, however, show that such a random search of the configurational space would take too long a time, much larger than the typical span of time taken in actual biological systems. In fact, there is no reason to believe that the protein wanders randomly in its configurational space. At any given time, the various groups constituting the protein feel strong forces. These forces originate from the protein itself as well as being induced by the solvent. If they are strong enough, they must also have the power to direct the motion of the representative point in the configurational space along some preferential pathways.

On the other hand, one cannot expect to find a unique trajectory leading from the initial to the final state. There is always a random component to the forces acting on the protein, originating from the thermal agitation of the solvent molecules. Nevertheless, if the average forces acting on the protein are strong enough, then, in spite of the thermal fluctuations, the motion of the protein will, on the average, be biased into a narrow range of preferred pathways.

Once the final desired product is attained, it must be stable enough to maintain its structure long enough to fulfill its biological tasks. What are the factors that contribute to the stability of these structures?

One of the earlier observations was that in both protein folding and protein association, no new covalent bonds are formed (we exclude here the cases involving disulfide bonds). This observation makes the question posed above more acute. It also suggests that the combined effects of relatively weak interactions can produce a large stabilizing force.

The second important observation is that the stability of the biomolecular structures is maintained almost exclusively in aqueous media. Addition of a large quantity of a cosolvent usually leads to destabilization of the structure. Is it possible that the solvent is responsible for the stability?

The answer to the last question is not yet known. It is believed, however, that the solvent does play an important, if not the dominant, role in maintaining the stability of these structures. Finding out how the solvent affects the stability is a very difficult task. Liquid water, as a major component of biological fluids, has been the subject of intense investigation by both experimentalists and theoreticians [1–4]. Although many questions are still open, we now understand some of the most unusual properties of pure water on the molecular level. Inserting a protein into water introduces a multitude of new interactions between the solvent molecules and the various groups along the protein molecule. Therefore, the study of solvent effects on processes involving proteins is still a formidable task.

It is with this background that the *hydrophobic effect hypothesis* has evolved [5–7]. This hypothesis is based on two experimental observations: (a) In both protein folding and protein association, hydrophobic ($H\phi O$) groups tend to accumulate in the interior of the final product; (b) The process of transferring a small nonpolar solute, such as methane, ethane, and the like, from water into an organic liquid, say ethanol or hexane, involves a large negative change in Gibbs energy. Typical values are in the range -3–-6 kcal/mol, for the process we schematically describe as

$$\{\text{Nonpolar solute in water}\} \rightarrow \{\text{nonpolar solute in organic solvent}\} \qquad (1)$$

This phenomenon is sometimes referred to as the *HϕO effect*. Qualitatively, it says that nonpolar molecules "dislike" the aqueous environment (hence the term *hydrophobia*), and therefore given the opportunity to distribute between water and organic liquid, they prefer the latter. The driving force for the process indicated in Eq. (1) can thus be ascribed to the tendency of the nonpolar solutes to avoid exposure to the aqueous environment.

The hydrophobic effect hypothesis, put forward by Kauzmann in 1959 [5], assumes that nonpolar, or *HϕO*, side chains along the polypeptide backbone have the same character as nonpolar molecules. Therefore, the removal of such side chains from the aqueous environment could also contribute a large negative Gibbs energy to the folding and association processes—much as in the transfer process indicated in Eq. (1).

In essence, the *HϕO* effect hypothesis is based on the apparent analogy between the transfer process (Eq. (1)) and only *one element* of the more complicated process involving nonpolar side chains of proteins. This analogy is indeed very appealing; no wonder that it was accepted by the majority of biochemists. Although Kauzmann himself suggested this hypothesis as a *tentative* conclusion, it was later cited as being a well-established fact, and the hypothetical character of this effect was ignored. It is now very common to find in the biochemical literature references to the *HϕO* effect as the major, or the most important, driving force in biochemical processes.

There are, however, two weak aspects to this hypothesis: (a) It ignores other solvent-induced effects involving *HϕO* as well as hydrophilic (*HϕI*) groups; (b) The analogy between the process (Eq. (1)) and the transfer of side chains, though appealing intuitively, is far from being perfect.

In order to accept the *HϕO* effect hypothesis, one needs to first establish that the analogy between the two processes of transfer leads also to quantitatively similar changes in Gibbs energies. Second, one must show that these changes in Gibbs energies are larger than other elements of the solvent-induced effects. Clearly, without having a list of all possible solvent-induced effects, one cannot make any statement regarding the relative importance of one element over another. But what are the other elements of the solvent-induced effects? We shall devote a large part of this chapter to answering this question.

In the context of the study of other solvent-induced effects, another effect involving hydrophilic (Hϕ*I*) groups has been considered. This is the formation of a direct hydrogen bond (HB) between two functional groups such as hydroxyl, carbonyl, or amine groups. It was recognized that the formation of such an HB in vacuum or in a nonpolar solvent could contribute some −6–−7 kcal/mol to the total driving force of the process.

The same HB formation when taking place in water, however, was described by a stoichiometric reaction of the form [5,8–10].

$$\succ\!\!-C=O\cdots w+\succ\!\!-OH\cdots w\rightarrow\succ\!\!-C=O\cdots HO\!\!-\!\!\prec+\,w\cdots w \qquad (2)$$

Thus, the two functional groups (here the carbonyl and the hydroxyl) are originally solvated by water molecules. When these two functional groups form one direct HB, they must get rid of the solvating water molecules. The water molecules released into the bulk of the water must also form HBs between themselves [as indicated by *w . . . w* on the rhs of Eq. (2)]. A glance at Eq. (2) might suggest that there are two HBs formed and two HBs broken in this process. Therefore, it was argued, such a process cannot contribute to the total driving force as much as it would have contributed had the same process taken place in an organic liquid. We shall refer to this argument as the *HφI* effect hypothesis. We shall later see that there are other effect involving *HφI* groups. For the present, we note, as we did for the *HφO* effect, that this hypothesis, though intuitively very appealing, is not established on solid grounds. We shall return to discuss this effect in Sec. VII.

The dismissal of the contribution of HBs, based on the *HφI* effect hypothesis, has indirectly enhanced the belief in the *HφO* effect hypothesis. If HBs do not contribute much to the driving force in a biochemical process, what else besides the *HφO* effect could be contributing? This is probably the main reason for the almost universal acceptance of the *HφO* effect hypothesis as the major, or the dominant, component of the driving forces in biochemical processes.

In this chapter, we present an analysis of solvent-induced effects on biochemical processes in terms of the various side chains or functional groups of the protein. We shall identify solvent effects that involve both *HφO* and *HφI* side chains. After having obtained a list of all possible solvent effects, we shall examine their orders of magnitude, based on data that is presently available. We shall see that neither the *HφO* nor the *HφI* hypothesis, in the sense of Eqs. (1) and (2), is supported by the available data. We shall also point out methods of estimating the various solvent effects form experimental data. It is only after having obtained these data that one can turn to examine the relative importance of the various contributions to the overall solvent-induced effects.

II. PROTEIN FOLDING AND PROTEIN–PROTEIN ASSOCIATION

We discuss here the thermodynamics of two fundamental processes: folding and association of proteins. Other, more complex, processes may be viewed as combinations of these two processes.

Consider first the process of folding of a protein, from an initial unfolded (**U**) state to a final folded (**F**) state. We write the equilibrium transformation as

$$\mathbf{U} \rightleftarrows \mathbf{F} \tag{3}$$

The chemical equilibrium condition is

$$\mu_{\mathbf{U}} = \mu_{\mathbf{F}} = \mu_{\mathbf{P}} \tag{4}$$

where $\mu_{\mathbf{U}}$ and $\mu_{\mathbf{F}}$ are the chemical potentials of the species **U** and **F**, respectively. Because of the existence of chemical equilibrium, $\mu_{\mathbf{U}}$ and $\mu_{\mathbf{F}}$ are also equal to the chemical potential of the protein (**P**) in the system.

For each of the species $\boldsymbol{\alpha}$ (**U, F,** or **P**) we can always write

$$\mu_{\alpha}{}^{l} = \mu_{\alpha}^{*l} + kT \ln \rho_{\alpha}{}^{l}\, \Lambda_{\alpha}{}^{3} \tag{5}$$

where $\rho_{\alpha}{}^{l}$ is the number density of the species $\boldsymbol{\alpha}$, $\Lambda_{\alpha}{}^{3}$ is the momentum partition function, and μ_{α}^{*l} is the pseudochemical potential of $\boldsymbol{\alpha}$ in the liquid phase l [11]. The latter may be interpreted as the work (say, at constant pressure P and temperature T) to insert $\boldsymbol{\alpha}$ at a fixed point in the liquid; that is, the translational degrees of freedom of $\boldsymbol{\alpha}$ as a whole are frozen. In this particular reaction, since $\Lambda_{\alpha}{}^{3}$ depends only on the mass of $\boldsymbol{\alpha}$, we have $\Lambda_{\mathbf{U}}{}^{3} = \Lambda_{\mathbf{F}}{}^{3} = \Lambda_{\mathbf{P}}{}^{3}$. Also, we note that $\rho_{\alpha}\Lambda_{\alpha}{}^{3}$ is a dimensionless quantity, and Eq. (5) is valid for any system containing $\boldsymbol{\alpha}$ at any concentration.

Similarly, for an ideal gas phase, we write the chemical potential of the species $\boldsymbol{\alpha}$ as

$$\mu_{\alpha}{}^{g} = \mu_{\alpha}^{*g} + kT \ln \rho_{\alpha}{}^{g}\Lambda_{\alpha}{}^{3} \tag{6}$$

where μ_{α}^{*g} is the pseudochemical potential of $\boldsymbol{\alpha}$ in an ideal gas phase (i.e., there are no interactions among the molecules).

The solvation Gibbs energy of the species $\boldsymbol{\alpha}$ is defined by

$$\Delta G_{\alpha}^{*} = \mu_{\alpha}^{*l} - \mu_{\alpha}^{*g} \tag{7}$$

and the equilibrium constant for the reaction show in Eq. (3) follows from Eqs. (4) and (5):

$$\mathrm{K}^{l} = \left(\frac{\rho_{\mathbf{F}}{}^{l}}{\rho_{\mathbf{U}}{}^{l}}\right)_{\mathrm{eq}} = \exp[-\beta(\mu_{\mathbf{F}}^{*l} - \mu_{\mathbf{U}}^{*l})] \tag{8}$$

where $\beta = (kT)^{-1}$, with k Boltzmann's constant. Similarly, the equilibrium constant in the ideal gas phase is defined by

$$K^{g} = \left(\frac{\rho_{\mathbf{F}}{}^{g}}{\rho_{\mathbf{U}}{}^{g}}\right)_{\mathrm{eq}} = \exp[-\beta(\mu_{\mathbf{F}}^{*g} - \mu_{\mathbf{U}}^{*g})] \tag{9}$$

Note that Eqs. (5), (7), (8), and (9) differ from the corresponding equations based on conventional thermodynamics. In the latter, we use similar equations applicable only in the limit of very dilute solutions, in which case one can define the standard chemical potential of the species $\boldsymbol{\alpha}$ by

$$\mu_\alpha = \mu_\alpha^{0l} + kT \ln \rho_\alpha^{l} \tag{10}$$

where the relation between μ_α^{*l} and μ_α^{0l} is

$$\mu_\alpha^{0l} = \mu_\alpha^{*l} + kT \ln \Lambda_\alpha^3 \tag{11}$$

Note that Eq. (11) is meaningful only when ρ_α is small enough (the ideal dilute limit) and μ_α^{0l} depends on the units chosen for Λ_α. Since the quantity μ_α^{*l} is more general (independent of the concentration ρ_α^{l}, and independent of the units chosen for ρ_α and Λ_α^3, these being chosen so that $\rho_\alpha\Lambda_\alpha^3$ becomes dimensionless), we shall use the pseudochemical potentials rather than the corresponding conventional *standard* chemical potentials.

In order to proceed and analyze the content of the quantity μ_α^{*l} (and hence of ΔG_α^*, K^l, and K^g), we need to *define* more precisely the species **U** and **F** in Eq. (3). There are infinite ways of defining these species; these may be conveniently collected in three groups.

(a) Let **U** and **F** represent *one* conformation of the unfolded and the folded states, respectively. Clearly, there are many ways of choosing these representative conformations. For the purpose of the study of the solvent effects in the following sections, we may choose **U** as the fully extended conformation of the protein and **F** as the average crystallographic structure of the folded form. Having made this choice, we have effectively frozen in the internal rotational degrees of freedom of the molecule.

Note that this definition will make ρ_U and ρ_F infinitely small. Since we shall always be interested in concentration ratios, however, the various equilibrium constants as well as the solvation Gibbs energies are finite quantities. This particular choice is very convenient in studying all possible solvent-induced effects on reaction (3), as we shall see in subsequent sections.

Assuming that the internal partition function is separable from the configurational partition function, we can write the chemical potential of the species $\boldsymbol{\alpha}$ (= **U** or **F**) as

$$\mu_\alpha^{l} = W(\boldsymbol{\alpha}|w) + U(\boldsymbol{\alpha}) + kT \ln \rho_\alpha^{l}\Lambda_\alpha^3 q_\alpha^{-1} \tag{12}$$

where $U(\boldsymbol{\alpha})$ is the energy change for the process of bringing all atoms of the protein from fixed positions in their ground states to the final conformation of $\boldsymbol{\alpha}$ in its ground state. q_α includes the rotational–vibrational and electronic partition functions (note that internal rotations are not included). $W(\boldsymbol{\alpha}|w)$ is the coupling work, or the Gibbs energy of interaction of $\boldsymbol{\alpha}$ with the solvent molecules. Here w stands for water, but it can be any solvent or mixture of solvents. The quantity $W(\boldsymbol{\alpha}|w)$ will be the central quantity in our study of solvent effects.

(b) The second case involves theoretical grouping of conformational states. Suppose that instead of a single conformation representing the unfolded state, we choose to include in the species **U** all conformations that fulfill certain conditions [12,13] (say belonging to a certain region in the Ramachandran map—or any other arbitrary selection of a group of conformations).

For simplicity, let M_U be the number of discrete conformations that we have chosen as belonging to the species **U**. For each specific conformation i, we can write, as in (12)

$$\mu_i^l = W(i\,|w) + U(i) + kT \ln \rho_i^l \Lambda_i^3 q_i^{-1} \tag{13}$$

At equilibrium, we have

$$\mu_i^l = \mu_U = \mu_U^{*l} + kT \ln \rho_U^l \Lambda_U^3 \tag{14}$$

where μ_U is the chemical potential of the species **U**, which includes all the M_U conformations that we have chosen to define this species. Rewriting Eq. (13) as

$$\Lambda_i^3 \rho_i^l = q_i \exp[-\beta U(i) - \beta W(i\,|w)] \exp[\beta \mu_U^l] \tag{15}$$

summing over all i belonging to the **U** species, and noting that $\Lambda_i^3 = \Lambda_U^3$ for all $i = 1, 2, \ldots, M_U$, we obtain

$$\begin{aligned}\Lambda_U^3 \rho_U^l &= \Lambda_U^3 \sum_{i=1}^{M_U} \rho_i^l = \sum_{i=1}^{M_U} q_i \exp[-\beta U(i) - \beta W(i\,|\,w)] \exp[\beta \mu_U^l] \\ &= q_U \exp[\beta \mu_U^l]\end{aligned} \tag{16}$$

Comparing with Eq. (14), we find

$$\mu_U^{*l} = kT \ln q_U^{-1} \tag{17}$$

where

$$q_U = \sum_{i=1}^{M_U} q_i \exp[-\beta U(i) - \beta W(i\,|\,w)] \tag{18}$$

The solvation Gibbs energy of **U** in this case is

$$\begin{aligned}\exp[-\beta\, \Delta G_U^*] = \exp[-\beta[\mu_U^{*l} - \mu_U^{*g}]] &= \frac{\sum q_i \exp[-\beta U(i) - \beta W(i\,|\,w)]}{\sum q_i \exp[-\beta U(i)]} \\ &= \sum_i Y_i \exp[-\beta W(i\,|\,w)]\end{aligned} \tag{19}$$

where Y_i is the mole fraction of the ith conformation in the ideal gas phase. Thus, $\exp[-\beta\, \Delta G_U^*]$ is an average of the quantities $\exp[-\beta W(i\,|\,w)]$, with the distribution of species given by Y_i. The average is over all the conformations i we have chosen to define the species **U**. Similar treatment applies for the species **F**.

The equilibrium constants K^l and K^g are now related by

$$\begin{aligned}K^l &= \exp[-\beta(\mu_F^{*l} - \mu_U^{*l})] \\ &= \exp[-\beta(\mu_F^{*g} - \mu_U^{*g})] \exp[-\beta(\Delta G_F^* - \Delta G_U^*)] \\ &= K^g \exp[-\beta(\Delta G_F^* - \Delta G_U^*)]\end{aligned} \tag{20}$$

We define the *solvent-induced* contribution to reaction (3) as

$$\delta G(\mathbf{U} \to \mathbf{F}) = \Delta G_{\mathbf{F}}^{*} - \Delta G_{\mathbf{U}}^{*} \tag{21}$$

This quantity connects the chemical equilibrium constants K^{l} and K^{g}, for the same process defined in Eq. (3). The same process implies also the same definitions of the species **U** and **F** in the liquid and in the gaseous phases. Since each of the solvation Gibbs energies on the rhs of Eq. (2) is an average of the form of Eq. (19), $\delta G(\mathbf{U} \to \mathbf{F})$ is also an average quantity of the same kind.

(c) The third case involves the experimentally determined grouping of conformational states. In the previous case, we have arbitrarily chosen two groups of conformations to *define* the species **U** and **F**. A limiting case of this grouping is when **U** and **F** consist of one conformation each; this is case (a) discussed above. Experimentally, the species are defined not by *choosing* the conformational states, but by choosing the experimental method to measure the concentrations of the two species. Different experimental methods can, in principle, pick up different sets of conformational states. To demonstrate this point, suppose that we can follow the adsorption spectrum of two chromophores *A* and *B* attached to the protein at two different places (Fig. 1). If *A* and *B* have different signals according to whether they are surrounded by water or by the protein itself, then we may use either *A* or *B* to follow the melting curve, that is, the transition from the **F** to the **U** form.

In the folded state **F** (indicated by structure I in Fig. 1), both probes *A* and *B* are buried within the interior of the protein. In the unfolded state **U** (indicated by IV in Fig. 1), the two probes are exposed to water. Let II and III be two intermediate states; in one, *A* is exposed to the solvent but not *B*, and in the second, *B* is exposed to the solvent but not *A*.

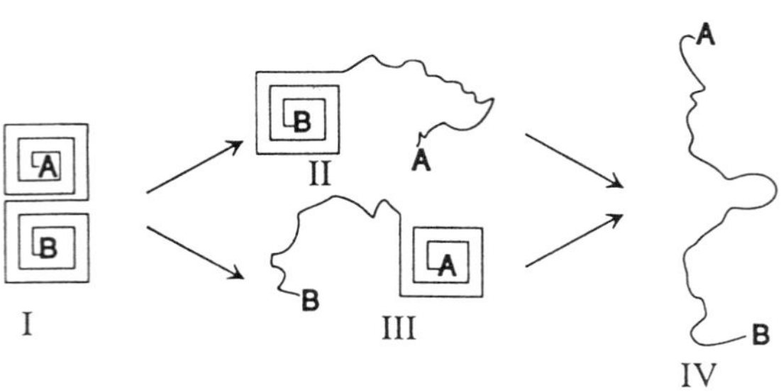

FIGURE 1 The denaturation of the folded form I into the unfolded form IV proceeds through two intermediary structures, II and III. The probe group *A* is exposed to the solvent in II and IV, whereas the probe group *B* is exposed to the solvent in III and IV. If we follow the spectrum of *A*, the mole fractions of **F** and **U** "read" by *A* are $x_{\mathbf{F}} = x_{\mathrm{I}} + x_{\mathrm{III}}$ and $x_{\mathbf{U}} = x_{\mathrm{II}} + x_{\mathrm{IV}}$, respectively. The corresponding mole fractions "read" by *B* are $x_{\mathbf{F}} = x_{\mathrm{I}} + x_{\mathrm{II}}$ and $x_{\mathbf{U}} = x_{\mathrm{III}} + x_{\mathrm{IV}}$, respectively.

If we follow the unfolding process by following the change of spectra of *A*, the mole fractions of the **U** and **F** forms will include

$$x_{\mathbf{F}} = x_{\mathrm{I}} + x_{\mathrm{III}}, \qquad x_{\mathbf{U}} = x_{\mathrm{II}} + x_{\mathrm{IV}} \tag{22}$$

But using *B* as our probe, the mole fractions, as "seen" by this probe, are different, that is,

$$x_{\mathbf{F}} = x_{\mathrm{I}} + x_{\mathrm{II}}, \qquad x_{\mathbf{U}} = x_{\mathrm{III}} + x_{\mathrm{IV}} \tag{23}$$

We see that the grouping of states is determined by the method we use to define the species **U** and **F**.

Once we recognize the specific grouping of states imposed by the specific experimental method, all the equations of the preceding subsection apply to this case, provided we understand the summation in Eqs. (16), (17), etc., as carried out over all states pertaining to the group of states chosen by the experimental method, not by our arbitrary decision as made in case (b).

In the analysis of the solvent effects carried out in subsequent sections, we shall use only one conformer to represent each of the species. It should be remembered that in order to establish the connection with the corresponding experimental quantities, a proper average over all conformations belonging to each species should be taken.

We now turn briefly to the association reaction. Let **P** and **L** be any two proteins, or a protein and a ligand, that form the complex **PL**. The reaction is

$$\mathbf{P} + \mathbf{L} \rightleftarrows \mathbf{PL} \tag{24}$$

The equilibrium condition is

$$\mu_{\mathbf{P}} + \mu_{\mathbf{L}} = \mu_{\mathbf{PL}} \tag{25}$$

from which we obtain the various equilibrium constants in the liquid and in an ideal gas phase.

These are defined as

$$K^{\mathrm{l}} = \left(\frac{\rho_{\mathbf{PL}}{}^{\mathrm{l}}}{\rho_{\mathbf{P}}{}^{\mathrm{l}}\rho_{\mathbf{L}}{}^{\mathrm{l}}}\right)_{\mathrm{eq}}, \qquad K^{\mathrm{g}} = \left(\frac{\rho_{\mathbf{PL}}{}^{\mathrm{g}}}{\rho_{\mathbf{P}}{}^{\mathrm{g}}\rho_{\mathbf{L}}{}^{\mathrm{g}}}\right)_{\mathrm{eq}} \tag{26}$$

and the connection between these two quantities is

$$K^{\mathrm{l}} = K^{\mathrm{g}} \exp[-\beta\delta G(\mathbf{P} + \mathbf{L} \rightarrow \mathbf{PL})] \tag{27}$$

where δG is given by

$$\delta G(\mathbf{P} + \mathbf{L} \rightarrow \mathbf{PL}) = \Delta G^{*}_{\mathbf{PL}} - \Delta G^{*}_{\mathbf{P}} - \Delta G^{*}_{\mathbf{L}} \tag{28}$$

The analogy between relations (27) and (20) is quite evident. Note that the quantities K^{l} and K^{g} as defined in Eq. (26) depend on the concentration units. The ratio of the two equilibrium constants, however, is independent of the chosen units.

We shall see in the next sections that the two quantities $\delta G(\mathbf{U} \rightarrow \mathbf{F})$ and $\delta G(\mathbf{P} + \mathbf{L} \rightarrow \mathbf{PL})$ have the same types of ingredients. Therefore, it is sufficient to examine the details of one reaction only and then apply this information to other, more complex processes.

III. DIRECT AND INDIRECT INTERACTIONS

We now return to the folding reaction (Eq. (3)), with the understanding that **U** and **F** are represented by one configuration each (this is case (a) discussed in the preceding section.) In this case, we can fix the position and orientation of the molecule.

The total change in the Gibbs energy is

$$\Delta G(\mathbf{U} \rightarrow \mathbf{F}) = \Delta U(\mathbf{U} \rightarrow \mathbf{F}) + \delta G(\mathbf{U} \rightarrow \mathbf{F}) \tag{29}$$

Since the molecules are devoid of translational and rotational degrees of freedom, the quantity ΔU on the rhs of Eq. (29) is simply the change in the potential energy of the protein associated with the transition $\mathbf{U} \rightarrow \mathbf{F}$. $\delta G(\mathbf{U} \rightarrow \mathbf{F})$ includes all possible effects due to the solvent. Traditionally, one assumes the solvent has a negligible effect on the internal degrees of freedom of the solute molecules. This assumption is plausible for simple solutes in a simple solvent. It is clearly an oversimplification for hydrogen-bonding solutes in water. To estimate the contribution to δG from solvent effects on the internal degrees of freedom would require a full quantum mechanical treatment of the entire solute and solvent system. Since it is impossible to carry out such a calculation at present, we shall adopt here the traditional approach; that is, we shall view the hydrogen-bonding capability as part of the potential function and treat the whole system of protein and water classically. Even with this drastic approximation, we still cannot estimate δG, which in our example involves averages over all the configurations of the solvent molecules; more specifically, the classical statistical mechanical expression for $\delta G(\mathbf{U} \rightarrow \mathbf{F})$ is [12,13]

$$\delta G(\mathbf{U} \rightarrow \mathbf{F}) = -kT \ln \frac{\langle \exp[-\beta B_{\mathrm{F}}] \rangle_0}{\langle \exp[-\beta B_{\mathrm{U}}] \rangle_0} \tag{30}$$

where B_α is the total interaction energy between a solute $\boldsymbol{\alpha}$ (here $\boldsymbol{\alpha}$ is either **U** or **F**) and all solvent molecules at some specific configuration, which we denote by $\mathbf{X}^N = \mathbf{X}_1 \cdots \mathbf{X}_N$; that is,

$$B_\alpha = \sum_{i=1}^{N} U(\mathbf{X}_\alpha, \mathbf{X}_i) \tag{31}$$

where $U(\mathbf{X}\boldsymbol{\alpha}, \mathrm{X}i)$ is the solute–solvent interaction energy at the configuration $\mathbf{X}\boldsymbol{\alpha}, \mathbf{X}i$. The averages $\langle\ \rangle_0$ in Eq. (30) are over all possible configurations of the solvent molecules in the $T, P,$ **N** ensemble. Note that if **U** and **F** are viewed as in cases

(b) and (c) of Sec. II, then a second average must be taken over all conformations of the protein [13]. Even without the additional average over conformations, however, the averages over solvent configurations involve integration over some $6N$ coordinates, with N on the order of Avogadro's number. Such averages cannot be computed even for simple liquids. What we shall be able to do in subsequent sections is to analyze δG in terms of its ingredients and then see if we can estimate some of these ingredients by either experimental or theoretical means.

There is a very fundamental difference between the two quantities on the rhs of Eq. (29). The first quantity involves *direct* interactions between atoms or groups of atoms *within* the protein molecule. This energy difference may be expressed symbolically as

$$\Delta U(\mathbf{U} \to \mathbf{F}) = U(\mathbf{F}) - U(\mathbf{U}) \tag{32}$$

where $U(\mathbf{F})$ and $U(\mathbf{U})$ are the energies of the protein in the states **F** and **U**, respectively, measured with respect to some common reference state (say, all the atoms at fixed positions and at infinite separation from each other). The potential energy function $U(\mathbf{R}^M)$ (where $\mathbf{R}^M = \mathbf{R}_1 \cdots \mathbf{R}_M$ is the *configuration* of the protein as defined by the locations $\mathbf{R}_i$ of all of its M atoms) is an extremely complicated function even for a small protein. Yet this function is by far the simpler of the two quantities on the rhs of Eq. (29). The main reason for this is that we know the ingredients that constitute the potential energy of the protein. These ingredients include bond energies, van der Waals, hydrogen bonds, and electrostatic interactions of which we know something either from experimental or from theoretical sources. Applying the transferability principle, we can construct the function $U(\mathbf{R}^M)$ by summing all the ingredients that contribute to the total potential function.

The quantity δG is "orders of magnitude" more difficult to study. Here we still do not know all the pertinent ingredients. Even if we knew these, they would not combine additively to form δG. This difficulty is the main reason for our meager understanding of solvent effects.

To demonstrate this difficulty, consider the following simple example. Let $\mathbf{R}_C$ and $\mathbf{R}_T$ be two conformers of *n*-butane, say, cis (C) and trans (T) as depicted in Fig. 2. Assuming that the bond energies are the same in these two conformations, the energy difference can be described by

$$\Delta U(\mathrm{C} \to \mathrm{T}) = U(\mathbf{R}_T) - U(\mathbf{R}_C) \tag{33}$$

and is due mainly to the difference in the van der Waals interactions between pairs of nonbonded groups. Furthermore, we can assume these interactions are *pairwise* additive and draw dashed lines to indicate the 1–3, 2–4, and the 1–4 pair interactions (Fig. 3). Of these three interactions, only the 1–4 pair has been significantly changed in the transition C → T. Hence, we can assign the total change in the potential function $\Delta U(\mathrm{C} \to \mathrm{T})$ to the difference in the 1–4 pair interaction in the two conformers (Fig. 4).

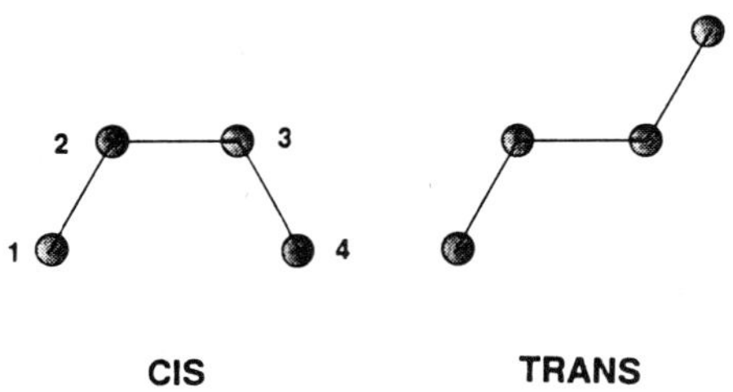

FIGURE 2 Schematic illustration of two conformations of *n*-butane: cis (C) and trans (T).

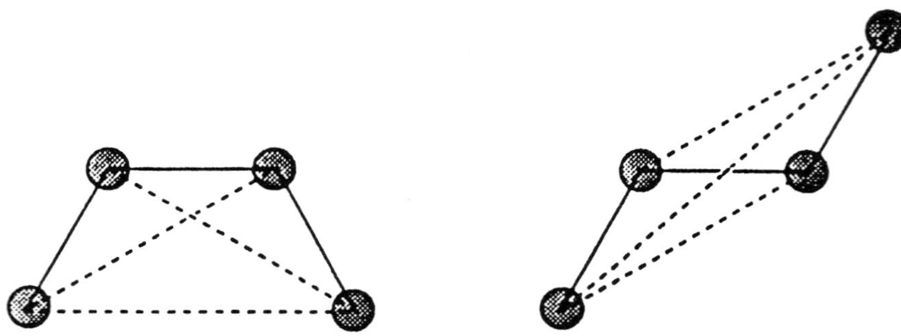

FIGURE 3 Dashed lines indicate interactions between pairs of nonbonded groups in Fig. 2.

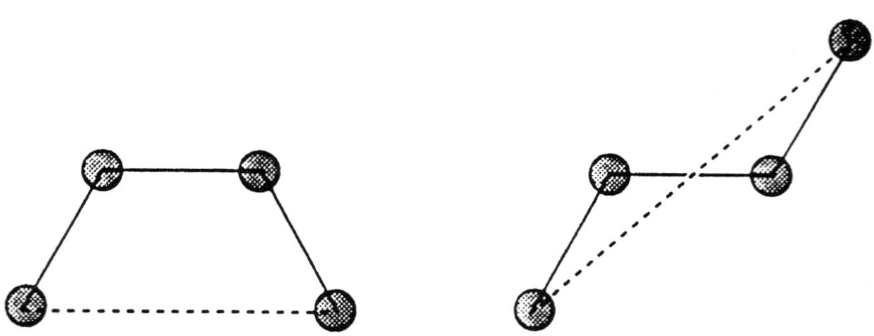

FIGURE 4 The main difference between interactions of nonbonded groups in the two conformations of Fig. 2 is due to the 1–4 pair.

Similar assignment of pairwise interactions is not possible for $\delta G(\mathrm{C} \rightarrow \mathrm{T})$. Here, in general, we *cannot* draw lines indicating solvent-induced interaction between pairs of atoms or groups of atoms. In other words, the pair of groups, say, 1 and 4 in Fig. 4, will contribute approximately the same to ΔU whether they are part of the butane or of any other molecule. The direct interaction, represented by the dashed line connecting 1 and 4, is independent of the molecular context. This is not true for δG. This function depends on the entire configuration of the solute molecule and cannot be split into pairwise additive indirect interactions.

Another example is the aggregation of M simple solute molecules to form a compact droplet, shown in Fig. 5. The Gibbs energy change for the process of bringing the M solutes from fixed positions at infinite separation to the final configuration $\mathbf{R}^M$ is written as

$$\begin{aligned}\Delta G(\infty \to \mathbf{R}^M) &= G(\mathbf{R}^M) - G(\infty)\\ &= U(\mathbf{R}^M) + \delta G(\infty \to \mathbf{R}^M)\\ &= \sum_{i<j}^{N} U(\mathbf{R}_{ij}) + \delta G(\infty \to \mathbf{R}^M)\end{aligned} \tag{34}$$

In the last equality in Eq. (34), we have assumed pairwise additivity of the total *direct* interaction among the M solute particles. Hence, one can draw pairwise interaction lines connecting pairs of particles. The same cannot be done for $\delta G(\infty \to \mathbf{R}^M)$, which cannot be split into a sum of pairwise additive contributions. We have the identity [7,13]

$$\delta G(\infty \to \mathbf{R}^M) = \Delta G^*(\mathbf{R}^M) - M\,\Delta G_1^* \tag{35}$$

where on the rhs of Eq. (35) we have the difference in the solvation Gibbs energy of the aggregate $\mathbf{R}^M$ and of M single solutes. $\Delta G^*(\mathbf{R}^M)$ depends on the configuration of the *entire* aggregate. If the solutes are hard spheres, then $\Delta G^*(\mathbf{R}^M)$ is simply the work of creating a cavity in the solvent suitable for accommodating the aggregate in the configuration $\mathbf{R}^M$. This work cannot be further split into pairwise indirect interactions terms, each of which is independent of the context in which it is found within the aggregate.

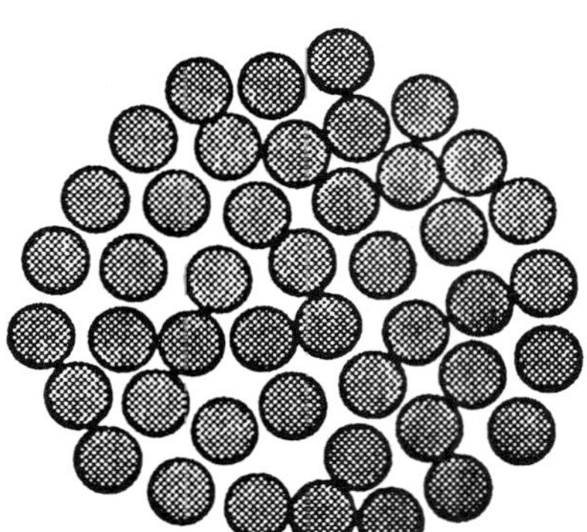

FIGURE 5 A compact aggregate formed by M simple solutes in a solvent.

The same argument applies to $\delta G(\mathbf{U} \rightarrow \mathbf{F})$. Even for the simple case where **U** and **F** are chosen to have only one conformation each (case (a) of Sec. II), we have the ratio of the two average quantities, as shown in Eq. (30). There is no way to break this quantity into pairwise additive contributions. Nevertheless, we shall see in Sec. V that $\delta G(\mathbf{U} \rightarrow \mathbf{F})$ can be broken into sums of contributions. Each of these contributions, although still involving the same kind of averaging of solvent configurations, can be studied by using small model compounds.

We now briefly turn to the association process. Here again, we consider the process of bringing **P** and **L** from fixed positions and orientations and at infinite separation to form the complex **PL.** Again, we assume that **P**, **L**, and **PL** are represented by a single conformation each. Since the molecules are devoid of translational and rotational degrees of freedom, the total Gibbs energy of the process can be written as

$$\Delta G(\mathbf{P} + \mathbf{L} \rightarrow \mathbf{PL}) = \Delta U(\mathbf{P} + \mathbf{L} \rightarrow \mathbf{PL}) + \delta G(\mathbf{P} + \mathbf{L} \rightarrow \mathbf{PL}) \tag{36}$$

Here again, ΔU is the *direct* interaction energy between **P** and **L** at the final configuration **PL**. This can be written to a good approximation as a sum of pairwise additive terms,

$$\Delta U(\mathbf{P} + \mathbf{L} \rightarrow \mathbf{PL}) = \sum_{\substack{i \in P \\ j \in L}} U(\mathbf{R}_{ij}) \tag{37}$$

where $U(\mathbf{R}_{ij})$ is the direct interaction energy between atom i on **P** and atom j on **L** (the internal interactions among all atoms within **P** or within **L** are presumed to be unchanged in the binding process). Each of these pair interactions can be studied independently of the environmental context in which it occurs. It is therefore meaningful to describe ΔU by lines connecting all atoms (or groups) i on **P** to all atoms (or groups) j on **L**. Each line represents a quantity of energy. The sum of these energies constitutes ΔU.

Similar assignment of "lines" representing indirect interactions cannot be done for δG. For the association process, we have the identity [13]

$$\delta G(\mathbf{P} + \mathbf{L} \rightarrow \mathbf{PL}) = \Delta G^*_{\mathbf{PL}} - \Delta G^*_{\mathbf{P}} - \Delta G^*_{\mathbf{L}} \tag{38}$$

which in analogy with Eq. (30) states that δG is the difference of the solvation Gibbs energy between the final state (**PL**) and the initial state (**P** and **L** at infinite separation). As in the case of the folding process, δG in Eq. (38) cannot be broken into pairwise additive contributions. Therefore, it is meaningless to draw lines that are presumed to represent pairwise indirect interactions, as is very often done in the biochemical literature. As an example, the binding Gibbs energy for the process depicted in Fig. 6 is described in terms of charge–charge, hydrogen-bonding, and hydrophobic interactions. The correct description should be that ΔG consists of two terms, direct and indirect interactions (Eq. (36)). The *direct* interaction consists of one charge–charge, one hydrogen-bond, and one van der Waals interaction (assuming that all other pairs are far apart and do not contribute to ΔU). The total *indirect* interaction δG cannot be described in a similar manner by assigning solvent-induced interactions to each pair of groups in the binding region of Fig. 6. We shall see in the following that δG depends on the configuration of the entire molecules, **P**, **L**, and **PL**. In some cases, significant contributions to δG can arise from groups (either $H\phi O$ or $H\phi I$) that do not belong to the binding region (i.e., groups that do not interact directly as the six groups depicted in Fig. 6).

In conclusion, for each of the processes discussed above we can separate the direct from the indirect contributions to ΔG. The former can, to a good approximation, be viewed as a sum of pairwise terms. The latter cannot be broken into a sum of similar terms, and other means of studying its ingredients are needed. We have also simplified our treatment by choosing one conformation for each species. This simplification is done for convenience in the study of all solvent effects. It should be remembered, however, that for any real process we must also take a second average over all possible conformations that belong to the relevant species [13].

IV. DRIVING FORCE, FORCE, AND STABILITY

In this section, we introduce a few additional concepts that are useful in the study of the thermodynamics of biochemical processes. Since our interest is mainly focused on the solvent contributions, it is convenient to freeze the translational and rotational degrees of freedom of the solutes involved in the process. The total *driving force* for the folding and association reactions are

$$\Delta G(\mathbf{U} \to \mathbf{F}) = \Delta U(\mathbf{U} \to \mathbf{F}) + \delta G(\mathbf{U} \to \mathbf{F}) \tag{39}$$

$$\Delta G(\mathbf{P} + \mathbf{L} \to \mathbf{PL}) = \Delta U(\mathbf{P} + \mathbf{L} \to \mathbf{PL}) + \delta G(\mathbf{P} + \mathbf{L} \to \mathbf{PL}) \tag{40}$$

When $\Delta G < 0$ for one of these processes, we say that there is a driving force to reach the final state from the initial state. For instance, if we have one protein in the **U** form in the solvent, a negative $\Delta G(\mathbf{U} \to \mathbf{F})$ implies that the **F** form is *rela-*

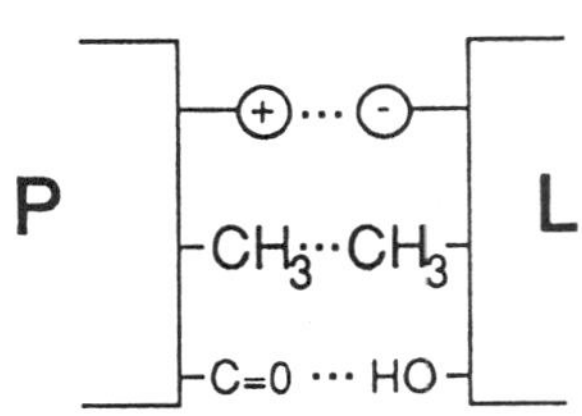

FIGURE 6 Schematic illustration of the direct interactions between groups on **P** and groups on **L**.

tively more stable than the **U** form. This does not mean that **U** will be totally converted to the **F** form. In fact, this never occurs. All we can say is that ΔG governs the probability ratio of finding the protein in the two states. Let $P_{\mathbf{F}}{}^{l}$ and $P_{\mathbf{U}}{}^{l}$ be the probabilities of finding the protein (at a fixed position in a system having fixed temperature, pressure, and composition) in the **F** and **U** states, respectively. Then

$$\left(\frac{P_{\mathbf{F}}{}^{l}}{P_{\mathbf{U}}{}^{l}}\right)_{eq} = \exp[-\beta\,\Delta G(\mathbf{U} \rightarrow \mathbf{F})] \tag{41}$$

The more negative ΔG is, the more populated is the **F** state relative to the **U** state.

Denoting by $P_{\mathbf{F}}{}^{g}$ and $P_{\mathbf{U}}{}^{g}$ the corresponding probabilities in the gaseous phase, we have the general relation between the probability ratio in the two phases:

$$\left(\frac{P_{\mathbf{F}}{}^{l}}{P_{\mathbf{U}}{}^{l}}\right)_{eq} = \exp[-\beta\,\Delta U]\exp[-\delta G] = \left(\frac{P_{\mathbf{F}}{}^{g}}{P_{\mathbf{U}}{}^{g}}\right)_{eq} \exp[-\delta G] \tag{42}$$

Thus, $\Delta U(\mathbf{U} \rightarrow \mathbf{F})$ determines the probability ratio in the ideal gas phase, whereas $\Delta G(\mathbf{U} \rightarrow \mathbf{F})$ determines the probability ratio in the liquid phase.

The probability ratios are useful for the study of the energetics of the processes. In order to establish the connection with the density ratio, or the equilibrium constants, we need to free the molecules from their fixed position and orientation.

If the two species **U** and **F** have one conformation each, then the density ratio is different from the probability ratio. The relation between the two is (see Eq. (20))

$$K^{l} = \left(\frac{\rho_{\mathbf{F}}{}^{l}}{\rho_{\mathbf{U}}{}^{l}}\right)_{eq} = \exp[-\beta(\mu_{\mathbf{F}}^{*l} - \mu_{\mathbf{U}}^{*l})] = \frac{q_{\mathrm{rot},\mathbf{F}}}{q_{\mathrm{rot},\mathbf{U}}}\exp[-\beta\,\Delta G(\mathbf{U} \rightarrow \mathbf{F})] = \frac{q_{\mathrm{rot},\mathbf{F}}}{q_{\mathrm{rot},\mathbf{U}}}\left(\frac{P_{\mathbf{F}}{}^{l}}{P_{\mathbf{U}}{}^{l}}\right)_{eq} \tag{43}$$

Thus, liberating the protein from the fixed position and orientation modifies the probability ratio given in Eq. (42) by the ratio of the rotational partition functions of the two forms. Here, the momentum partition functions of the two forms are the same. In the association process, both rotational and translational partition functions should be taken into account.

For the experimental equilibrium constant, we usually need to consider also averages over all conformations that belong to the **U** and the **F** states (see case (c) in Sec. II). In this case, the rotational partition function of the entire molecule, as well as the internal rotational partition functions, will modify the density ratio of the two forms.

When speaking about stability, care must be exercised in distinguishing between the stability of the system and the relative stability of the two species. A thermodynamic system, characterized by the temperature T, pressure P, and composition $\mathbf{N}$ is said to be stable when the Gibbs energy of the system reaches its minimum value with respect to some internal variable. The internal variable could be the location of a movable partition, separating two compartments, or the mole fraction of the two forms. For instance, the condition of stability of the system having $\mathbf{N}$ water molecules at a given T and P, and protein molecules that can attain one of two states **U** or **F** is

$$\frac{\partial G(P,T,\mathbf{N};x_\mathrm{F})}{\partial x_\mathrm{F}} = 0 \tag{44}$$

The minimum is with respect to the mole fraction x_F, when P, T, $\mathbf{N}$ are held constant; see Fig. 7.

The stability condition leads to the equality of the chemical potentials (Eq. (4)) and to the corresponding equilibrium constant (Eq. (8)). Whenever we *prepare* a system with a mole fraction x_F that is different from the equilibrium mole fraction that satisfies Eq. (44), the system is unstable with respect to the variable x_F. The system will tend to reach the minimum value of $G(T,P,\mathbf{N};x_\mathrm{F})$ by letting x_F attain the equilibrium value. Two particular cases are $x_\mathrm{F} = 1$ and $x_\mathrm{U} = 1$, that is, when we start with a system containing the pure **F** or pure **U** form. In both cases, letting the protein solution attain equilibrium with respect to the conversion between the two forms, the Gibbs energy will be reduced and the eventual value of x_F will be the equilibrium value, that is, the value of x_F that minimizes the Gibbs energy, Eq. (44).

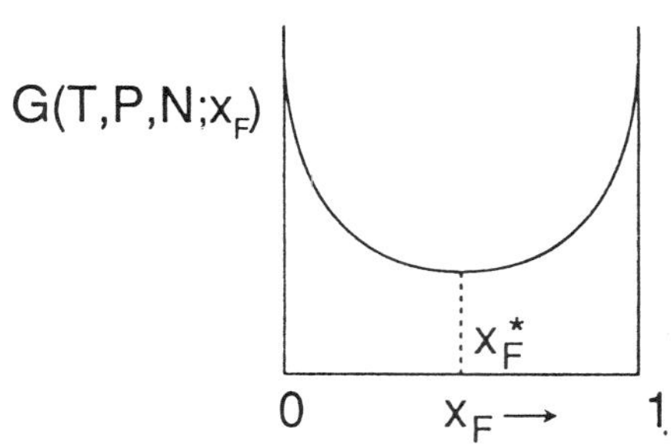

FIGURE 7 The Gibbs energy as a function of the mole fraction x_F has a single minimum at x^*_F.

A different view of the stability of a protein solution system is the following: Let $\mathbf{R}^M$ be any specific conformation of the protein (i.e., the locations $\mathbf{R}_i$ of all its atoms are fixed). The Gibbs energy of the system can now be viewed as a function of the internal parameters $\mathbf{R}^M$. We thus construct the function $G(T, P, \mathbf{N}; \mathbf{R}^M)$ and ask, what is the minimum of G as a function of the variables $\mathbf{R}^M$, keeping P, T, $\mathbf{N}$ constant?

In this case, we cannot expect a single minimum at some configuration $\mathbf{R}^{*M}$. Starting from any arbitrary initial conformation $\mathbf{R}^M$, the system, in general, does not proceed to attain the configuration $\mathbf{R}^{*M}$ for which the Gibbs energy has a minimum. This is in contrast to the case of the function $G(T, P, \mathbf{N}; x_\mathbf{F})$ depicted in Fig. 7. Starting from any $x_\mathbf{F}$ different from $x_\mathbf{F}^*$, the system will tend monotonically to $x_\mathbf{F}^*$ and the relative fluctuations about $x_\mathbf{F}^*$ are small. Starting with any conformation $\mathbf{R}^M$ other than $\mathbf{R}^M$ (presuming that $\mathbf{R}^{*M}$ is the conformation for which $G(T, P, \mathbf{N}; \mathbf{R}^M$ has an absolute minimum) the system will not proceed monotonically to $\mathbf{R}^{*M}$. Even after $\mathbf{R}^{*M}$ has been reached, fluctuations about $\mathbf{R}^{*M}$ will be large.

If there is an absolute minimum of $G(T, P, \mathbf{N}; \mathbf{R}^M)$ at $\mathbf{R}^{*M}$, all we can say is that the conformation $\mathbf{R}^{*M}$ is relatively more stable than any other conformation $\mathbf{R}^{*M}$. We can also say that there exists a *thermodynamic force* or driving force leading from $\mathbf{R}^M$ to $\mathbf{R}^{*M}$. A special case is when $\mathbf{R}^M$ can attain only two conformations, say **U** and **F** as discussed in Sec. II. In this case, we can speak of the relative stability of the two conformations. This quantity is related to the measurable relative stability of the two species ΔG^0 ($\mathbf{U} \rightarrow \mathbf{F}$) [or $\Delta G^*(\mathbf{U} \rightarrow \mathbf{F})$] as studied by protein chemists [14–17]. Sometimes the dependence of ΔG^0 on the temperature is referred to as the stability curve. Note, however, that this is different form the stability curve in the sense of Fig. 7. In Fig. 7, we have the function $G(T, P, \mathbf{N}; x_\mathbf{F})$, where $T, P, \mathbf{N}$ are fixed and $x_\mathbf{F}$ is some internal parameter that can be varied. On the other hand, ΔG^0 as a function of the temperature is better referred to as the *relative stability* of the two species as a function of temperature, rather than as a stability curve.

As we have defined it, the thermodynamic driving force has dimensions of *energy*. This is not a *force* in the mechanical sense. One can ask what is the force acting on the molecule, or on part of it, that sets it in motion? The general answer to this question can be obtained from the gradient of the function $G(\mathbf{R}^M)$ with respect to one of the variables $\mathbf{R}_i$.

The force exerted on an atom, or on a group, at $\mathbf{R}_i$ is given by [18]

$$F_i = -\nabla_i G(\mathbf{R}^M) = -\nabla_i U(\mathbf{R}^M) - \nabla_i\, \delta G(\mathbf{R}^M) \tag{45}$$

where ∇_i is the gradient operator with respect to the vector $\mathbf{R}_i$. The first term on the rhs of Eq. (45) is the *direct* force exerted on the *i*th group, by all the atom of the protein itself, being at some specific configuration $\mathbf{R}^M$. Knowing the potential function $U(\mathbf{R}^M)$, one can easily find the force acting on any atom. The second term is the solvent-induced, or the *indirect*, force acting on the *i*th group. It is straightforward to show that the indirect force is given by

$$F_i^{(\mathrm{SI})} = -\nabla_i \, \Delta G^*(\mathbf{R}^M) \tag{46}$$

A solvent-induced force exists if and only if an infinitesimal change in the vector $\mathbf{R}_i$ produces a change in the solvation Gibbs energy of the *entire* molecule at $\mathbf{R}^M$. This force is more difficult to study since it is derived from an average over all possible configurations of the solvent molecules. The statistical mechanical expression for the force is [13]

$$F_i^{(\mathrm{SI})} = \int -\nabla_i U(\mathbf{R}_i, \mathbf{X}_\mathrm{W}) \rho(\mathbf{X}_\mathrm{W}/\mathbf{R}^M) \, d\mathbf{X}_\mathrm{W} \tag{47}$$

where $-\nabla_i U(\mathbf{R}_i, \mathbf{X}_\mathrm{W})$ is the *direct* force exerted by a *water* molecule at $\mathbf{X}_\mathrm{W}$ on the group at $\mathbf{R}_i$ and $\rho(\mathbf{X}_\mathrm{W}/\mathbf{R}^M)$ is the local density of water molecules at $\mathbf{X}_\mathrm{W}$ given a specific conformation of the solute $\mathbf{R}^M$.

In order to simulate the motion of the protein in a solvent, we need to know the total (both direct and indirect) force exerted on each part of the molecule at each configuration $\mathbf{R}^M$. The direct force is unique, in the sense that given the function $U(\mathbf{R}^M)$ we can operate the exact force at each atom of the protein. The indirect force is only the *average* force induced by the solvent. The exact, or instantaneous, force is unpredictable since it depends on the configuration of the water molecules in the neighborhood of *i* at the moment. The function $\delta G(\mathbf{R}^M)$ provides only the *average* solvent-induced forces.

The difference between a driving force, the average force, and the instantaneous force can be illustrated with a simple example.

Consider two simple solutes, say two argon atoms, at fixed positions in a solvent. Let R be the distance between the centers of the two atoms. For each R we can construct the Gibbs energy function $G(T, P, \mathbf{N}; R)$, or simply $G(R)$. A typical curve is shown in Fig. 8. In Fig. 8a, the function $G(R)$ in vacuum is simply the pair potential $U(R)$. If R_e is the distance for which $U(R)$ has a minimum, then for any initial distance $R \neq R_\mathrm{e}$ the system will have a driving force to reach R_e. Also, the *force* at any point, say at R_1 or at R_2, is also in the same direction as the driving force, that is, toward R_e.

The situation is quite different when the same two solutes are in a solvent. In Fig. 8b we have again a distance R_e for which $G(R)$ has the lowest minimum. Starting from any initial distance R, there is always a driving force from R to R_e. The *force*, however, may be toward R_e if we start at R_1, away from R_e if we start at R_2, and no force at R_3. Note also that if we start from the initial distance R_2 and release the two solute particles, the instantaneous motion depends on the instantaneous configuration of all the solvent molecules in the surroundings of the two solutes. This motion is unpredictable from the $G(R)$ curve. All we can say from the $G(R)$ curve is that the *average* force at R_1 is attractive, and repulsive at R_2.

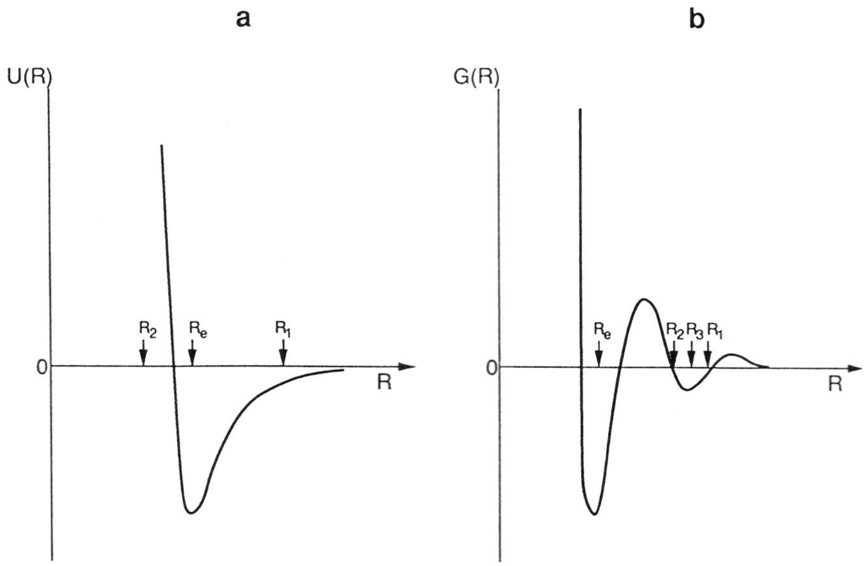

FIGURE 8 The direct pair potential (a) and the Gibbs energy function $G(R)$ (b) as a function of the distance R between two simple spherical atoms.

This force does not necessarily have to be in the same direction as the thermodynamic force.

The generalization to the case of proteins is quite straightforward. Instead of the single variable in $G(R)$, we now have the multiple variable function $G(\mathbf{R}^M)$ ($T, P, \mathbf{N}$ are presumed to be constants). Again, there might be a single configuration $\mathbf{R}^{*M}$ for which this function has an absolute minimum. Therefore, for any initial configuration $\mathbf{R}^M$, there is a thermodynamic force toward $\mathbf{R}^M$. Releasing the molecule at any initial configuration $\mathbf{R}^M$, however, we cannot tell what are the instantaneous forces that are exerted on all the atoms of the protein. Taking the gradient of $G(\mathbf{R}^M)$ as in Eq. (45) will tell us only the *average* force exerted on atom i, given the configuration $\mathbf{R}^M$. This force can be in the direction toward $\mathbf{R}^{*M}$ or away from $\mathbf{R}^M$. Also, knowing the function $G(\mathbf{R}^M)$ does not tell us whether or not the minimum *can* be reached from any initial configuration $\mathbf{R}^M$. Thus, although there exists a thermodynamic driving force to reach $\mathbf{R}^{*M}$, the actual arrival at $\mathbf{R}^{*M}$ depends on the initial configuration and on the temperature.

We conclude this section with the definitions of the energy and entropy changes associated with the process (either the folding or the association). These are defined by

$$\Delta E = \frac{\partial(\beta\,\Delta G)}{\partial\beta} \tag{48}$$

$$-T\,\Delta S = \Delta G - \Delta E \tag{49}$$

The thermodynamic energy change ΔE is sometimes confused with the direct energy change ΔU. The connection between the two quantities follows from Eq. (39) or (40), that is,

$$\Delta E = \Delta U + \frac{\partial(\beta\, \delta G)}{\partial \beta} \tag{50}$$

where we have assumed that ΔU is independent of the temperature (this is true for case (a) Sec. II, but not true for cases (b) and (c), where ΔU is an average quantity).

Thus, the thermodynamic energy change has one component due to the direct interaction energy. The second component is due to the temperature dependence of the solvation Gibbs energies of all the solutes involved in the process. The latter may be shown to be related to structural changes induced in the solvent by the process [3].

The entropy change corresponding to the process under discussion is obtained from Eq. (49):

$$-T\,\Delta S = \delta G - \frac{\partial(\beta\, \delta G)}{\partial \beta} \tag{51}$$

Thus, if ΔU is truly temperature-independent (as in case (a) of Sec. II), then all the entropy change associated with the process results from changes in the solvation Gibbs energies of the various solutes involved in the process.

If ΔU is an average over all conformations of the species (as in cases (b) and (c) of Sec. II), then it is temperature-dependent, and instead of Eq. (51) we have

$$-T\,\Delta S = \Delta U + \delta G - \frac{\partial[\beta\, \Delta U - \beta\, \delta G]}{\partial \beta} \tag{52}$$

V. INVENTORY OF SOLVENT-INDUCED EFFECTS

We consider here two idealized processes of folding and association. For these particular processes, we shall derive a complete list of all possible solvent effects. This analysis will also indicate the way to obtain an inventory of solvent effects for processes involving real proteins.

Consider a linear hydrocarbon backbone having altogether M side chains of two kinds only: methyl groups to represent $H\phi O$ groups, and hydroxyl or carbonyl groups to represent $H\phi I$ groups. We also assume that these side chains are far apart from each other, so that when the molecule is in the **U** form, they are independently solvated. This is to say, the solvation spheres around these groups do not overlap. (For more precise definition of this independence, see Eq. (57).)

The **F** form is presumed to have a compact conformation, where some of the side chains (both $H\phi O$ and $H\phi I$) might be buried in the interior of the molecule. These are the side chains that are unexposed to the solvent. The remaining side chains are located on the surface of the molecule. We assume that the latter are either independently solvated or pairwise correlated. No higher-order correlations among the side chains in the **F** form are taken into account. The process is schematically depicted in Fig. 9.

There are many differences between this model polymer and a real protein. The polypeptide backbone is different from a hydrocarbon backbone. There are some 20 different side chains. Correlations between pairs, triplets, and higher orders exist both in the **U** and **F** forms; partially buried side chains in the **F** form are solvated differently from fully exposed side chains, etc. All these make the analysis of the solvent effects on real proteins extremely difficult. Once we have learned, however, how to obtain the list of solvent effects in our model compound, and perhaps also when we are able to say something about the order of magnitude of the various effects, it will be rather straightforward to extend the analysis to real proteins. Such an analysis will also indicate which experiments are needed to obtain the required estimates of solvent effects in real proteins.

For our idealized polymer, we write the general expression for the quantity $\delta G(\mathbf{U} \rightarrow \mathbf{F})$ as

$$\delta G(\mathbf{U} \rightarrow \mathbf{F}) = \Delta G_{\mathbf{F}}^{*} - \Delta G_{\mathbf{U}}^{*} \tag{53}$$

Following the assumption of independence of all side chains in the **U** form, we write

$$\Delta G_{\mathbf{U}}^{*} = \Delta G_{\mathbf{U}}^{*\mathrm{BB}} = \sum_{i=1}^{M} \Delta G_{\mathbf{U}}^{*i/\mathrm{BB}} \tag{54}$$

where $\Delta G_{\mathbf{U}}^{*\mathrm{BB}}$ is the solvation Gibbs energy of the backbone (BB) of the polymer only. Next, we introduce all the M side chains, one at a time (because of the inde-

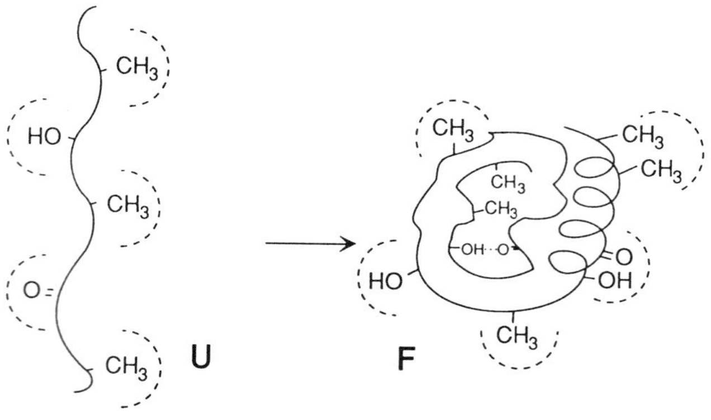

FIGURE 9 Schematic process of folding an idealized linear molecule. In the **U** form, all side chains are independently solvated. In the **F** form, some side chains are buried in the interior of the molecule, some are independently solvated, and some are engaged in pairwise correlation. The solvation sphere about each of the side chains is indicated by the dashed curve.

pendence assumption, it does not matter in which order we do this). The contribution of the ith side chain is $\Delta G_{\mathbf{U}}^{*i/BB}$, which is the *conditional* solvation Gibbs energy of the ith side chain, given the backbone [12,13].

Next, we consider the solvation Gibbs energy of the **F** form. Following our simplifying assumptions made above, we write

$$\Delta G_{\mathbf{F}}^{*BB} = \Delta G_{\mathbf{F}}^{*BB} + \sum_{(\text{independent})} \Delta G_{\mathbf{F}}^{*i/BB} + \sum_{(\text{correlated})} \Delta G_{\mathbf{F}}{}^{i,i/BB} \tag{55}$$

Here $\Delta G_{\mathbf{F}}^{*\,BB}$ is the solvation Gibbs energy of the "backbone" of the **F** form. This backbone includes all the side chains that are buried within the polymer and are not exposed to the solvent. The exposed side chains are to two types. Those that are independently solvated contribute the first sum on the rhs of Eq. (55). The second sum includes all the pairs of correlated side chains. Normally, this includes pairs of side chains at a distance shorter than 6–7 Å.

For this idealized process, we can follow the "fate" of each side chain in the folding process. The three possible fates are the following:

(i) A side chain, independently solvated in the **U** form, is still exposed to the solvent and independently solvated in the **F** form. Such side chains contribute a term $\Delta G_{\mathbf{U}}^{*i/BB}$ in Eq. (54) and exactly the same term in Eq. (55). Therefore, in forming the difference δG in Eq. (53), the contributions of all these side chains will cancel out. Intuitively, this is clear since a side chain that is equally exposed to the solvent in the two states **U** and **F** cannot contribute to the solvent-induced driving force. The group of all such side chains is denoted by E (for external).

(ii) A side chain that is exposed to the solvent in the **U** form becomes buried, or unexposed to the solvent, in the **F** form. In this case, there is a term $-\Delta G_{\mathbf{U}}^{*i/BB}$ corresponding to such side chains in Eq. (54) but not in Eq. (55) [Note that side chains that are buried within the **F** form are part of the "backbone" and therefore collectively contribute to $\Delta G_{\mathbf{F}}^{*BB}$ in Eq. (55).] When forming the difference (Eq. (53)), we shall have one term $-\Delta G_{\mathbf{U}}^{*i/BB}$ for each such side chain. All the side chains in this group are referred to as the I group (I for internal).

(iii) A pair of side chains that are independently solvated in the **U** form becomes correlated in the **F** form. Here we lose the solvation of the side chains in the **U** form, but gain the solvation of the pair as a single entity, in the **F** form. The side chains in this group are referred to as the J group.

We now form the difference (Eq. (53)) and collect the various contributions according to the fates of all the side chains. The result is (for more details, see Refs. 12 and 13)

$$\begin{aligned}\delta G(\mathbf{U} \rightarrow \mathbf{F}) = {} & [\Delta G_{\mathbf{F}}^{*BB} - \Delta G_{\mathbf{U}}^{*i/BB}] - \left[\sum_{i \in I} \Delta G_{\mathbf{U}}^{*i/BB}\right] \\ & + \left[\sum_{i,j \in J} \Delta G_{\mathbf{F}}^{*i,j/BB} - \Delta G_{\mathbf{U}}^{*i,j/BB} - \Delta G_{\mathbf{U}}{}^{j/BB}\right]\end{aligned} \tag{56}$$

The total solvent effect on the folding process is broken here into three major types of terms: first, the change in the solvation Gibbs energies of the backbones in the two forms; second, the loss of the *conditional* solvation Gibbs energies of all side chains in the I group; and finally, the change in the solvation Gibbs energy of all side chains in the J group. Note that no contribution due to side chains in the E group appears in Eq. (56). Their existence has no effect on the overall solvent effect $\delta G(\mathbf{U} \rightarrow \mathbf{F})$. These side chains might well contribute to the dynamical forces along the folding pathway and perhaps will also determine the structure of the **F** form. In the present consideration, however, we are interested in $\delta G(\mathbf{U} \rightarrow \mathbf{F})$ for a given conformation of **U** and **F**.

In our simple model polymer, we have only two kinds of side chains, $H\phi O$ and $H\phi I$ groups. Hence, the complete inventory of all possible solvent effects, for this particular model, is the following:

1. Change in the solvation Gibbs energy of the backbone (note the different definitions of the "backbones" of **U** and **F**)
2. Loss of the conditional solvation Gibbs energy of an $H\phi O$ side chain
3. Loss of the conditional solvation Gibbs energy of an $H\phi I$ side chain
4. Correlation between two $H\phi O$ side chains
5. Correlation between two $H\phi I$ side chains
6. Correlation between an $H\phi O$ and an $H\phi I$ side chain

There are six different types of contributions to $\delta G(\mathbf{U} \rightarrow \mathbf{F})$. All of these terms can be estimated by a combination of theoretical and experimental means. Thus, for this particular model polymer, we have not only a complete inventory of solvent-induced effects, but also an idea of the order of magnitude of each of these effects. We shall summarize, briefly, the methods of evaluating these effects. The solvation Gibbs energy of the backbone of the **U** form may be estimated by using the experimental data on the solvation Gibbs energies of small hydrocarbons [11]. It is known that, to a good approximation, the solvation Gibbs energies are linear in the number of carbon atoms n in the molecule. Extrapolation to larger n would give the solvation Gibbs energy of the backbone of the **U** form. As for the **F** form, presuming that it is compact and nearly spherical, the best way of estimating its solvation Gibbs energy is to use scaled particle theory [19].

The loss of the conditional solvation Gibbs energy of a methyl group hung on a saturated hydrocarbon backbone is of the order 0.3 kcal/mol. This value may be obtained from experimental data on small model compounds [12,13], the methodology of which is described elsewhere [13]. Similarly, the conditional solvation Gibbs energy of a hydrophilic group is of the order -6 kcal/mol [13].

Correlation between two side chains depends on the distance and relative orientation. One can easily show that if the two side chains are far enough apart so that their solvation spheres are independent, then

$$\Delta G^{*i,j/\mathrm{BB}} = \Delta G^{*i/\mathrm{BB}} + \Delta G^{*j/\mathrm{BB}} \tag{57}$$

This is actually the definition of independence of the *i*th and *j*th group. For any finite distance on the order of a few molecular diameters, the correlation between the two groups can contribute either positively or negatively to δG. For two methyl groups attached to a benzene ring, we have experimental data showing that the correlation is on the order of -0.3 kcal/mol. [13]. Somewhat larger values are obtained for the correlation between two methane molecules in water [13]. Since these are not attached to any backbone, however, their magnitude cannot be used for estimating the correlation between two methyl groups on a polymer.

Correlation between two $H\phi I$ groups can be estimated by both theoretical and experimental means. The presently available information indicates that two $H\phi I$ groups at a distance of about 4–5 Å and favorably oriented to form a one-water bridge with a water molecule can contribute as much as -3 kcal/mol to δG [13].

Finally, the correlation between an $H\phi O$ and an $H\,\phi I$ group can also be estimated from experimental data. No definite figures are available. It is likely, however, that at short distances such correlations contribute positively to δG. The reason is that an $H\phi I$ group has a solvation Gibbs energy of about -6 kcal/mol, whereas for the methyl group, the solvation Gibbs energy is about $+0.3$ kcal/mol. When two such groups come close to each other, it is likely that the $H\phi O$ group interferes with the solvation of the $H\phi I$ group, thereby reducing the net solvation Gibbs energy of the pair. Hence, the contribution to δG due to such a pair is likely to be positive.

The analysis carried out above for a simple model polymer has two virtues in the study of real proteins. First, the complete inventory of solvent effects listed above can be easily extended to the case of real proteins. Second, the above analysis indicates the way to obtain the relevant data for real proteins.

Clearly, the inventory of all solvent effects in real proteins is much larger than the one given above for the model polymer. Instead of having to estimate the Gibbs energy of solvation of a linear hydrocarbon, we need to estimate the solvation Gibbs energy of the polypeptide backbone—or approximately polyglycine. The main difference between the two is that the latter have the C=O and NH groups along the backbone. These can also be estimated from small model compounds [12,13]. The solvation Gibbs energy of the backbone of the **F** form can be based on scaled particle theory, as indeed has been done earlier [19].

The loss of the conditional solvation Gibbs energies of all of the 20 different side chains can be estimated by the same methodology used for a methyl group on a hydrocarbon backbone. The main difficulty here is to find small model compounds that can mimic the environment about the side chains as in the real protein. Likewise, partially buried side chains can be viewed as new side chains, the solvation of which can be studied by similar model compounds.

The list of all possible pairwise correlations is far larger than the three cases listed above (i.e., the $H\phi O$–$H\phi O$, the $H\phi O$–$H\phi I$, and the $H\phi O$–$H\phi I$ correlations). Here, each one of the 20 side chains can correlate with any one of the 20 side chains. Even without accounting for partially buried side chains, the total number of different pairs is too large. In addition, side chains exposed to the solvent in both the **U** and the **F** form may be engaged in pairwise, tripletwise, etc., correlations. At present, almost nothing is known about the higher-order correlations among such chains. A preliminary study of triplet and quadruplet correlations among $H\phi I$ groups has been published recently [20].

Clearly, it would be very inefficient to study every possible solvent-induced effect. Not all of these necessarily occur in real proteins. Instead, the more practical approach would be to study each protein for which the crystallographic structure is known. For each protein, one can list all side chains that are independent, pairwise correlated, etc., and then study these by using small model compounds to determine the relative contribution of each solvent effect.

We have until now discussed the ingredients of the solvent effects on the Gibbs energy change for the folding process. It is easy to verify that the same inventory of effects occurs in the association between two or more proteins. Using the simple model compound introduced in the beginning of this section, we can think of two proteins, **P** and **L**, each having a fixed conformation, forming a complex **PL**. Here again, we can classify all side chains as belonging to three groups. E consists of all side chains that are unaffected by the association process; therefore, these also do not contribute to δG. The group I contains all side chains the solvation of which is lost in the process. These are the side chains that are located in the binding region in **PL** and are therefore unexposed to the solvent. Finally, the J group consists of all side chains the solvation of which is partially changed in the process. These are likely to occur near the boundaries of the binding region of the complex **PL**.

Having classified all side chains into these three groups, we can proceed to write the analogue of Eq. (56) for the association process. This reads

$$\begin{aligned}\delta G(\mathbf{P}+\mathbf{L}\rightarrow\mathbf{PL}) = {} & [\Delta G_{\mathbf{PL}}^{*\mathrm{BB}} - \Delta G_{\mathbf{P}}^{*\mathrm{BB}} - \Delta G_{\mathbf{L}}^{*\mathrm{BB}}] \\ & - \left[\sum_{i\in I} \Delta G_{\mathbf{P}}^{*i/\mathrm{BB}} + \sum_{i\in I} \Delta G_{\mathbf{L}}^{*\mathrm{BB}}\right] \\ & + \left[\sum_{i,j\in J} \Delta G_{\mathbf{PL}}^{*i,j/\mathrm{BB}} - \Delta G_{\mathbf{P}}^{*i/\mathrm{BB}} - \Delta G_{\mathbf{L}}^{*j/\mathrm{BB}}\right]\end{aligned} \tag{58}$$

The first term is the difference in the solvation Gibbs energies of the backbones between the initial and the final states. Note that the backbones of **P**, **L**, and **PL** are different for the different species. The second term includes the loss of the conditional solvation Gibbs energies of all the side chains in **P** and **L** that belong to the I group. (We assume for simplicity that all of these are independently solvated in the initial state—**P** and **L**. This assumption is not necessary, however. We can also include in the second term on the rhs of Eq. (58) the loss of the solvation Gibbs energies of pairs of correlated side chains.) The last term on the rhs of Eq. (58) includes those side chains (i belongs to **P** and j belongs to **L**) that are initially independently solvated and become correlated in the final state **PL**. Again, higher-order correlations are neglected in this simplified model. As in the folding process, Eq. (58) can be used to construct an inventory of all possible solvent effects for our simple model polymers, having only two kinds of side chains from which we can obtain the general list for a real protein. It is clear that the same types of ingredients that contribute to the $\delta G(\mathbf{U} \rightarrow \mathbf{F})$ will appear also in $\delta G(\mathbf{P} + \mathbf{L} \rightarrow \mathbf{PL})$.

VI. THE MISSING INFORMATION AND HOW TO OBTAIN IT

The preceding section dealt with folding and association of a hypothetical polymer. This hypothetical polymer was constructed in such a way that it contains only side chains, the conditional solvation Gibbs energies of which are known. Therefore, we could say something about the order of magnitude of the various ingredients that constitute δG. More important, the study of the hypothetical polymer points the way to study the ingredients of the solvent effects, δG, for processes involving real proteins. In this section, we discuss the missing information needed to study solvent effects in real folding and association processes. We shall also suggest methods of obtaining the missing information.

As we have pointed out in Sec. V, it would be inefficient to study *all* possible solvent effects and then apply this information to estimate the solvent contribution to the driving force for any specific process. There are simply too many ingredients, some of which might never occur in real proteins. Therefore, the better approach would be to choose a specific folding or association process, list all the ingredients that contribute to δG of that process, and then seek to estimate their magnitude using whatever experimental or theoretical means are available. In the rest of this section, we shall discuss the different ingredients and the methods of their study.

A. The Solvation Gibbs Energy of the Large Linear Polypeptide Having No Side Chains

This quantity is important for the study of solvent effects on protein folding. There are several ways to compute this quantity. First, as pointed out in the previous section, we can start with the linear hydrocarbon chain, estimate its solvation Gibbs energy in water by assuming the linearity in n, the number of carbon atoms in the molecule [21], that is,

$$\Delta G_n^* = a + bn \tag{59}$$

where a and b may be determined from experimental data on small hydrocarbon molecules [21]. Next, we can add the special features of the polypeptide chain, such as the peptide bond and the NH and C=O groups. The latter may also be studied by using small model compounds, say small polypeptides and the corresponding small hydrocarbons [13].

The second method is to measure the solvation Gibbs energies of small polypeptides of different lengths and then extrapolate linearly to the required length of the particular backbone of interest. In any case, it should be remembered that the NH and the C=O groups are close enough so that their solvation cannot be treated as being independent.

B. Solvation of the Backbone of the F Form

This is needed both in the folding and in the association processes. Here, the backbone includes all side chains that are not exposed to the solvent. From the point of view of the solvent, we have a large nonpolar solute, and the main part of the interaction with the solvent is the repulsive interaction between the solute and a solvent molecule. Note that the NH and C=O exposed to the solvent in the **F** form are not included in the "backbone." These can be studied separately by using model compounds in the same manner as we study the conditional solvation of the entire side chains. Since one is unlikely to find a real nonpolar solute the size of a protein, there is no experimental method to measure its solvation Gibbs energy. The best we can do, at present, is to use scaled particle theory to estimate the hard part of the solvation and then add the relatively small contribution of the soft interaction. This procedure has been carried out recently for a few proteins, the crystallographic structures of which are known [19].

C. Loss of the Conditional Solvation Gibbs Energies of the Various Side Chains

For any side chain that has been transferred to the interior of the protein, we have one term, $-\Delta G_{\mathbf{F}}^{*i/\mathrm{BB}}$ in the case of protein folding. This term can be estimated by using small model compounds. The methodology of this procedure and some numerical examples have been discussed elsewhere [13]. The presently available data include a few side chains only, and the conditional solvation Gibbs energies involve a hydrocarbon backbone as the "condition." What is needed are conditional solvation Gibbs energies of the side chains adjacent to a small model compound that has a polypeptide environment similar to the one in a real protein. Knowing how sensitive the conditional solvation Gibbs energies are to the variation in the "conditions," we can expect the value of the conditional solvation Gibbs energies near a polypeptide to differ considerably from the corresponding values next to a hydrocarbon backbone. Perhaps the closest to the required data are those published by Fauchere and Pliska [22], who reported the differences in the conditional solvation Gibbs energies between two liquids. The required quantities are the conditional solvation Gibbs energies of the side chains in water or in physiological aqueous solutions. To obtain these quantities, we need to measure the distribution coefficients of two compounds, denoted as αR and αH, between water and a gaseous phase. αR and αH are two model compounds having the side chains R and hydrogen (H), respectively. The carrier molecule α should be small enough so that the distribution coefficients of both αR and αH can be measured. It should be long enough so that the environment around the side chain in αR is similar to environment around R in real proteins. The conditional solvation Gibbs energy of R is thus determined from [13]

$$\Delta G^{*R/\alpha} = \Delta G^{*\alpha R} - \Delta G^{*\alpha \mathrm{H}} \tag{60}$$

where on the rhs we have two measurable quantities. More details on the nature of the approximations involved in this procedure can be found elsewhere [13].

Having a list of all the quantities $\Delta G^{*R/\alpha}$ for all the side chains R (including perhaps partially buried side chains and the conditional solvation Gibbs energies of the backbone NH and C=O groups) will allow us to estimate the contribution of each side chain in the I group to the driving force in the protein folding and association processes. Such a list could well be referred to as a (conditional) hydrophobicity scale. The difference between this scale and other (unconditional) scales should be noted, however [12].

D. Pairwise Correlations

Much is known about the solvent-induced part of the potential of average force between two nonpolar solutes in water. This hydrophobic interaction problem has been studied theoretically, experimentally, and by simulation methods [23]. Very little is known about the correlation between the same two nonpolar solutes, or groups, next to a backbone. Only very recently it was found that pairwise correlation between $H\phi I$ groups could contribute significantly to the driving forces for biochemical processes [12,20].

The methodology of obtaining the required data is similar to that used for the conditional solvation Gibbs energies of single side chains. For pairwise correlations, there are two kinds of experiments that may be carried out to obtain the required data.

The first system, shown schematically in Fig. 10 involves a carrier molecule α with two side chains, say R_1 and R_2. In one conformation, denoted A, the two side chains are presumed to be far apart and therefore independently solvated. In the second conformation, denoted B, the two groups are close enough and therefore correlated. The contribution of this pair correlation to δG is given by

$$\delta G(R_1, R_2) = \Delta G_B^* - \Delta G_A^* \tag{61}$$

where on the rhs we have two experimentally measurable quantities. Again, if the carrier molecule α can be chosen to provide an environment about R_1 and R_2 sim-

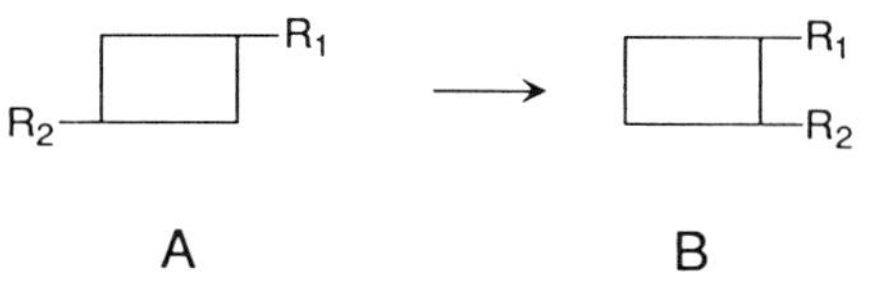

FIGURE 10 Model compounds used to measure the contribution of pairwise correlation to the total solvent effect. In A, the two side chains are far apart, hence independently solvated. In B, the two side chains are close enough to be correlated. The carrier molecule is schematically drawn as a rectangle.

ilar to that of a real protein, then δG as measured by Eq. (61) is a good approximation for the pairwise correlation between R_1 and R_2.

Another method that provides essentially the same information is shown in Fig. 11. Here we start with two molecules, denoted C and D, each of which has one side chain. We form the new compounds E and F and measure the correlation between the two side chains R_1 and R_2 by [13]

$$\delta G(R_1, R_2) = \Delta G_E^* + \Delta G_F^* - \Delta G_A^* - \Delta G_B^* \tag{62}$$

where again we have experimental quantities on the rhs. The advantage of the this method is that the two side chains R_1 and R_2 are strictly independent in the initial state, whereas in the previous method the two side chains are initially far apart, but not at infinite separation from each other.

The available data is very scarce. We know the quantity δG for two methyl and for two ethyl groups attached to a benzene carrier. There is also information, from both theoretical and experimental sources, on the correlation between two $H\phi I$ groups, such as two hydroxyl groups [13]. No information is available on any pair of side chains attached to a carrier that is similar to that of a protein backbone.

E. Higher-Order Correlations

In principle, one can extend the same methodology to obtain triplet or higher-order correlations among side chains. Either the analogue of Fig. 10 or of Fig. 11 could be used to measure the extent of correlation among three or more side chains. A rough estimate of the triplet and quadruplet correlations among $H\phi I$ groups has been reported recently [20]. It was concluded that triplet and quadruplet correlations among $H\phi I$ groups could contribute significantly to both the stability of the protein as well as to the forces between two proteins. Nothing is known about the corresponding correlations among $H\phi O$ groups.

To conclude, we have seen that much experimental and theoretical work is needed to study the ingredients of the solvent effects in biochemical processes. It is clear the required data is quite different from that currently used to estimate protein stability and driving forces in biochemical processes. The latter are based on

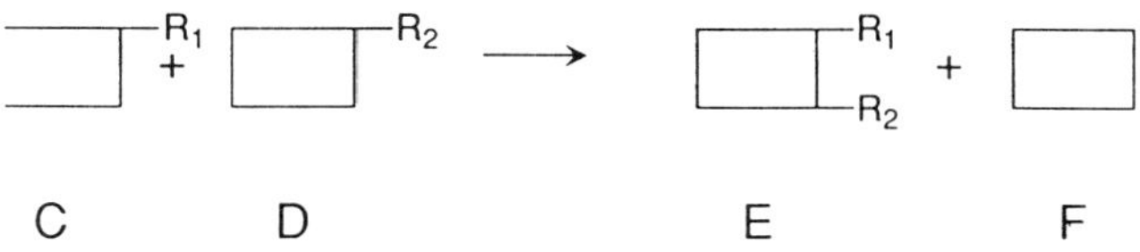

FIGURE 11 A different "reaction" to measure the contribution of pair correlation. See text for details.

various hydrophobicity scales that were constructed with the hope that they would provide the contribution of each single amino acid to the overall driving force. The analysis carried out in this chapter clearly shows that single-amino-acid properties cannot tell the whole story. An important part of the driving force comes from correlations among the side chains of the various amino acids. We hope that the outline presented in this section will encourage experimentalists to undertake the required measurements of the ingredients that are still missing in the study of the solvent effects on biochemical processes.

VII. CONCLUDING REMARKS

In considering the overall driving force for any process involving large macromolecules, one has to take into account changes in the internal partition functions of the solutes involved in the process, changes in the direct interactions, and solvent effects. These components are intertwined and in general cannot be studied in isolation. Nevertheless, if we focus on the study of all possible solvent effects, it is sufficient to consider a simplified process in which one conformation of each species is assumed. In this case, there exists a clear-cut separation of the changes in internal partition functions, direct interactions, and solvent effects. Under these assumptions, the latter is still a very complicated problem, various side chains interact differently with the solvent, and their contribution to the overall driving force depends not only on their type, but also on their immediate neighbors in both the initial and the final states. For this reason, it is premature to speculate on the dominant driving force in any biochemical process. Once we have a list of all the ingredients of solvent effects and an idea about the order of magnitude of each, it then makes sense to make a statement regarding the relative importance of each of the components. From the limited information that is presently available, we can make the following tentative statements.

A small $H\phi O$ side chain in the I group (i.e., a side chain that loses its solvation in the process) contributes about -0.3 kcal/mol, whereas an $H\phi I$ group such as OH or C=O contributes about $+6$ kcal/mol. It should be stressed that although the loss of solvation of an $H\phi I$ group contributes positively to δG of the process, it can still be decisive in the determination of the specific structure of the folded protein or the specific binding site in the association process [13].

For a pair of $H\phi O$ side chains that become correlated, the contribution to δG is between -0.3 and -0.6 kcal/mol. For a pair of $H\phi I$ side chains, at favorable distance and orientation, the contribution to δG is on the order of -3 kcal/mol [13].

Thus, based on the available information on side chains attached to hydrocarbon carriers, we can conclude that in each of these cases the $H\phi I$ groups are relatively more important. We believe that this conclusion will hold true also for side chains attached to a polypeptide backbone.

The foregoing conclusion still leaves open the question of the global or the overall contributions of *all* $H\phi O$ groups relative to *all* $H\phi I$ groups. It is in principle possible that, although on the basis of one-to-one comparison, the $H\phi I$ effects are larger than the corresponding $H\phi O$ effects, the cumulative effects of all $H\phi O$ groups be larger than that of all $H\phi I$ groups. For instance, if an $H\phi I$ group contributes 10 times more than an $H\phi O$ group in the I group, then it is possible that in a specific protein having 10 times more of the $H\phi O$ groups the overall $H\phi O$ effect will be larger than the overall $H\phi I$ effect (as far as side chains in the I group are concerned). This is unlikely to be the case. Whatever the relative numbers of $H\phi O$ and $H\phi I$ side chains in a specific protein, the total number of $H\phi I$ groups (including $H\phi I$ side chains as well as the NH and C=O groups of the backbone) will always be larger than the total number of $H\phi O$ side chains. It is therefore very likely that the overall $H\phi I$ effects are more important than the overall $H\phi O$ effects. Of course, the relative magnitudes of these effects is dependent on the specific protein under consideration. The relative magnitude of the contribution of each side chain, or each pair of side chains, etc., is independent of the specific protein.

The conclusion that $H\phi I$ rather than $H\phi O$ effects might be more important in biochemical processes, although relatively new, has provoked some criticism. If the $H\phi I$ effects are more important than the $H\phi O$ effects, the critics say, then the protein should have folded "inside outside," that is, burying the $H\phi I$ groups rather than the $H\phi O$ groups into the interior of the protein (and, similarly, in the association process, more $H\phi I$ groups should have lost their solvation). The rationale behind this view is that in reality we do indeed *observe* that $H\phi O$ side chains are more likely to lose their solvation and become unexposed to the solvent. This was the observation that led Kauzmann to propose that the $H\phi O$ effect (or $H\phi O$-bond, as it was originally referred to) is probably one of the most important driving forces in the process of folding of a protein. This conclusion is still prevalent in the biochemical literature [25]. The *experimental observation* in itself, however, cannot be used as an argument in favor of the role of $H\phi O$ effects in biochemical processes. That $H\phi O$ side chains tend to occupy the interior of the protein does not necessarily mean that they are responsible for the driving force that led them into the interior of the protein. This argument reminds me of a similar situation regarding mixing of two different ideal gases [26]. Here we also *observe mixing*. A careful analysis, however, shows that the mixing, by itself, does not contribute to the driving force for that process. It is the expansion of each species into the entire volume that provides the driving force, and as a result of which we also observe mixing. Likewise, the process of folding of a protein is an extremely complicated one, driven by many different factors, one of which is the *HφO* effect. The observation that *HφO* groups tend to appear inside is a result of the combined effects of all these factors. One needs to make an independent analysis, like the one carried out in the preceding sections, to find out which of the factors are more or less important.

While *HφO* effects have gained a great deal of attention from both theoreticians and experimentalists, there exists no biochemical processes involving proteins where these effects have been shown to play a dominant role. There is not even a theoretical argument to support such a contention.

On the other hand, *HφI* effects have not enjoyed the attention they deserve. Though most people admit that direct hydrogen bonding is an important aspect of the driving forces in biological systems, other solvent effects involving *HφI* groups were not searched for. In fact, even direct (inter- or intramolecular) hydrogen bonding, when viewed in the context of the solvent effects, was dismissed, apparently as a result of the hypothesis discussed in Sec. I (following Eq. (2)). This hypothesis seems to suggest that two HBs are lost and two gained in the process of formation of one direct HB between two *HφI* groups. The fallacy of this argument has been discussed in detail elsewhere [27]. Briefly, the presentation in Eq. (2) of an apparent stoichiometric reaction is erroneous for two reasons: First, what is lost on the lhs of Eq. (2) is the conditional solvation Gibbs energies of the two *HφI* groups. This loss amounts to 4.5–12 kcal/mol depending on the number of "arms" (along which HBs may be formed) lost in the process. On the other hand, one HB energy is gained on the rhs of the equation. Therefore, the net contribution to the driving force due to direct hydrogen bonding is between -1.5 and $+6$ kcal/mol [27]. Second, whatever the water molecule does when released in the process (Eq. (2)) does not affect the driving force of the process. Therefore, the appearance of the term $w \ldots w$ in Eq. (2) is potentially misleading. Thus, formation of a direct HB between two *HφI* groups could contribute either negatively or positively to the overall driving force of the process.

Another effect involving *HφI* groups that could contribute significantly to the driving force was recently examined and referred to as *HφI interaction* [24]. This effect occurs whenever two *HφI* groups such as OH, C=O, or NH are at a distance of about 4.5–5 Å and properly oriented in such a way that a hydrogen-bonded bridge can be formed with one water molecule. In Sec. V, we referred to this phenomenon as pair correlation between two *HφI* groups. Experimental evidence and theoretical estimates indicate that this effect can contribute as much as -3 kcal/mol to the driving force. This is purely a solvent-induced effect involving *HφI* groups, and it does not involve any direct interaction between the groups.

Similarly, it was also shown that correlations among three and four *HφI* groups can contribute significantly to both the stability of the protein and the forces operating between proteins [20]. It is clear that the same type of solvent-induced effects should contribute also to the stability of double-strand nucleic acids and to the binding of proteins to DNA. In all these processes, the solvent can rely on the abundance of *H*ϕ*I* groups on the surface of biological macromolecules to gain control of both the extent and the specificity of vital biochemical processes.

REFERENCES

1. Eisenberg, D., and Kauzmann, W., *The Structure and Properties of Water*, Oxford University Press, New York (1969).
2. Franks, F. (ed.), *Water: A Comprehensive Treatise*, Volume 1, Plenum Press, New York (1972).
3. Ben-Naim, A., *Water and Aqueous Solutions: Introduction to a Molecular Theory*, Plenum Press, New York (1974).
4. Stillinger, F. H., *Science, 209*: 451 (1980).
5. Kauzmann, W., *Adv. Protein Chem. 14*: (1959).
6. Tanford, C., *The Hydrophobic Effect: Formation of Micelles and Biological Membranes*, Wiley Interscience, New York (1973).
7. Ben-Naim, A., *Hydrophobic Interactions*, Plenum Press, New York (1980).
8. Schellman, J. A., *C. R. Trav. Lab. Carlsberg. Ser. Chim., 29*: 223 (1955).
9. Jencks, W. P., *Catalysis in Chemistry and Enzymology*, McGraw-Hill, New York (1969).
10. Hine, J., *J. Am. Chem. Soc., 94*: 5766 (1972).
11. Ben-Naim, A., *Solvation Thermodynamics*, Plenum Press, New York (1987).
12. Ben-Naim, A., *Biopolymers, 29*: 567 (1990).
13. Ben-Naim, A., *Statistical Thermodynamics for Chemists and Biochemists*, Plenum Press, New York, 1992.
14. Becktel, W. J., and Schellman, J. A., *Biopolymers, 26*: 1859 (1987).
15. Schellman, J. A., *Ann. Rev. Biophys. Biophys. Chem., 16*: 115 (1987).
16. Privalov, P. L., and Khechinashvilli, N. N., *J. Mol. Biol., 86*: 665 (1974).
17. Privalov, P. L., and Gill, S. J., *Adv. Protein Chem., 39*: 191 (1988).

17a. Privalov, P. L., *Adv. Protein Chem., 33*: 167 (1979).

18. Ben-Naim, A., *J. Phys. Chem., 94*: 6893 (1990).
19. Ben-Naim, A., Ting, K. L., and Jernigan, R. L., *Biopolymers, 28*: 1309 (1989).
20. Ben-Naim, A., *J. Chem. Phys., 93*: 8196 (1990).
21. Ben-Naim, A., and Marcus, Y., *J. Chem. Phys., 83*: 4744 (1985).
22. Fauchere, J. L., and Pliska, V., *Eur. J. Med. Chem., 18*: 369 (1983).
23. Ravishanker, G., and Beveridge, D. L., in *Theoretical Chemistry of Biological Systems*, Naray-Szabo, G. (ed.), Elsevier, Amsterdam (1986).
24. Ben-Naim, A., *J. Chem. Phys., 92*: 7412 (1989).
25. Dill, K. A., *Biochemistry, 29*: 1501 (1985).
26. Ben-Naim, A., *Am. J. Physics, 55*: 725 (1987).
27. Ben-Naim, A., *J. Phys. Chem., 95*: 1437 (1991).

10

Thermodynamic Mechanisms for Enthalpy–Entropy Compensation

Ernest Grunwald
Brandeis University, Waltham, Massachusetts
Lorrie L. Comeford
Salem State College, Salem, Massachusetts

I. INTRODUCTION

Historically, the term *enthalpy–entropy compensation* summarizes the conspicuous tendency for changes in ΔH° and ΔS°, due to the introduction of a substituent or a change of solvent, to have the same sign. Reactivity, as expressed by the equilibrium constant, is measured by ΔG°. Because $G = H - TS$, reactivity is also measured by the *difference* between ΔH° and $T\,\Delta S^\circ$; that is, $\Delta G^\circ = \Delta H^\circ - T\,\Delta S^\circ$. When the changes in ΔH° and $T\,\Delta S^\circ$ have the same sign and magnitude, their effects on reactivity cancel exactly. In most actual cases, however, the magnitudes are not equal, and compensation is partial. The phenomenon also encompasses thermodynamic data for activation obtained by the methods of chemical kinetics. In that case, the variables are $\Delta H^\ddagger$ and $T\,\Delta S^\ddagger$ [1–3].

In essence, enthalpy–entropy compensation stems from the microscopic complexity that exists in all chemical systems. *Thermodynamic* composition is specified with minimalist simplicity in terms of components, while *chemical* composition is a rich blend of molecular species and subspecies. In principle, when the external conditions change, the equilibria among the molecular species shift and—again in principle—this causes a contribution to the change in enthalpy that is compensated exactly by a like contribution to the change in entropy. This general effect is often too small to be detected, but there are cases when it is significant, and some cases when it is dramatic. Some of the latter cases are found in protein chemistry [3] and will be treated by Lumry in Chapter 1.

We are not specialists on proteins, and so the present chapter will deal with compensation in a broad perspective. Our general discussion will apply both to proteins and to the substrates and ions with which they interact.

Enthalpy–entropy compensation is most often found for liquid solutions. It is particularly significant for aqueous solutions, where it is closely tied to the fact that entropies and enthalpies of many aqueous solutes are "anomalous." After presenting a general discussion, we shall turn specifically to the thermodynamics of nonpolar solutes in water and infer a description of the molecular species in the hydration shells of these solutes, to rationalize the anomalies.

II. EXPERIMENTAL EXAMPLES

The demonstration of compensation is most straightforward when ΔG° is constant throughout a reaction series while ΔH° and $T\,\Delta S^\circ$ vary significantly. Examples of this sort are shown in Figs. 1–3. Figure 1 shows the solvent effect on the proton transfer equilibrium (next page):

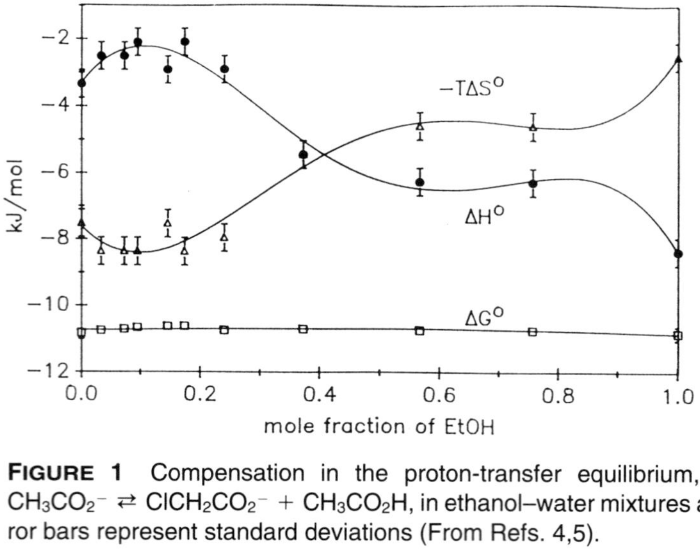

FIGURE 1 Compensation in the proton-transfer equilibrium, $ClCH_2CO_2H + CH_3CO_2^- \rightleftarrows ClCH_2CO_2^- + CH_3CO_2H$, in ethanol–water mixtures at 298 K. The error bars represent standard deviations (From Refs. 4,5).

$$ClCH_2COOH + CH_3COO^- \rightleftarrows ClCH_2COO^- + CH_3COOH$$

in ethanol–water mixtures [4,5]. ΔH° and $T\,\Delta S^\circ$ each vary by more than 4 kJ/mol, while ΔG° varies by less than 0.2 kJ/mol. The compensation is especially convincing because the individual curves for ΔH° and $T\,\Delta S^\circ$ are complex, showing experimentally significant maxima and minima.

Figure 2 shows data [6] for the binding of the zwitterions of a series of amino acids with Cryptand-222, a bicyclic ligand whose complex with glycine is depicted as

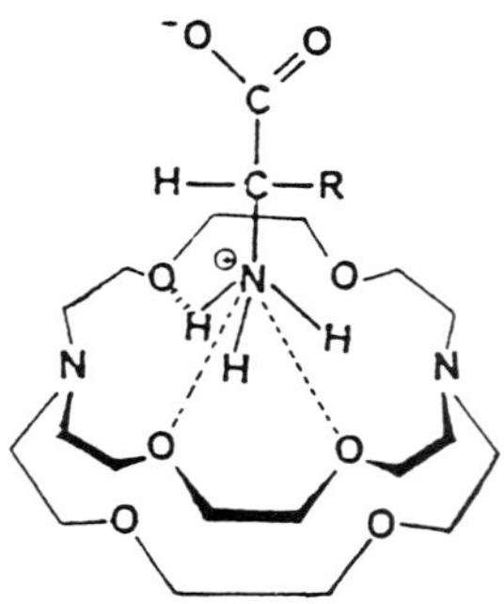

(1)

Again, the curve for ΔG° is flat. In this case, ΔH° is plotted as a function of ΔS°, and the line predicted by $\Delta H^\circ = T\,\Delta S^\circ + \Delta G^\circ$ (a constant) is observed. ΔG° for binding to the NH_3^+ group varies by less than 4 kJ/mol, while ΔH° varies by 37 kJ/mol. Although the ΔH° data for DL-phenylalanine are puzzling, the near constancy of ΔG° guarantees the existence of compensation.

Figure 3 shows a kinetic example of the compensation effect, the radical decomposition of phenylazotriphenylmethane, $Ph{-}N{=}NC(Ph)_3$, in a series of normal non-hydrogen-bonding solvents [7]. Once again, the compensation is nearly perfect. The free energy of activation, $\Delta G^\ddagger$, varies by less than 1 kJ/mol, while $\Delta H^\ddagger$ and $T\,\Delta S^\ddagger$ vary by almost 20 kJ/mol. The figure shows that $\Delta H^\ddagger$ and $T\,\Delta S^\ddagger$ are *not* simple functions of the dielectric constant. On the other hand, the aromatic solvents form a nearly constant group that differs from malonic ester and cyclohexane.

III. INTERACTION MECHANISMS AND COMPENSATION VECTOR DIAGRAMS

Before showing less obvious examples involving partial compensation, we need to present some notation. The operator Δ (as in ΔH) is the familiar reaction operator and denotes a change in the reaction zone that converts reactants to products. The plots of Figs. 1–3, on the other hand, show the effects of perturbations originating outside the reaction zone—due to an added substituent R or a change of solvent M. The effects of such perturbations will be denoted by the perturbation

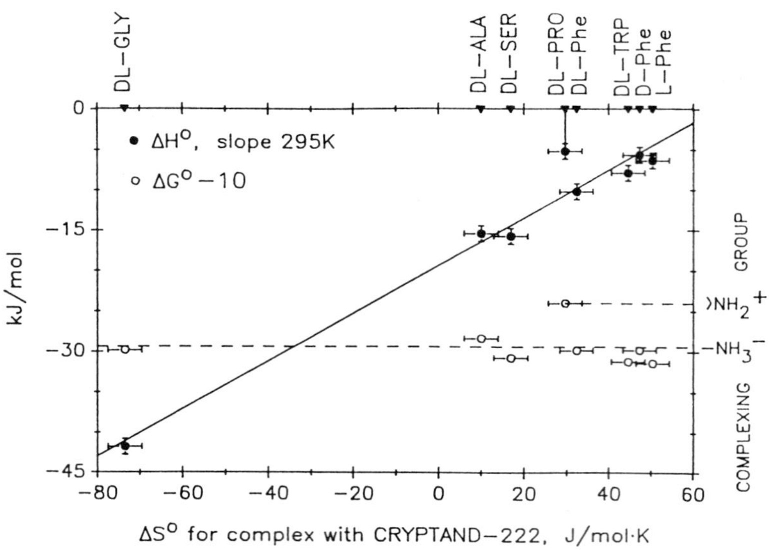

FIGURE 2 Binding of the zwitterions of a series of amino acids with Cryptand-222 in methanol at 298 K. (From Ref. 6.)

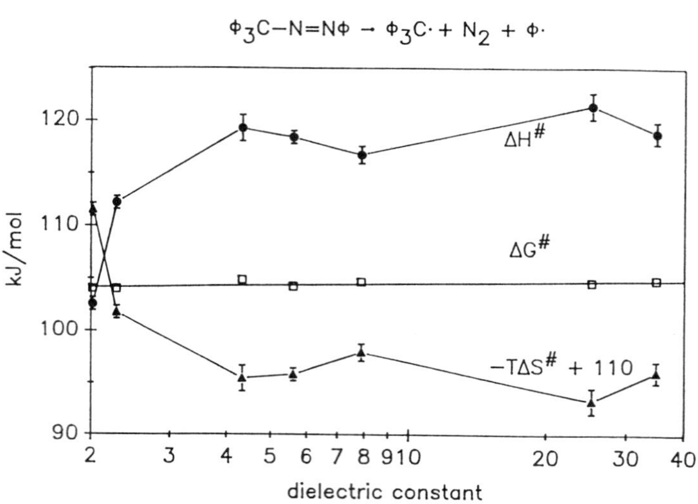

FIGURE 3 Compensation of the kinetic medium effect in the radical decomposition of phenylazotriphenylmethane at 298 K. Solvents, in order of increasing dielectric constant: cyclohexane, benzene, anisole, chlorobenzene, malonic ester, benzonitrile, nitrobenzene. (From Ref. 7.)

operator δ, as in $\delta_R \Delta G$ or $\delta_M \Delta S$. Perturbations often involve two or more independent interaction mechanisms. For example, a polar substituent might polarize a reaction zone by dipole interaction and also by quantum-mechanical resonance. In first approximation, such perturbations are additive:

$$\delta\Delta G = \delta_1\Delta G + \delta_2\Delta G \tag{2a}$$

$$\delta\Delta H = \delta_1\Delta H + \delta_2\Delta H \tag{2b}$$

$$\delta\Delta S = \delta_1\Delta S + \delta_2\Delta S \tag{2c}$$

Because the two perturbations are independent, they will show different relationships of ΔH to ΔS. For example, $\delta_1\Delta H - T\,\delta_1\Delta S$ might be zero, in perfect compensation, while $\delta_2\Delta H$ and $T\,\delta_2\Delta S$ might not even have the same sign. Since we measure the overall effect, there are not enough data to unravel the contributions from individual mechanisms, and mechanisms with good compensation might be missed. This is the reason the data often seem confusing.

Before presenting some of the more complex data, it will be useful to represent $\delta\Delta H$ and $\delta\Delta S$ on a vector diagram. The two vectors V_1 and V_2 in Fig. 4a have the components ($\delta_1\Delta H$, $\delta_1\Delta S$) and $\delta_2\Delta H$, $\delta_2\Delta S$). These vectors are not resolved experimentally, and their sum is the property that is measured. Figures 4b and 4c show such a plot for an entire reaction series. In Fig. 4b, $\delta_1\Delta H$ and $\delta_1\Delta S$ show good compensation, while $\delta_2\Delta H$ and $\delta_2\Delta S$ are uncorrelated. The vector sums ($\delta\Delta H$, $\delta\Delta S$) are plotted in Fig. 4c. These sums show a linear trend due to ($\delta_1\Delta H$, $\delta_1\Delta S$), with considerable scatter due to ($\delta_2\Delta H$, $\delta_2\Delta S$). Indeed, a skeptic might doubt that there is any compensation at all.

IV. EXAMPLES OF PARTIAL COMPENSATION

Figure 5 shows an example of partial compensation based on the binding of a series of cations by gramicidin A [8].

This linear polypeptide was dispersed in lipid lysophosphocholine, thereby causing the gramicidin chains to fold into structures that contain channels that specifically bind cations from aqueous solution. In the figure, the cations are arranged so that the strength of binding, as measured by $\Delta G°$, increases from left to right. $\Delta H°$ and $T\,\Delta S°$ always have the same sign but differ in magnitude, and so the compensation is partial. Moreover, $\Delta H°$ and $\Delta G°$ (and, less certainly, $\Delta S°$) vary monotonically, so that $\Delta G°$ remains a roughly constant fraction of $\Delta H°$, in accordance with Leffler's empirical equation, Eq. (3) [1]; the slope, β, is called the *isokinetic temperature*:

$$\delta\Delta H = \beta\delta\Delta S \tag{3a}$$

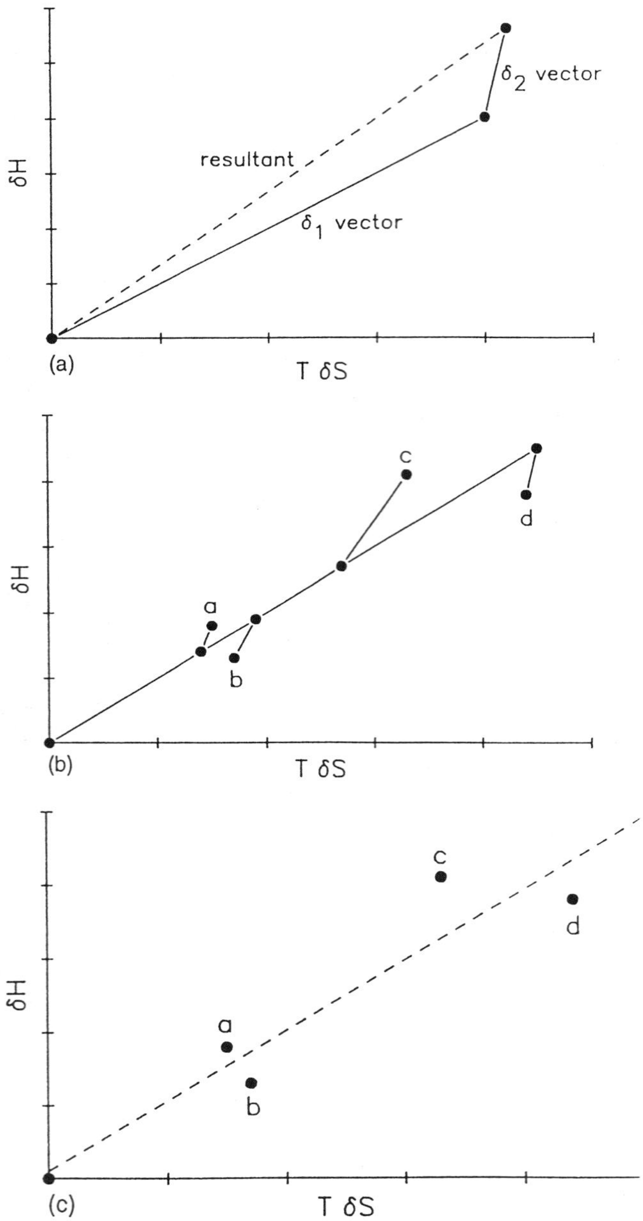

FIGURE 4 Enthalpy–entropy vectors based on two perturbation mechanisms: (a) a single vector; (b) a series of vectors; one of the mechanisms shows exact compensation; (c) the resultants obtained in (b) plotted with a correlation line.

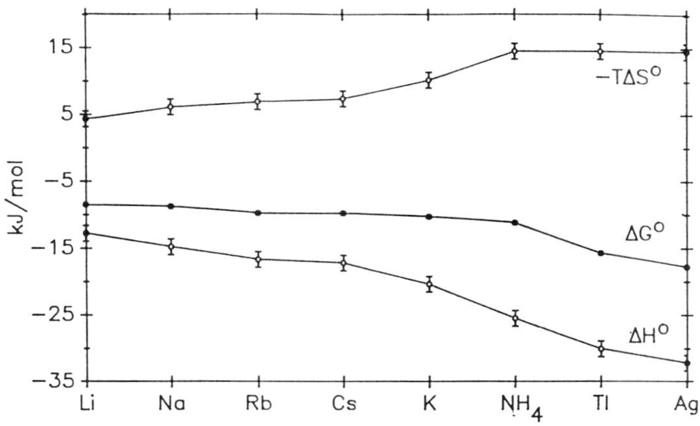

FIGURE 5 Binding of aqueous cations to gramicidin A at 313 K. The gramicidin is dispersed in lysophosphocholine. (From Ref. 3.)

$$\delta\,\Delta G = \frac{\beta - \mathrm{T}}{\beta}\,\delta\Delta H \tag{3b}$$

Figure 6 shows an example where even partial compensation has been hard to prove, the acid dissociation of substituted benzoic acids in water [9]. There are two methods for measuring ΔH° for chemical reactions: by calorimetry, and via the temperature derivative of the equilibrium constant using $\Delta H^\circ = RT^2\, d \ln K/dT$. The second method is in wider use, but is less accurate than calorimetry unless the equilibrium measurements are unusually accurate [10]. The data in Fig. 6 are based on calorimetry, and the error bars are realistic [9]. Compensation occurs when $\delta\Delta H$ and $-T\delta\Delta S$ lie on opposite sides of the horizontal line through the origin. This is true for most of the data.

V. THERMODYNAMIC COMPENSATION

The preceding examples illustrate that there is substantial experimental evidence for enthalpy–entropy compensation. Until recently, however, there has been no general theoretical evidence to demonstrate that this is a lawful phenomenon. This changed when one of us showed there are cases when compensation is predicted by thermodynamics [11]. We shall now describe a kind of compensation that is virtually ubiquitous for chemical systems, although the magnitude of the compensation is often too small to be detected. The description will be direct, involving the partial free energy, enthalpy, and entropy of individual components and molecular species, rather than the functions resulting from Δ and δ operations.

A good starting point for the description of thermodynamic compensation is a change of phase at the transition temperature, such as $H_2O(s) \rightleftarrows H_2O(l)$ at T_{mp}.

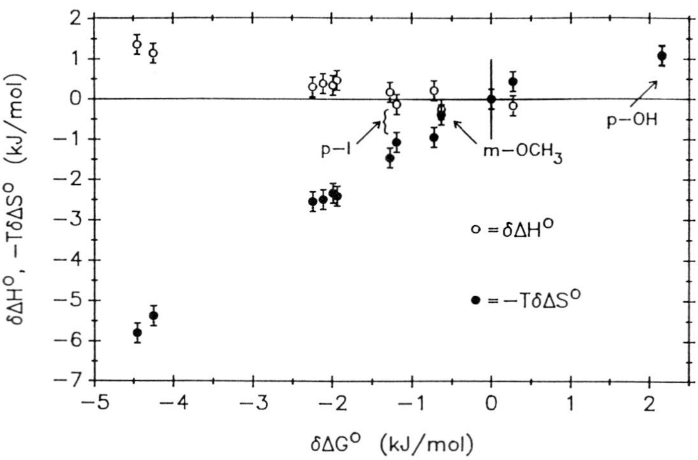

FIGURE 6 Substituent effects in the acid dissociation of benzoic acid in water at 298 K. The substituents are m-F, Cl, Br, I, OH, OCH_3, CH_3, NO_2, and p-Br, I, OH, and NO_2. (From Refs. 9,10.)

Because the phases are pure substances at equilibrium, $\Delta_{fus}G^\circ = 0$. Therefore, $\Delta_{fus}H^0 = T_{mp}\,\Delta_{fus}S^\circ$: ΔH° and $T\,\Delta S^\circ$ are locked into equality because $\Delta G^\circ = 0$. This mechanism is general: Thermodynamic compensation occurs whenever, at constant T and P, the free-energy change δG in a process is zero.

A. Molecular Species

The virtually ubiquitous kind of compensation to be described is based on the fact that many thermodynamic components are not single molecular species, but mixtures of species in rapid equilibrium. Traditionally, a thermodynamic component is a tangible substance that is well characterized and that can be obtained reproducibly in macroscopic quantities. A formal component is a pure substance of known chemical formula.

Molecular species are derived from formal components and are introduced to account for chemical activity and reactivity. A molecular species is an ensemble of related molecules that are characterized by a definite molecular formula, equilibrium geometry, modes of motion, and spectral properties.

Molecular species can be further divided into subspecies. A subspecies is any distinguishable subset of a molecular species whose variables of state, such as T and P, can be defined.

For example, liquid acetic acid, a thermodynamic component, consists of the following molecular species: *cis*-CH_3COOH monomer, its nonpolar head-to-head dimer, *trans*-CH_3COOH monomer, and polar hydrogen-bonded chains based on the *trans* monomer (Eq. (4)) [12]:

(4)

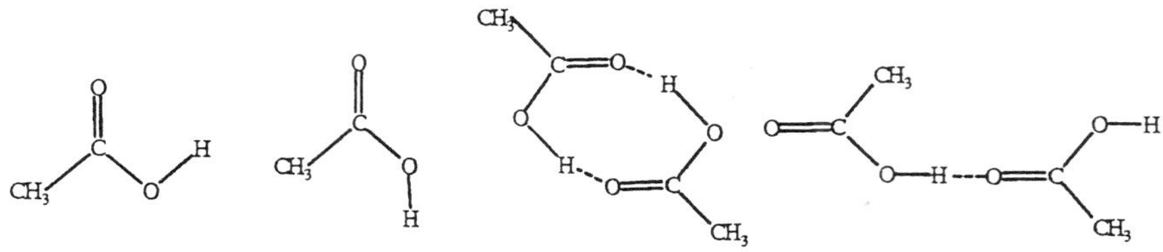

cis, $\mu = 1.4$D | trans, $\mu = 3.9$D | dimer of the cis isomer | dimer of the trans isomer

These molecular species exist in dynamic equilibrium. The equilibrium is sensitive to the addition of solutes because polar solutes favor the formation of polar solvent species, nonpolar solutes favor that of nonpolar species, and hydrogen-bonding solutes interfere with the hydrogen-bonding equilibria [13].

The chemical thermodynamics of molecular species is nonequilibrium thermodynamics, in the sense that the fractions of interconverting species are treated as variables. For definiteness, consider the composition tree (T-1). Here and in the following, number subscripts denote thermodynamic components, and capital letter subscripts denote molecular species. Component 1 is modeled as an equilibrium mixture of two species A and B; the fraction α is a variable. Component 2 is modeled as a single species U.

Composition Tree (T-1)
T and P are constant.

Component $(1, n_1, m_1)$		Component $(2, n_2, m_2)$
$A\ (n_A, m_A)$	$B\ (n_B, m_B)$	$U\ (n_U, m_U)$
$m_A = (1 - \alpha)m_1$	$m_B = \alpha m_1$	$m_U = m_2$
$K = \alpha_{eq}/(1 - \alpha_{eq})$		

where n_1, n_2 = formula-weight numbers; n_A, n_B, n_U = mole numbers; m = molality; and α = progress variable for reaction.

Because the two molecular species of component 1 need not be at equilibrium, the state of the system is defined by $\{T, P, n_1, n_2, \alpha\}$. In order to describe displacements of the system from equilibrium, we shall introduce the following functions: the *molar strain* $\alpha - \alpha_{eq}$; and the molar stress, $\partial G/\partial\alpha = y$. These are functions of the (nonequilibrium) state.

Enthalpy–entropy compensation follows from a general thermodynamic theorem, called *molar strain tolerance*, whose proof is outside the scope of this chapter. It is well known that at equilibrium, the free energy G of a closed system at constant T and P is at a minimum; hence, $(\partial G/\partial\alpha)_{eq} = 0$. Molar strain tolerance states that a similar relationship applies to the *partial* free energy of *any* component and *any* molecular species. Thus, for any ith component and any progress variable α in a closed system at equilibrium at constant T and P, molar strain tolerance is expressed by Eq. (5a); enthalpy–entropy compensation follows as the corollary (5b):

$$\frac{\partial G_i}{\partial \alpha} = \frac{\partial G_i^\circ}{\partial \alpha} = 0, \qquad \text{when } \alpha \text{ is at (or near) equilibrium} \tag{5a}$$

Therefore:

$$\frac{\partial H_i}{\partial \alpha} = T\,\frac{\partial S_i}{\partial \alpha}; \qquad \frac{\partial H_i^\circ}{\partial \alpha} = T\,\frac{\partial S_i^\circ}{\partial \alpha} \tag{5b}$$

Physically, the partial derivatives in (5) represent an infinitesimal process in which α changes by an amount $\partial\alpha$ at (or near) its equilibrium value α_{eq}.

Similarly, for any Jth molecular species and any progress variable β, in a closed system at equilibrium at constant T and P, molar strain tolerance is expressed by (6a), and the corollary enthalpy–entropy compensation by (6b):

$$\frac{\partial G_J}{\partial \beta} = \frac{\partial G_J^\circ}{\partial \beta} = 0, \qquad \text{when } \beta \text{ is at (or near) equilibrium} \tag{6a}$$

Therefore:

$$\frac{\partial H_J}{\partial \beta} = T\,\frac{\partial S_J}{\partial \beta}; \qquad \frac{\partial H_J^\circ}{\partial \beta} = T\,\frac{\partial S_J^\circ}{\partial \beta} \tag{6b}$$

The infinitesimal change $\partial\beta$ occurs at (or near) the equilibrium value, β_{eq}.

Compensation according to Eqs. (5) and (6) is a general phenomenon, because *any* component corresponds to one or more distinct molecular species, and *any* molecular species can be analyzed into subspecies. Thus, *any* specification of molar composition in principle involves progress variables such as α or β. Their equilibrium values, such as α_{eq} or β_{eq}, in principle vary with the external conditions. As illustrated in Fig. 7, any change in external conditions in which the closed system starts and ends at equilibrium can therefore be dissected into two steps, one with and one without thermodynamic compensation. The change depicted in Fig. 7 is divided into a first step in which the temperature changes from T to T' at constant α—an *isomolar* change. This is followed by a step in which α changes from $\alpha_{eq}(T)$ to $\alpha_{eq}(T')$ at constant T'. Because $\alpha_{eq}(T)$ differs from the equilibrium value at T', the second step is an isothermal step in which α returns to equilibrium in a closed system at constant T and P. It can be shown that this step proceeds with molar strain tolerance. That is, $\partial G_1^\circ/\partial\alpha = 0$; hence, $\delta G_1^\circ = 0$, and $\delta H_1^\circ = T'\,\delta S_1^\circ$. By contrast, the free energy change in the isomolar first step is not constrained to be zero, and there is no requirement for enthalpy–entropy compensation.

Reflection will show that the dissection illustrated in Fig. 7 is generally possible: Any finite change in a variable of state can be represented as a sum of infinitesimal changes, each of which is carried out so that the system is initially and finally at molar equilibrium. Each infinitesimal step can be divided into an iso-

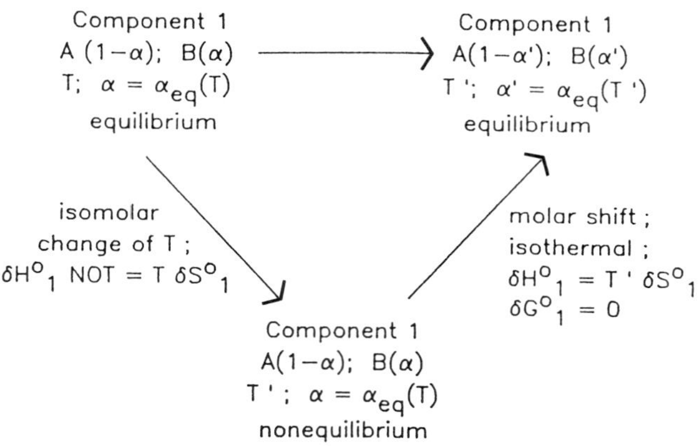

FIGURE 7 Dissection of a change of state into an isomolar step and a step with thermodynamic enthalpy–entropy compensation. See text.

molar step in which the system departs from molar equilibrium followed by a molar-shift step in which the resulting system returns to equilibrium under conditions of molar strain tolerance. Any finite change in a variable of state, in principle, therefore includes a contribution with exact enthalpy–entropy compensation. There remains the practical issue (to which we shall return) of whether this contribution is large enough to be detected.

VI. MATHEMATICAL FORMULATION

It is convenient to continue with composition tree (T-1) and to derive an expression for the partial free energy G_2 of component 2. Formal chemical thermodynamics does not subdivide components into molecular species and subspecies. When such a subdivision is made (e.g., $Na^+ + Cl^-$ for NaCl), it is viewed as "extrathermodynamic" and one assumes that the species resulting from subdivision exist at chemical equilibrium before and after the change of state. On that basis, the partial free energy $G_2 = \partial G/\partial n_2$ corresponds to an infinitesimal addition, ∂n_2, in such a way that chemical equilibrium (eq) is maintained. That is, $G_2 = (\partial G/\partial n_2)_{n_1,T,P,\mathrm{eq}}$. Thermodynamically, the condition implied by the subscript "eq" is that the molar stress $y = \partial G/\partial \alpha$ stays constant at zero.

To derive an expression for G_2 of component 2 in (T-1), we begin by representing the free energy G as a function of $\{n_1, n_2, T, P, \alpha\}$. Since T, P, and n_1 will be constant, we shall not show these variables explicitly in the following. That is, $G = G(n_2, \alpha)$ and dG is given by Eq. (7a). Letting $\alpha = \alpha(n_2, y)$ then yields Eq. (7b); substitution of (7b) in (7a) then gives (7c) for $(\partial G/\partial n_2)_y$. Finally, evaluation at equilibrium ($\partial G/\partial \alpha = y = 0$) yields the desired result in (7d):

$$dG = \left(\frac{\partial G}{\partial n_2}\right)_{\alpha} dn_2 + \left(\frac{\partial G}{\partial \alpha}\right)_{n_2} d\alpha \tag{7a}$$

$$d\alpha = \left(\frac{\partial \alpha}{\partial n_2}\right)_y dn_2 + \left(\frac{\partial \alpha}{\partial y}\right)_{n_2} dy \tag{7b}$$

$$\left(\frac{\partial G}{\partial n_2}\right)_y = \underbrace{\left(\frac{\partial G}{\partial n_2}\right)_\alpha}_{\text{isomolar}} + \underbrace{\left(\frac{\partial G}{\partial \alpha}\right)_{n_2}\left(\frac{\partial \alpha}{\partial n_2}\right)_y}_{\text{molar-shift term}} \tag{7c}$$

$$G_2 = \left(\frac{\partial G}{\partial n_2}\right)_{y=0} = \underbrace{\left(\frac{\partial G}{\partial n_2}\right)_{\alpha=\alpha_{eq}}}_{\text{isomolar}} \tag{7d}$$

Note that the mathematics of Eq. (7c) naturally divides the partial derivative into an isomolar term and a molar-shift term, similar to the separation into consecutive steps in Fig. 7. The derivative $\partial\alpha/\partial n_2$ expresses the molar shift, that is, the shift in α per formula weight of the added component. The molar-shift term vanishes at equilibrium, and the expression for G_2 is purely isomolar.

To derive an expression for the partial molar entropy, we begin with the thermodynamic equation, $S_2 = -(\partial^2 G/\partial n_2\, \partial T)$, in which the infinitesimal changes in n_2 and T take place with maintenance of $A = B$ equilibrium, in component (1) in (T-1). To derive an expression for $\partial^2 G/\partial n_2\, \partial T$, Eq. (7c) is differentiated term by term with respect to T, with y and n_2 constant. (P and n_1 are constant throughout.) S_2, which is defined for the solution at equilibrium, is then obtained by setting y equal to zero. The partial differentiation of Eq. (7c) with respect to T is a little tricky because, for any function $\phi(T,n_2,\alpha)$,

$$\left(\frac{\partial \phi}{\partial T}\right)_{y,n_2} = \left(\frac{\partial \phi}{\partial T}\right)_{\alpha,n_2} + \left(\frac{\partial \phi}{\partial \alpha}\right)_{n_2,T}\left(\frac{\partial \alpha}{\partial T}\right)_{y,n_2}$$

Details of the derivation have been published elsewhere [11b]. The result, after mathematical simplification, is given by

$$\left(\frac{\partial^2 G}{\partial n_2 \partial T}\right)_y = \left(\frac{\partial^2 G}{\partial n_2 \partial T}\right)_\alpha - \left(\frac{\partial^2 G}{\partial \alpha^2}\right)_{n_2,T}\left(\frac{\partial \alpha}{\partial T}\right)_y\left(\frac{\partial \alpha}{\partial n_2}\right)_y + \left(\frac{\partial G}{\partial \alpha}\right)_{n_2}\left(\frac{\partial^2 \alpha}{\partial n_2 \partial T}\right)_y \tag{8a}$$

Letting $y = 0$ and defining $(S_2)_\alpha = -(\partial^2 G/\partial n_2 \partial T)_{\alpha=\alpha_{eq}}$, S_2 is given by

$$S_2 = \underbrace{(S_2)_\alpha}_{\text{isomolar}} + \underbrace{\left(\frac{\partial^2 G}{\partial \alpha^2}\right)_{T,n_2}\left(\frac{\partial \alpha}{\partial n_2}\right)_{T,y=0}\left(\frac{\partial \alpha}{\partial T}\right)_{n_2,y=0}}_{\text{molar-shift term}} \tag{8b}$$

The molar-shift term does not vanish because $\partial^2 G/\partial \alpha^2 > 0$, G being at a minimum at equilibrium.

Finally, to obtain H_2, one substitutes Eqs. (7d) and (8b) into $H_2 = G_2 + TS_2$. The result is Eq. (9), in which $(H_2)_\alpha = (G_2)_\alpha + T(S_2)_\alpha$, and $(G_2)_\alpha = G_2$ according to (7d):

$$\underbrace{H_2 = (H_2)_\alpha}_{\text{isomolar}} + \underbrace{T\left(\frac{\partial^2 G}{\partial \alpha^2}\right)_{T,n_2}\left(\frac{\partial \alpha}{\partial n_2}\right)_{T,y=0}\left(\frac{\partial \alpha}{\partial T}\right)_{n_2,y=0}}_{\text{molar-shift term}} \tag{9}$$

Thermodynamic compensation is implicit in Eqs. (7c), (8b), and (9), because G_2 is rigorously an isomolar property, while H_2 and TS_2 are made up of an isomolar and a molar-shift term. The molar-shift terms compensate exactly.

In addition to the exact compensation produced by the molar-shift terms, there may also be partial compensation between the isomolar terms, $(H_2)_\alpha$ and $T(S_2)_\alpha$, but thermodynamics does not inform us on that. As stated earlier, there is, however, a widespread propensity for compensation, which should apply here.

A. Standard Partial Enthalpies and Entropies in Dilute Solutions

We shall find that the molar-shift terms in Eqs. (8b) and (9), in dilute solutions, are characteristic of the solute but independent of its concentration. The standard partial enthalpy of a dilute solute H_2° simply equals H_2. The standard partial entropy $S_2^\circ = S_2 - R \ln m_2$ and contains all concentration-independent contributions to S_2. The molar-shift terms therefore are part of H_2° and S_2°:

$$H_2^\circ = (H_2^\circ)_\alpha + T\left(\frac{\partial^2 G}{\partial \alpha^2}\right)_{T,n_2}\left(\frac{\partial \alpha}{\partial n_2}\right)_{T,y=0}\left(\frac{\partial \alpha}{\partial T}\right)_{n_2,y=0} \tag{10a}$$

$$\underbrace{S_2^\circ = (S_2^\circ)_\alpha}_{\text{isomolar}} + \underbrace{\left(\frac{\partial^2 G}{\partial \alpha^2}\right)_{T,n_2}\left(\frac{\partial \alpha}{\partial n_2}\right)_{T,y=0}\left(\frac{\partial \alpha}{\partial T}\right)_{n_2,y=0}}_{\text{molar-shift term}} \tag{10b}$$

B. Molar-Shift Mechanism

We shall use composition tree (T-2), a modification of (T-1). Component 1 is the solvent, and component 2 is a dilute solute. Each component consists of two molecular species. Because the solute is dilute, its molecular species, U and V, do not mutually interact, and the fraction β_{eq} is independent of m_2.

Composition Tree (T-2)
T and *P* are constant.

Solvent $(1, n_1, m_1)$		Dilute Solute $(2, n_2, m_2)$	
$A\ (n_A, m_A)$	$B\ (n_B, m_B)$	$U\ (n_U, m_U)$	$V\ (n_V, m_V)$
$m_A = (1 - \alpha)m_1$	$m_B = \alpha m_1$	$m_U/m_2 = (1 - \beta)$	$m_V/m_2 = \beta$
$K_1 = \alpha_{eq}/(1 - \alpha_{eq})$		$K_2 = \beta_{eq}/(1 - \beta_{eq});\ \partial\beta_{eq}/\partial m_2 = 0$	

Although β_{eq} is independent of m_2, the relative stability of the *solvent* species A and B is sensitive to the presence of solute: In principle, $G_A°$ and $G_B°$ are functions of m_2. Accordingly, the equilibrium fraction α_{eq} may vary with m_2.

Given these boundary conditions, the molar shift terms in Eqs. (8b) and (9) can be derived rigorously. The derivation has been published [14] and will not be repeated here. The result is

$$T\left(\frac{\partial^2 G}{\partial \alpha^2}\right)_{T,n_2}\left(\frac{\partial \alpha}{\partial n_2}\right)_{T,y=0}\left(\frac{\partial \alpha}{\partial T}\right)_{n_2,y=0} = m_1\left(\frac{\partial \alpha_{eq}}{\partial m_2}\right)_{T,P}(\Delta H_{A\to B})_{P,m_2} \tag{11}$$

molar-shift term for H_2 and H_2^0

To estimate $\partial\alpha_{eq}/\partial m_2$ we shall use a "nearest-neighbor" model. Let $\alpha_{eq}(0)$ and $\alpha_{eq}(m_2)$ denote the equilibrium value of α in the pure solvent and in the solution, respectively. Let $\alpha_{solv.shell}$ denote the equilibrium value of α for solvent molecules that are adjacent to solute molecules, and let s denote the mean number of solvent molecules next to a solute molecule. Using these definitions, we may write

$$\alpha_{eq}(m_2) = \frac{\alpha_{solv.shell} m_2 s + \alpha_{eq}(0)(m_1 - m_2 s)}{m_1} \tag{12a}$$

$$m_1 \frac{\partial \alpha_{eq}}{\partial m_2} = s[\alpha_{solv.shell} - \alpha_{eq}(0)] \tag{12b}$$

Substitution in Eq. (11) then yields the working equation

$$H_2 = (H_2)_\alpha + s[\alpha_{solv.shell} - \alpha_{eq}(0)]\, \Delta H_{A\to B} \tag{13a}$$

$$H_2° = (H_2°)_\alpha + s[\alpha_{solv.shell} - \alpha_{eq}(0)]\, \Delta H_{A\to B} \tag{13b}$$

To show that the molar-shift term in (13) can be significant, we shall adopt the values $\Delta H_{A\to B} = 10$ kJ/mol and $\alpha_{eq}(0) \approx 0.3$, which characterize the two-state model for liquid water described in the next section (H_2O with 4 and 5 nearest neighbors) [15]. According to the authors' model of hydrophobic hydration shells [14], $\alpha_{solv.shell} \approx 0$ and $s \approx 10$. The molar shift term for H_2 then is -30 kJ/mol, which is highly significant experimentally.

The derivation is readily extended from the overall solute component to its individual molecular species. We begin by dissecting H_2 into contributions from its molecular species U and V [15]:

$$H_2 = (1-\beta)H_U + \beta H_V = (1-\beta)(H_U)_\alpha + \beta(H_V)_\alpha + m_1\left(\frac{\partial \alpha_{eq}}{\partial m_2}\right)\Delta H_{A\to B} \tag{14}$$

To identify the contributions from U and V, we shall dissect $\partial\alpha/\partial m_2$ into terms proportional to $\partial\alpha/\partial m_U$ and $\partial\alpha/\partial m_V$. Let $y_1 = \partial G/\partial\alpha$ and $y_2 = \partial G/\partial\beta$. Specify that $y_1 = 0$, but allow y_2 to vary. Then, at constant T and P, α is a function either of m_U and m_V, or of y_2 and m_2. Transformation of variables yields the general Eq. (15a), which reduces to (15b):

$$\left(\frac{\partial \alpha}{\partial m_2}\right)_{y_2} = \left(\frac{\partial \alpha}{\partial m_U}\right)_{m_V}\left(\frac{\partial m_U}{\partial m_2}\right)_{y_2} + \left(\frac{\partial \alpha}{\partial m_V}\right)_{m_U}\left(\frac{\partial m_V}{\partial m_2}\right)_{y_2} \tag{15a}$$

$$\left(\frac{\partial \alpha}{\partial m_2}\right) = (1 - \beta)\left(\frac{\partial \alpha}{\partial m_U}\right)_{m_V} + \beta\left(\frac{\partial \alpha}{\partial m_V}\right)_{m_U} \tag{15b}$$

Equations (14) and (15b) show that H_2 and $\partial\alpha/\partial m_2$ are mole-fractional averages, averaged over the solute species. Substitution of (15b) in (14) then leads to (15c), in which β is a variable; equating the coefficients of β and $(1 - \beta)$ then yields the desired expressions for the molecular species in (15d):

$$\begin{aligned}(1 - \beta)H_U + \beta H_V = {} & (1 - \beta)\left[(H_U)_\alpha + m_1\left(\frac{\partial \alpha}{\partial m_U}\right)_{m_V} \Delta H_{A\to B}\right] \\ & + \beta\left[(H_V)_\alpha + m_1\left(\frac{\partial \alpha}{\partial m_V}\right)_{m_U} \Delta H_{A\to B}\right]\end{aligned} \tag{15c}$$

$$H_U = (H_U)_\alpha + m_1\left(\frac{\partial \alpha}{\partial m_U}\right)_{m_V} \Delta H_{A\to B}; \qquad H_V = (H_V)_\alpha + m_1\left(\frac{\partial \alpha}{\partial m_V}\right)_{m_U} \Delta H_{A\to B} \tag{15d}$$

C. Solvation Mechanism

In condensed phases including liquids, the cage effect can generate environmental isomers [15a]. For example, as indicated in Fig. 8, a solvent molecule in a dilute solution may exist in a cage consisting entirely of solvent molecules, or in one in which the solvent molecule is adjacent to a solute molecule.

Composition Tree (T-3)

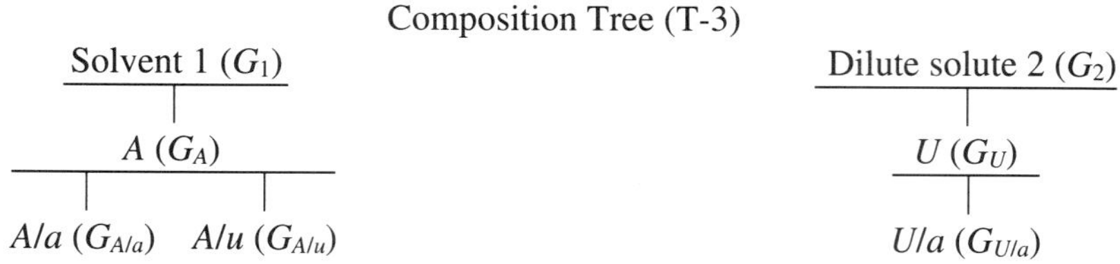

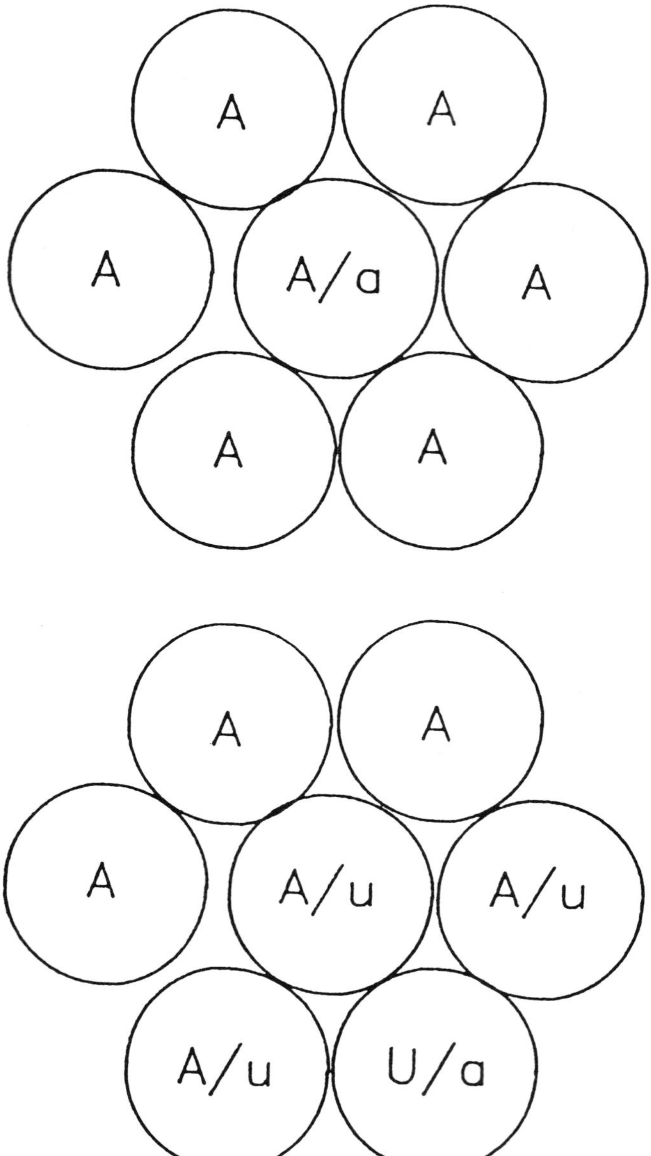

FIGURE 8 An example of environmental isomers of the solvent in dilute solutions. A = solvent molecule; U = solute molecule. Environments: $/a$ = adjacent only to solvent molecules; $/u$ = adjacent to one solute molecule. The other neighbors are solvent molecules.

The simplest composition tree that makes such a distinction is (T-3). Here and in the following, the symbol "/" means "exists in environment" and is followed by a symbol for the environment. Thus *A*, the solvent molecular species, consists of two subspecies, *A*/*a* and *A*/*u*, as defined in Fig. 8. *U*, the solute molecular species, is dilute enough to be entirely *U*/*a*; that is, the *U* molecules are adjacent only to solvent molecules.

The composition tree (T-3) requires that *A*/*a* and *A*/*u* be distinguishable, but this is true—ideally if not actually—in most cases. By solving the time-dependent part of the Schrödinger wave equation for molecules in liquid cages, Grunwald and Steel [15a] have shown that distinguishability exists if the interaction energy of the *A* molecule with an /*a* cage differs from that with a /*u* cage by an order of 20 I/mol or more.

Given distinguishability, we can develop a thermodynamic mechanism for enthalpy—entropy compensation, which we shall call the *solvation mechanism.* Let n_1 and n_2 denote the formula-weight numbers of the components, and z_2 the mean number of *A* molecules adjacent to a *U* molecule. The free energy of the solution may then be written in the form

$$G = (n_1 - z_2 n_2)G_{A/a} + n_2 G_{U/a} + z_2 n_2 G_{A/u} \tag{16a}$$

$$= n_1 G_{A/a} + n_2 G_{U/a} + z_2 n_2 (G_{A/u} - G_{A/a}) \tag{16b}$$

Similarly,

$$H = (n_1 - z_2 n_2)H_{A/a} + n_2 H_{U/a} + z_2 n_2 H_{A/u} \tag{17a}$$

$$S = (n_1 - z_2 n_2)S_{A/a} + n_2 S_{U/a} + z_2 n_2 S_{A/u} \tag{17b}$$

Thermodynamic analysis then shows that at equilibrium at constant *T*, *P*:

$$G_{A/u} = G_{A/a} \tag{18}$$

$$S_2 = S_{U/a} + z_2(S_{A/u} - S_{A/a}) \tag{19a}$$

$$H_2 = H_{U/a} + z_2(H_{A/u} - H_{A/a}) \tag{19b}$$

and

$$\mathrm{S}_{\mathrm{A/u}} \neq \mathrm{S}_{\mathrm{A/a}} \tag{19c}$$

The result is enthalpy–entropy compensation of the shift-of-cage term: $z_2(H_{A/u} - H_{A/a}) = Tz_2(S_{A/u} - S_{A/a})$.

The shift-of-cage term propogates into the *standard* partial enthalpy $H_2°$ and entropy $S_2°$. We therefore have a compensating contribution from solvation to $\Delta H°$ and $T\Delta S°$ for reaction. The magnitude of this contribution will often be negligible, but it could easily range up to tens of kv/mol for macromolecules.

While it is still risky to describe a specific example of compensation in terms of the solvation mechanism, we shall speculate that the solvation mechanism may be significant in the example shown in Fig. 3. London dispersion forces are distinctively strong in the solvation of the decomposing azo compound and of its radicaloid transition state, and $z_{2,eq}$ is relatively large. Except for this feature, the solvents, reactant, and transition state in this example would be expected to form "regular" solutions.

VII. APPLICATION TO NONPOLAR SOLUTES IN WATER

The concept that different nearest-neighbor environments can generate isomeric subspecies is especially useful for water and aqueous solutions. This kind of isomerism depends on the fact that translation and rotation for molecules in a liquid are hindered motions describable approximately as libration and torsional motion in a potential well (really a cage). The existence of a potential barrier increases the spacing between the energy levels for the hindered translation and rotation and thus, via statistical thermodynamics, changes the molar thermodynamic properties of the caged molecular species. When the interactions among the molecules permit the formation of isomeric cage configurations, with similar stabilities but distinctly different potential barriers, the sets of molecules residing in different cages acquire different thermodynamic properties and become environmental isomers. The interaction of a molecule with its surrounding molecular cage is not the same as molecular complex formation. Caged molecules moving under the influence of a specific hindering potential remain independent kinetic units, while the molecules in a molecular complex move as a single kinetic unit.

For liquid water at ordinary temperatures, a model in which pure water consists of two "states" in 1:1 equilibrium [16–20] approximately reproduces a broad spectrum of its properties [16–22]. It is now almost certain that the two "states" are environmental isomers, *W*/4*w* and *W*/5*w*, in which the water molecule *W* is nearest neighbor to 4 and 5 water molecules, respectively [21–24]. The identification of the environments as *W*/4*w* and *W*/5*w* rests on the fact that the experimental mean number of nearest water neighbors in liquid water is between 4 and 5 [26]. It is supported theoretically by molecular-dynamics simulations employing semiempirical intermolecular potentials, in which over 95% of the water molecules are found with roughly equal probabilities in /4*w* or /5*w* environments [24]. The condition that the two kinds of water molecules must be distinguishable is supported by the possible resolution of infrared and Raman spectral bands of liquid water into pairs of subbands with consistent properties [16,17,23]. *W*/4*w* is essentially a tetrahedrally hydrogen-bonded water molecule as depicted in the classical Bernal–Fowler model [25]. *W*/5*w* represents a water molecule in a denser, less polar environment, so that the interaction of *W* with the /5*w* environment is characterized by a higher enthalpy and a higher entropy than that of *W* with the /4*w* environment [21]. Empirical parameters [19b,20] for the interconversion of the isomers of liquid water are given in composition tree (T-4).

Composition Tree (T-4)

Water component $(1, n_1, m_1)$		Dilute solute $(2, n_2, m_2)$
W		U
$\langle W/4w \rangle\ (1 - \alpha)$	$W/5w\ (\alpha)$	$U/(W/4w)$

$\langle W/4w \rangle = [W/4w + W/(4w + u)]$
in fast exchange

$K_{4,5} = \alpha_{eq}/(1 - \alpha_{eq}) = 0.44$ (298 K); $\Delta V_{4,5} = -3{-}{-4}$ ml/mol;
$\Delta H_{4,5} = 10.5$ kJ/mol; $\Delta S_{4,5} = 28.2$ J/mol K. (Refs. 19b, 20.)

When a non-hydrogen-bonding solute U is introduced into the liquid water lattice, one expects that the U molecules if possible will fit interstitially between the hydrogen-bonded connections. Space requirements for the solute then operate so as to produce preferential solvation by the less dense $W/4w$ isomer. Solvated U is therefore shown in (T-4) as $U/(W/4w)$, and water molecules adjacent to U are consistently shown as $W/(4w + u)$. Solvation by the denser $W/5w$ isomer has been neglected.

A further assumption in (T-4) is that $W/4w$ and $W/(4w + u)$ are in fast exchange, producing a single exchange-averaged species $\langle W/4w \rangle$ whose fraction $1 - \alpha_{eq}$ and standard free energy, enthalpy, and entropy in dilute solution vary linearly with m_2. This assumption is supported by the fact that the solutes we shall consider, when dissolved in non-hydrogen-bonding solvents, do not show enthalpy–entropy compensation by the solvation mechanism.

A. Delphic Dissection of Standard Partial Entropies

When one of us first used the composition model shown in (T-4) to analyze thermodynamic properties of nonpolar solutes in water [14a], he called the procedure a "delphic dissection" because at the time it bordered on the oracular. It seems fitting to retain this name, though the scientific basis for (T-4) is now much stronger. On the other hand, the related adjective *isodelphic* is now replaced by *isomolar*, and the *lyodelphic* term in the earlier work is now the *molar-shift* term.

Our delphic dissection will be limited to standard partial entropies; enthalpies and heat capacities have been analyzed elsewhere [14]. The entropy data will be taken from Refs. 27–29. The dissection is based on the composition model given in (T-4). Using Eqs. (10b), (11), and (12b), the standard partial entropy of the nonpolar solute component is expressed by Eq. (20a); introduction of numerical data from (T-4) and letting $\alpha_{solv.shell} = 0.000$ (because the solvation shell is all $W/4w$) then yields Eq. (20b):

$$S_2^\circ(w) = (S_2^\circ)_\alpha + s[\alpha_{solv.shell} - \alpha_{eq}(0)]\,\frac{\Delta H_{4,5}}{T} \tag{20a}$$

$$\underset{\text{isomolar}}{S_2^\circ(w) = (S_2^\circ)_\alpha} \ \underset{\text{molar-shift term}}{-10.8\ s\ \text{J/mol K}} \tag{20b}$$

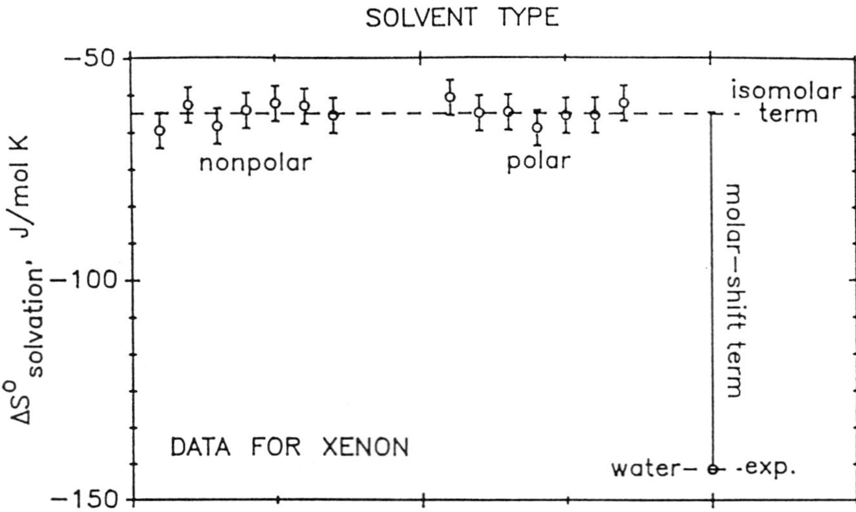

FIGURE 9 Evaluation of the molar-shift entropy term for xenon in water. The data for $\Delta S_{\text{solvation}}{}^{\circ}$ are based on the process, Xe(g,latm) → Xė(dilute solute, standard state of unit mole fraction). (From Refs. 27, 28.)

TABLE 1 Analysis of Entropies for Nonpolar Solutes in Water[a]

Solute	$\Delta S_2{}^{\circ}(w)$	$\Delta(S_2{}^{\circ})_{\alpha}$[b]	Molar-shift entropy	$\frac{\partial \alpha_{eq}}{\partial m_2}$	s[c]	σ (Å)[d]
He	−101	−43	−58	−0.0298	5.4	2.29
Ne	−110	−45	−65	−0.0336	6.1	2.69
Ar	−128	−55	−73	−0.0376	6.8	3.63
Kr	−135	−57	−78	−0.0400	7.2	4.11
Xe	−143	−63	−80	−0.0412	7.5	4.82
H_2	−106	−48	−58	−0.0296	5.4	2.97
N_2	−131	−51	−80	−0.0408	7.4	3.74
O_2	−130	−57	−73	−0.0376	6.8	3.60
CO	−128	−55	−73	−0.0376	6.8	3.74
CH_4	−133	−60	−73	−0.0376	6.7	4.14
C_2H_6	−151	−68	−83	−0.0427	7.7	5.27
C_3H_8	−163	−70	−93	−0.0477	8.6	6.20
n-C_4H_{10}	−174	−78	−96	−0.0493	8.9	6.88

[a]All entropies in units of J mol^{-1} K^{-1}.
[b]Mean ΔS° for solvation in non-hydrogen-bonding solvents.
[c]Eq. (20b).
[d]Gas-kinetic collision diameter of solute.

In delphic dissection, $(S_2°)_\alpha$ is predicted, and the molar-shift term is obtained from the experimental data by difference. $(S_2°)_\alpha$ can be obtained by prediction because physical theories (such as the regular-solution theory) equate formal components to molecular species and thus predict isomolar variables. Regular-solution theory predicts that for nonpolar solutes in non-hydrogen-bonding solvents, $\Delta S_2°$ for solvation ($g \rightarrow l$) is independent of the solvent [30], in agreement with observation. Typical data for xenon are plotted in Fig. 9. Since $\Delta S_2° = \Delta(S_2°)_\alpha$, extrapolation of the horizontal regular-solvent line yields $\Delta(S_2)_\alpha$ in the solvent water.

Results for 13 nonpolar gases are listed in Table 1. The molar shift entropies are large, ranging from -58 to -96 J/mol K, and would contribute factors ranging from 10^{-3} to 10^{-5} to an equilibrium constant. The compensating molar shift enthalpies, T times the entropies, range from -17 to -29 kJ/mol.

The test for validity of this dissection is that the s parameters listed in Table 1 must be physically plausible as nearest-neighbor numbers. In fact they are. Independent predictions for s based on the model of random packing of spheres and gas-kinetic collision diameters σ gave results within a few percent of those derived by dissection [14]. The results for S are also compatible with results of computer simulations [14b]. Indeed, one may be allowed to claim that the mystifying entropy anomaly for nonpolar solutes in water has been transformed into a predictable molar-shift term.

VIII. CONCLUDING REMARKS

Interest in enthalpy–entropy compensation in chemical processes and reactions has been strong ever since Leffler's original discovery of this propensity [1]. The interest is not misplaced. Compensation phenomena, as we have shown, result from interactions among molecular species and elucidate the molecular species, their subspecies, their molecular environments, and sometimes even their rates of exchange between specific environments.

Twenty years ago, Benzinger pioneered the concept of a connection between enthalpy–entropy compensation and conformational diversity among the molecular species in a thermodynamic component [31]. The present chapter builds on that concept. We agree, however, with Rhodes [32] that the specific compensation mechanism suggested by Benzinger may be insubstantial. Instead, we hope to have shown that predictable and substantial compensation can appear in liquid solutions because of solute-induced shifts of equilibria among molecular species in the solvent.

DEDICATION

We dedicate this chapter to John E. Leffler on the occasion of his retirement from Florida State University.

REFERENCES

1. J. E. Leffler, *J. Org. Chem. 20*: 1202 (1955).
2. J. E. Leffler and E. Grunwald, *Rates and Equilibria of Organic Reactions*, Wiley, New York, 1963, reprinted by Dover, 1989.
3. (a) R. Lumry and S. Rajender, *Biopolymers 9*: 1125–1227 (1970). (b) R. Lumry and R. B. Gregory, in *The Fluctuating Enzyme*, G. R. Welch, ed., Wiley-Interscience, New York, 1986.
4. F. J. Millero, C.-H. Wu, and L. G. Hepler, *J. Phys. Chem. 73*: 2453 (1969).
5. E. Grunwald and B. J. Berkowitz, *J. Amer. Chem. Soc. 73*: 4939 (1951).
6. A. F. Danil de Namor, M. C. Ritt, D. F. V. Lewis, M. J. Schwing-Weill, and F. A. Neu, *Pure & Appl. Chem. 63*: 1435 (1991).
7. M. G. Alder and J. E. Leffler, *J. Am. Chem. Soc. 76*: 1425 (1954).
8. J. F. Hinton, J. Q. Fernandez, C. S. Dikoma, W. L. Whaley, R. E. Koeppe, and F. S. Millett, *Biophys. J. 54*: 527 (1988).
9. T. Matsui, H. C. Ko, and L. G. Hepler, *Can. J. Chem. 52*: 2906 (1974).
10. P. D. Bolton, K. A. Fleming, and F. M. Hall, *J. Am. Chem. Soc. 94*: 1033 (1972).
11. E. Grunwald, *J. Am. Chem. Soc. 106*: 5414 (1984). (a) The compensation law is treated on p. 5416. (b) p. 5415.
12. E. Grunwald and T.-P. I, *Pure & Appl. Chem. 51*: 53 (1979).
13. E. Grunwald, S. P. Anderson, A. Effio, S. E. Gould, and K. C. Pan, *J. Phys. Chem. 80*: 2935 (1976); E. Grunwald, K.-C. Pan, and A. Effio, *ibid. 80*: 2937 (1976).
14. (a) E. Grunwald, *J. Amer. Chem. Soc. 108*: 5726 (1986); (b) E. Grunwald and L. Comeford, in *Environmental Influences and Recognition in Enzyme Chemistry*, J. F. Liebman and A. Greenberg, eds., VCH Publishers, New York, 1988, Chapter 3.
15. E. Grunwald, *Prog. Phys. Org. Chem. 17*: 31 (1990). (a) Distinguishability of liquid environments: E. Grunwald and C. Steel, *J. Phys. Chem. 97*: 13326 (1993).
16. G. E. Walrafen, *J. Chem. Phys. 40*: 3249 (1964); *ibid. 47*: 114 (1967); *ibid. 48*: 244 (1968).
17. J. D. Worley and I. M. Klotz, *J. Chem. Phys. 45*: 2868 (1966).
18. C. A. Angell, *J. Phys. Chem. 75*: 3698 (1971).
19. (a) T. Litowitz and E. Carnevale, *J. Appl. Phys. 26*: 816 (1955). (b) C. M. Davis and J. M. Jarzynski, in *Water and Aqueous Solutions*, R. A. Horne, ed., Wiley-Interscience, New York, 1972, Chapter 10. A two-state model with constant statewise molar volumes does not predict the density maximum of liquid water.
20. S. W. Benson, *J. Am. Chem. Soc. 100*: 5640 (1978).
21. E. Grunwald, *J. Am. Chem. Soc. 108*: 5719 (1986).
22. P. A. Giguere, *J. Chem. Phys. 87*: 4835 (1987).
23. G. E. Walrafen, M. S. Hokmabadi, W.-H. Yang, Y. C. Chu, and B. Monosmith, *J. Phys. Chem. 93*: 2909 (1989).
24. F. Sciortino, A. Geiger, and H. E. Stanley, *Nature 354*: 218 (1991).
25. S. D. Bernal and R. H. Fowler, *J. Chem. Phys. 1*: 515 (1933).
26. A. H. Narten and H. A. Levy, *Science 165*: 447 (1969).
27. (a) E. Wilhelm and R. Battino, *Chem. Reviews 73*: 1 (1973). (b) E. Wilhelm, R. Battino, and R. J. Wilcock, *Chem. Reviews 77*: 219 (1977).

28. S. F. Dec and S. J. Gill, *J. Solution Chem. 14*: 417, 827 (1985).
29. B. B. Benson and D. Krause, *J. Chem. Phys. 64*: 689 (1976).
30. J. H. Hildebrand, J. M. Prausnitz, and R. L. Scott, *Regular and Related Solutions*, Van Nostrand Reinhold, New York, 1970.
31. T. H. Benzinger, *Nature 229*: 100 (1971).
32. W. Rhodes, *J. Phys. Chem. 95*: 10246 (1991).

11

Preferential Interactions of Water and Cosolvents with Proteins

SERGE N. TIMASHEFF
Brandeis University, Waltham, Massachusetts

I. INTRODUCTION

It has been common practice over many decades for biologists and biochemists to control the stability of enzymes and other proteins, as well as of assembled organelles, by the addition of cosolvent at high concentration (typically at 0.3–10 M). This chapter will discuss issues involved in this concentration range. It seems evident from the concentration range utilized that the interactions must be weak, characterized by free energy changes, ΔG, not far from zero and frequently positive. Compounds most commonly used for stabilization are sucrose and other sugars, glycerol and other polyols (sorbitol, mannitol), as well as good salting-out salts, such as ammonium sulfate. The most common denaturants are urea and guanidine hydrochloride. The stabilizing cosolvents are also depressors of protein solubility, to the point of being precipitants. Conversely, not all precipitants are structure stabilizers. For example, both polyethyleneglycol (PEG) and 2-methyl-2,4-pentanediol (MPD), among the most widely used agents to precipitate proteins and crystallize them for structural studies, are protein denaturants.

The mechanism of action of these cosolvents on protein stability and solubility has been the subject of extensive studies over the past three decades. These have led to the conclusion that the macroeffects induced in proteins by these cosolvents are the direct consequence of their preferential interactions with the proteins. Preferential interactions—the redistribution of solvent composition in the presence of a protein—are commonly measured by equilibrium solution methods, such as dialysis equilibrium, vapor pressure equilibration, or light scattering. The observed interactions may be preferential binding (excess of cosolvent over the bulk solvent composition) or preferential exclusion (deficiency of cosolvent in the vicinity of the protein). The last corresponds to an excess of water, that is, to preferential hydration. These measured changes in solvent composition are the direct consequence of the mutual perturbation of the chemical potentials of the protein and cosolvent by each other. As such, they are true thermodynamic parameters and should be referred to as *thermodynamic binding*. This is in contrast to the occupancy of a site on a protein by a ligand, which, by itself, does not give a full description of the interaction process because it neglects the role of the principal solvent, water. This chapter will be devoted to the thermodynamic definition of preferential interactions, how they affect protein stability, their relation to site occupancy, and the physical and chemical factors that modulate them.

II. COSOLVENT CONTROL OF PROTEIN SOLUTION STABILITY AND STATE OF DISPERSION

A. Binding of Cosolvent and Displacement of Reaction Equilibria

The effects of cosolvents on proteins can be regarded in terms of three principal types of reactions: (a) protein denaturation–stabilization; (b) protein self-association and assembly into organelles; (c) protein solubility and precipitation (but not necessarily crystallization, since this involves nonequilibrium physical processes, such as crystal nucleation, growth, etc.). The last two are, in fact, manifestations of the same phenomenon.

Let us take a general equilibrium reaction, $R \rightleftharpoons Pr$, which could be, for example, the denaturation equilibrium, $N \rightleftharpoons D$. Let the equilibrium be affected by a cosolvent X, so that $K = [Pr]/[R]$ is $f[X]$. Following Scatchard [1], let us define components 1, 2, and 3 as water, protein, and cosolvent, respectively. Then, according to the Wyman linkage relation [2,3],

$$-\left(\frac{\partial \Delta G^\circ}{\partial \mu_3}\right)_{T,P,m_2} = \left(\frac{\partial \ln K}{\partial \ln a_3}\right)_{T,P,m_2} = \left(\frac{\partial m_3}{\partial m_2}\right)_{T,P,\mu_3} = \left(\frac{\partial m_3}{\partial m_2}\right)_{T,P,\mu_3}^{Pr} = \left(\frac{\partial m_3}{\partial m_2}\right)_{T,P,\mu_3}^{R} \tag{1}$$

where ΔG° is the standard free energy change of the $R \rightleftharpoons Pr$ reaction; μ_i is the chemical potential of component i, $\mu_i = \mu_i^\circ + RT \ln a_i$; a_i is the activity of component i, $a_i = m_i\gamma_i, m_i$ is the molal concentration of component i, γ_i is its activity coefficient; T is the thermodynamic (Kelvin) temperature, and P is pressure. The quantity $(\partial m_3/\partial m_2)_{T,P,\mu_3}$ is the perturbation of the concentration of cosolvent by addition of the protein to the system at chemical (dialysis) equilibrium of the cosolvent. It is normally referred to as *binding*. Equation (1) states that the two processes, the $R \rightleftharpoons Pr$ equilibrium and the binding of the ligand (cosolvent) to the protein, are linked functions [4]; that is, a change in either process must necessarily be accompanied by a change in the other. Therefore, if $\Delta(\partial m_3/\partial m_2)_{T,P,\mu_3}$ is positive, the cosolvent drives the reaction to the right (in the case of unfolding, it is a denaturant). If it is negative, the cosolvent drives the reaction to the left (in the case of unfolding, it is a stabilizer).

As stated above, the right-hand side of Eq. (1) is the change in "binding" of the cosolvent to the protein during the course of the reaction, where *binding* is defined as the change in the concentration of component 3 upon addition of component 2 to the system at constant chemical potential of component 3. With the minor approximation that $(\partial m_3/\partial m_2)_{T,P,\mu_3} \simeq (\partial m_3/\partial m_2)_{T,\mu_1,\mu_3}$ [5], it is clear that this parameter can be equated to the ligand binding measured by dialysis equilibrium and related techniques, $\bar{\nu}_3$ in Scatchard's notation [6], expressed as moles of ligand bound per mole of protein.

B. What Is Binding?

Rigorously, the quantity usually referred to as binding, $\bar{\nu}_3 \equiv (\partial m_3/\partial m_2)_{T,\mu_1,\mu_3}$, means the change in concentration of ligand in the domain of the protein needed to restore chemical equilibrium. This requires a redistribution of solvent components in the vicinity of the protein, which can be accomplished by a change in the concentration of ligand, of water, or of both. Therefore, the binding parameter $(\partial m_3/\partial m_2)_{T,\mu_1,\mu_3}$ is the resultant of the balance between the interactions of ligand and water with the protein. It follows this parameter can be decomposed into the individual contributions B_i of the number of molecules of each component found in the domain of each protein molecule that is interacting with it by, for example, occupying sites on it. As will be shown below, the form of this relation is [7]

$$\bar{\nu}_3 \equiv \left(\frac{\partial m_3}{\partial m_2}\right)_{T,\mu_1,\mu_3} = B_3 - \frac{m_3}{m_1} B_1 \tag{2}$$

From this equation it is evident that if the ratio B_3/B_1 is equal to the bulk solvent composition, that is, $B_3/B_1 = m_3/m_1$, then the dialysis equilibrium result will be $\bar{\nu} = 0$; that is "binding" will be stated to be zero even though loci on the surface of the protein are occupied by the ligand. For $B_3/B_1 > m_3/m_1$, $\bar{\nu}_3$ will be positive; for $B_3/B_1 < m_3/m_1$, $\bar{\nu}_3$ will be negative. Thus, the thermodynamic measurement of binding gives *not* the number of sites occupied by the ligand of interest (component 3), but the balance relative to bulk solvent composition between the occupancy of sites by the ligand and water [8]. This balance can be positive, negative, or zero. As a result, the thermodynamic parameter $(\partial m_3/\partial m_2)_{T,\mu_1,\mu_3}$, measured by dialysis equilibrium, is known as *preferential binding*, since it expresses the preference of the protein for water or ligand, that is, its relative affinities for water and ligand (cosolvent, denaturant). Tanford [9] has shown formally that the change in ligand binding in the Wyman linkage equation refers to change in preferential binding, not site occupancy, a fact implicit in the Wyman treatment, as will become evident later.

C. Cosolvent Effects on Equilibria Relative to Water

The Wyman linkage relation (Eq. (1)) describes the effect of a ligand on a chemical equilibrium at a given concentration of the ligand. In the most general case, the extent to which a cosolvent affects an equilibrium can obviously vary with cosolvent concentration, and actually the directions in which the equilibrium is displaced can be opposite at different concentrations of the same ligand. Frequently, it is desired to measure the effect that a ligand has on a macromolecular reaction relative to pure water. Thermodynamically, for the reaction State I $\rightleftharpoons$ State II, this is expressed by the change in its standard free energy when the system is transferred from water to the cosolvent solution:

$$\delta\,\Delta G^\circ = \Delta G^\circ\,(\text{cosolvent}) - \Delta G^\circ\,(\text{water}) \tag{3}$$

For a solution component to have an effect on a reaction, a change must occur in the interactions of that component with the entity (protein) undergoing the reaction. This interaction can be expressed through the change in the chemical potential of the protein when transferred from water to the cosolvent system, defined as the standard free energy of transfer:

$$\Delta\mu_{2,\text{tr}} = \mu_2\,(\text{cosolvent}) - \mu_2\,(\text{water}) \tag{4}$$

For a reaction, this must be defined for the two end states:

$$\delta\,\Delta\mu_{2,\text{tr}} = \Delta\mu_{2,\text{tr}}\,(\text{State II}) - \delta\mu_{2,\text{tr}}\,(\text{State I}) \tag{5}$$

Combining Eqs. (3) and (5) reduces the system to a two-step process in which the starting material is protein in State I in water and the product is protein in State II in the cosolvent system. This transformation can be accomplished along two pathways:

I. (1) Conversion of State I to State II in water;
   (2) Transfer of State II from water to the cosolvent system;

II. (1) Transfer of State I from water to the cosolvent system;
   (2) Conversion of State I to State II in the cosolvent system.

Since the two paths are thermodynamically equivalent, this gives rise to a classical thermodynamic box [8]:

$$\begin{array}{ccc} \text{State I (cosolvent)} & \overset{\Delta G^\circ(\text{cs})}{\rightleftharpoons} & \text{State II (cosolvent)} \\ \updownarrow \ \Delta\mu_{2,\text{tr}}\ (\text{I}) & & \updownarrow \ \Delta\mu_{2,\text{tr}}\ (\text{II}) \\ \text{State I (water)} & \overset{\Delta G^\circ(\text{w})}{\rightleftharpoons} & \text{State II (water)} \end{array}$$

Therefore, $\Delta\mu_{2,\text{tr}}\ (\text{I}) + \Delta G^\circ(\text{cs}) = \Delta\mu_{2,\text{tr}}\ (\text{II}) + \Delta G^\circ(\text{w})$, where cs and w stand for cosolvent and water, respectively.

Let us express this specifically for the three processes of interest: denaturation (stabilization), precipitation (solubility), self-assembly (organelle stability).

1. Denaturation ($N \rightleftharpoons D$):

$$\Delta G_{m_3}{}^{\circ,N-D} - \Delta G_{\text{w}}{}^{\circ,N-D} = \Delta\mu_{2,\text{tr}}{}^{D} - \Delta\mu_{2,\text{tr}}{}^{N} \tag{6a}$$

2. Solubility (protein in solution $\rightleftharpoons$ precipitate), neglecting protein activity coefficients:

$$\ln \frac{S_{2,m_3}}{S_{2,\text{w}}} = \frac{1}{RT} (\Delta\mu_{2,\text{tr}}{}^{\text{pr}} - \Delta\mu_{2,\text{tr}}{}^{\text{sol}}) \tag{6b}$$

where S_2 is protein solubility, the subscript w is water, and the superscripts pr and sol mean precipitate and solution

3. Self-assembly ($P_n + P \rightleftharpoons P_{n+1}$):

$$\Delta G_{m_3}{}^{\circ(\text{growth})} - \Delta G_{\text{w}}{}^{\circ(\text{growth})} = \Delta\mu_{2,\text{tr}}{}^{P_{n+1}} - \Delta\mu_{2,\text{tr}}{}^{P} \tag{6c}$$

where the superscript P means protein subunit freely dispersed in solution, and P_{n+1} means the same subunit when incorporated into the assembled structure, such as a subunit enzyme complex or an organelle, say an actin fiber

III. RELATION BETWEEN PREFERENTIAL INTERACTIONS AND TRANSFER FREE ENERGY

A. Thermodynamic Definition of Binding

As discussed above, the effect of a cosolvent on a chemical reaction can be expressed through the change either in preferential interactions or in the transfer free energy during the course of the reaction. The first has as reference state the solution of the given composition, the second pure water. Therefore, the change in preferential binding tells the direction in which a cosolvent will displace an equilibrium at the given cosolvent concentration at which the slope of the Wyman plot (Eq. (1)) is being measured, and the strength of that displacement. The change in the transfer free energy defines the effect the ligand has on the free energy of the reaction relative to pure water.

The redistribution of solvent components in the vicinity of a protein measured by dialysis equilibrium, $(\partial m_3/\partial m_2)_{T,\mu_1,\mu_3}$, is a necessary consequence of the mutual perturbations of the chemical potentials of the protein and ligand by each other [2,6,10–12]:

$$\left(\frac{\partial m_3}{\partial m_2}\right)_{T,P,\mu_3} = -\frac{(\partial \mu_3/\partial m_2)_{T,P,m_3}}{(\partial \mu_3/\partial m_3)_{T,P,m_2}} = -\left(\frac{\partial \mu_2}{\partial \mu_3}\right)_{T,P,m_2} \tag{7}$$

The denominator on the right-hand side is the expression of the cosolvent nonideality.

$$\left(\frac{\partial \mu_3}{\partial m_3}\right)_{T,P,m_2} = RT\left(\frac{1}{m_3} + \frac{\partial \ln \gamma_3}{\partial m_3}\right) \tag{8}$$

The quantity $(\partial \mu_2/\partial m_3)_{T,P,m_2} = (\partial \mu_3/\partial m_2)_{T,P,m_3}$, which expresses the mutual perturbations of the chemical potentials of the protein and cosolvent by each other, is known as the *preferential interaction parameter.*

Equation (7) has the practical meaning that depending on whether, at a given solvent composition m_3, the interaction of the protein with the cosolvent-containing solution is thermodynamically favorable, unfavorable, or indifferent (respectively, negative, positive, or zero value of the preferential interaction parameter), the parameter $(\partial m_3/\partial m_2)_{T,\mu_1,\mu_3}$ will be positive, negative, or zero, resulting in turn in experimentally measured preferential binding, preferential exclusion, or no interaction. Preferential exclusion of the cosolvent manifests itself in dialysis equilibrium experiments by giving *negative* stoichiometries of binding. Quite evidently, this means that the solvent vicinal to the protein contains an excess of water over its presence in the bulk solvent; that is, the protein is preferentially hydrated. Application of the Gibbs–Duhem equation to Eq. (7) permits expression of the dialysis equilibrium results in terms of the preferential binding of water [13]:

$$\left(\frac{\partial m_1}{\partial m_2}\right)_{T,\mu_1,\mu_3} = -\frac{m_1}{m_3}\left(\frac{\partial m_3}{\partial m_2}\right)_{T,\mu_1,\mu_3} = \frac{m_1}{m_3}\left(\frac{\partial \mu_2}{\partial \mu_3}\right) \tag{9}$$

B. Binding Is Replacement of Water by Ligand at a Site

The result that, at dialysis equilibrium, the interaction of the protein with the mixed solvent, expressed as interaction with component 3, can be favorable or unfavorable follows because the amount of water and ligand found in the immediate domain of the protein is defined by the relative affinities of water and ligand for the protein. By definition, in an aqueous medium, the binding of one molecule of ligand to a site on a protein molecule must displace molecules of water from that site, if for no other reason than the physical law that two bodies cannot occupy the same space at the same time. This is illustrated in Fig. 1. A contact between water molecules and protein at the site must manifest itself by some characteristic free energy of interaction that may be expressed through an equilibrium constant, K_w. Therefore, the interaction of the protein with an aqueous mixed solvent at a given site must be written as

$$P + m\,H_2O \overset{K_w}{\rightleftharpoons} P \cdot H_2O_m \tag{10a}$$

$$P + L \overset{K_L}{\rightleftharpoons} PL \tag{10b}$$

where P is protein with the binding site empty, PL is the protein with the site occupied by ligand, and $P \cdot H_2O_m$ is protein with the site occupied by m molecules of water. It is clear that, at the site, there is competition between water and ligand molecules [13]. Therefore, the free energy of binding at a site, measured by dialysis equilibrium, ΔG^b, is the difference between the intrinsic free energies of interaction of the protein with ligand, ΔG^{32}, and with water, ΔG^{12} [13,14]:

$$\Delta G^b = \Delta G^{32} - \Delta G^{12} \tag{11}$$

Since $(\partial\mu_2/\partial m_3)_{T,P,m_2}$ for one site is equal to $(\partial\,\Delta G^b/\partial m_3)_{T,P,m_2}$, Eq. (11), when combined with Eqs. (7) and (9) and followed by summation over all sites on a protein molecule, gives the relation between the dialysis equilibrium "binding" result and total site occupancies by ligand and water molecules on the protein [7,13,14]:

$$\left(\frac{\partial m_3}{\partial m_2}\right)_{T,P,\mu_3} = B_3 - \frac{m_3}{m_1} B_1$$

This is seen to be Eq. (2).

C. The Wyman Slope Is the Change in Thermodynamic Interaction

In his derivation of Eq. (1), Wyman derived "binding" from Eq. (7) [2], that is, from the variation of the chemical potential of the ligand induced by the protein. Therefore, Wyman's definition of binding was general and not equated with site

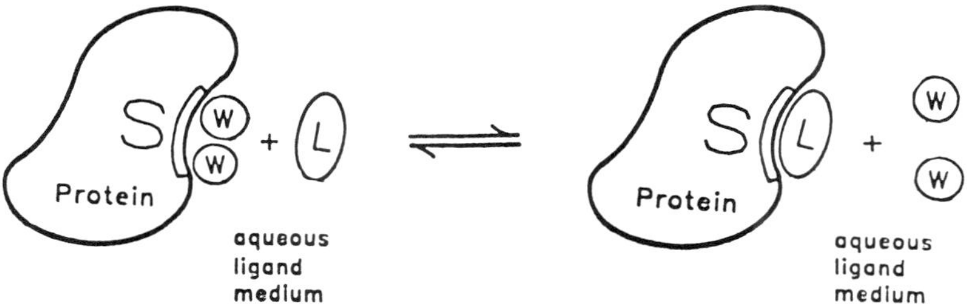

FIGURE 1 Scheme of protein–ligand interaction: competition between the binding of ligand (L) and water (W) molecules at the site (S), as defined by Eq. (11).

occupancy. The Wyman equation (Eq. (1)) can be restated then in a thermodynamically more significant form:

$$\left(\frac{\partial \ln K}{\partial \ln a_3}\right)_{T,P,m_2} = \frac{(\partial\mu_3/\partial m_2)_{T,P,m_3}{}^{R} - (\partial\mu_3/\partial m_2)_{T,P,m_3}{}^{Pr}}{(\partial\mu_3/\partial m_3)_{T,P,m_2}} \tag{12}$$

which leads to

$$\left(\frac{\partial \Delta G^\circ}{\partial m_3}\right)_{T,P,m_2} = \Delta RT\left(\frac{\partial \ln \gamma_2}{\partial m_3}\right)_{T,P,m_2} \tag{13}$$

Therefore, at any concentration m_3 of cosolvent, its effect on the chemical equilibrium is defined by the change of its perturbation of the excess chemical potential of the protein during the course of the reaction.

D. Relation Between Transfer Free Energy and Preferential Interaction

The transfer free energy of a protein from water to a cosolvent-containing system is the total free energy change induced in the system by addition of the cosolvent to an aqueous medium. Therefore, it is related to the variation of the preferential interaction of that protein with the cosolvent by [15]

$$\Delta\mu_{2,\mathrm{tr}} = \int_0^{m_3}\left(\frac{\partial\mu_2}{\partial m_3}\right)_{T,P,m_2} dm_3 \tag{14}$$

Combining Eqs. (5), (12), and (14) shows that the change in transfer free energy during the course of a reaction, $\delta\,\Delta\mu_{2,\mathrm{tr}}$, can be obtained directly from integration of the modified Wyman equation (Eq. (13)):

$$\delta\,\Delta\mu_{2,\mathrm{tr}} = RT\int_0^{m_3}\left(\frac{\partial \ln K}{\partial m_3}\right)_{T,P,m_2} dm_3 \tag{15}$$

Examination of these relations shows that knowledge of the preferential interaction at a single solvent composition does not reveal the effect of a cosolvent on a reaction relative to water. The relation between the preferential interaction parameter and transfer free energy is illustrated in Fig. 2, based on data of the preferential interactions of β-lactoglobulin with $MgCl_2$ [16]. Let us scrutinize system 2. In it, the measured preferential interaction parameter, shown in the lower part of the figure, follows a cosolvent concentration dependence pattern in which it starts at a high positive value, then decreases linearly, and finally becomes negative above 2.5 molal. This means that at low concentration the cosolvent is preferentially excluded from the protein; that is, there is preferential hydration. This exclusion decreases progressively until, above 2.5 molal, the cosolvent becomes preferentially bound. The corresponding variation of $\Delta\,\mu_{2,\mathrm{tr}}$ with m_3 (upper part of the figure, curve 2) shows an increasingly unfavorable thermodynamic state of the protein as the salt concentration increases. The rate of increase of $\Delta\mu_{2,tr}$ with cosolvent concentration gradually decreases, however. At 2.5 molal, this parameter reaches a maximum, after which it starts to fall. It is important to note that at the high cosolvent concentrations (>2.5 m) the protein is still in an unfavorable environment relative to water and is salted out, though now it binds cosolvent preferentially. This difference in the patterns of variation of these two thermodynamic parameters is the direct consequence of the difference in their reference states: $\Delta\mu_{2,tr}$ always refer to pure water, while $(\partial\mu_2/\partial m_3)_{T,P,m_2}$ refers to water in a solvent that contains cosolvent at a concentration, m_3 . The preferential interaction parameter is the rate of change of $\Delta\mu_{2,tr}$; that is, it is a comparison of the state of the protein in a given concentration of cosolvent relative to its state at an infinitesimally lower concentration. As a result, the trends of the two parameters may be opposite in direction, and they may actually assume opposite signs (see Table 1, BSA in 3 M $MgCl_2$ and β-lactoglobulin in 80% 2-chloroethanol), an observation that may seem paradoxical at first.

IV. HOW TRANSFER FREE ENERGY MODULATES PROTEIN REACTIONS

As shown by Eq. (1), the effect of cosolvent on a reaction in a solvent of a given composition is defined by the change in preferential binding of the cosolvent to the protein during the course of the reaction. Its effect relative to water is given by the difference between the transfer free energies in the two end states of the reaction. Examination of a large number of cosolvents by dialysis equilibrium has shown that all of the stabilizing and salting out agents are preferentially excluded from native globular proteins, that is, their contact with protein is thermodynamically unfavorable—$(\partial\mu_2/\partial m_3)_{T,P,m_2}$ is positive. Conversely, denaturants (6 M GuaHCl, 8 M urea) mostly interact favorably with the unfolded state of the protein—$(\partial\mu_2/\partial m_3)_{T,P,m_2}$ is negative. A selected list of typical results is given in Table 1.

A. Precipitation

Since the effect of a cosolvent relative to water is determined by $\delta\ \Delta\mu_{2,tr}$, it is evident that in order to define this effect the interactions between protein and cosolvent must be known for the two end states of the reaction. There is a fundamental difference, however, between the denaturation (unfolding) and coalescence (precipitation, self-assembly) reactions. This is illustrated in Fig. 3. By definition, in the protein precipitation and self-assembly reactions the structure of the protein molecules is not changed during the course of the reaction; in denaturation, the structure of the protein is changed. This means that in the coalescence reactions the chemical nature of the contacts between protein and solvent components is essentially the same in the two end states. The resulting very weak interactions can

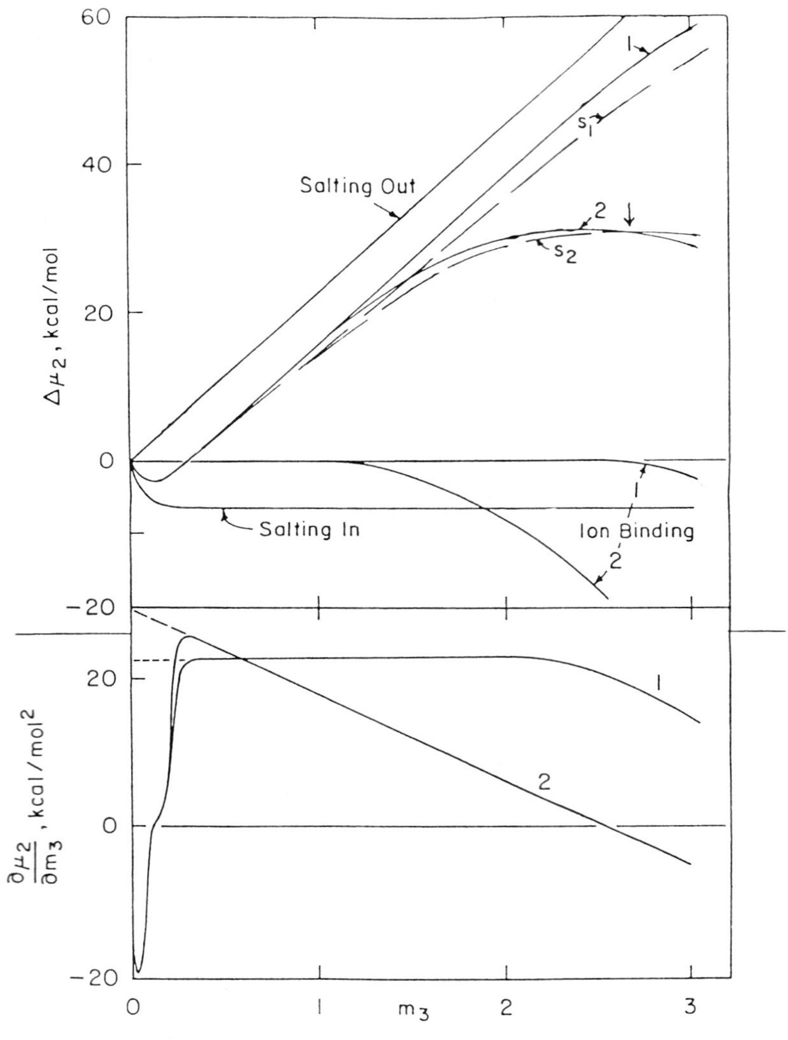

FIGURE 2 Relation between transfer free energy $\Delta\mu_{2,tr}$ and the preferential interaction parameter $(\partial\mu_2/\partial m_3)_{T,P,m_2}$ as a function of cosolvent concentration m_3. In case 1 (top), the preferential interaction parameter remains positive and constant over a large concentration range. Concomitantly, the transfer free energy increases monotonely at a steady rate. In case 2 (bottom), the preferential interaction parameter is a steadily decreasing function, passing from a high positive value at low cosolvent concentration to a negative value at high cosolvent concentration. The transfer free energy is positive at all cosolvent concentrations; it passes through a maximum at the cosolvent concentration at which $(\partial\mu_2/\partial m_3)_{T,P,m_2}$ is equal to zero

be assumed, as a reasonable approximation, to be statistically distributed over the surface of a protein molecule. Coalescence into a precipitate or a self-assembled structure reduces the protein–solvent contact area due to the protein–protein contacts, as illustrated in Fig. 3a [16]. In salting-out systems, the protein–cosolvent interactions are thermodynamically unfavorable in disperse solution; that is, $\partial m_3/\partial m_2$ is negative. This unfavorable situation, the magnitude of which is taken to be proportional to the protein–solvent contact area, must decrease in the coalesced state [31]. The net consequence is that $(\partial m_3/\partial m_2)^{pr} - (\partial m_3/\partial m_2)^{sol} = \Delta\bar{\nu}_3$ is positive and the equilibrium is pushed toward precipitation or self-assembly. Therefore, in the case of salting out or self-assembly, knowledge of the preferential interactions of the dispersed protein is usually sufficient to analyze the full reaction if the protein structure remains unaltered, as it is defined to be.

Thermodynamically this is described by the chemical potential diagram shown in Fig. 4 [16]. The standard chemical potentials of a protein in aqueous solution in the precipitated (pr) state, $\mu_{2,w}^{\circ(pr)}$, and in the solution (sol) state, $\mu_{2,w}^{\circ(sol)}$, are depicted by the horizontal lines on the left. Their difference gives the standard free energy of solution of the protein in water, $\Delta G_{2,w}^{\circ(sol)} = -RT \ln S_{2,w}$, within the approximation that protein activity coefficients may be neglected. This quantity contains all the specific and nonspecific interactions involved in the interprotein contacts, as well as all changes in entropy related to the incorporation of a single protein molecule into the large structure. As the salting-out material is added to the system, the environment of the protein in both the solution and the precipitate becomes less favorable than in water and its standard chemical potentials in the two states, $\mu_{2,m_3}^{\circ(pr)}$ and $\mu_{2,m_3}^{\circ(sol)}$, depicted by the horizontal lines at the right, become more positive, the magnitude of the change being equal to the transfer free energy in each state. Since $\Delta\mu_2^{\circ(pr)} < \Delta\mu_2^{\circ(sol)}$, $\Delta G_{2,m_3}^{\circ(sol)}$ must be more positive than $\Delta G_{2,w}^{\circ(sol)}$ and the solubility in the salting-out medium must be reduced relative to that in water.

B. Structure Stabilization–Destabilization

In the case of denaturation (unfolding), the situation is much more complex. In this reaction, *by definition*, the structure of the protein changes. As a consequence, the

---

and decreases at higher values of m_3, but remains positive; that is, the thermodynamic state relative to pure water remains unfavorable. In the upper part of the figure, the solid lines refer to protein dispersed in solution, the dashed lines refer to the precipitate. The shapes of the thermodynamic functions at low m_3 reflect the salting-in region at low salt concentration. At higher concentrations of cosolvent, extensive ion binding was superposed on the salting out in case 2, and a low level of ion binding in case 1. This, in fact, could be the binding pattern of any weakly interacting ligand or cosolvent.

TABLE 1 Thermodynamics of Protein–Solvent Interactions in Structure Stabilizing and Destabilizing Systems

Protein	Solvent	$\partial m_3/\partial m_2$ (mol/mol)	$\partial \mu_2/\partial m_3$ (kcal/mol of protein/mol of ligand)	$\Delta \mu_{2,tr}$[a] (kcal/mol)	Ref.
Lysozyme	8 *M* urea	12.0	−0.56	−7.0	17[b]
β-Lg	Cl EtOH, 20%,v/v	108	−17.1	−41.6	18
β-Lg	Cl EtOH, 40%,v/v	166	−9.8	−129	18
β-Lg	Cl EtOH, 65%,v/v	0	0	−203	18
β-Lg	Cl EtOH, 80%,v/v	−145	+1.45	−185	18
β-Lg	glycerol, 30%,v/v	−14.7	+1.50	+8.7	15
RNase A	1 *M* glucose	−3.1	+1.70	+1.5	19
RNase A	1 *M* sucrose	−7.6	+4.3	+4.1	20
CTGen, native	PEG-1000, 20%, v/v	−1.0	+11.0	+2.7	21[b]
CTGen, denatured	PEG-1000, 20%, v/v	1.3	−13.0	−3.1	21[b]
Lysozyme	PEG-4000, 2.5%, v/v	−1.0	+85	+0.5	22
RNase A	MPD, 50%, v/v	−111	+8.7	+35	23
BSA	2 *M* betaine	−72.5	+17.0	+38	24
BSA	2 *M* lysine HCl	−45.8	+19.7	+34	25
BSA	2 *M* Na Glutamate	−68.8	+33.0	+72	25
BSA	1 *M* Na_2SO_4 pH 5.6	−35.4	+37.2	+37	26[b]
BSA	1 *M* $MgSO_4$ pH 4.5	−26.5	+17.7	+18	27[b]
BSA	1 *M* MgCl pH 5.6	−3.0	+7.7	+7.7	26
BSA	1 *M* $MgCl_2$ pH 3.0	−11.3	+29.4	+40	16
BSA	3 *M* $MgCl_2$ pH 3.0	9.6	−20.0	+55	16
BSA	1 *M* Gua HCl	17.5	−13.8	−15.4	28

TABLE 1 Continued

Protein	Solvent	$\partial m_3/\partial m_2$ (mol/mol)	$\partial \mu_2/\partial m_3$ (kcal/mol of protein/mol of ligand)	$\Delta\mu_{2,tr}$[a] (kcal/mol)	Ref.
BSA	6 *M* Gua HCl	44	−3.1	−95	29
BSA	0.5 *M* Gua_2SO_4	2.55	−4.6	−2.3	28
BSA	1 *M* Gua_2SO_4	−16.4	+8.7	+4.4	28
β-Lg	0.5 *M* glycine pH 5.1	1.22	−1.30	−1.3	30
β-Lg	1.0 *M* glycine pH 5.1	−6.5	+3.20	−0.18	30
β-Lg	2.0 *M* glycine pH 5.1	−23.8	+5.40	+4.1	30

[a]Calculated by integration of $(\partial \mu_2/\partial m_3)_{T,P,m_2}$, according to Eq. (14).
[b]Calculated from a single point value of $(\partial \mu_2/\partial m_3)_{T,P,m_2}$, with the assumption that this parameter is constant.

nature of protein–solvent interactions may also change; for example, new extensive nonpolar regions and peptide groups become exposed to solvent and the distance between charges (ionizable residues) increases greatly. The cosolvent effects on the reaction must depend then on a fine balance between patterns of interaction of the native and unfolded proteins. This may involve combinations of exclusion and binding that change during the course of the reaction. Therefore, the effect of a cosolvent on the $N \rightleftharpoons D$ equilibrium requires knowledge of preferential interactions with both the *N* and *D* states. Several general cases will be described [32].

Case I: The cosolvent is preferentially excluded from proteins by a strictly nonspecific mechanism in which the cosolvent is totally inert toward chemical groups on the protein surface and acts only by general exclusion from interfaces [20,32]. This is illustrated in Fig. 3b. Since on unfolding the protein molecule becomes more asymmetric and the protein–solvent interface increases, the unfavorable contacts also increase. Preferential exclusion is increased. By Eq. (1), $\Delta\bar{\nu}_3$ is negative and the equilibrium is displaced toward the native state. Such a situation is found with sugars [20] and strong salting-out salts, such as $(NH_4)_2SO_4$ and $MgSO_4$ [32]. In these systems, $\Delta\mu_{2,tr}$ increases monotonely with cosolvent concentration, since $(\partial \mu_2/\partial m_3)_{T,P,m_2}$ remains constant (see Fig. 2) [16].

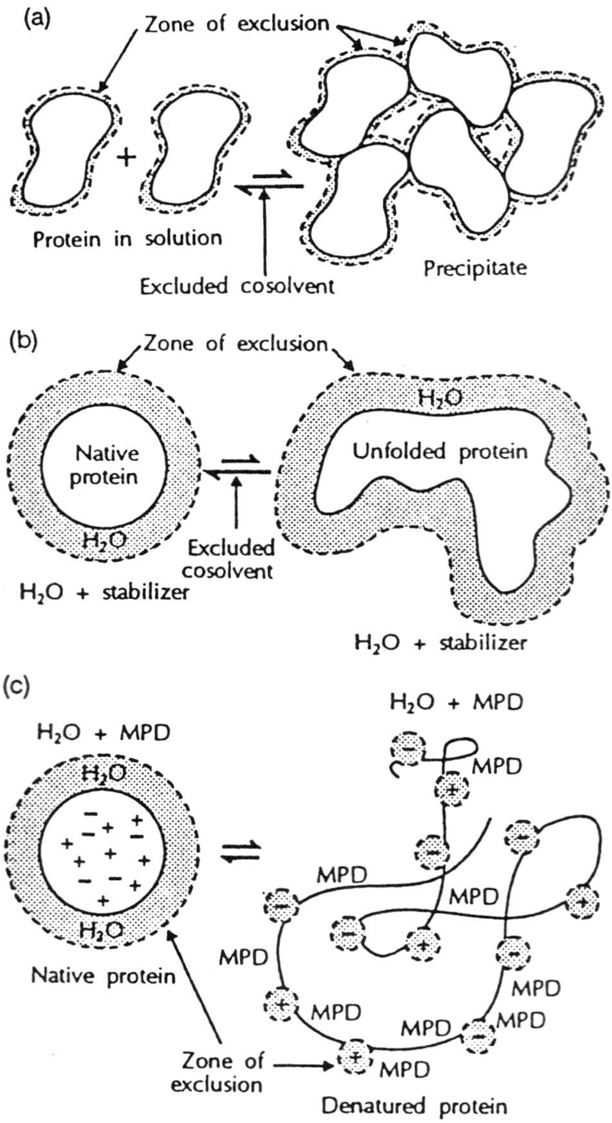
(a)
Zone of exclusion
Protein in solution
Excluded cosolvent
Precipitate
(b)
Zone of exclusion
Native protein
H_2O
Excluded cosolvent
H_2O + stabilizer
Unfolded protein
H_2O
H_2O + stabilizer
(c)
H_2O + MPD
H_2O
H_2O
Native protein
Zone of exclusion
H_2O + MPD
MPD
Denatured protein

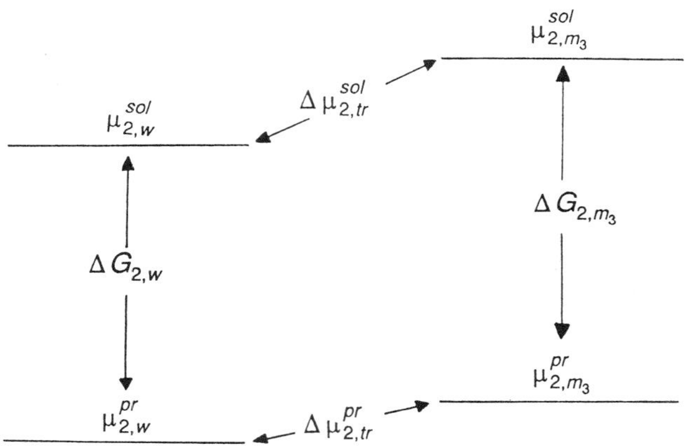

FIGURE 4 Free energy diagram of protein solubility in water (*w*) and a salting-out aqueous cosolvent (m_3) system.

Case II: The cosolvent is preferentially excluded from protein in the native state, but is preferentially bound to the unfolded state, as is true of PEG [21]. In this case, $(\partial m_3/\partial m_2)_{T,\mu_1,\mu_3}$ changes from negative to positive during the course of the reaction and, by Eq. (1), the equilibrium is displaced to the right. This situation is illustrated for MPD in Fig. 3c [32].

Case III: The cosolvent binds weakly or not at all to the native protein, but binds extensively to the unfolded form. This is the classical situation with strong denaturants, such as urea and guanidine hydrochloride. For these, $\Delta\bar{\nu}_3$ is positive and $\delta\ \Delta\mu_{2,\mathrm{tr}}$ is negative, and the denaturation is driven by the favorable interactions of the denaturant with the unfolded form of the protein [29,33].

Case IV: A final case is one in which the denaturant is indifferent to protein in the unfolded form, but is excluded from the native form. This has been identified by Schellman [33] for ribonuclease A in 6 *M* GuaHCl, for which $(\partial m_3/\partial m_2)_{T,\mu_1,\mu_3} = 0$, but $\delta\ \Delta G^{\circ,N-D}$ is negative [34]. By Eq. (6a), $\delta\ \Delta\mu_{2,\mathrm{tr}}$ must be negative, which leads to the conclusion that $(\partial m_3/\partial m_2)_{T,\mu_1,\mu_3}$ for the native protein

FIGURE 3 Patterns of protein–solvent interactions. (a) Schematic representation of the salting-out reaction. The preferential exclusion of cosolvent per protein subunit is reduced in the precipitated state. (b) Schematic representation of the denaturation reaction when the predominant interaction of the cosolvent with the protein is nonspecific exclusion due to, for example, the increase in water surface tension by the cosolvent. Since in the asymmetric denatured state the exclusion is greater per protein molecule than in the compact native state, the equilibrium is displaced to the left. (c) The denaturation reaction in which the mode of protein–solvent interactions is different in the native and unfolded states. This is shown for the water–MPD (2-methyl-2,4-pentanediol) system; MPD exclusion due to the high charge density on the native protein is replaced by MPD exclusion from individual charged sites and MPD binding to newly solvent-exposed nonpolar regions in the denatured state.

is negative and that the unfolding of ribonuclease A in 6 *M* GuaHCl is due not to the stabilization of the unfolded form, but to the destabilization of the native form.

The first two cases are illustrated by the chemical potential diagram shown in Fig. 5 [32]. For the protein denaturation reaction in water (shown in the middle), $\Delta G_w^\circ = -RT \ln K_w$ is the difference between the chemical potentials of the denatured ($\mu_{2,w}^{\circ,D}$) and native ($\mu_{2,w}^{\circ,N}$) forms of the protein in water at constant temperature and pressure. Addition of any precipitant cosolvent to the protein solution in the native state raises the chemical potential of the protein by the free energy of transfer of the native protein from water to the particular solvent system. After denaturation, the chemical potentials of the proteins shift to the upper levels on the diagram. In the case of a stabilizer, exemplified by $MgSO_4$ [32] (Fig. 5), the exclusion is greater in the unfolded state than in the native state, with the consequence that $\Delta\mu_{2,MgSO_4}^{\circ,D} > \Delta\mu_{2,MgSO_4}^{\circ,N}$ and the free energy of denaturation, $\Delta G_{MgSO_4}^\circ$, becomes more positive than in water. For a structure destabilizer, the transfer free energy in the denatured state must be more favorable than in the native state. This has been found to be true [21] for chymotrypsinogen in 20% PEG-1000, which is strongly excluded from the native protein and acts as a precipitant. Yet, it is a denaturant [21,35]. The standard free energies of denaturation

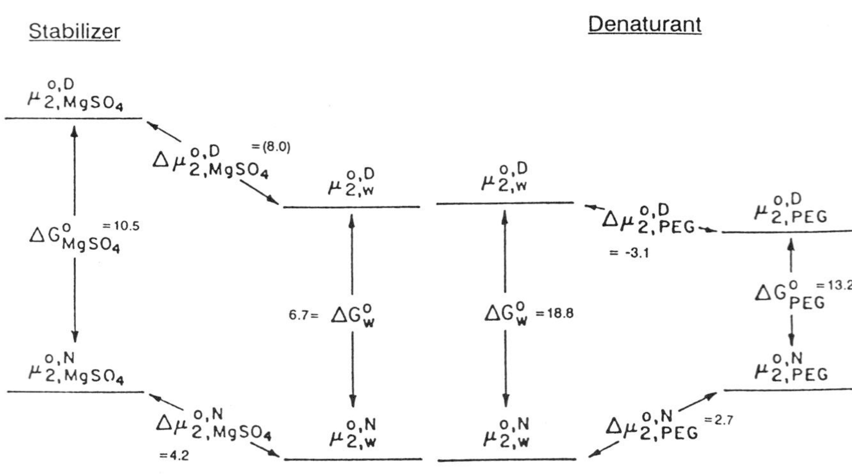

FIGURE 5 Free energy diagram of the effect of cosolvents on protein stability. The stabilization system is ribonuclease A at pH 2.8 in 1.2 *M* $MgSO_4$. The numbers are free energy values in kcal/mol measured at 20°C [32] ($\Delta\mu_2^{\circ,D}$ has not been measured; hence, the expected value is given in parentheses). The denaturation system is chymotrypsinogen in 20% PEG-1000. The numbers are free energy values in kcal/mol measured at 20°C; the thermodynamic square closes to within ±0.2 kcal/mol [21].

of chymotrypsinogen in water and in this solvent system and the preferential interactions in the native and denatured states, determined by Lee and Lee [21], are given in the free energy box of Fig. 5, which closes to within 0.2 kcal/mol. In this system, therefore, the observed [21] lowering of the transition temperature by 7.5°C is the consequence of the decrease in transfer free energy by 5.8 kcal/mol (from an unfavorable to a favorable interaction) when the protein is transformed from the native folded state to the unfolded one.

C. Why Precipitants Are Not Necessarily Stabilizers

The above illustrations of the relations between preferential interactions and structural stabilization of proteins bring out clearly the reasons why the preferential interactions of native proteins with solvent additives do not necessarily correlate with protein stability, while such a correlation is true apparently without exception with protein solubility. Preferential hydration always induces salting out, regardless of the mechanism of the preferential interaction [16]. This is a necessary consequence of the identity of protein structure in the dispersed and precipitated states. In the denaturation equilibrium, the situation is exactly the converse. By definition, the protein structure must be different in the two end states of the process [32]. Therefore, as illustrated in Fig. 3b, when the protein–solvent interactions are unfavorable and independent of the chemical nature of the protein surface, the effect will always be that of stabilization. On the other hand, when the interactions are defined to a major extent by the chemical nature of the protein surface, addition of a cosolvent may either stabilize or destabilize the protein, even though it acts as a strong precipitant and, possibly, crystallizer of the native protein; and measurement of the interactions of the protein in either end state gives no direct insight into the effect of the solvent on protein stability.

V. PREFERENTIAL INTERACTIONS AND BINDING AT SITES

From the above analysis, it is clear the parameter that determines the direction into which an equilibrium will be driven by a cosolvent, or any other ligand, at a given solution composition is the change in its preferential interaction with the protein, $\Delta(\partial m_3/\partial m_2)_{T,P,\mu_3}$, during the course of the reaction. Frequently, this is equated with the change in site occupancy by the ligand. Although it does not introduce any significant errors when the binding is strong, this practice may lead to serious errors of interpretation, and it *must* do so in the case of weakly interacting ligands (used at >1 *M*), not only in the magnitude of the effect, but possibly even in its sign, namely, the direction in which the reaction will be driven. Let us examine why this is so. To do this, let us compare classical site occupancy binding theory with preferential interaction theory.

A. Classical Site Binding Theory

In the classical theory [36], binding is defined as the occupancy of a site S on the protein by the ligand, so that if $[PL]$ is the probability of site S on the protein being occupied by ligand and $[P]$ is that of the site being free of ligand, then an equilibrium constant is written as $K = [PL]/[P][L]$, where $[L]$ is the activity of the free ligand. This equilibrium constant is seen to be identical with K_L of Eq. (10b). In this treatment, water is neglected; that is, it is implicitly regarded as a totally neutral, inert species, which, as a matter of fact, does not occupy space. In this approach for multiple binding to a protein that has a number of sites, n, for the binding of ligand, L, each with a binding constant, K_i, the equilibrium reaction is

$$P + nL \rightleftharpoons \sum_{i=1}^{n} PL_i \tag{16}$$

The probability, θ, that a given site is occupied by ligand L, which defines site occupancy (extent of binding per site, ν in Scatchard notation), is given by the standard Langmuir isotherm [37]:

$$\theta = \frac{K_i[L]}{1 + K_i[L]} \tag{17}$$

The free energy of binding is [38,39]

$$\Delta G_b = -RT \ln\left(1 + \sum_{i=1}^{n} K_i[L]^i\right) \tag{18}$$

For a reaction State I $\rightleftharpoons$ State II, for example, protein unfolding, if all the sites are regarded as thermodynamically identical in both end states of the reaction [38], then

$$\delta\,\Delta G_b = -RT\,\Delta n \ln(1 + K_i[L]) \tag{19}$$

where Δn is defined as the number of sites newly exposed (or eliminated) during the course of the reaction. It is frequently equated with the slope of the Wyman linkage relation (Eq. (1)).

B. Inadequacy of the Site Binding Treatment

By definition, classical binding theory can give only interaction stoichiometries. Its inadequacy when applied to weak interactions characteristic of cosolvents becomes strikingly vivid when one deals with cosolvents that are preferentially excluded from the protein, namely, those for which the results of a dialysis equilibrium experiment give a negative value of $\bar{\nu}_3$—a negative binding stoichiometry. This is true of all stabilizing and salting-out (precipitant) cosolvents, namely, sugars, polyols, amino acids, methyl amines, and salting-out salts (such as ammonium sulfate) (see Table 1). Reading these results in terms of classical binding for, e.g., RNase A in 1 *M* sucrose, one would have to say, "In 1 *M* sucrose, one molecule of RNase A binds *minus* 7.6 molecules of sugar." Application of Eq. (17) would lead to the absurdity of a *negative* equilibrium constant. This is clearly the consequence of neglecting the displacement of water from the site, that is, of using only Eq. (10b) to describe the interaction.

This difficulty of equating thermodynamic binding with occupancy of a site is further brought out in an illustration by Schellman [40], who has addressed the question in a series of important theoretical papers [33,40–44]. In this example, a protein molecule is immersed into a mixed solvent and is, by definition, thermodynamically indifferent to its components. Then, the fraction of the time that a site X on the protein will be occupied by water or cosolvent molecules must be proportional to their relative concentrations in the solvent. Since cosolvent concentration is high, its presence at site X will be a frequent occurrence. If binding is equated to site occupancy, then each time the cosolvent (component 3) is situated at site X, it must be regarded as binding to that site. According to the classical treatment, this binding would be described by an equilibrium constant that corresponds to a finite free energy of binding, $\Delta G^\circ = -RT \ln K_L$. Yet, the site occupancy by water and the ligand being in the same proportion as their presence in the solvent, the observation in a dialysis equilibrium experiment must be that of identical concentrations of ligand inside and outside the bag, that is, zero binding, in contradiction to the first conclusion.

C. Preferential Binding as Exchange at Sites: Weak and Strong Binding

The preceding difficulties are eliminated when the classical binding is replaced by the theory of preferential binding, that is, of the complete thermodynamic interaction, which considers the total reaction and takes into account that whenever a ligand molecule L comes into contact with the protein at site X, it must displace water molecules from that site, as described by Eqs. (10a), (10b), and (11) [13,14]. This exposition will combine Schellman's treatment for exchange at a site [33,40,42–44] with arguments advanced by the present author [7,8,13,14,45]. If we take now protein in pure water as the reference state in which all sites are hydrated, Eqs. (10a) and (10b) reduce to an exchange equilibrium

$$P \cdot H_2O_m + L \rightleftharpoons PL + mH_2O \tag{20}$$

The equilibrium constant for the case of the exchange of one water molecule for one ligand molecule at site X is [43,44]

$$K_{ex} = \frac{K_L}{K_w} = \frac{[PL][H_2O]}{[PH_2O][L]} \tag{21}$$

and $-RT \ln K_{ex}$ is seen to be ΔG^b of Eq. (11). The total free energy of the interaction of the unsolvated protein with the mixed solvent has been shown by Schellman to be [43]

$$\Delta G_s = RT \ln(1 + K_w A_w + K_L A_L) \tag{22}$$

where the activity A_i of component i and the equilibrium constants are expressed in mole fraction units. Comparison with Eq. (18) shows that, for the complete reaction that starts with a bare site, a term for the binding of water ($K_w A_w$) is stated explicitly. In the real system normally encountered in the laboratory, in which the reference state is the hydrated protein in pure water, all sites are initially occupied by water, since no bare sites are possible. The change in the free energy of the interaction due to the introduction of the cosolvent into the system, ΔG_b, is obtained, therefore, by subtracting the water-unhydrated protein interaction term, so that [43,44]

$$\Delta G_b = -RT \ln(A_w + K_{ex} A_L) \tag{23}$$

Comparison with Eq. (18) shows that the difficulty with the classical site occupancy treatment is its equating of ΔG^b of Eq. (11) with ΔG^{32}, since it automatically sets $\Delta G^{12} = 0$. The frequent use of Eq. (18) rather than Eq. (23) to calculate the free energy of binding (and hence the transfer free energy) for weakly interacting systems [44a,56a] leads to values of ΔG_b that can be incorrect not only in magnitude, but also in sign.

For the model of the exchange of one water molecule for one ligand molecule at a single independent site, Schellman [44] has shown that the preferential interaction can be related to interactions with water and ligand in terms of the exchange equilibrium constant by

$$\left(\frac{\partial m_3}{\partial m_2}\right)_{T,P,\mu_3}^{(\text{per site})} = \frac{(K_{ex} - 1/m_1)m_3}{1 + K_{ex}m_3} = \frac{(K_{ex} - 1/m_1)}{K_{ex}}\,\theta \tag{24}$$

where K_{ex} is in molal activity units. This can be easily modified for the replacement of several water molecules at the site by properly redefining K_{ex}. The probability of a site being occupied by ligand is given by

$$\theta = \frac{K_{ex}m_3}{1 + K_{ex}m_3} \tag{25}$$

Equation (25) is identical in form with Eq. (17). The usual binding constant, however, has been replaced by the exchange constant [44]. Equation (24) shows explicitly how the value of K_{ex} defines whether the preferential binding at a given site will be positive or negative, whereas site occupancy is always positive. Since $m_1 = 55.56$, when $K_{ex} > 0.018$, preferential interaction is positive; when $K_{ex} < 0.018$, the value is negative. When $K_{ex} = 0.018$ on the molal scale and 1 on the mole fraction scale, the result is thermodynamic indifference of the protein toward the solvent components. The same balance is expressed for the summation over all sites on a protein by Eq. (2). Equation (24) also identifies the lower limit of the binding constant above which interaction with water may be neglected, and identification of K_{ex} with K_L and of θ with $(\partial m_3/\partial m_2)_{T,P,\mu_3}$ (per site), that is, application of the classical binding theory, does not lead to detectable numerical error [44] even though conceptually it remains incorrect. Taking that limit at $K_{ex} > 50/m_1$ leads to the conclusion that at $K_{ex} > 1$, $K_{ex} \simeq K_b$. This analysis, expressed in terms of the general Eq. (2), also explains why biochemists, in equating results of dialysis equilibrium with site occupancy, obtain the correct numerical values of binding equilibrium constants. The binding of biochemical ligands to receptor sites is usually strong ($K_b \geq 10^4\ M^{-1}$), so that the concentrations used are low, $m_3 \leq 10^{-4}$. Since $m_1 = 55.56$, the second term on the right-hand side of Eq. (2) becomes negligible and $(\partial m_3/\partial m_2)_{T,\mu_1,\mu_3} \simeq B_3$. On the other hand, when interactions are weak, as in the case of cosolvent stabilizers or denaturants, the concentrations used are high, 1–8 *M*, and the water binding term in Eq. (2), that is, site occupancy by water molecules, becomes significant [7]. Hence, in these systems it is operationally, as well as conceptually, incorrect to equate the equilibrium dialysis results with site occupancy, or the slope of a Wyman plot (Eq. (1)) with the change in the number of ligand molecules, for example, denaturant molecules, that occupy sites on a protein during unfolding.

D. Preferential Binding as the Balance Between Water and Ligand Binding to a Protein: Meaning of Zero "Binding"

The complexity of the relationship between site occupancy and preferential binding is clearly illustrated by the interaction of four proteins with the 2-chloroethanol–water solvent system [18,46]. The results of dialysis equilibrium experiments as a function of the cosolvent concentration are shown in Fig. 6. For all four proteins, the pattern of interaction is initially an increasing preferential binding up to a cosolvent concentration at which it reaches a maximum. As 2-chloroethanol concentration is raised further, the preferential binding decreases until it crosses zero at a high cosolvent concentration, above which the interaction becomes one of preferential hydration. Classical binding theory would lead to the conclusion that the interaction between the protein and the ligand is complex, with inhibition of binding at high ligand concentration. Let us examine what the true interaction pattern is in terms of preferential binding that takes both water and 2-chloroethanol into account [45–47]. Figure 7, which simulates these data, is a schematic presentation of the decomposition of the dialysis equilibrium results into contributions from site occupancy on the protein by water and cosolvent, in accordance with Eq. (2) [45]. For simplicity, it is assumed that hydration (B_1) remains constant and site occupancy by the cosolvent (B_3) increases linearly with its con-

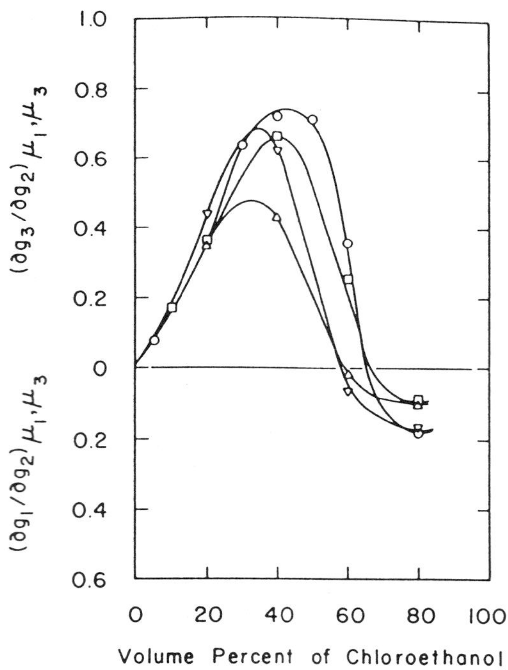

FIGURE 6 Dependence on solvent composition of the preferential interaction of solvent components with various proteins in the water–2-chloroethanol system. Key: (0) β-lactoglobulin A, (△) lysozyme, (∇) bovine serum albumin, and (□) insulin [46]. The preferential interactions are expressed in grams 2-chloroethanol (g_3) or water (g_1) per gram protein (g_2). This is converted to the form of Eq. (2) by $g_i = m_iM_i/1000$.

centration in the solution. This is expressed by the two straight lines in part B of Fig. 7. The upper part of the figure compares the composition of the solvent in contact with the protein with that of the bulk solvent (dialyzate). It is evident that the relative proportions pass from excess cosolvent on the protein to excess water, after going through a neutral point at which the concentrations on the protein and in the bulk are identical. Combination of B_1 and B_3 according to Eq. (2) gives the bell-shaped dependence of $(\partial m_3/\partial m_2)_{T,P,\mu_3}$, that is, preferential interaction, on solvent composition, drawn in Fig. 7B. This curve is seen to mimic the experimental results obtained with the four proteins. It is important to note that at the solvent composition at which the water and cosolvent distributions on the protein and in the bulk solvent are identical, preferential (thermodynamic) binding, as measured by dialysis equilibrium, is zero, even though site occupancy by the cosolvent is high. In other words, a measurement of zero "binding" by dialysis equilibrium at a single concentration of ligand does not mean binding is not occurring. It only means that, at the given solvent composition and in the corresponding state of site occupancy by water and ligand, the protein is indifferent to infinitesimal changes in contact with the two solvent components; that is, $(\partial\mu_2/\partial m_3)_{T,P,m_2}$ is zero. The total interaction between protein and solvent with reference to pure water may, in fact, be characterized by a large free energy. In the present example, the value of this free energy, $\Delta\mu_{2,tr}$, at the point of zero preferential binding is given by the integral under the $(\partial m_3/\partial m_2)_{T,P,\mu_3}$ curve, after transformation to $(\partial\mu_2/\partial m_3)_{T,P,m_2}$, between the limits of pure water and the crossover point (0.50 fractional composition of cosolvent), by using Eqs. (7) and (14). As is clear from Fig. 7 (see also Table 1), this is a large negative value; that is, there is a large favorable interaction of the four proteins with 2-chloroethanol at the point of zero preferential binding.

E. Meaning of Thermodynamic Indifference

This analysis shows how a zero binding measured by dialysis equilibrium does not necessarily mean total thermodynamic indifference of the protein to the solvent components. For this to be true, the solvent compositions on the protein and in the bulk must be identical at all concentrations of the cosolvent and $B_3 = (m_3/m_1)B_1$ would be true at all m_3. Then $(\partial m_3/\partial m_2)_{T,P,\mu_3}$, $(\partial\mu_2/\partial m_3)_{T,P,m_2}$, and $\Delta\mu_{2,tr}$ would all have zero values, though sites on the protein would be occupied by ligand strictly for statistical reasons. The most common origin of the zero point by far is a balance between the interactions of the protein with water and the cosolvent at a definite cosolvent concentration, as defined by Eqs. (2) and (11) [18,27,46]. At this point, as illustrated by Figs. 6 and 7, $(\partial m_3/\partial m_2)_{T,P,\mu_3}$ and $(\partial\mu_2/\partial m_3)_{T,P,m_2}$ will be zero, but $\Delta\mu_{2,tr}$ will assume a finite negative value for favorably interacting cosolvents and a positive value for unfavorably interacting ones.

F. Relation Between Global Preferential Interactions and Exchange at Sites

In the above discussion, preferential interaction was related to site occupancy by two relations, Eqs. (2) and (24). The first relates the total preferential interactions of a protein molecule to the total occupancy of sites on the protein by water or the cosolvent summed over all attractive and repulsive interactions [7,13,14]. The second defines specifically the exchange of a water molecule for a cosolvent molecule at any given site on the protein [43,44]. Let us relate the general equation (Eq. (2)) with the specific one for each site (Eq. (24)). The surface of a protein molecule contains a spectrum of sites that can interact with water and with ligand with different affinities. Each of these sites can be described by its particular value of K_{ex}. There can be no vacant sites. All the sites must be occupied either by water or by cosolvent, be it simply by van der Waals contacts. For the sake of simplicity, let us classify the sites into three general sets. The first set of sites will be totally

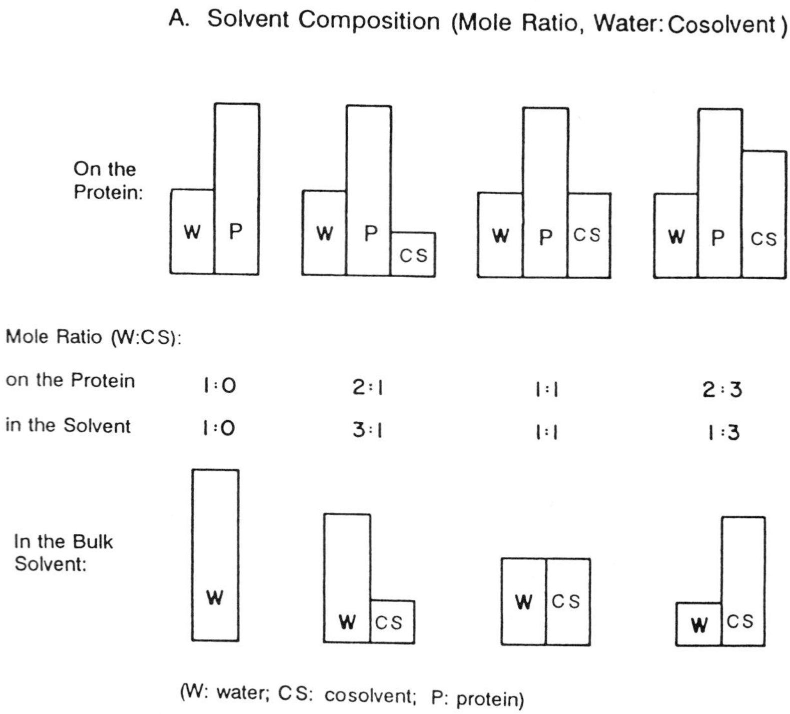

FIGURE 7 Comparison of the preferential interactions of and site occupancy by solvent components in a mixed solvent system, water (*w*) (component 1) and cosolvent (CS) (component 3), as a function of bulk solvent composition. The calculations mimic the experimental results of Fig. 6. (A) Composition of solvent occupying sites on the protein (P) in equilibrium with bulk solvent of various compositions. These compositions are given in mole ratios between the two sets of diagrams. In this calculation, it has been assumed that hydration is independent of solvent composition, while site occupancy by cosolvent is directly proportional to its contents in the medium. (B) Resulting dependence of interaction parameters on solvent composition, B_1, and B_3 are site occupancies by water and cosolvent, in

indifferent to contact with water or cosolvent, with $K_{ex} = 0.018\ m^{-1}$. The second set consists of sites that have essentially no affinity for the ligand relative to water, that is, ligand is excluded from them, and $K_{ex} \ll 0.018\ m^{-1}$. The third set are sites with various affinities for the ligand (cosolvent) molecules, $K_{ex} > 0.018\ m^{-1}$. Each site will contribute to the thermodynamic interaction, $(\partial m_3/\partial m_2)_{T,P,\mu_3}$, so that for a total of n sites on the surface of the protein molecule,

$$\bar{\nu}_3 \equiv \left(\frac{\partial m_3}{\partial m_2}\right)^{\text{total}}_{T,P,\mu_3} = \sum_{i=1}^{n} \left(\frac{\partial m_3}{\partial m_2}\right)^{\text{per site}}_{T,P,\mu_3} \tag{26}$$

B. Interaction Parameters

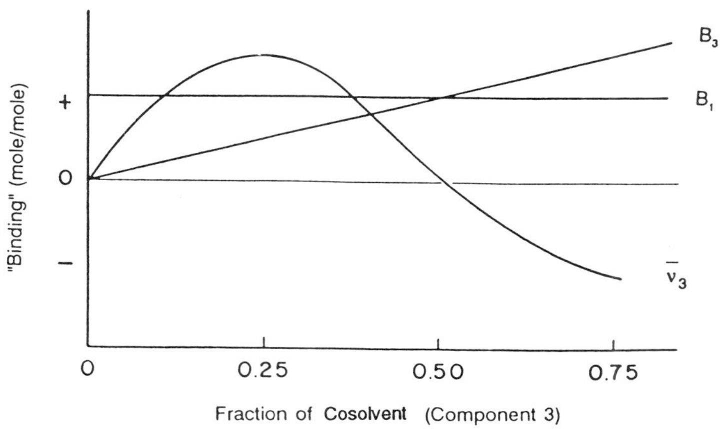

FIGURE 7 Continued
accordance with the above assumption. As a consequence of the relation between preferential binding, $\bar{\nu}_3 \equiv (\partial m_3/\partial m_2)_{T,P,\mu_3}$, and the site occupancies (B_1 and B_3) expressed by Eq. (2), the positive site occupancies result in a bell-shaped concentration dependence of the preferential interaction. We note that at high cosolvent concentration the preferential binding of cosolvent to protein is negative, even though its site occupancy on the protein increases monotonely. We also note that at the solvent composition at which the mole ratios of solvent components that occupy sites on the protein or are present in the bulk solvent are identical, the preferential interaction is zero. This is the point of thermodynamic indifference of the protein to solvent components even though site occupancy is high.

and Eq. (2) will be related to Eq. (24) by

$$\left(\frac{\partial m_3}{\partial m_2}\right)_{T,P,\mu_3} = B_3 - \frac{m_3}{m_1} B_1 = \sum_{i=1}^{n} \frac{(K_{ex}{}^{(i)} - 1/m_1)m_3}{1 + K_{ex}{}^{(i)}m_3} \tag{27}$$

If we analyze each class of sites, the indifferent sites make no contribution to the thermodynamic interaction. Therefore, they can be neglected. The sites from which ligand is excluded [48,49] make no contribution to B_3. These nonexchangeable sites are occupied by $B_1{}^{\mathrm{Nex}}$ water molecules. Therefore, the contribution of these sites to $(\partial m_3/\partial m_2)_{T,P,\mu_3}$ is given by $-(m_3/m_1)B_1{}^{\mathrm{Nex}}$ [49a]. This leads to the full relationship between site occupancy and preferential interaction [49a]:

$$\left(\frac{\partial m_3}{\partial m_2}\right)_{T,P,\mu_3} = \sum_{1}^{l} B_3{}^{\mathrm{ex}} - \frac{m_3}{m_1}\left(\sum_{1}^{l} B_1{}^{\mathrm{ex}} + B_1{}^{\mathrm{Nex}}\right) \tag{28}$$

where the summations are over all l sites with different exchange constants between water and cosolvent molecules, that is, different effective affinities of cosolvent for hydrated protein. If the assumption is made that, for a weakly interacting ligand, such as urea, glycerol, $MgCl_2$, or other cosolvents, the values of K_{ex} are not too broadly distributed about a mean value for those sites at which exchange can occur, and that, therefore, they can be regarded as a group of identical sites, then Eq. (28) can be simplified to

$$\left(\frac{\partial m_3}{\partial m_2}\right)_{T,P,\mu_3} = B_3 - \frac{m_3}{m_1}(B_1^{ex} + B_1^{Nex}) \tag{29}$$

In similar manner, the transfer free energy $\Delta\mu_{2,tr}$ consists of contributions from interactions at all the different classes of sites, so that, for the simple case of two classes of sites, one identical exchangeable, the other nonexchangeable, we have [49a]

$$\Delta\mu_{2,tr} = \Delta\mu_2^{ex} + \Delta\mu_2^{Nex} = \int_0^{m_3}\left(\frac{\partial\mu_2}{\partial m_3}\right)_{T,P,m_3}^{ex} dm_3 + \int_0^{m_3}\left(\frac{\partial\mu_2}{\partial m_3}\right)_{T,P,m_2}^{Nex} dm_3 \tag{30}$$

G. Direct Site Occupancy Measurements Cannot Define the Thermodynamic Interaction

This discussion permits us to approach now the relation between thermodynamic interaction and experimentally measured binding at sites. Protein–ligand contacts can be measured by a number of nonthermodynamic techniques that report the site occupancy by a specific ligand. This can be the perturbation of a spectral property of the protein or of the ligand, for example, UV absorption of fluorescence. It can also be the evolution or absorption of heat measured by microcalorimetry. Careful experiments, followed by the use of standard analyses such as a Scatchard plot, can generate a value of n, the number of sites available to ligand occupancy, as well as the extent of occupancy θ, the number of ligand molecules bound to sites, $B_3 = n\theta$, and a binding constant, which will be an average value of K_{ex}. It should be noted that K_L of Eq. (10b) is never measurable for weakly interacting systems such as cosolvents and denaturants. Application of Eq. (24) can then give a value of the preferential interactions over the sites available to cosolvent (e.g., urea) binding. This, however, as shown above, *does not* define the full thermodynamic interaction. This would be true only if the sum total of sites were available to occupancy by both water and ligand, that is, for $n = B_1 + B_3$ ($B_1^{Nex} = 0$). The prevailing situation is one in which B_1^{Nex} is finite. Now, since $n \equiv B_1^{ex} + B_3 = B_3/\theta$, $n \leq B_1 + B_3$, and the total number of sites determined by such a site-probing method cannot give the total number of sites involved in interactions with solvent. The complete thermodynamic interaction, $(\partial m_3/\partial m_2)_{T,P,\mu_3}$, is available only from solution equilibrium thermodynamic measurements, such as dialysis equilibrium, isopiestic vapor pressure equilibrium [50], or related techniques, for example, light scattering [18,51], small-angle x-ray scattering [13,52], and sedimentation equilibrium [12,53]. Numbers of water and ligand molecules interacting, that is, B_1 and B_3, can be measured by non-equilibrium techniques, for example, B_1, by NMR [54,55], B_3 by spectral perturbation or microcalorimetry. Since these techniques respond only to contact between protein and ligand, their application must always lead to a positive number of ligand molecules interacting with protein, whether the thermodynamic (preferential) interaction is positive, negative, or indifferent. Therefore, information obtained by binding (site contact) techniques, such as fluorescence perturbation or microcalorimetry, is *by definition* incomplete for a thermodynamic characterization of the interaction. Such directly measured values of B_3, when used in conjunction with dialysis equilibrium results, could be used, however, to deduce values of B_1^{Nex} by Eq. (29), after calculation of $B_1^{ex} = n - B_3$.

Identical considerations apply to the characterization of protein reactions, such as denaturation, by the techniques that report only on cosolvent–protein contacts. If site binding is measured for both the native and denatured protein by, say, microcalorimetry, the information gained is $\Delta B_3 = B_3^{Denat} - B_3^{Nat}$. Yet, whether this number is positive or negative, it tells us nothing about the direction in which the equilibrium will be displaced in the absence of a knowledge of ΔB_1 and especially ΔB_1^{Nex}, since both the Wyman relation (Eq. (1)) and the calculation of a change in transfer free energy (Eq. (6a)) are based on a knowledge of the change in the complete thermodynamic interaction [49a],

$$\Delta\left(\frac{\partial m_3}{\partial m_2}\right)_{T,P,\mu_3} = \Delta B_3 - \frac{m_3}{m_1}(\Delta B_1^{ex} + \Delta B_1^{Nex}) \tag{31}$$

H. Weak Effects as Results of Strong Interactions at Sites

The above analysis permits resolution of the dilemma that measurements of the direct binding of denaturants, such as guanidine hydrochloride or urea, by site contact reporting techniques frequently indicate strong interactions, with ΔG values as negative as -1.0 kcal/mol, while their effects on protein stability are exercised at high concentration, 4–8 *M*, which suggests weak interactions. For example, Pfeil et al. [56], using a microcalorimetric titration technique, have reported a binding constant (which, in fact, is the exchange constant K_{ex} of Eq. (21)) of GuaHCl to unfolded apocytochrome c of 1.16 M^{-1}. In 6 *M* GuaHCl, this leads to $\Delta G_b = -1.1$ kcal/mol. Similar values have been reported for the binding of GuaHCl to other denatured, as well as native, proteins [56a]. The resolution of the apparent dilemma that reasonably strong binding leads to much weaker effects than expected will be given by the examination of two specific systems, one in which the cosolvent is preferentially excluded, the other in which it is preferentially bound. The first is the interaction of β-lactoglobulin with 30% glycerol [15];

the other is the interaction of lysozyme with 8 *M* urea [17]. To establish numerically the relation between preferential binding and site occupancy, defined by Eqs. (2) and (29), it is necessary to have some mechanistic knowledge of the interaction, such as the total number of sites available to binding by either or both ligands, as well as the extent of binding (site occupancy) by one of the ligands. Since such knowledge is essentially nonexistent, a simple model will be utilized. In this model, it is assumed that the surface of the protein contains l independent sites that can bind either water or the ligand, $l = B_1^{ex} + B_1^{Nex} + B_3$. It will be assumed further that, in the absence of data on B_3, site occupancy by the ligand, $B_1 = B_1^{ex} + B_1^{Nex}$, the number of water molecules that occupy sites, is given by the value of the hydration measured by some experimental technique such as isopiestic vapor pressure equilibrium [57] or NMR [54,55].

For β-lactoglobulin in 30% (5.81 m) glycerol at pH 2.0, the measured value of $(\partial m_3/\partial m_2)_{T,\mu_1,\mu_3}$ was -14.7 molecules of glycerol per molecule of protein. This corresponds to $\Delta\mu_{2,tr} = 8.7$ kcal/mol [15]. The value of B_1 calculated by the nmr method of Kuntz [54,55] was 303 molecules H_2O per molecule of protein. When combined with $(\partial m_3/\partial m_2)_{T,\mu_1,\mu_3} = -14.7$, this, by Eq. (2), gives $B_3 = 17$ molecules of glycerol that occupy sites on a protein molecule. Application of Eq. (24) within the model of n identical independent sites that can bind either water or glycerol, $n = B_1 + B_3$, results in $K_{ex} = 0.0097$ m^{-1} (0.54 in mole fraction units) and $(\partial m_3/\partial m_2)_{T,\mu_1,\mu_3}$ per site of -0.046, with a total number of sites of 320. The free energy of interaction per site calculated by Eq. (23) was +24.3 cal/mol, which for the n sites resulted in 7.8 kcal/mol, or essentially the value of $\Delta\mu_{2,tr}$ deduced from experimental data by Gekko and Timasheff [15].

For lysozyme in 8 *M* urea, the value of $(\partial m_3/\partial m_2)_{T,\mu_1,\mu_3}$, measured by dialysis equilibrium, was found to be 12.0 molecules of urea per molecule of protein [17]. This, by Eqs. (7) and (14), results in $\Delta\mu_{2,tr}$ of -7.0 kcal/mol, if $(\partial\mu_2/\partial m_3)_{T,P,m_2}$ is assumed to be constant. If this is equated with the free energy of binding of 12 ligand molecules, as would follow from the naive application of the classical binding treatment, the result would be a binding free energy of -580 cal/mol per urea bound. This is a surprisingly high value for the binding of a ligand that requires a concentration much greater than 1 *M* ligand. It indicates that the classical treatment is inadequate and that the more complete analysis in terms of preferential interactions with the application of Eqs. (2), (27), and (28) is required.

The recent publication of measurements of the interaction of unfolded proteins with urea and GuaHCl by calorimetric titration [56a] has made available values of the number of protein–denaturant contacts as a function of denaturant concentration. These are, in fact, values of B_3 of Eqs. (2) and (28). Scatchard-type analysis of the isotherms [56a] has led to the total number of exchangeable sites, $n = B_3 + B_1^{ex}$, and the exchange equilibrium constants, K_{ex}. Combination of these experimental values for lysozyme in 8 *M* urea, $B_3 = 72$ molecules of urea making contact with one protein molecule, and $n = 229$ sites available to urea–water exchange, with the measured value of $(\partial m_3/\partial m_2)_{T,\mu_1,\mu_3} = 12.0$ [17] gives, by Eq. (2), $B_1^{\text{total}} = 266$ and, by Eq. (25), $K_{\text{ex}} = 0.036\ \text{m}^{-1}$ (2.0 in mole fraction units), since $\theta = B_3/n = 0.31$. Equation (24) then results in $(\partial m_3/\partial m_2)_{T,\mu_1,\mu_3}$ per exchangeable site of 0.155. Over the 229 exchangeable sites, this makes a contribution of 35.4 moles urea per mole protein. This is in great excess over the experimental value of 12.0 [17]. What is the source of the difference? It is the water present on the sites at which it does not exchange with urea, that is, B_1^{Nex} of Eq. (28), which is equal to $B_1^{\text{total}} - (n - B_3)$, or 109 water molecules per molecule of protein. Their contribution to the measured value of $(\partial m_3/\partial m_2)_{T,P,\mu_3}$ is $-(m_3/m_1)B_1^{\text{Nex}} = -24.6$ molecules of urea per molecule of protein. Summing up the two contributions by Eq. (28) gives $(\partial m_3/\partial m_2)_{T,P,\mu_3} = 11.0$, which is essentially the "binding" value measured by dialysis equilibrium, that is, the true value of the thermodynamic interaction.

How is the balance between protein–ligand or protein–water contacts at exchangeable and nonexchangeable sites reflected in the thermodynamic parameters of the protein–urea interaction? Taking the exchangeable sites, the free energy of interaction per site, calculated with Eq. (23), is −58.4 cal/mol. Over the 229 exchangeable sites, this sums up to −13.4 kcal/mol, or more negative by −6.7 kcal/mol than the experimentally measured value of the transfer free energy. This favorable contribution to the free energy of interaction is equal to −180 cal/mol per urea molecule actually bound. The value of $-RT \ln K_{\text{ex}} = -404$ cal/mol (mole fraction standard state) is the quantity that would be reported as the free energy of binding from measurements by nonthermodynamic techniques, such as microcalorimetry, that report strictly the number of sites on a protein occupied by a ligand. What is the cause of the difference between the measured value of $\Delta\mu_{2,\text{tr}} = -7.0$ kcal/mol and that predicted from interaction at the exchangeable sites, that is, −13.4 kcal/mol? It is the unfavorable free energy of interaction between the protein and the urea–water solvent system at those sites at which water cannot exchange with urea.

In fact, the low value of "binding" measured by dialysis equilibrium, which stems from large values of site occupancy by water and cosolvent (urea), is a reflection of compensation of the corresponding interaction free energies. The relatively weak value of the experimentally measured $\Delta\mu_{2,\text{tr}} = -7.2$ kcal/mol is the resultant of quite strong favorable and unfavorable interactions between the protein and the cosolvent (urea). The binding of the 72 urea molecules at the exchangeable sites makes a contribution of −13.4 kcal/mol, while its exclusion from the 109 nonexchangeable sites contributes 6.7 kcal/mol of unfavorable interaction free energy. At these sites, contact with water is favored over that with urea by 61 cal per site. The thermodynamic effect of protein–cosolvent interactions on a reaction is determined by the values of $\partial m_3/\partial m_2$ and $\Delta\mu_{2,\text{tr}}$, *summed over all interacting sites*. These parameters are the resultants of compensation between a variety of favorable and unfavorable interactions that may be relatively strong. The above calculations, based on available experimental data, illustrate how a reasonably strong binding ($\Delta G^\circ = -RT \ln K_{ex} = -400$ cal/mol per site) may lead only to a weak thermodynamic effect (ΔG net per bound urea $= \Delta\mu_{2,tr}/B_3 = -97$ cal/mol). When the binding is coupled with exclusion at other sites, the thermodynamic effect per site becomes very weak indeed ($\Delta\mu_{2,tr}/(B_3 + B_1^{ex} + B_1^{Nex}) = -21$ cal/mol), hence the need of high concentrations of ligand (e.g., 8 *M* urea, 6 *M* GuaHCl). These apparent dilemmas present themselves only when the interactions are weak, that is, with denaturing and stabilizing cosolvents that are required at >1 *M* concentration. When interactions are strong, as is true of the binding of various biochemical factors, $(m_3/m_1)B_1^{Nex}$ becomes negligibly small and the thermodynamic consequences of the interaction take place at the expected concentrations.

These considerations bring out the fundamental thermodynamic distinction between global preferential binding over the entire protein molecule, $(\partial m_3/\partial m_2)_{T,P,\mu_3}$, site occupancy, B_3, and water–ligand exchange measured at sites. The first is a true equilibrium thermodynamic parameter that defines the total effect of adding the ligand (cosolvent) to the aqueous protein solution. As such, it can be used to define thermodynamically the course of a reaction by the Wyman equation (Eq. (1)) or through the calculation of $\delta\Delta\mu_{2,tr}$. Site occupancy, on the other hand, is an incomplete description of the interactions. It deals only with particular protein–ligand contacts. Even if expressed via K_{ex}, in terms of ligand exchange with water on protein sites, it cannot describe the complete thermodynamic interaction without a knowledge of the number of nonexchangeable water molecules bound to the protein with the application of Eqs. (2) and (28). Furthermore, site occupancy cannot be used to define the course of a reaction, because K_{ex} leads to θ and B_3 but gives no information about the total $(\partial m_3/\partial m_2)_{T,P,\mu_3}$, and ΔG_b gives no information about $\Delta\mu_{2,tr}$. For example, for the two systems discussed, fluorescence or calorimetric titration would give binding values of 72 and 17 molecules of ligand bound per protein molecule for the urea and glycerol systems, respectively, with negative values of the standard free energies of binding, while the thermodynamic binding stoichiometries, $(\partial m_3/\partial m_2)_{T,P,\mu_3}$, are 12.0 [15] and -14.7 [17], respectively, and the transfer free energies are negative for urea and positive for glycerol.

VI. MEANING OF SITES IN WEAK BINDING

When interactions are weak, as in the case of cosolvents, it seems precarious to equate the number of binding sites deduced, B_1 or B_3, with actual loci on which molecules of water or the cosolvent become complexed. These numbers are better looked at as effective quantities. They are a consequence of expressing as numbers of whole molecules that make contact with the protein a wide spectrum of interactions that may occur between the protein and water or the cosolvent, be it actual occupancy of a site, a transient contact that is energetically or entropically nonneutral, a momentary perturbation of the translational or rotational motions of a cosolvent molecule by contact with protein, or even an interaction with a water or cosolvent molecule that is itself in contact with the protein. These interactions are best expressed in terms of purely thermodynamic parameters, such as the mutual perturbation of chemical potentials by solution components: $(\partial\mu_3/\partial m_2)_{T,P,m_3} = (\partial\mu_2/\partial m_3)_{T,P,m_3}$. This perturbation is manifested through a perturbation of solvent composition vicinal to the protein. The numerical value of B_1 certainly does not mean a definite number of water molecules immobilized on the protein as in the iceberg model. The same is true of B_3. For some cosolvents, a correlation may be deduced with chemically reasonable loci for interaction (e.g., peptide groups for urea, charge clusters for salts). This, however, is interpretation of binding data. It may be useful in cautious predictive processes. Direct results of dialysis equilibrium when the ligand concentration is high (>1 *M*) may never be used to predict binding stoichiometries.

The same arguments hold true for cosolvent-dependent equilibria, such as urea-induced denaturation. Following Eqs. (1) and (31), all the parameters must be taken as differentials between the two end states, that is, $\Delta(\partial m_3/\partial m_2)_{T,\mu_1,\mu_3}$, ΔB_1, ΔB_3. Since during such a reaction any and all of these parameters may change, the slope of a Wyman plot can *never* give Δn, that is, the number of newly exposed sites, for, say, the binding of urea. The slope is an effective value of Δn, which is the resultant of the changes in all the discussed modes of water and cosolvent interaction with the protein. The protein reaction may abolish sites or generate new ones at which there may be competition between water and cosolvent, or it may induce changes in that competition at already existing sites or in nonexchangeable sites due to a change in the environment, or it may lead to changes in interactions between sites (cooperativity) and between water and cosolvent molecules in proximity to the protein. Therefore, in a reaction in which the protein structure changes, a detailed analysis is needed of the changes in B_1^{ex}, B_1^{Nex}, and B_3 during the course of the reaction in order to arrive at a model of the interactions at a molecular level.

VII. WHY ARE SOME COSOLVENTS PREFERENTIALLY EXCLUDED FROM PROTEIN?

Attractions between protein and cosolvent molecules are easy to understand. The concept of exclusion is psychologically more difficult to accept. Thermodynamically, it is a direct consequence of the positive value of $(\partial\mu_2/\partial m_3)_{T,P,m_2}$. What are the causes of this unfavorable interaction? Examination of a large number of preferentially excluded molecules has led to their classification into two general categories [32]: (a) those whose exclusion is due to forces totally independent of the chemical nature of the protein surface; that is, the protein and cosolvent are chemically inert toward each other; (b) those which recognize particular chemical groups on, or chemical characteristics of, the protein surface.

The first category is composed essentially of two mechanisms, steric exclusion and increase in water surface free energy by the cosolvent. Steric exclusion as a source of the excess of water on a protein surface in mixed solvents was first proposed by Kauzmann in 1949, as quoted by Schachman and Lauffer [58]. The effect is pictured schematically in Fig. 8A, which shows a protein molecule in contact with a cosolvent molecule that is much larger than water. Since the protein and cosolvent are chemically inert toward each other, they can approach until they

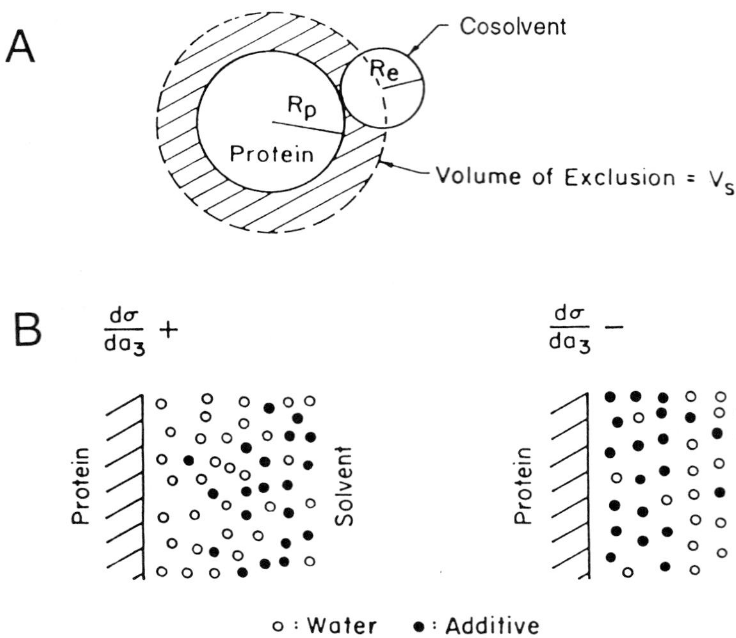

FIGURE 8 Schematic representation of mechanisms of cosolvent exclusion that are independent of the chemical nature of the protein surface. (A) Steric exclusion: R_e, the radius of the cosolvent, being much greater than that of water, a shell with volume V_s is created around the protein. This shell being enriched with water, the observation is preferential exclusion of the cosolvent. (B) Solvent component distribution at an interface due to the perturbation of the surface tension of water (σ) by the cosolvent. When the cosolvent raises the surface tension of water, it is depleted from the surface layer, which results effectively in preferential hydration; when it lowers σ, it is accumulated in the surface layer (a_3 is the activity of the cosolvent).

make contact at a center-to-center distance equal to the sum of their radii, $R_p + R_e$, if the molecules are taken as spheres [35]. Since the cosolvent molecule is larger than a water molecule, a shell is created about the protein molecule that is penetrable to water but not to cosolvent. The net effect of this cosolvent exclusion is preferential hydration. Within this very simple model, the volume of the shell, V_s, is equal to the volume of the water of preferential hydration, that is, $V_3 = (M_2/\rho_w)(\partial g_1/\partial g_2)_{T,P,\mu_3}$, where preferential hydration is expressed in gram per gram units and ρ_w is the density of water. Then, by Eqs. (7) and (9),

$$\left(\frac{\partial \mu_2}{\partial m_3}\right)_{T,P,m_3}^{\text{Ster. excl.}} = \frac{m_3}{1000} V_s \rho_s \left(\frac{\partial \mu_3}{\partial m_3}\right)_{T,P,m_2} \tag{32}$$

Since, by definition, the shell contains an excess of water, $(\partial g_1/\partial g_2)_{T,\mu_1,\mu_3}$ and V_s are positive and the preferential interaction parameter is positive; that is, the observation in a dialysis equilibrium experiment is preferential exclusion of the cosolvent. This mechanism has been shown to account for the preferential exclusion of PEG from native globular proteins [22,35,59,59a].

Exclusion by perturbation of the surface free energy of water follows from the Gibbs adsorption isotherm (1878) [60], which defines the excess concentration of a solute at an interface. If we consider a solvent–protein interface, with s being the molar surface area of the protein and σ the surface tension [20,61], then

$$\left(\frac{\partial m_3}{\partial m_2}\right)_{T,P,\mu_3}^{\text{Surf. tens.}} = -\frac{s a_3}{RT} \left(\frac{\partial \sigma}{\partial a_3}\right)_{TP} \tag{33}$$

The consequences of this relation for protein–cosolvent interactions are expressed schematically in Fig. 8B. It is evident cosolvents that increase the surface tension of water will be depleted in the interface; that is, the experimental observation will be preferential exclusion of the cosolvent. Sugars [61], most amino acids [63], and most salts [64], in particularly the salting-out ones, raise the surface tension of water. Therefore, by Eq. (33), they are all preferentially excluded from the surface of a protein, whether globular native or unfolded denatured. It is clear, therefore, that this is the predominant source of preferential exclusion.

The second category, in which the cosolvent exclusion is related to the chemical recognition of particular sites on a protein molecule, consists of all the situations in which the affinity of a protein site is greater for water than for the cosolvent or the cosolvent is repelled, as would be true for, say, like charges. The most prominent such cause of exclusion is the solvophobic effect. Solvophobic compounds, which include glycerol [15] and other polyols [65], such as sorbitol and mannitol, affect the interactions between solvent molecules in such a way that the activity of water is raised even more than normally upon contact with nonpolar regions on a protein surface. Such cosolvent molecules migrate away from the protein surface in order to reduce the unfavorable free energy effect. The net effect is observed preferential hydration. A cosolvent frequently used to crystallize proteins is MPD (2-methyl-2,4-pentanediol) [66]. It appears to be repelled from charges [23]. Since a globular protein molecule contains a mosaic of charges on its surface, MPD is preferentially excluded from it.

VIII. CONCLUSION: COMPETITION, COMPENSATION, BINDING–EXCLUSION BALANCE

The above discussion on the nature of ligand binding to proteins shows that when dealing with weak interactions binding becomes a complicated concept. Any particular way of looking at binding must be analyzed critically in terms of what exactly is being measured. As pointed out in the discussion of protein–urea interactions, the experimentally observed quantity is the result of a fine balance between binding and exclusion, site competition, and compensation between different modes of interaction (binding or exclusion) of a ligand with different sites on a protein.

In an aqueous medium, the binding of a cosolvent molecule to a protein involves the displacement of water molecules; that is, there is competition between the cosolvent and water for a particular site and any binding constant measured is an exchange constant. It is possible to detect the resulting site occupancy on the protein by any technique that reports on the formation of contacts, be it the evolution of heat measured by microcalorimetry, or the perturbation of a spectral signal, such as fluorescence emission. The total thermodynamic interaction between a protein and a cosolvent system is given, however, by the sum of all such interactions. Therefore, if the protein surface contains sites that have an affinity for water but are indifferent to the ligand, these sites will be hydrated and their net contribution the total thermodynamic interaction will be one of an effective repulsion of the ligand; that is, the free energy of the interaction of the ligand with such sites will be positive. This positive contribution will compensate the negative free energy due to ligand occupancy of other sites. The net effect may be positive, negative, or zero, depending on the magnitudes of the various compensating interactions. In the examples discussed here, it was shown that the net urea interaction with the protein is positive, since site occupancy overcomes exclusion. This is not surprising, as urea is a denaturant. In the case of glycerol, a stabilizer, the net effect is one of preferential exclusion. Yet, a number of glycerol molecules do occupy sites on the protein. This binding–exclusion balance is what constitutes the true thermodynamic interaction that is measured in complete equilibrium thermodynamic experiments such as dialysis equilibrium or light scattering. In reactions, the effect of a cosolvent is measured by the change in its thermodynamic interactions with the protein. This effect therefore reflects the fine balance between all the competitions, compensations, and binding–exclusion balances in the two end states of the reaction.

REFERENCES

1. Scatchard, G., Physical chemistry of protein solutions. I. Derivation of the equations for the osmotic pressure, *J. Am. Chem. Soc. 68*:2315–2319 (1946).
2. Wyman, J., Jr., Linked functions and reciprocal effects in hemoglobin: A second look, *Adv. Protein Chem. 19*:223–286 (1964).
3. Wyman, J., and S. J. Gill, "Binding and Linkage. Functional Chemistry of Biological Macromolecules," University Science Books: Mill Valley, California (1990).
4. Wyman, J., Jr., Heme proteins, *Adv. Protein Chem. 4*:407–531 (1948).
5. Stigter, D., Interactions in aqueous solutions. IV. Light scattering of colloidal electrolytes, *J. Phys. Chem. 64*:842–846 (1960).
6. Scatchard, G., The attractions of proteins for small molecules and ions, *Ann. N.Y. Acad. Sci. 51*:660–672 (1949).
7. Inoue, H., and S. N. Timasheff, Preferential and absolute interactions of solvent components with proteins in mixed solvent systems, *Biopolymers 11*:737–743 (1972).
8. Na, G. C., and S. N. Timasheff, Interaction of calf brain tubulin with glycerol, *J. Mol. Biol. 151*:165–178 (1981).
9. Tanford, C., Extension of the theory of linked functions to incorporate the effects of protein hydration, *J. Mol. Biol. 39*:539–544 (1969).
10. Kirkwood, J. G., and R. J. Goldberg, Light scattering arising from composition fluctuations in multi-component systems, *J. Chem. Phys. 18*:54–57 (1950).
11. Stockmayer, W. H., Light scattering in multi-component systems, *J. Chem. Phys. 18*:58–61 (1950).
12. Casassa, E. F., and H. Eisenberg, Thermodynamic analysis of multicomponent solutions, *Adv. Prot. Chem. 19*:287–395 (1964).
13. Timasheff, S. N., The application of light scattering and small-angle X-ray scattering to interacting biological systems, in *Electromagnetic Scattering*, M. Kerker, Ed., Pergamon Press, New York, pp. 337–355 (1963).
14. Timasheff, S. N., and M. J. Kronman, The extrapolation of light scattering data to zero concentration, *Arch. Biochem. Biophys. 83*:60–75 (1959).
15. Gekko, K., and S. N. Timasheff, Mechanism of protein stabilization by glycerol: Preferential hydration in glycerol–water mixtures, *Biochemistry 20*:4667–4676 (1981).
16. Arakawa, T., R. Bhat, and S. N. Timasheff, Preferential interactions determine protein solubility in three-component solutions: The $MgCl_2$ system, *Biochemistry 29*:1914–1923 (1990).
17. Prakash, V., C. Loucheux, S. Scheufele, M. J. Gorbunoff, and S. N. Timasheff, Interactions of proteins with solvent components in 8 *M* urea, *Arch. Biochem. Biophys. 210*:455–464 (1981).
18. Inoue, H., and S. N. Timasheff, The interaction of β-lactoglobulin with solvent components in mixed water–organic solvent systems, *J. Am. Chem. Soc. 90*:1890–1897 (1968).
19. Arakawa, T., and S. N. Timasheff, Stabilization of protein structure by sugars, *Biochemistry 21*:6536–6544 (1982).
20. Lee, J. C., and S. N. Timasheff, The stabilization of proteins by sucrose, *J. Biol. Chem. 256*:7193–7201 (1981).

21. Lee, J. C., and L. L. Y. Lee, Thermal stability of proteins in the presence of poly(ethylene glycols), *Biochemistry 26*:7813–7819 (1987).
22. Lee, J. C., and L. L. Y. Lee, Preferential solvent interactions between proteins and polyethylene glycols, *J. Biol. Chem. 256*:625–631 (1981).
23. Pittz, E. P., and S. N. Timasheff, Interaction of ribonuclease A with aqueous 2-methyl-2,4-pentanediol at pH 5.8, *Biochemistry 17*:615–623 (1978).
24. Arakawa, T., and S. N. Timasheff, Preferential interactions of proteins with solvent components in aqueous amino acid solutions, *Arch. Biochem. Biophys. 224*:169–177 (1983).
25. Arakawa, T., and S. N. Timasheff, The mechanism of action of Na glutamate, lysine HCl, and PIPES in the stabilization of tubulin and microtubule formation, *J. Biol. Chem. 259*:4979–4986 (1984).
26. Arakawa, T., and S. N. Timasheff, Preferential interactions of proteins with salts in concentrated solutions, *Biochemistry 21*: 6545–6552 (1982).
27. Arakawa, T., and S. N. Timasheff, Mechanism of protein salting in and salting out by divalent cation salts: Balance between hydration and salt binding, *Biochemistry 23*:5912–5923 (1984).
28. Arakawa, T., and S. N. Timasheff, Protein stabilization and destabilization by guanidinium salts, *Biochemistry 23*: 5924–5929 (1984).
29. Lee, J. C., and S. N. Timasheff, The calculation of partial specific volumes of proteins in guanidine hydrochloride, *Arch. Biochem. Biophys. 165*:268–273 (1974).
30. Arakawa, T., and S. N. Timasheff, The abnormal solubility behavior of β-lactoglobulin: Salting-in by glycine and NaCl, *Biochemistry 26*:5147–5153 (1987).
31. Lee, J. C., and S. N. Timasheff, In vitro reconstitution of calf brain microtubules: Effects of solution variables, *Biochemistry 16*:1754–1764 (1977).
32. Arakawa, T., R. Bhat, and S. N. Timasheff, Why preferential hydration does not always stabilize the native structure of globular proteins, *Biochemistry 29*:1924–1931 (1990).
33. Schellman, J. A., Solvent denaturation, *Biopolymers 17*:1305–1322 (1978).
34. Greene, R. F., Jr., and C. N. Pace, Urea and guanidine hydrochloride denaturation of ribonuclease, lysozyme, α-chymotrypsin and β-lactoglobulin, *J. Biol. Chem. 249*: 5388–5393 (1974).
35. Arakawa, T., and S. N. Timasheff, Mechanism of poly(ethylene glycol) interaction with proteins, *Biochemistry 24*:6756–6762 (1985).
36. Tanford, C., "Physical Chemistry of Macromolecules," Wiley: New York, pp. 526–548 (1961).
37. Langmuir, I., The constitution and fundamental properties of solids and liquids, *J. Am. Chem. Soc. 38*:2221–2295 (1916).
38. Schellman, J. A., Stability of hydrogen bonded peptide structures in aqueous solution, *Compt. rend. lab. Carlsberg*, Sér. *Chim. 29*:230–259 (1955).
39. Aune, K. C., and C. Tanford, Thermodynamics of the denaturation of lysozyme by guanidine hydrochloride. II. Dependence on denaturant concentration at 25°C, *Biochemistry 8*:4586–4590 (1969).
40. Schellman, J. A., The thermodynamic stability of proteins, *Ann. Rev. Biophys. Biophys. Chem. 16*:115–137 (1987).
41. Schellman, J. A., Macromolecular binding, *Biopolymers 14*:999–1018 (1975).

42. Schellman, J. A., and R. B. Hawkes, The measurement of protein stability, in *Protein Folding*, R. Jaenicke, Ed., Elsevier: New York, pp. 331–344 (1980).
43. Schellman, J. A., Selective binding and solvent denaturation, *Biopolymers* *26*:549–559 (1987).
44. Schellman, J. A., A simple model of solvation in mixed solvents. Applications to the stabilization and destabilization of macromolecular structures, *Biophys. Chem.* *37*:121–140 (1990).
44a. Hu, C.-Q., J. M. Sturtevant, J. A. Thomson, R. E. Erickson, and C. N. Pace, Thermodynamics of ribonuclease T1 denaturation, *Biochemistry 31*:4876–4882 (1992).
45. Lee, J. C., K. Gekko, and S. N. Timasheff, Measurements of preferential solvent interactions by densimetric techniques, *Methods Enzymol. 61*:26–49 (1979).
46. Timasheff, S. N., and H. Inoue, Preferential binding of solvent components to proteins in mixed water–organic solvent systems, *Biochemistry 7*:2501–2513 (1968).
47. Timasheff, S. N., Protein–solvent interactions and protein conformation, *Acc. Chem. Res. 3*:62–68 (1970).
48. Kupke, D. W., Density and volume change measurements, in *Physical Principles and Techniques of Protein Chemistry*, S. J. Leach, Ed., Part C, Academic Press: New York, pp. 1–75 (1973).
49. Reisler, E., Y. Haik, and H. Eisenberg, Bovine serum albumin in aqueous guanidine hydrochloride solutions. Preferential and absolute interactions and comparison with other systems, *Biochemistry 16*:197–203 (1977).
49a. Timasheff, S. N., Water as ligand: Preferential binding and exclusion of denaturants in protein unfolding, *Biochemistry 31*:9857–9864 (1992).
50. Hade, E. P. K., and C. Tanford, Isopiestic compositions as a measure of preferential interactions of macromolecules in two-component solvents. Application to proteins in concentrated aqueous cesium chloride and guanidine hydrochloride, *J. Am. Chem. Soc. 89*:5034–5040 (1967).
51. Timasheff, S. N., H. M. Dintzis, J. G. Kirkwood, and B. D. Coleman, Light scattering investigation of charge fluctuations in isoionic serum albumin solutions, *J. Am. Chem. Soc. 79*:782–791 (1957).
52. Timasheff, S. N., Small-angle x-ray scattering measurements of biopolymer molecular weights in interacting systems, *Advan. Chem. 125*:327–342 (1973).
53. Kielley, W. W., and W. F. Harrington, A model for the myosin molecule, *Biochim. Biophys. Acta 41*:401–421 (1960).
54. Kuntz, I. D., Hydration of macromolecules. III. Hydration of polypeptides, *J. Am. Chem. Soc. 93*:514–516 (1971).
55. Kuntz, I. D., Jr., and W. Kauzmann, Hydration of proteins and polypeptides, *Adv. Protein Chem. 28*:239–345 (1974).
56. Pfeil, W., K. Welfle, and V. E. Bychkova, Guanidine–hydrochloride titration of the unfolded apocytochrome-C studied by calorimetry, *Studia Biophysica 140*:5–12 (1991).
56a. Makhatadze, G. I., and Privalov, P. L., Protein interactions with urea and guanidinium chloride, *J. Mol. Biol. 226*:491–505 (1992).
57. Bull, H. B., and Breese, K., Protein hydration. I. Binding sites, *Arch. Biochem. Biophys. 128*:488–496 (1968).

58. Schachman, H. K., and M. A. Lauffer, The hydration, size and shape of tobacco mosaic virus, *J. Am. Chem. Soc. 71*:536–541 (1949).
59. Lee, J. C., and L. L. Y. Lee, Interaction of calf brain tubulin with poly(ethylene glycols), *Biochemistry 18*:5518–5526 (1979).
59a. Bhat, R., and Timasheff, S. N., Steric exclusion is the principal source of the preferential hydration of proteins in the presence of polyethylene glycols, *Protein Science 1*:1133–1143 (1992).
60. Gibbs, J. W., On the equilibrium of heterogeneous substances, *Trans. Conn. Acad. 3*:343–524 (1878).
61. Timasheff, S. N., J. C. Lee, E. P. Pittz, and N. Tweedy, The interaction of tubulin and other proteins with structure-stabilizing solvents, *J. Colloid Interfac. Sci. 55*:658–663 (1976).
62. Landt, E., The surface tensions of solutions of various sugars, *Z. Ver. Dtsch. Zuckerindustrie 81*:119–124 (1931).
63. Bull, H. B., and K. Breese, Surface tension of amino acid solutions: A hydrophobicity scale of the amino acid residues, *Arch. Biochem. Biophys. 161*:665–670 (1974).
64. Melander, W., and C. Horvath, Salt effects on hydrophobic interactions in precipitation and chromatography of proteins: A interpretation of the lyotropic series, *Arch. Biochem. Biophys. 183*:200–215 (1977).
65. Gekko, K., and T. Morikawa, Thermodynamics of polyol-induced thermal stabilization of chymotrypsinogen, *J. Biochem. 90*:51–60 (1981).
66. King, M. V., B. S. Magdoff, M. B. Adelman, and D. Harker, Crystalline forms of bovine pancreatic ribonuclease: Techniques of preparation, unit cells and space groups, *Acta Crystallogr. 9*:460–465 (1956).

12

Thermodynamic Nonideality and Protein Solvation

DONALD J. WINZOR
The University of Queensland, Brisbane, Queensland, Australia

PETER R. WILLS
University of Auckland, Auckland, New Zealand

I. INTRODUCTION

Although thermodynamic nonideality is usually regarded by physical chemists and biochemists as a complicating factor that should be avoided if possible, the biological world does not seem to share such fears of this phenomenon. Indeed, in view of the highly crowded state of most biological environments, it could be argued that nature has perfected the art of taking advantage of thermodynamic nonideality. For example, some plants respond to drought and other forms of stress by accumulating high levels (0.1 *M* or greater [1,2]) of small solutes such as proline [3], glycine betaine [4], and sucrose [5]. Similarly, certain organisms that live under conditions of high osmotic pressure accumulate low-molecular-weight components to raise the osmotic pressure of the cytoplasm [6]. A favored explanation of these observations is that the high concentration of nonelectrolyte or osmolyte affords a defense mechanism against stress by stabilizing protein structure [7–10].

The stabilization of protein structure by high concentrations of small nonelectrolytes such as carbohydrates and other polyhydric alcohols has usually been considered in terms of preferential solvation, that is, in terms of preferential occupancy of the protein domain by solvent (water) at the expense of small solute [9,11–20]. The evidence obtained, however, by density measurements that is used to support this interpretation of thermodynamic nonideality effected by small nonelectrolytes has also proven amenable to statistical-mechanical interpretation in terms of the excluded volume concept, whereby the nonideality emanates from the space-filling effects of the added inert solute [21,22]. In this chapter, we demonstrate the absolute equivalence of these two interpretations of thermodynamic nonideality arising from the inclusion of small nonelectrolytes in protein solutions. We derive, correct to first order in solute concentration variables, the formal relationships that define the variation of solution density with protein concentration. Since interpretation of thermodynamic nonideality via statistical-mechanical concepts is more rewarding from the viewpoint of assessing the composition dependence of activity coefficients [21,22], published results of density and thermal denaturation studies of protein stability are reinterpreted in order to provide further insight into the experimental evaluation of parameters required for predicting space-filling effects on the basis of excluded volume considerations.

II. QUANTITATIVE INTERPRETATION OF PARTIAL SPECIFIC VOLUMES

We commence this section by summarizing the traditional interpretation of partial specific volume measurements in terms of protein solvation [9,11–20]. Since this subject is the topic of Chapter 11, however, the approach is only treated in sufficient detail to allow direct comparisons to be made with the interpretation based on the statistical-mechanical concept of excluded volume. We shall restrict our discussion to cases where the solvent and solution are incompressible, this being a very good approximation for dilute protein solutions.

A. Traditional Approach

From the usual experimental viewpoint, the apparent partial specific volume, ϕ'_A, of macromolecular component in a solution comprising protein (component A) and solvent (component s) is defined as [23]

$$\lim_{c_A \to 0} \left(\frac{\partial \rho}{\partial c_A} \right)_{T,\mu_s} = (1 - \phi'_A \rho_0)_{T,\mu_s} \tag{1}$$

where ρ denotes the density of the protein-containing solution and ρ_0 denotes the density of the phase with which the protein solutions with weight concentrations c_A are in dialysis equilibrium. In instances where the solvent is supplemented with a molar concentration C_M of dialyzable nonelectrolyte (component M), the corresponding density measurements on the dialyzed protein solutions yield an apparent partial specific volume defined by the relationship [24,25]

$$(1 - \phi_A'\rho_0)_{T,\mu_s,\mu_M} = (1 - \bar{v}_A\rho_0) + \xi_M^0(1 - \bar{v}_M\rho_0) \tag{2a}$$

$$\xi_M = \frac{M_M}{M_A}\left(\frac{\partial m_M}{\partial m_A}\right)_{T,\mu_s,\mu_M} \tag{2b}$$

where ρ_0 continues to denote the density of the diffusate, m_X is the molality of component X, and $\bar{v}_A$ is the true partial specific volume of protein, $(\partial V/\partial c_A)_{T,P,m_M}$, defined under conditions of constant pressure and molality of small solute. $\bar{v}_M$ is the partial specific volume of the small solute imparting thermodynamic nonideality, and ξ_M is the thermodynamic parameter describing the interaction of small solute with protein: the superscript ξ_M^0 in Eq. (2a) denotes the limiting value as $c_A \rightarrow 0$. The quantity ξ_M measures the mass of small solute to be removed from solution, per unit of mass protein added, in order to maintain constancy of the chemical potential of the small solute in the protein and diffusate phases [24,25]. (The pressure is also varied to keep the solvent chemical potential constant.) A positive value of ξ_M signifies an excess of small solute in the immediate domain of the protein, that is, the binding of solute to protein. A negative value of ξ_M implies preferential interaction of the protein with solvent. That ξ_M has been negative for most small solutes studied [9,11–13,15,17,19] signifies their preferential exclusion from the protein domain, a situation termed *preferential solvation* because the protein domain is occupied by solvent in preference to the mixture of solvent and small solute found in the bulk of the solution. In conformity with this concept, the parameter ξ_M may be converted to a corresponding solvation (hydration) parameter [26,27], $-\xi_M/m_M M_M$, which corresponds to a mass of solvent in the protein domain per unit mass of protein.

The parameter ξ_M is also a measure of the thermodynamic nonideality of the protein solution effected by the addition of small solute. From the condition

$$d\mu_M = \left(\frac{\partial \mu_M}{\partial m_A}\right)_{T,P,m_M} dm_A + \left(\frac{\partial \mu_M}{\partial m_M}\right)_{T,P,m_A} dm_M + \left(\frac{\partial \mu_M}{\partial P}\right)_{T,m_A,m_M} dP = 0 \tag{3}$$

the interaction parameter ξ_M may be expressed in terms of other thermodynamic variables by means of the relationship (Eq. (15) of [24])

$$-\left(\frac{\partial m_M}{\partial m_A}\right)_{T,\mu_s,\mu_M} = \frac{(\partial \mu_M/\partial m_A)_{T,P,m_M} + M_M\bar{v}_M(\partial \Pi/dm_A)_{T,\mu_s,\mu_M}}{(\partial \mu_M/\partial m_M)_{T,P,m_A}} \tag{4}$$

The partial differentials of the small solute chemical potential (μ_M) with respect to solute molality (m_M or m_A) are most conveniently written in terms of activity coefficient, y_M, as

$$\left(\frac{\partial \mu_M}{\partial m_M}\right)_{T,P,m_A} = RT\left[\frac{1}{m_M} + \left(\frac{\partial \ln y_M}{\partial m_M}\right)_{T,P,m_A}\right] \tag{5}$$

$$\left(\frac{\partial \mu_M}{\partial m_A}\right)_{T,P,m_M} = RT\left(\frac{\partial \ln y_M}{\partial m_A}\right)_{T,P,m_M} \tag{6}$$

to give

$$\xi_M = -\frac{M_M}{M_A} m_M \frac{(\partial \ln y_M/\partial m_A)_{T,P,m_M} + (M_M \bar{v}_M/RT)(\partial \Pi/\partial m_A)_{T,\mu_s,\mu_M}}{1 + m_M(\partial \ln y_M/\partial m_M)_{T,P,m_A}} \tag{7}$$

which illustrates clearly that the interaction parameter has a leading term that depends on the first order of the concentration of small solute M. The effect that it measures may therefore be considered to represent a deviation from ideality in solutions containing a protein and small solute.

A potential source of error in the application of Eq. (7) is its reliance upon partial derivatives, with respect to protein molality, of chemical potentials for which the constraints do not coincide with those used experimentally. Some of the partial derivatives demand constancy of pressure and molalities of other species, whereas in the experimental situation the chemical potentials of solvent and small solute are maintained constant. Generally, there is a lack of appreciation of the conditions imposed by the differing constraints that apply in different experimental situations, simply because any quantitative differences appear as first-order concentration corrections that may be ignored for ideal solutions (the situation most frequently under consideration). Any consideration of the effects of thermodynamic nonideality, however, must take differences of this magnitude into account. A discussion of the choice of concentration scale assists in the development of a more direct approach to the analysis of thermodynamic nonideality in equilibrium dialysis studies of the type under consideration.

B. Choice of Concentration Scale

Consider initially an experiment in which a solution of protein (A) is brought into dialysis equilibrium with solvent (s). Since aqueous solutions are being considered, it will be assumed that both solvent and solution are incompressible, a simplification that allows the partial molar volumes of all constituents to be regarded as constants. In thermodynamic studies, the chemical potential (μ_i) of a solute component is considered to be the sum of a standard state value (μ_i^0) and a part that, under ideal conditions, depends logarithmically upon solute concentration. Although the solute concentration may be measured on the mole-fraction, molal, molar, or weight-based scale, the choice of concentration scale serves only to dictate the value of the standard state chemical potential [28,29]. For very dilute solutions, all concentration scales are related to each other in approximately linear fashion, whereupon the variation of chemical potential can be taken into account through the logarithmic term, regardless of the concentration variable used [28,29]. In considerations of thermodynamic nonideality, however, the interest is on effects that arise when this approximation fails. Consequently, it becomes necessary to account for all the additional terms that appear in the thermodynamic equations, either as the result of molecular interactions or as a result of the nonlinear relationship between concentration scales. Polynomial expansions in concentration allow this to be done in a self-consistent manner.

The addition of protein to solvent at constant temperature gives rise to one of two situations, depending on the nature of the experiment. In equilibrium dialysis and classical osmometry, for example, the chemical potential of solvent in the protein-containing phase (α) and the solvent phase (β) remains equal to that of solvent at atmospheric pressure (P); that is, $(\mu_s^\alpha)_{T,P+\Pi} = (\mu_s^\beta)_{T,P}$. On the other hand, the laboratory constraints of fixed temperature and pressure that apply to the determination of partial specific volumes from density measurements on undialyzed solutions lead to a situation in which μ_s^α differs from the value that applied before the addition of protein.

In partition studies such as equilibrium dialysis, the expression relating the thermodynamic activity z_i of a solute component to its chemical potential may be written [30] as

$$(\mu_i)_{T,\mu_s} = (\mu_i^0)_{T,\mu_s} + RT \ln z_i \tag{8a}$$

$$z_i = \gamma_i C_i \tag{8b}$$

where the thermodynamic activity of solute is a molar quantity and is therefore most appropriately expressed as the product of its molar (or comparable weight-based) concentration C_i and the corresponding activity coefficient γ_i. By adopting this concentration scale, it is possible to express the activity coefficient of any solute component in terms of virial coefficients, $\mathbf{B}_{ij}$ and so on, which appear in the usual expression for the osmotic pressure, namely,

$$\frac{\Pi}{RT} = \sum C_i + \sum_i \sum_j \mathbf{B}_{ij} C_i C_j + \cdots \tag{9}$$

where the index i encompasses all solute species, and $j \geq i$. (Note that the range of the index j in the summation differs from that used by us on previous occasions. The current notation is consistent with that used by Hill and other authors.) The result obtained is

$$\ln \gamma_i = 2\mathbf{B}_{ii} C_i + \sum_{j \neq i} \mathbf{B}_{ij} C_j + \cdots \tag{10}$$

which has the advantage that the coefficients $\mathbf{B}_{ij}$, etc., find simple statistical-mechanical interpretation in terms of the physical interactions between pairs, triplets, etc., of molecules [30,31]. In this formulation, γ_i is a molar activity coefficient of solute component i, but it is obtained in terms of a standard state with a lower pressure (P) than that ($P + \Pi$) of the solution.

Alternatively, in circumstances where the chemical potential of solute is being defined under conditions of fixed temperature and pressure, the expressions analogous to Eq. (8) must now be written as

$$(\mu_i)_{T,P} = (\mu_i^0)_{T,P} + \mathbf{R}T \ln a_i \tag{11a}$$

$$a_i = y_i m_i \tag{11b}$$

$$m_i = \frac{N_i}{M_s N_s} \tag{11c}$$

where the thermodynamic activity of solute, a_i, is a molal quantity and is therefore described most simply as the product of an activity coefficient y_i and the molal concentration m_i. In the definition of molality [Eq. (11c)], N_A and N_s denote the respective numbers of solute and solvent molecules, while M_s is the molar mass of solvent expressed in kg/mol to retain customary convention that the units of m_i are mol/kg solvent. The counterparts of Eqs. (9) and (10) are

$$\frac{-[(\mu_s)_{T,P} - (\mu_s^0)_{T,P}]}{RT\rho_s} = \sum m_i + \sum_i \sum_j \mathbf{C}_{ij} m_i m_j + \cdots \tag{12}$$

$$\ln y_i = 2\mathbf{C}_{ii} m_i + \sum_{j \neq i} \mathbf{C}_{ij} m_j + \cdots \tag{13}$$

where y_i is a molal activity coefficient that is obtained by defining chemical potential in terms of a standard state at the same pressure but with a different chemical potential of solvent. The coefficients $\mathbf{C}_{ij}$ are not normally expressible in simple statistical-mechanical terms, but for incompressible solutions the two sets of osmotic virial coefficients conform to the identities

$$\mathbf{C}_{ii} = (\mathbf{B}_{ii} - M_i \bar{v}_i)\rho_s \tag{14a}$$

$$\mathbf{B}_{ii} = \frac{\mathbf{C}_{ii}}{\rho_s} + M_i \bar{v}_i \tag{14b}$$

$$\mathbf{C}_{ij} = \left(\mathbf{B}_{ij} - \frac{M_i \bar{v}_i + M_j \bar{v}_j}{2}\right)\rho_s \tag{14c}$$

$$\mathbf{B}_{\mathrm{ij}} = \frac{\mathbf{C}_{ij}}{\rho_s} + \frac{M_i \bar{v}_i + M_j \bar{v}_j}{2} \tag{14d}$$

where ρ_s is the solvent density. These relationships allow the two thermodynamic activities to be expressed as polynomial expansions of either concentration scale. Specifically,

$$\ln z_i = \ln C_i + 2\mathbf{B}_{ii}C_i + \sum_{j\neq i}\mathbf{B}_{ij}C_j + \cdots \tag{15a}$$

$$= \ln m_i + \ln \rho_s + (2\mathbf{C}_{ii} + M_i\bar{v}_i\rho_s)m_i + \sum_{j\neq i}(\mathbf{C}_{ij} + M_j\bar{v}_i\rho_s)m_j + \cdots \tag{15b}$$

$$\ln a_i = \ln m_i + 2\mathbf{C}_{ii}m_i + \sum_{j\neq i}\mathbf{C}_{ij}m_j + \cdots \tag{15c}$$

$$= \ln C_i - \ln \rho_s + (2\mathbf{B}_{ii} - M_i\bar{v}_i)C_i + \sum_{j\neq i}(\mathbf{B}_{ij} - M_j\bar{v}_j)C_j + \cdots \tag{15d}$$

Although z_i may be expressed in terms of either molar or molal concentration and an appropriate activity coefficient, its definition as a molar activity remains unchanged; the activity coefficients therefore differ [Eqs. (15a,b)]. Furthermore, Eqs. (8) and (10) provide a valid description of the dependence of solute chemical potential upon protein concentration only when solutions are compared at the same chemical potential of solvent. Likewise, Eqs. (11) and (13) require the chemical potentials of the protein in different solutions to be compared at the same pressure.

The distinction between the alternative definitions of the thermodynamic activities z_i and a_i is a subtlety that must be taken into account in the interpretation of partial specific volume measurements. We therefore make a direct attack on analysis of the effect of small solute on the thermodynamic activity of the protein in such measurements.

C. Direct Thermodynamic Interpretation

As mentioned in Sec. II.A, the apparent partial specific volume ϕ'_A is measured under conditions of constant chemical potentials of solvent and small solute. It is now clear that under these circumstances it is most convenient to express concentrations on the molar scale. We shall therefore derive all expressions in such terms [22].

For incompressible solutions, the density may be expressed as the sum of the weight concentrations of all species, *including solvent*; that is, $\rho = \Sigma C_i M_i$. The density ρ of a protein solution (α) and the density ρ_0 of the diffusate (β) with which it is in dialysis equilibrium are therefore given by

$$\rho = M_A C_A^\alpha + M_M C_M^\alpha + \rho_s(1 - M_M\bar{v}_M C_M^\alpha - M_A\bar{v}_A C_A^\alpha) \tag{16a}$$

$$\rho_0 = M_M C_M^\beta + \rho_s(1 - M_M\bar{v}_M C_M^\beta) \tag{16b}$$

where ρ_s refers to the density of unsupplemented solvent. These expressions are readily combined to yield

$$\frac{\rho - \rho_0}{M_A C_A^\alpha} = (1 - \bar{v}_A \rho_s) + M_M(1 - \bar{v}_M \rho_s)\frac{C_M^\alpha - C_M^\beta}{M_A C_A^\alpha} \tag{17}$$

Because of the identity of small solute activities (z_M) in the two phases, the concentration difference may be written as

$$C_M^\alpha - C_M^\beta = C_M^\beta\,(\gamma_M^\beta/\gamma_M^\alpha - 1) \tag{18}$$

Activity coefficients are now replaced by expressing them as power series in concentration, whereupon

$$\gamma_M = \exp\,(\mathbf{B}_{AM}C_A + 2\mathbf{B}_{MM}C_M + \cdots) \tag{19a}$$

$$= 1 + \mathbf{B}_{AM}C_A + 2\mathbf{B}_{MM}C_M + \cdots \tag{19b}$$

where we have retained terms up to first order in solute concentration. With these substitutions, the ratio of activity coefficients becomes

$$\frac{\gamma_M{}^\beta}{\gamma_M{}^\alpha} = 1 - 2\mathbf{B}_{MM}(C_M^\alpha - C_M^\beta) - \mathbf{B}_{AM}C_A^\alpha + \cdots \tag{20}$$

Because $C_M^\alpha - C_M^\beta \to 0$ in the limit of zero protein concentration, Eqs. (17), (18), and (20) may be combined to obtain

$$1 - \phi_A'\rho_0 \equiv \lim_{C_A \to 0} \frac{\rho - \rho_0}{M_A C_A^\alpha}$$

$$= (1 - \bar{v}_A\rho_s) - (1 - \bar{v}_M\rho_s)\frac{\mathbf{B}_{AM}}{M_A}M_M C_M + \cdots \tag{21}$$

where the value of C_M may now be taken as that of the diffusate.

Provided the concentration of small solute is confined to the region wherein the expansion of activity coefficients may be truncated after the linear term, the linear dependence of $1 - \phi_A'\rho_0$ upon C_M has a slope of $(1 - \bar{v}_M\rho_s)M_M(\mathbf{B}_{AM}/M_A)$. The thermodynamic nonideality of the system may thus be expressed experimentally in terms of the second virial coefficient ($\mathbf{B}_{AM}$), which, for an inert nonelectrolyte as small solute, may be identified with the protein–small solute covolume. Specifically,

$$\mathbf{B}_{AM} = U_{AM} = \frac{4\pi\mathbf{N}(R_A + R_M)^3}{3} \tag{22}$$

where $\mathbf{N}$ is Avogadro's number, and R_A and R_M are the respective exclusion radii of A and M, both assumed spherical.

Having outlined this direct procedure [22] for interpreting partial specific volume measurements in terms of thermodynamic nonideality, we are now in a position to make comparisons with the classical approach in terms of protein solvation.

D. Equivalence of Treatments

In order to establish equivalence of the preferential solvation and statistical-mechanical treatments of thermodynamic nonideality, the relationship between ξ_M and the second virial coefficient will be obtained by manipulating Eq. (2a) to a form more in keeping with Eq. (21). The first step is to introduce the solvent density (ρ_s) into Eq. (2a) by expressing the solution density as a power series in solute concentration, the molal equivalent of Eq. (16) being

$$\rho = \rho_s\left(1 + \sum_i M_i m_i\right)\Big/\left(1 + \sum_i M_i \bar{v}_i m_i \rho_s\right) \tag{23a}$$

$$= \rho_s\left[1 + \sum_i (1 - \bar{v}_i \rho_s) M_i m_i + \cdots\right] \tag{23b}$$

where the subscript i spans all species, including solvent. On taking into account that ξ_M^0 is obtained in the limit of zero solute concentration ($m_A \rightarrow 0$), substitution of Eq. (23b) into Eq. (2a) gives

$$1 - \phi_A' \rho_0 = (1 - \bar{v}_A \rho_s) + (1 - \bar{v}_M \rho_s)\ [\ (\xi_M^0\ - \bar{v}_A \rho_s M_M m_M) + \cdots] \tag{24}$$

The next step is to express the partial derivatives appearing in Eq. (7) as polynomial expansions in solute concentrations, but it is only necessary to retain the leading terms in each to obtain the description of ξ_M correct to first order in concentration. With that proviso, the three partial derivatives in Eq. (7) become

$$\lim_{C_A \rightarrow 0} \left(\frac{\partial \ln y_M}{\partial m_A}\right)_{T,\,P,\,m_M} = \mathbf{C}_{AM} + \cdots \tag{25}$$

$$\lim_{C_A \rightarrow 0} \left(\frac{\partial \ln y_M}{\partial m_M}\right)_{T,\,P,\,m_A} = 2\mathbf{C}_{MM} + \cdots \tag{26}$$

$$\lim_{C_A \rightarrow 0} \left(\frac{\partial \Pi}{\partial m_A}\right)_{T,\mu_s,\mu_M} = RT\rho_s[1 + \cdots] \tag{27}$$

The last derivative follows from Eq. (9), while the other two emanate from Eq. (15c) on noting that $\ln y_M = \ln a_M - \ln m_M$. Substitution of Eqs. (25)–(27) into Eq. (7) then gives

$$\lim_{C_A \rightarrow 0} \xi_M = -\frac{M_M}{M_A}\,\frac{(\mathbf{C}_{AM} + M_M \bar{v}_M \rho_s) m_M + \cdots}{1 + 2\mathbf{C}_{MM} m_M + \cdots} \tag{28a}$$

$$\xi_M^0 = -\frac{M_M}{M_A}\,[(\mathbf{C}_{AM} + M_M \bar{v}_M \rho_s) m_M + \cdots]\,[1 - \cdots] \tag{28b}$$

whereupon Eq. (24) becomes

$$1 - \phi_A' \rho_0 = (1 - \bar{v}_A \rho_s) - (1 - \bar{v}_M \rho_s)\frac{M_M}{M_A}[(\mathbf{C}_{AM} + M_M v_M \rho_s + M_A\,\bar{v}_A \rho_s) m_M + \cdots] \tag{29}$$

Conversion from molal to molar concentration on the basis that

$$m_M(C_A = 0) = \frac{C_M}{\rho_s}[1 + M_M\bar{v}_M\rho_s C_M + \cdots] \tag{30}$$

and substitution of $[\mathbf{B}_{AM} - (M_M\bar{v}_M + M_A\bar{v}_A)/2]\rho_s$ for $\mathbf{C}_{AM}$ [Eq. (14c)] then yields Eq. (21), the expression derived in Sec. C.

Finally, the formal relationship between ξ_M and $\mathbf{B}_{AM}$ is obtained either directly from Eqs. (14c) and (28b) or by equating the leading terms in Eqs. (21) and (24). Either approach shows that

$$\xi_M^0 = -\left(\frac{M_M}{M_A}\right)(\mathbf{B}_{AM} - M_A\bar{v}_A)\rho_s m_M + \cdots \tag{31}$$

whereupon the solvation parameter, $-\xi_M^0/m_M M_M$, becomes $(\mathbf{B}_{AM} - M_A\bar{v}_A)\rho_s/M_A$, correct to zero order in concentration.

From a thermodynamic viewpoint, $\mathbf{B}_{AM}$ and ξ_M^0/m_M provide different means of expressing deviations from thermodynamic ideality that arise from interactions between molecules of the protein (A) and small solute (M). Analysis of results in terms of $\mathbf{B}_{AM}$ allows readier access to the composition dependence of the thermodynamic activity coefficient of the protein; and when the only significant interactions between molecules of protein and small solute are of the excluded volume type, $\mathbf{B}_{AM}$ represents the covolume, (U_{AM}) [Eq. (22)]. In this case, ξ_M is seen to be a measure of the difference between the volume in the region of the protein that is inaccessible to the small solute (U_{AM}) and the corresponding volume that is inaccessible to solvent $(M_A\bar{v}_A)$. As Eisenberg [24,25,32] has noted, interpretation of ξ_M as a measure of preferential solvation is equivocal. Like the excluded volume interpretation of $\mathbf{B}_{AM}$, it entails the consideration of a thermodynamic parameter in terms of postulated molecular interactions. Although such resort to model-dependent interpretations of thermodynamic data represents a departure from classical protocol, the approach is justified in the event that the model provides a helpful explanation of the thermodynamic behavior of the system under consideration.

In order to gain further insight into the prediction of covolumes for protein–small nonelectrolyte systems, we now employ the statistical-mechanical approach for reanalysis of published densimetric studies of thermodynamic nonideality in protein solutions containing high concentrations of a range of small nonelectrolytes.

III. VIRIAL COEFFICIENTS FROM DENSITY MEASUREMENTS

There have been many studies in which partial specific volume measurements have been used in conjunction with equilibrium dialysis to establish generality of the concept that the protein domain is preferentially occupied by solvent at the expense of small inert solutes [9,11–20]. We now reassess that literature by reverting to a more fundamental thermodynamic parameter, viz., the second virial coefficient $\mathbf{B}_{AM}$, as a means of describing the nonideality of protein solutions effected by the presence of the small solute.

A. Protein–Small Nonelectrolyte Systems

The use of Eq. (21) for the determination of second virial coefficients from partial specific volume measurements [22] is illustrated in Fig. 1, which reappraises results for the nonideality of ribonuclease and chymotrypsinogen that stems from the presence of sucrose [12], lactose [17], and glucose [17]. These plots are essentially linear, as Eq. (21) predicts for low enough concentrations; and furthermore, there is clearly reasonable agreement between results for the two disaccharides, sucrose and lactose. Second virial coefficients obtained from the slopes, $-\mathbf{B}_{AM}(1 - \bar{v}_M\rho_s)(M_M/M_A)$, are summarized in Table 1, which also includes values of $\mathbf{B}_{AM}$ that emanate from corresponding analyses of results obtained with other polyhydric alcohols (14,15) as the small nonelectrolyte. In general, the dependence of $1 - \phi'_A\rho_0$ upon molar concentration of small solute in the diffusate, C_M, was essentially linear in the experimental concentration range examined, the slight curvature of the plot for glycerol being taken to imply inadequacy of

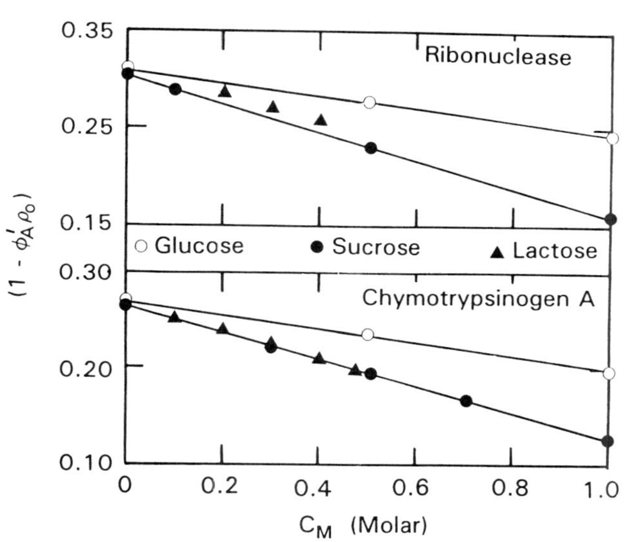

FIGURE 1 Evaluation of the second virial coefficient $\mathbf{B}_{AM}$ from the effect of carbohydrate concentration (C_M) on the apparent partial specific volumes of ribonuclease and chymotrypsinogen A. Results obtained with sucrose [12], glucose [17] and lactose [17] as small solute are plotted in accordance with Eq. (21).

TABLE 1 Determination of Second Virial Coefficients ($\mathbf{B}_{AM}$) for Proteins and Small Nonelectrolytes from Partial Specific Volume Measurements[a,b]

	$\mathbf{B}_{AM}$(L/mol)					
Protein	Sucrose[c]	Lactose	Glucose	Mannitol	Sorbitol	Glycerol
Ribonuclease A	15.5 [12]	12.8 [17]	12.7 [17]	—	—	11.0 [14]
Lysozyme	—	—	13.0 [17]	—	—	—
α-Chymotrypsin	24.1 [12]	—	—	—	—	21.9 [14]
Chymotrypsinogen A	26.8 [12]	28.7 [17]	27.0 [17]	—	—	23.9 [14]
Bovine albumin	—	64.2 [17]	60.1 [17]	64.5 [15]	67.6 [15]	55.8 [15]
β-Lactoglobulin	—	—	—	—	—	16.0 [14]

[a] Calculated via Eq. (21) from the limiting slopes of plots of $(1 - \phi'_A\rho_0)$ versus C_M.
[b] Numbers in brackets denote sources of the partial specific volume measurements.
[c] Reported in [22].

Eq. (21) truncated after the linear term for describing results over such a large range (0–5.5 *M*) of C_M. The value of $\mathbf{B}_{AM}$ reported in Table 1 was therefore inferred from the limiting slope as $C_M \rightarrow 0$, which disregarded data beyond 3 *M* glycerol.

B. Osmolytes as Inert Solute

Small charged but electrically neutral solutes such as aliphatic amino acids have also been examined for their effects on partial specific volume measurements [9,33]. Analysis of results for two such osmolytes, glycine and betaine, are shown in Fig. 2, which also includes the second virial coefficients obtained from these plots for lysozyme and bovine serum albumin in accordance with Eq. (21). The earlier inference [33] that osmolytes are also excluded preferentially from the protein domain is manifested in the present analysis as a general similarity between values of $\mathbf{B}_{AM}$ obtained for a given protein with osmolytes and nonelectrolytes as small solute.

C. Excluded Volume Interpretation

The values of $\mathbf{B}_{AM}$ are thermodynamic coefficients describing the first-order concentration dependence of the nonideality due to the presence of nonelectrolyte or osmolyte in the protein solutions. As such, they may clearly be used in calculations of the activity coefficient of the protein component in the mixtures via Eq. (19). If the only interactions between protein and the small solute are of the

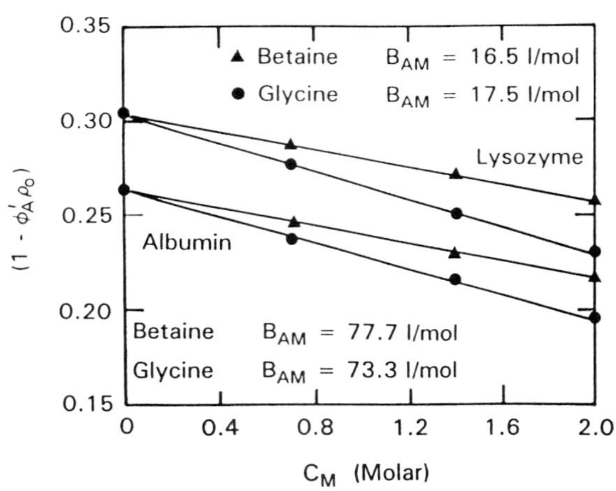

FIGURE 2 Evaluation of the second virial coefficient $\mathbf{B}_{AM}$ from published data [9] for the effects of osmolyte concentration (C_M) on the apparent partial specific volumes of lysozyme and bovine serum albumin, the results being plotted in accordance with Eq. (21).

excluded volume type, then $\mathbf{B}_{AM}$ is simply the covolume U_{AM}, in which case it may well prove possible to predict adequately the composition dependence of thermodynamic activity of the protein from knowledge of the species' effective radii [Eq. (22)]. Indeed, to a first approximation the radius of the small solute should be of negligible magnitude in relation to that of the protein. It has therefore been suggested [21,22] that the molar volume of the hydrated protein should provide a reasonable estimate of U_{AM}. That suggestion is explored further in Table 2, which compares the values of $\mathbf{B}_{AM}$ obtained from densimetric measurements with the corresponding estimate of the molar volume of the hydrated protein that is inferred from hydrodynamic measurements. On the basis of these results, it appears that the volume derived from the Stokes radius of the protein overestimates $\mathbf{B}_{AM}$. Because allowances for thermodynamic nonideality usually amount to relatively small corrections being applied to experimental results, this use of hydrodynamic radii seems likely to provide an acceptable estimate of U_{AM} and hence of the activity of a globular protein in moderately concentrated solutions of small solute. Furthermore, adoption of the most pessimistic outlook would still lead to the conclusion that such allowances must lead to better quantitative description of the system than an analysis that ignores the phenomenon of thermodynamic nonideality altogether.

This section has demonstrated a means whereby results previously attributed to preferential protein solvation may also be interpreted on the statistical-mechanical basis of excluded volume. Furthermore, an advantage of the latter mode of interpretation is the quantification in a manner that also allows evaluation of the thermodynamic activity coefficient. An obvious criticism, however, that may be leveled at Table 2 is its reliance upon comparison between a thermodynamic parameter, $\mathbf{B}_{AM}$, and a hydrodynamic parameter, the Stokes volume. The validity for so doing is examined further in Secs. IV and V.

IV. CONSIDERATION OF SMALL SOLUTES AS EFFECTIVE SPHERES

The validity of adopting the statistical-mechanical concept of excluded volume for interpretation of the second virial coefficient ($\mathbf{B}_{AM}$) describing thermodynamic nonideality of protein–small solute systems would clearly be strengthened by demonstration that the small solutes used in such studies can be modeled, even approximately, as effective thermodynamic spheres. One source of relevant information is the concentration dependence of the activity coefficient of the small solute that is obtained by the classical procedure of isopiestic vapor pressure measurements.

A. Interpretation of Isopiestic Measurements

Provided that the solute whose activity is being assessed is nonvolatile, the final equilibrium state in isopiestic experiments reflects a situation in which the chem-

TABLE 2 Comparison of the Magnitudes of Second Virial Coefficients for Protein–Small Solute Interactions with the Hydrodynamic Molar Volumes (V_h) of the Hydrated Proteins

Protein	M_A	$\bar{v}_A$(ml/g)	$s^0_{20,w}$(S)[a]	V_h(L/mol)[b]	$\mathbf{B}_{AM}$(L/mol)[c]
Ribonuclease	13,700	0.693	2.02 [34]	15.7	11.0–15.5
Lysozyme	14,000	0.703	1.91 [35]	17.8	16.5–17.5
α-Chymotrypsin	25,000	0.738	2.75 [36]	23.7	21.9–24.1
Chymotrypsinogen A	25,500	0.734	2.75 [36]	26.1	23.9–28.7
Bovine albumin	66,000	0.734	5.01 [37][d]	108.1	55.8–77.7

[a]Numbers in brackets denote sources of the sedimentation coefficients.
[b]Hydrodynamic volume deduced from the previous three entries.
[c]Range of values from Table 1 and Fig. 2.
[d]Value of $s^0_{25,w}$ reported by Baldwin [37].

ical potential of solvent (water) in the mixture is equal to that of the vapor phase and also that of solvent in the reference solution. By using a series of reference solutions with known, but different, solvent chemical potentials, μ_s, it is possible to determine the solute activity at constant temperature and pressure, a_M, in accordance with Eqs. (11)–(13).

Figure 3 summarizes the molal concentration dependence of activity coefficients (y_M) reported for sucrose and glycerol [38], and also for betaine and proline [40], on the basis of isopiestic measurements. The first point to note is the conformity of these results with Eq. (15c), not only in regard to the predicted linear dependence but also the expectation that the slope, $2\mathbf{C}_{MM}$, be positive. Second, combination of the slopes with the density of water and the partial molar volumes of the solutes leads, by use of Eq. (14c), to the estimates of the second virial coefficient, $2\mathbf{B}_{MM}$, reported in Table 3. Third, the final column in Table 3 records the radius R_M obtained on the basis that the second virial coefficient corresponds to the covolume U_{MM} for self-interaction of a spherical solute, that is, on the basis that $2\mathbf{B}_{MM} = U_{MM} = 32\pi\mathbf{N}R_M^3/3$.

That the small solute considered in Fig. 3 exhibited the positive deviations from Raoult's law that are required for the representation of solute molecules as inert geometric particles should not be taken to imply generality of the observation. To emphasize this point, the results of isopiestic measurements on urea [38] and glycine [39] are presented in Fig. 4. Clearly, interpretation of the results for these solutes, which are exhibiting negative deviations from Raoult's law, in terms of excluded volume would lead to the calculation of a negative covolume and

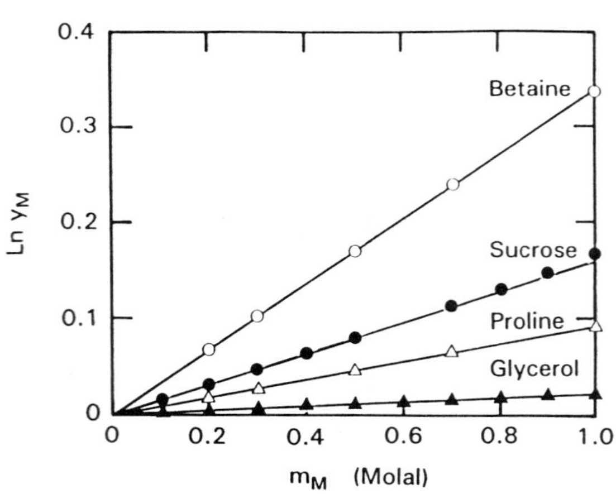

FIGURE 3 Concentration dependence of activity coefficients reflecting self-interaction of sucrose and glycerol [38], and also of betaine and proline [40]. These results, obtained by the isopiestic method, are plotted in accordance with Eq. (13).

TABLE 3 Self-Covolumes and Effective Thermodynamic Radii of Small Solutes from Isopiestic and Freezing Point Measurements

Solute	$2\mathbf{B}_{MM} = U_{MM}$ (L/mol)[a]	R_M (nm)
Sucrose	0.58 (I) ; 0.65 (F)	0.31, 0.32
Glucose	0.26 (I)[b]; 0.30 (F)	0.24, 0.25
Glycerol	0.17 (I)	0.20
Betaine	0.5.3 (I)	0.30
Proline	0.26 (I)	0.23

[a]Letter in parentheses denotes method of evaluation: I, from isopiestic data (Fig. 3); F, from freezing point depression data (Fig. 5).
[b]Based on a value of 1.020 for the osmotic coefficient, $(\mu_s - \mu_s^0)/RTm_M$, for 1 molal glucose [42] and assumed linearity of its dependence upon m_M.

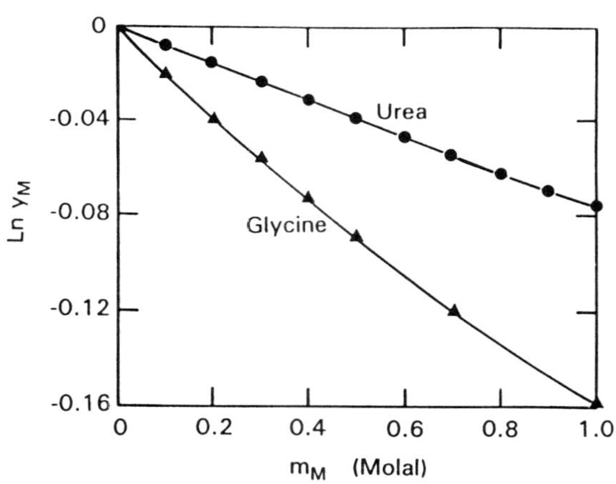

FIGURE 4 Nonconformity of activity coefficients reflecting the self-interactions of urea [38] and glycine [39] with interpretation in terms of the excluded volume concept. Results of these isopiestic studies are plotted in accordance with Eq. (13), as in Fig. 3.

hence negative value for the effective thermodynamic radius. For urea, the deviation from classical excluded volume behavior reflects the reversible self-association of this solute as the result of intermolecular hydrogen bonding [43]: dimerization of M, for example, leads to a second virial coefficient with magnitude $\mathbf{B}_{MM} - K_2$, where K_2 is the molar dimerization constant [44]. On the other hand, the most plausible explanation of the result for glycine is that this small osmolyte is exhibiting behavior more in keeping with Debye–Hückel theory for electrolytes [45].

B. Freezing Point Depression Data

In common with their isopiestic counterparts, freezing point depression measurements also refer to systems where the solvent chemical potential in the liquid phase containing small solute, M, is identical with that in a second phase comprising solvent alone. This chemical potential of solvent is again being defined at fixed pressure, and hence its dependence upon solute concentration is most appropriately defined in molal terms [as in Eq. (12)]. On the basis of the reasonable approximation that the molar latent heat of fusion, λ_F, is independent of temperature over the relatively small range covered by the freezing point measurements, the second virial coefficient $\mathbf{C}_{MM}$ may be evaluated from the expression

$$\frac{\lambda_F(1/T - 1/T_0)}{RM_s m_M} = 1 + \mathbf{C}_{MM} m_M + \cdots \tag{32}$$

where T_0 and T are the respective freezing points of pure solvent and a solution with molal concentration m_M of solute.

The application of Eq. (32) to freezing point data [46] for sucrose and glucose is presented graphically in Fig. 5. Conversion of the values of $\mathbf{C}_{MM}$ obtained from the limiting slopes to covolumes ($U_{MM} = 2\mathbf{B}_{MM}$) via Eq. (14) yields magnitudes of $2\mathbf{B}_{MM}$, and hence R_M, that are in good agreement with the corresponding values obtained from analysis of the isopiestic measurements (Table 3). Results for two other disaccharides, maltose and lactose, are indistinguishable from those for sucrose, there being a corresponding identity of plots for the monosaccharides glucose and fructose.

These values of effective thermodynamic radii for small solutes are open to potential criticism on the grounds that they refer to solute in water rather than a buffer supplemented with supporting electrolyte. Moreover, the relevance of an effective thermodynamic radius determined by freezing point depression to studies at (say) 25°C may also be questioned. On that score, that the estimates of $2\mathbf{B}_{MM}$ and, hence R_M, for sucrose and glucose obtained by freezing point depression are in fair agreement with those from isopiestic measurements at 25°C suggests the relative unimportance of the temperature factor; but clearly, it does not comment on the validity of the other potential source of criticism. It is therefore important to have available a means of measuring activity coefficients under less restrictive conditions. Gel chromatography affords such a method [47,48].

C. Frontal Gel Chromatography of Sucrose

The more recent characterization of the thermodynamic nonideality of sucrose in acetate-chloride buffer (pH 3.9, I 0.2) by frontal gel chromatography on Sephadex G-10 [48] affords an example of activity coefficient measurement under conditions pertinent to a specific environment. In such experiments, the molar thermodynamic activity of solute (z_M) is being measured, because the distribution of solute between the mobile and stationary phases is effected under conditions of

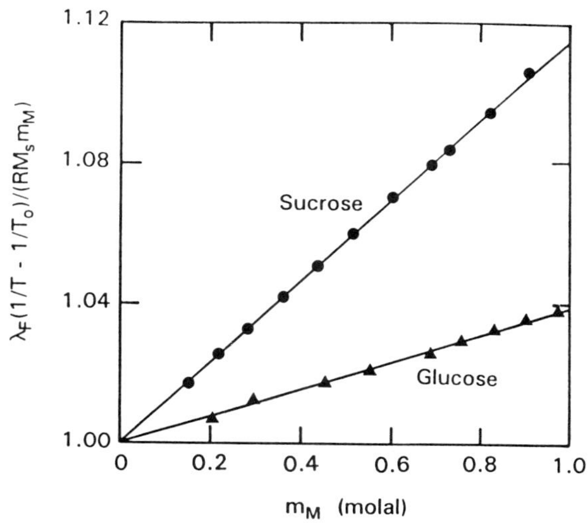

FIGURE 5 Evaluation of the second virial coefficients for self-interaction of sucrose and glucose by analysis of freezing point depression data [46] in terms of Eq. (32).

constant solvent chemical potential. Solute concentrations are therefore expressed in molar terms.

On the grounds that the partitioning of solute M between mobile (α) and stationary (β) phases establishes the identity of its chemical potential in both phases, the inference that the two thermodynamic activities are the same may be written [47–49]

$$\gamma_M^\beta C_M^\beta = \gamma_M^\alpha C_M^\alpha \tag{33}$$

It follows that the ratio of solute concentrations in the two phases is the reciprocal of the corresponding ratio of activity coefficients, that is,

$$\sigma_M = \frac{C_M^\beta}{C_M^\alpha} = \frac{\gamma_M^\alpha}{\gamma_M^\beta} \tag{34}$$

where σ_M, the partition coefficient pertaining to solute concentration C_M^α, may be determined experimentally by frontal gel chromatography [50] from the relationship [51,52]

$$\sigma_M = \frac{V - V_0}{V_t - V_0} \tag{35}$$

V denotes the median bisector of the elution profile from a frontal gel chromatography experiment with applied solute concentration C_M^α on a column with void volume V_0 and total accessible volume V_t.

The activity coefficients of the solute [Eq. (10)] in the mobile (α) and stationary (β) phases are given by

$$\ln \gamma_M^\alpha = 2\mathbf{B}_{MM} C_M^\alpha + \cdots \tag{36a}$$

$$\ln \gamma_M^\beta = 2\mathbf{B}_{MM} C_M^\beta + \mathbf{B}_{MX} C_X^\beta + \cdots \tag{36b}$$

where the additional term in Eq. (36b) reflects thermodynamic nonideality of solute due to the presence of gel matrix at an effective concentration C_X^β, presumed independent of C_M^β. Combination of these expressions with Eq. (34) leads to the relationship

$$\ln \sigma_M = 2\mathbf{B}_{MM}(C_M^\alpha - C_M{}^\beta) - \mathbf{B}_{MX} C_X^\beta + \cdots \tag{37a}$$

The problem of evaluating the solute–matrix contribution to nonideality is avoided by noting that in the limit of infinite dilution ($C_M^\alpha \rightarrow 0$), the expression analogous to Eq. (37a) is

$$\ln \sigma_M^0 = -\mathbf{B}_{MX} C_X^\beta + \cdots \tag{37b}$$

Combining Eqs. (36) and (37) then gives

$$\ln \sigma_M = \ln \sigma_M^0 + 2\mathbf{B}_{MM} C_M^\alpha (1 - \sigma_M) + \cdots \tag{38}$$

as the expression for the concentration dependence of the experimental partition coefficient. The second virial coefficient for solute self-interaction, $2\mathbf{B}_{MM}$, may therefore be obtained from the slope of a plot of $\ln \sigma_M$ *versus* $C_M^\alpha (1 - \sigma_M)$.

The application of this procedure to frontal gel chromatography results for sucrose in acetate-chloride buffer (pH 3.9, I 0.2) is illustrated in Fig. 6, from which a value of 0.58 L/mol is obtained for $2\mathbf{B}_{MM}$. Similar values have been inferred from the isopiestic and freezing point depression measurements (Table 3).

D. Validity of the Proposition

That the small solutes considered in Figs. 3, 5, and 6 all exhibited positive deviations from Raoult's law lends initial justification to the interpretation of their self-interaction on the statistical-mechanical basis of excluded volume. On the other hand, the thermodynamic nonideality of solutes such as urea and glycine (Fig. 4) cannot be so interpreted. Although the spherical geometry assigned to the solutes exhibiting positive deviations from Raoult's law is a very poor representation of their physical shapes, such criticism does not detract from the use of these radii as effective thermodynamic parameters, irrespective of the degree to which that thermodynamic model provides an acceptable description of the actual molecule.

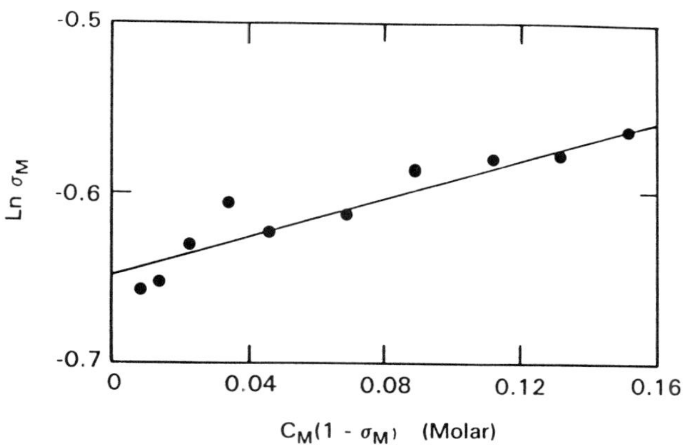

FIGURE 6 Evaluation of the second virial coefficient, $2\mathbf{B}_{MM}$, for sucrose in acetate-chloride buffer (pH 3.9, I 0.2) by analysis of frontal gel chromatographic data in terms of Eqs. (35) and (38). (Adapted from [48] with permission.)

V. EFFECTIVE THERMODYNAMIC RADII OF GLOBULAR PROTEINS

In most studies of thermodynamic nonideality in protein solutions, the stance has been taken that the Stokes radius of a globular protein provides a reasonable estimate of its effective hydrated radius for use in covolume calculations [21,22,47–49,53–58]. As noted elsewhere [59], there would be no theoretical objection to this substitution of a hydrodynamic parameter for a thermodynamic quantity if the protein were in fact an impenetrable sphere. Since, however, the Stokes radius of a protein merely refers to the size of a sphere with equivalent hydrodynamic characteristics, there is clearly no guarantee that the sphere with equivalent thermodynamic characteristics need necessarily have the same dimensions [60]. Unfortunately, there is little information available to provide comment on that comparison. The size deduced from the partial specific volume is an underestimate of the required parameter because it refers solely to the volume occupied by the unsolvated species and does not take into account the region of the solvent domain that is included in the particles whose interactions give rise to the thermodynamic effects under consideration. The decision to substitute hydrodynamic for thermodynamic radii has thus largely reflected the availability of values for the former and a virtual absence of values for the latter. We therefore summarize experimental means by which estimates of an effective thermodynamic radius may be obtained.

A. Evaluation from Self-Covolume Measurements

By analogy with the preceding section on small solutes, an obvious way of determining an effective thermodynamic radius of a hydrated globular protein is to measure the second virial coefficient for self-interaction. Although osmotic pressure and light-scattering studies have the potential to provide the magnitude of $2\mathbf{B}_{AA}$, the main emphasis of such investigations has invariably been the evaluation of molecular weight by extrapolating Π/RTc_A or Hc_A/τ to zero weight-concentration, c_A, in order to eliminate the effects of thermodynamic nonideality. In ultracentrifuge studies, the second viral coefficient has traditionally been evaluated from the concentration dependence of the apparent weight-average molecular weight inferred from interferometric records of the sedimentation equilibrium distribution [61]; but a far more direct analysis is available [62].

1. Evaluation by Sedimentation Equilibrium

The direct method of analysis takes advantage of the omega function [62], defined by the relationship

$$\Omega(r) = \frac{c_A(r)}{c_A(r_F)} \exp \frac{M_A(1 - \bar{v}_A\rho_s)\omega^2(r_F{}^2 - r^2)}{2RT} \tag{39}$$

in which $c_A(r)$ and $c_A(r_F)$ denote the respective protein concentrations at radial distance r and a reference radial position r_F in a sedimentation equilibrium experiment conducted at temperature T and angular velocity ω. In the context of quantifying thermodynamic nonideality of a single nonassociating solute, the important feature of the omega function is its relationship to solute concentration [62,63]. Specifically,

$$\Omega(r) = \frac{z_A(r_F)c_A(r)}{z_A(r)c_A(r_F)} = \frac{\Omega_0 c_A(r)}{z_A(r)} = \frac{\Omega_0}{\gamma_A(r)} \tag{40}$$

where $z_A(r)$ and $z_A(r_F)$ refer to the thermodynamic activities at the respective radial positions [22,44]; the latter equalities follow from the demonstration [62,63] that the ratio $z_A(r_F)/c_A(r_F)$ is given by the ordinate intercept Ω_0 of a plot of $\Omega(r)$ *versus* $c_A(r)$. For moderate protein concentrations, the expression for the activity coefficient [Eq. (10)] may be truncated after the linear concentration term, whereupon its substitution into Eq. (40) gives

$$\Omega(r) = \Omega_0 \exp \frac{-2\mathbf{B}_{AA}c_A(r)}{M_A} = \Omega_0[1 - 2\mathbf{B}_{AA}C_A(r) + \cdots] \tag{41}$$

The predicted linear dependence of $\Omega(r)$ upon molar protein concentration $C_A(r)$ thus allows evaluation of the second virial coefficient as the ratio of the slope $(-2\mathbf{B}_{AA}\Omega_0)$ to the ordinate intercept.

Use of the omega function for quantifying thermodynamic nonideality of a nonassociating protein is illustrated in Fig. 7, which refers to a sedimentation equilibrium study of isoelectric ovalbumin [63]. Indeed, the linear concentration dependence of $\Omega(r)$ was used to evaluate Ω_0 and hence the activity $z_A(r_F)$ of the reference solution with a concentration $c_A(r_F)$ of 1.25 mg/ml [$C_A(r_F) = 27.8\ \mu M$]. Whereas, however, that value of $z_A(r_F)$ was then substituted into the basic sedimentation equilibrium equation to obtain $z_A(r)$, and hence $\gamma_A(r)$, for each protein concentration, such action is now seen to be redundant because the second virial coefficient describing the concentration dependence of $\gamma_A(r)$ may be obtained directly from Fig. 7 by application of Eq. (41). On the grounds that the resultant value of 500 ±20 L/mol is also the covolume, an effective thermodynamic radius of 2.92 (± 0.06) nm is obtained for the hydrated ovalbumin molecule.

2. Determination by Exclusion Chromatography

Second virial coefficients for self-interaction of proteins have also been determined by the frontal chromatographic method described in Sec. IV.C for sucrose, the only procedural difference being the selection of a sufficiently porous chromatographic matrix to allow partition of the protein under study; control-pore glass [47,64] and Fractogel TSK [65] have been used for this purpose. Results of chromatographic measurements of $2\mathbf{B}_{AA}$ for several proteins are summarized in Table 4. As noted above, values of $2\mathbf{B}_{AA}$ obtained under isoelectric conditions also describe the covolume U_{AA}. Under conditions where the protein bears a net charge (valence) Z_A, the second virial coefficient and covolume are related by the expression [22]

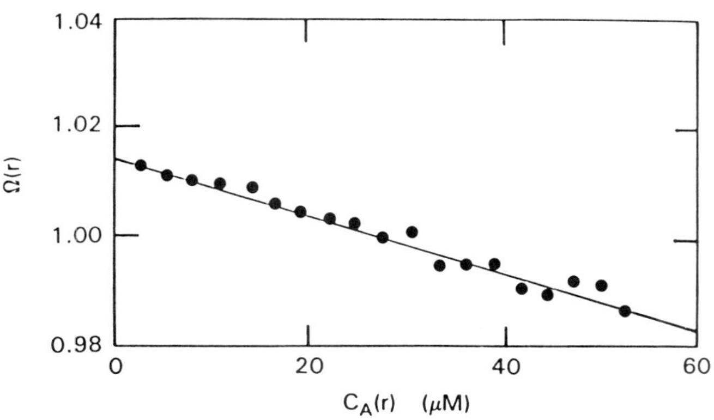

FIGURE 7 Use of a sedimentation equilibrium distribution to evaluate the second virial coefficient, $2\mathbf{B}_{AA}$, for isoelectric ovalbumin from a plot of the results in accordance with Eq. (41). (Adapted from [63] with permission.)

TABLE 4 Evaluation of Effective Thermodynamic Radii of Proteins from Second Virial Coefficients Obtained Chromatographically[a]

		Conditions					Radius (nm)	
Protein	Matrix[b]	pH	I	Z_A	$2\mathbf{B}_{AA}$ (L/mol)	U_{AA}[c,d] (L/mol)	R_A[d]	R_h
Ovalbumin	HW-55	4.6	0.20	0	462 ± 74	462	2.84 ± 0.14	2.92
Ovalbumin	CPG-75	7.4	0.16	−16	781 ± 36	474	2.86 ± 0.07	2.92
Ribonuclease	HW-55	9.5	0.20	0	114 ± 23	114	1.78 ± 0.11	1.84
Serum albumin	HW-55	4.6	0.20	0	854 ± 90	854	3.49 ± 0.12	3.51
Catalase	HW-55	7.6	0.25	−48	3850 ± 560	2950	5.27 ±0.32	5.27
α-Crystallin	CPG-240	7.3	0.17	−46	10700 ± 900	9700	7.83 ± 0.24	8.05

[a]Experimental data taken from Ref. 65.
[b]Either Fractogel TSK HW-55 or Control-Pore Glass (CPG) with stipulated pore size.
[c]Calculated from Eq. (42).
[d]Values differ slightly from those reported in [65], where a molar volume term was incorrectly incorporated into Eq. (42) [22,44].

$$2\mathbf{B}_{AA} = U_{AA} + \frac{Z_A^2(1 + 2\kappa R_A)}{2I(1 + \kappa R_A)^2} \tag{42}$$

in which the second term on the right-hand side describes the contribution of charge–charge interactions in terms of solute parameters and the inverse screening length κ associated with ionic strength I. Inspection of the penultimate column of Table 4, which lists the effective thermodynamic radii, R_A, deduced from U_{AA}, reveals substantial agreement between the two chromatographic estimates of R_A for ovalbumin, which essentially duplicate the value emanating from the sedimentation equilibrium study (Fig. 7). For all six proteins studied, there is also close correspondence between the effective thermodynamic radius (R_A) and its hydrodynamic counterpart, the Stokes radius (R_h).

B. Evaluation from Protein–Small Solute Covolume

The use of partial specific volume measurements to characterize thermodynamic nonideality reflecting the interaction of protein with small solute has already been described in Sec. III. We therefore restrict methodological discussion here to other methods that may also be used to evaluate the second virial coefficient $\mathbf{B}_{AM}$.

1. Determination by Equilibrium Dialysis or Gel Chromatography

Provided that the dialysis membrane or gel chromatographic matrix is chosen to confine the protein to one phase, the ratio of small-solute concentrations in the protein-containing and protein-free phases (C_M^α and C_M^β, respectively) may be substituted into the expression [66]

$$\mathbf{B}_{AM} = \frac{\ln(C_M^\beta / C_M^\alpha)}{C_A^\alpha} \tag{43}$$

to determine a value of $\mathbf{B}_{AM}$, the second virial coefficient describing thermodynamic nonideality due to mutual interaction of protein and small solute.

Application of Eq. (43) in its gel chromatographic context is illustrated in Fig. 8, which represents the trailing elution profile obtained in frontal gel chromatography of a mixture of ovalbumin (0.92 mM) and glucose (35.0 mM) on a column of Sephadex G-25 equilibrated with acetate buffer (pH 5.0, I 0.1), which was also the solvent for the mixture. The initial plateau of glucose, with applied concentration C_M^α, is maintained until a volume equal to the elution volume of ovalbumin (V_A) has been eluted, at which stage the concentration of glucose attains a new value, C_M^β. The concentration of glucose in this protein-free phase remains at C_M^β until the glucose is eluted at V_M. Combination of these two glucose concentrations (C_M^α = 35.0 mM, C_M^β = 36.9 mM) with the applied ovalbumin concentration (C_A^α = 0.92 mM) in Eq. (43) gives a value of 57 L/mol for $\mathbf{B}_{AM}$. A

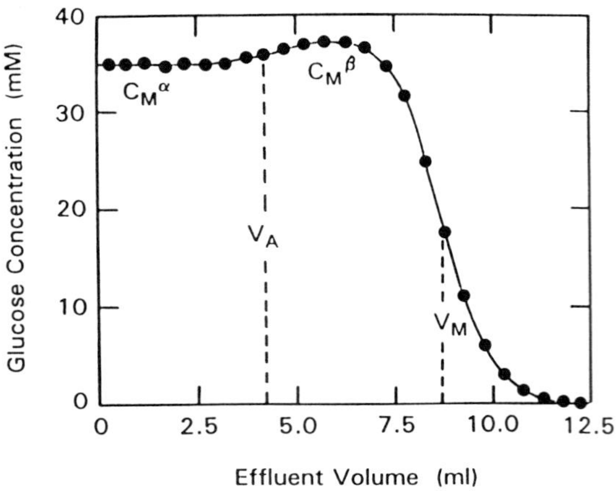

FIGURE 8 Evaluation of $\mathbf{B}_{AM}$ from Eq. (43) and the trailing elution profile obtained in frontal gel chromatography of a mixture of ovalbumin (0.92 m*M*) and glucose (35.0 m*M*) on a column (0.9 × 11.4 cm) of Sephadex G-25.

similar virial coefficient (63 L/mol) emanated from a second experiment with slightly lower ovalbumin concentration (0.84 m*M*).

2. Effective Thermodynamic Radii of Proteins

Statistical-mechanical consideration of the values of $\mathbf{B}_{AM}$ as covolumes for protein and small neutral solute allows an effective thermodynamic radius of the protein (R_A) to be estimated from Eq. (22) on the basis of the magnitudes of R_M reported in Table 3. Such an interpretation of $\mathbf{B}_{AM}$ values for the interactions of sucrose and glucose with several proteins is summarized in Table 5, which also includes the Stokes radius (R_h) and the radius of unsolvated protein that is calculated from the partial specific volume [$R_u = (3\bar{v}_A/4\pi M_A\mathbf{N})^{1/3}$]. The values of R_A obtained from Eq. (22) are generally smaller than the corresponding Stokes radii; indeed, they are closer to the calculated radii of the unsolvated proteins.

C. Relationship to the Stokes Radius

The protein radius that is operative in covolume effects is unquestionably a thermodynamic parameter that must be distinguished from its hydrodynamic counterpart, the Stokes radius. In principle, it may be argued that the magnitudes of these two parameters must differ [60]; but in practice, results obtained with the few proteins for which comparisons are available (Table 4) certainly suggest that the Stokes radius is a fairly good approximation of R_A for prediction of the covolume due to protein self-interaction. The fact that hydrated globular proteins therefore exhibit thermodynamic behavior consistent with their representation as equivalent impenetrable spheres [59] signifies that substitution of the Stokes radius for its

TABLE 5 Determination of Effective Thermodynamic Radii of Proteins from Covolumes for Interaction with Carbohydrates[a]

Protein	U_{AM} (L/mol)		R_A (nm)[b]		R_h[c] (nm)	R_u[d] (nm)
	Sucrose	Glucose	Sucrose	Glucose		
Ribonuclease	15.5	15.7	1.52	1.47	1.84	1.56
Chymotrypsinogen	26.8	27.0	1.89	1.96	2.18	1.95
Ovalbumin	—	60.0[e]	—	2.64	2.92	2.38
Lysozyme	—	13.0	—	1.49	1.92	1.58
Serum albumin	64.2	60.1	2.63	2.64	3.51	2.68

[a]Taken from Table 1.
[b]Radius calculated from Eq. (22) with R_M taken as 0.31 and 0.24 nm for sucrose and glucose, respectively (Table 3).
[c]Stokes radius taken from Table 4 or deduced from V_h (Table 2).
[d]Radius of unhydrated protein calculated on the basis that $M_A\bar{v}_A = 4\pi\mathbf{N}R_u^3/3$.
[e]Value obtained gel chromatographically (Fig. 7).

thermodynamic counterpart should also be an acceptable approximation in the prediction of the covolume contribution to thermodynamic nonideality reflecting interaction between dissimilar proteins. A glance at Table 5, however, shows that this inference does not extend to situations in which one of the species contributing to the covolume effect is a small solute.

The effective thermodynamic radii of proteins calculated on the basis of nonideality reflecting their interaction with small solutes are clearly smaller than their Stokes radii (Table 5). Although the unhydrated radius R_u has been regarded as the effective thermodynamic radius in studies of preferential solvation [17,19], that interpretation seems precluded on the grounds that it implies equal access by small solute and solvent to the hydrated protein domain, a situation that clearly contradicts the concept of preferential solvation as the source of the thermodynamic nonideality. Furthermore, demonstrations that isomeric enzyme equilibria may be displaced in favor of the more compact solvated state by the inclusion of small inert solutes [21,58,67,68] indicate that the radius of the solvated protein must be the governing parameter in excluded volume considerations. In seeking a plausible explanation of this finding that the effective hydrated radius of a protein in its interaction with small solutes is smaller than that pertaining to self-interaction, we suggest that protein asymmetry provides at least a partial answer. Whereas asymmetry of protein molecules must lead to hydrodynamic and effective thermodynamic radii that overestimate the volume of the hydrated protein, departure from spherical shape should be of little consequence in the delineation of the protein domain by means of its covolume interactions with a small solute. Excluded volumes measured for protein–small solute interactions are therefore likely to provide more accurate indications of actual hydrated protein volumes than the overestimates inferred from $2\mathbf{B}_{AA}$ or, indeed, Stokes radii.

The discrepancy between R_A and R_h for lysozyme cannot be explained solely on this basis, because the calculated value of R_A is also smaller than R_u (Table 5). Such underestimation of R_A may reflect invalid application of the present method for determining the effective thermodynamic radius of this protein due to reactivity of the space-filling solute, glucose, which should not be regarded as inert in the light of its demonstrated weak interaction with lysozyme [69]. By analogy with the effect of dimerization on the second virial coefficient for self-interaction [44], binding of the small solute to a single protein site leads to a second virial coefficient with magnitude $\mathbf{B}_{AM} - K_{AM}$, where K_{AM} is the association constant for the protein–small solute interaction.

VI. EFFECTS OF SMALL SOLUTES ON PROTEIN ISOMERIZATION

We have established the formal equivalence of the interpretations of thermodynamic nonideality based on preferential solvation and excluded volume concepts, and we have also indicated ways of evaluating parameters required for application of the latter. It therefore remains to substantiate the inference that statistical-mechanical interpretation is the more rewarding, for example, in predicting and employing to advantage the effects of thermodynamic nonideality as a probe for detecting and quantifying isomerization equilibria.

For a protein solution comprising an equilibrium mixture of isomeric states ($A \rightleftarrows B$), the thermodynamic isomerization constant X is given by

$$X = \frac{a_B}{a_A} \tag{44}$$

where, from Sec. II.B, a_B and a_A are thermodynamic activities of the respective isomers defined on the molal basis, because the chemical potentials of A and B are being defined under conditions of constant pressure. Although molality is therefore the more direct concentration scale to employ, we shall adopt the usual practice of using molar concentrations. Consider now the situation in which a very dilute solution of isomerizing protein is supplemented with a moderately high concentration of inert solute. Because $C_M \gg C_A + C_B$, the relationship for the composition dependence of the thermodynamic activity of each isomer [Eq. (15d)] simplifies to

$$\ln a_A = \ln C_A - \ln \rho_s + (U_{AM} - M_M\bar{v}_M - M_A\,\bar{v}_A)C_M + \cdots \tag{45a}$$

$$\ln a_B = \ln C_B - \ln \rho_s + (U_{BM} - M_M\,\bar{v}_M - M_B\,\bar{v}_B)C_M + \cdots \tag{45b}$$

Substitution of these expressions in Eq. (44) then gives

$$X \cong \frac{C_B}{C_A}\exp[(U_{BM} - U_{AM})C_M] \tag{46a}$$

On identifying the ratio C_B/C_A as the isomerization constant amenable to experimental measurement, X', Eq. (46a) may be rearranged as

$$X' = X\exp[(U_{AM} - U_{BM})C_M] \tag{46b}$$

Since Eq. (46b) turns out to be identical with an earlier expression [55] derived prior to the realization that the isomerization constant should be regarded as the ratio of two molal thermodynamic activities, the earlier inferences retain validity.

In the context of reversible protein unfolding, the unfolded state (B) is the larger isomer ($R_B > R_A$), and accordingly the covolume difference term in Eq. (46b) is negative. Consequently, the effect of an inert space-filling solute on an isomeric equilibrium between native and unfolded states is to decrease the measured isomerization constant X', or, in other words, to increase the concentration of compact isomer at the expense of the unfolded form.

A. pH-Induced Unfolding of Proteins

Description of the reversible unfolding/refolding of small, single-domain proteins as a two-state equilibrium between native and unfolded states is a widely established concept [70]. Insufficient attention, however, has been given to the problem of establishing that the reversible unfolding detected by, for example, UV difference spectroscopy is, indeed, an equilibrium phenomenon. The need for caution in this regard is evident from Fig. 9, where the open symbols summarize the pH dependence of the absorbance change at 287 nm for bovine serum albumin (Fig. 9a) and ribonuclease (Fig. 9b). Despite the similarity in form of these two titration curves, both of which are reversible, only that for ribonuclease reflects an equilibrium between native and acid-expanded states [68]: serum albumin exhibits reversible unfolding in which all molecules expand progressively as the pH is lowered from 5 to 2 [55,71]. For studies of protein folding and unfolding, it is therefore important to have available methods that provide comment on the applicability of the equilibrium model to the particular system being investigated. From Eq. (46) it is evident that thermodynamic nonideality arising from excluded volume effects has the potential to provide such a probe.

From the solid symbols in Fig. 9a, which refer to difference spectroscopy experiments on serum albumin in the presence of sucrose and glycerol, it is apparent that these solutes are completely ineffective in displacing the titration curve. In the event that the curve reflected pH dependence of the isomerization constant governing an equilibrium between acid-expanded (pH 2) and native (pH 5) states of albumin, calculations based on Eq. (46) show that the concentra-

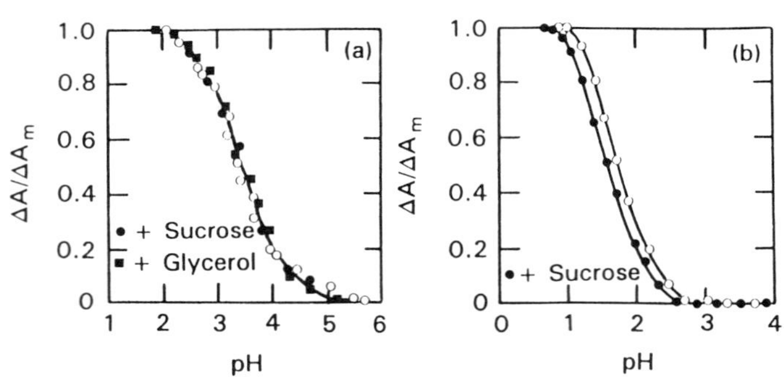

FIGURE 9 Use of thermodynamic nonideality to probe the existence of isomerization equilibria in the pH-induced unfolding of proteins as monitored by UV difference spectroscopy. (a) The effect of sucrose and glycerol on the fractional change in absorbance at 287 nm during acid expansion of bovine serum albumin. (b) The corresponding results for the acid unfolding of ribonuclease A. (Adapted from [68] with permission.)

tions of sucrose and glycerol should have sufficed to maintain the albumin in its native configuration even at pH 2. The earlier conclusions that the acid expansion of albumin should not be regarded in such equilibrium terms is therefore substantiated. On the other hand, the acid expansion of ribonuclease is describable in terms of the two-state equilibrium concept, since the inclusion of sucrose does lead to displacement of the spectral titration curve in a manner consistent with stabilization of the more compact native isomer (Fig. 9b). Combination of the value of X' with that of X inferred from the spectral change at the same pH in the absence of sucrose yields an estimate of 0.56 (± 0.15) for the ratio X'/X; and hence, via Eq. (46b), a value of -1.16 L/mol for the difference in covolume, $U_{AM} - U_{BM}$. The resultant conclusion that the acid-expanded state has a 7.5% larger volume than the native enzyme, or a 2.5% ± 1.5% larger radius, finds parallel in viscosimetric studies of ribonuclease that yielded a 1.6% difference in hydrodynamic radius [72]; indeed, the agreement is even closer if the radii of enzyme and sucrose are considered to be additive in the calculations of covolumes for the native and acid-expanded species [68].

By showing that the molecular-crowding effect of sucrose was able to displace the titration curve in the direction implicating stabilization of the more compact, native state of the enzyme, this use of thermodynamic nonideality as a probe has certainly established the equilibrium nature of the unfolding undergone by ribonuclease between pH 4 and pH 1. Furthermore, on the grounds that quantitative consideration of the displacement of the curve at any intermediate pH yields a calculated volume for the expanded state that matches the value measured experimentally, the study also justifies consideration of the unfolding as a two-state equilibrium between native and acid-expanded states of the enzyme. In that regard, an obvious advantage of the statistical-mechanical treatment of thermodynamic nonideality in terms of excluded volume is its generation of magnitudes for activity coefficients, which in turn allow the application of expressions such as Eq. (46) for predicting and elucidating the effects of inert solutes on protein interactions.

B. Ligand-Induced and Preexisting Isomerizations

The development, over two decades ago, of the Monod [73] and Koshland [74] theories of allostery has often triggered vigorous, though pointless, debate about the relative merits of these alternative explanations of sigmoidal binding or kinetic responses observed with various proteins and enzymes. Such arguments inevitably foundered through lack of methods for establishing whether the protein/enzyme comprised a single state or an equilibrium mixture of isomers in the absence of ligand (substrate). The advent of interest in effects of thermodynamic nonideality has finally led to means of distinguishing between isomerizations that are preexisting and ligand-induced [58,67,75].

Displacement of an isomerization equilibrium, be it preexisting or ligand-induced, by the space-filling effects of a small solute again provides the means of detecting the phenomenon; but in this instance, there is possibly an additional interaction to be taken into account, viz., the preferential or exclusive binding of ligand L to one isomeric state of the protein/enzyme. In considerations of ligand interactions, the stance has been taken that no volume change accompanies the actual binding phenomenon [54,56,75], for which the dependence of the apparent equilibrium constant on the concentration of space-filling solute is then governed solely by the effect of M on the thermodynamic activity of ligand, a_L. Under conditions where thermodynamic nonideality is dominated by the inert solute terms [those applying to Eq. (46)], the expression for the experimentally determinable isomerization constant, $\bar{X}'$, becomes

$$\bar{X}' = \bar{X} \exp[(U_{AM} - U_{BM})C_M] \tag{47a}$$

where $\bar{X}$ is the constitutive isomerization constant [76] with magnitude defined as

$$\bar{X} = \frac{C_B(1 + K_B a_L)^q}{C_A(1 + K_A a_L)^p} \tag{47b}$$

$$= X\frac{(1 + K_B a_L)^q}{(1 + K_A a_L)^p} \tag{47c}$$

K_B is the intrinsic binding constant [77] for the interaction of ligand with q sites on isomeric state B, and K_A the corresponding equilibrium constant for interaction with p sites on A. In the only experimental study of preferential ligand binding where thermodynamic nonideality has been taken into account [67], z_L has been substituted incorrectly for a_L in the expression for exclusive binding to a single site on A; and hence the deduced equilibrium constant (K_A) should be multiplied by the ratio a_L/z_L to obtain its thermodynamically rigorous molal counterpart. Regardless of that deficiency in interpretation, the most important point to emerge from those deliberations is the correct conclusion that the effects of thermodynamic nonideality and preferential ligand binding can act either in unison or in opposition, depending on which isomeric state of the protein/enzyme has the higher affinity for ligand.

1. Pyruvate Kinase: A Monod System

Sedimentation velocity studies in the presence and absence of sucrose have been used to establish preexistence of the isomerization equilibrium responsible for the allosteric behavior of rabbit muscle pyruvate kinase [58]. Whereas inclusion of the allosteric inhibitor, phenylalanine, with enzyme gave rise to a decrease of 0.3 S in the sedimentation coefficient of pyruvate kinase, the corresponding effect of substrate, phosphoenolpyruvate, was to increase the sedimentation coefficient by 0.03 S. Consideration of these findings to signify an isomerization constant (X) of 0.1

for the equilibrium between an active, compact state (A) and an expanded, inactive form (B) of the enzyme was substantiated by the finding that inclusion of sucrose (0.1 M) also brought about the change in sedimentation coefficient effected by phosphoenolpyruvate. Such demonstration that rabbit muscle pyruvate kinase undergoes isomerization in the absence of substrate clearly justifies adoption of the Monod model of allostery [73] for interpreting kinetic data obtained with this enzyme [78].

2. Detection of Isomerizations by Kinetic Studies

From expressions derived for the space-filling effects of small inert solutes on kinetic parameters for the separate situations in which univalent enzymes undergo isomerizations that are substrate-induced and preexisting, it has been concluded that experimental observation of an enhanced or diminished maximal velocity in the presence of small inert solute can only reflect the existence of a substrate-induced conformational change [75]. In keeping with this proposition, the inhibition of the invertase-catalyzed hydrolysis of sucrose by high concentrations of sucrose and ethanol has been interpreted as signifying isomerization of the enzyme–substrate complex to an expanded activated state [57]. Similarly, the inhibitory effects of sucrose and glycerol on the rate of thrombin inactivation by antithrombin have been attributed to the existence of a slight increase in volume and/or asymmetry associated with formation of a thrombin–antithrombin complex that subsequently undergoes covalent modification in an irreversible inactivation step [79]. On the other hand, the inclusion of sucrose led to a marked decrease in Michaelis constant but had no effect on the maximal velocity of ester hydrolysis of α-chymotrypsin at pH 3.5 and 50°C, conditions under which the enzyme comprises an equilibrium mixture of compact (native) and expanded states [67]. In all three investigations, the changes in kinetic parameters brought about by inclusion of the small solutes were used entirely plausibly to evaluate an effective thermodynamic radius of the expanded isomeric state of the enzyme or enzyme–substrate complex on the basis of excluded volume considerations.

C. Thermal Unfolding of Proteins

Having described examples that demonstrate the strength of the approach whereby thermodynamic nonideality is used as a probe of isomerizations, we feel honor bound to provide also the example that illustrates its limitations. For the thermal unfolding of ribonuclease, the difference between the Stokes radii of the native and thermally unfolded states is 0.3–0.5 nm [80,81], much larger than the 0.03 nm change associated with acid unfolding. Although there are several reports that the midpoint of the thermal transition, the melting temperature (T_m), is elevated by the inclusion of a small solute such as glycerol [14,82–84], detailed analysis of the most extensive set of results for glycerol [14] in accordance with Eq. (46) yields, however, a value of -0.61 L/mol for $U_{AM} - U_{BM}$ [68], which signifies an apparent difference of only 1.3% between the radii of the native and thermally unfolded states of the enzyme. Possible reasons for the disparity between this estimate and that of 16–27% inferred from hydrodynamic studies of the native and unfolded forms of ribonuclease [80,81] are clearly required.

One plausible explanation of the disparity is that the thermally unfolded form of ribonuclease does not meet the theoretical requirement inherent in Eq. (46) that it can be modeled satisfactorily as an impenetrable sphere. In that regard, the considerably expanded nature of the thermally denatured enzyme may well allow access of small solute to at least part of the solvated protein domain, thereby rendering the region of solute exclusion smaller than the hydrated molar volume of B. Alternatively, the present result could also reflect preferential binding of glycerol to the larger (unfolded) state of the enzyme. The dependence of $\ln(X'/X)$ upon C_M would then reflect a balance between displacement of the isomeric equilibrium toward native enzyme (A) as the result of excluded volume effects, and its displacement toward the expanded isomer (B) as the result of preferential binding of glycerol to the unfolded state of the enzyme [$K_B > K_A$ in Eq. (47)]. Such interpretation of the effects of glycerol in terms of dual actions as a ligand as well as a space-filling solute finds support in observations that comparable concentrations of a range of small solutes give rise to different shifts in T_m for ribonuclease [82–84]. Indeed, the isomerization equilibrium is displaced in favor of the denatured enzymic state by poly(ethylene glycol) [20].

This analysis of the effect of glycerol on the thermal unfolding of ribonuclease [68] has clearly not met with the degree of success that was enjoyed by the other applications of thermodynamic nonideality as a probe of protein isomerization. The inability to obtain an acceptable quantitative description of the results solely on the statistical-mechanical basis of excluded volume has not detracted, however, from the demonstration that glycerol is exhibiting a net excluded volume effect; and hence from the conclusion that the analysis provides justification for consideration of the thermal unfolding of ribonuclease in terms of an equilibrium transition between discrete states of the enzyme.

VII. CONCLUDING REMARKS

The major contribution of this investigation to research into protein solvation is undoubtedly the demonstration (Sec. II) of absolute equivalence between theoretical treatments of thermodynamic nonideality in terms of preferential solvation and on the statistical-mechanical basis of excluded volume. Some progress toward establishing such equivalence of treatments had been made during our initial foray into the analysis of thermodynamic nonideality of protein solutions engendered by the presence of small solutes [21], but attainment of that goal has had to await our appreciation of several thermodynamic subtleties inherent in the correct definition of solute chemical potential under different circumstances [22,44]. The present chapter not only amends our misdemeanors in the earlier statistical-mechanical treatment of thermodynamic nonideality [21], but also eliminates inconsistencies from the theoretical analysis of preferential solvation, which also displayed a corresponding lack of mathematical rigor. In that regard, it is of interest that similar corrections have had to be made to the derivation of the basic sedimentation equilibrium equation [44]. Since there is clearly widespread confusion and uncertainty about the thermodynamic requirements associated with selection of standard states and the appropriate concentration scales for correct definition of solute chemical potential, Sec. II.B (and its counterpart in the review of thermodynamic nonideality in sedimentation equilibrium [44]) may well provide some clarification. On that score we claim no originality, because our treatment of the subject merely reflects substitution of the simplest case (an incompressible solution comprising a single solute) into the definitive theoretical expressions that Hill formulated more than 30 years ago [30,85,86]. Inasmuch as that exercise has had a dramatic effect on our understanding of basic thermodynamics, there is a very distinct possibility that the content of Sec. II.A may also be a revelation to many others within the scientific fraternity.

To summarize the main theme of the chapter, we note that the excluded volume and preferential solvation treatments both exhibit the same degree of thermodynamic rigor in the sense that the basic experimental quantity determined is a measure of the second virial coefficient describing nonideality due to the mutual presence of protein and small solute. Indeed, the parameter is identified as such in the statistical-mechanical treatment (Sec. III). Thereafter, both treatments lose thermodynamic rigor as they seek interpretation of the second virial coefficient, according to different models, in an attempt to take better advantage of the thermodynamic information contained within the magnitude of the second virial coefficient. Our feeling that the statistical-mechanical interpretation is the more rewarding in that regard is obvious from the remainder of the chapter, which endeavors to justify that viewpoint by elaborating methods for experimental determination of the parameters required for its application (Secs. IV and V), and by outlining progress made thus far in its application (Sec. VI). We trust that any unjust effect of that bias is offset by the content of Chapter 11, which describes in greater detail the analysis of preferential protein solvation.

ACKNOWLEDGMENTS

It is a pleasure to acknowledge the hospitality and thought-provoking debate provided by Dr. S. N. Timasheff in his laboratory at Brandeis University during a five-week visit by D. J. W. as part of a study-leave program from the University of Queensland. Financial support from the Australian Research Council is also gratefully acknowledged.

REFERENCES

1. Stewart, G. R., and Lee, J. A., *Planta 120*: 279–289 (1974).
2. Munns, R., Brady, C. J., and Barlow, E. W. R., *Aust. J. Plant Physiol 6*: 379–389 (1979).
3. Aspinall, D., and Paleg, L., in *Physiology and Biochemistry of Drought Resistance in Plants,* L. G. Paleg and D. Aspinall (Eds.). Academic Press, Sydney, pp. 205–241 (1981).
4. Wynn Jones, R. G., and Storey, R., in *Physiology and Biochemistry of Drought Resistance in Plants,* L. G. Paleg and D. Aspinall (Eds.). Academic Press, Sydney, pp. 172–204 (1981).
5. Setter, T. L., and Greenway, H., *Aust. J. Plant Physiol. 6*: 47–60 (1979).
6. Yancey, P. H., Clark, M. E., Hand, S. C., Bowlus, R. D., and Somero, G. N., *Science 217*: 1214 –1222 (1982).
7. Pollard, A., and Wynn Jones, R. G., *Planta 144*: 291–298 (1979).
8. Paleg, L. G., Douglas, T. J., van Daal, A., and Keech, D. B., *Aust. J. Plant Physiol. 8*: 107–114 (1981).
9. Arakawa, T., and Timasheff, S. N., *Arch. Biochem. Biophys. 224*: 169–177 (1983).
10. Paleg, L. G., Stewart, G. R., and Bradbeer, J. W., *Plant Physiol. 75*: 974–978 (1984).
11. Lee, J. C., and Lee, L. L.-Y., *J. Biol. Chem. 256*: 625–631 (1981).
12. Lee, J. C., and Timasheff, S. N., *J. Biol. Chem. 256*: 7193–7201 (1981).
13. Gekko, K., and Timasheff, S. N., *Biochemistry 20*: 4667–4676 (1981).
14. Gekko, K., and Timasheff, S. N., *Biochemistry 20*: 4677–4686 (1981).
15. Gekko, K., and Morikawa, T., *J. Biochem. (Tokyo) 90*: 39–50 (1981).
16. Gekko, K., and Morikawa, T., *J. Biochem. (Tokyo) 90*: 51–60 (1981).
17. Arakawa, T., and Timasheff, S, N., *Biochemistry 21*: 6536–6544 (1982).
18. Arakawa, T., and Timasheff, S. N., *Biochemistry 21*: 6545–6552 (1982).
19. Arakawa, T., and Timasheff, S. N., *Biochemistry 24*: 6756–6762 (1985).
20. Lee, L. L.-Y., and Lee, J. C., *Biochemistry 26*: 7813–7819 (1987).
21. Winzor, D. J., and Wills, P. R., *Biophys. Chem. 25*: 243–251 (1986).
22. Wills, P. R., Comper, W. D., and Winzor, D. J., *Arch. Biochem, Biophys. 300*: 206–212 (1993).
23. Casassa, E. F., and Eisenberg, H., *Adv. Protein Chem. 19*: 287–395 (1964).
24. Cohen, G., and Eisenberg, H., *Biopolymers 6*: 1077–1100 (1968).
25. Eisenberg, H., and Reisler, E., *Biochemistry 8*: 4572–4578 (1969).
26. Timasheff, S. N., and Kronman, M. J., *Arch. Biochem. Biophys. 83*: 60–75 (1959).
27. Inoue, H., and Timasheff, S. N., *Biopolymers 11*: 737–743 (1972).
28. Tanford, C., *Physical Chemistry of Macromolecules,* Wiley, New York (1961).
29. Tombs, M. P., and Peacocke, A. R., *The Osmotic Pressure of Biological Macromolecules,* Clarendon Press, Oxford (1974).
30. Hill, T. L., *J. Chem. Phys. 30*: 93–97 (1959).
31. McMillan, W. G., and Mayer, J. E., *J. Chem. Phys. 13*: 276–305 (1945).
32. Eisenberg, H., *Eur. J. Biochem. 187*: 7–22 (1990).
33. Arakawa, T., and Timasheff, S. N., *Biophys. J. 47*: 411–414 (1985).
34. Creeth, J. M., *J. Phys. Chem. 62*: 66–74 (1958).

35. Sophianopoulos, A. J., Rhodes, C. K., Holcomb, D. N., and Van Holde, K. E., *J. Biol. Chem. 237*: 1107–1112 (1962).
36. Neet, K. E., and Brydon, S. E., *Arch. Biochem. Biophys. 136*: 223–227 (1970).
37. Baldwin, R. L., *Biochem. J. 65*: 503–512 (1957).
38. Scatchard, G., Hamer, W. J., and Wood, S. E., *J. Am. Chem. Soc. 60*: 3061–3070 (1938).
39. Smith, E. R. B., and Smith, P. K., *J. Biol. Chem. 117*: 209–216 (1937).
40. Smith, P. K., and Smith, E. R. B., *J. Biol. Chem. 132*: 57–64 (1940).
41. Robinson, R. A., and Stokes, R. H., *J. Phys. Chem. 65*: 1954–1958 (1961).
42. Stokes, R. H., and Robinson, R. A., *J. Phys. Chem. 70*: 2126–2131 (1966).
43. Schellman, J. A., *Compt. rend. trav. lab. Carlsberg, Sér. Chim. 29*: 223–229 (1955).
44. Wills, P. R., and Winzor, D. J., in *Ultracentrigufation in Biochemistry and Polymer Science,* S. E. Harding, A. J. Rowe and J. C. Horton (Eds.). Roy. Soc. Chem., Cambridge, pp. 311–330 (1992).
45. Debye, P., and Hückel, E., *Phys. Z. 24*: 185–206 (1923).
46. *CRC Handbook of Chemistry and Physics* (69th ed.), R. C. Weast (Ed.). CRC Press, Boca Raton, Florida (1988).
47. Nichol, L. W., Siezen, R. J., and Winzor, D. J., *Biophys. Chem. 9*: 47–55 (1978).
48. Shearwin, K. E., and Winzor, D. J., *Biophys. Chem. 31*: 287–294 (1988).
49. Nichol, L. W., Siezen, R. J., and Winzor, D. J., *Biophys. Chem. 10*: 17–26 (1979).
50. Winzor, D. J., and Sheraga, H. A., *Biochemistry 3*: 1263–1267 (1963).
51. Gelotte, B., *J. Chromatogr. 3*: 330–342 (1960).
52. Ackers, G. K., *J. Biol. Chem. 243*: 2056–2064 (1968).
53. Wills, P. R., Nichol, L. W., and Siezen, R. J., *Biophys. Chem. 11*: 71–82 (1980).
54. Nichol, L. W., Sculley, M. J., Ward, L. D., and Winzor, D. J., *Arch. Biochem. Biophys. 222*: 574–578 (1983).
55. Winzor, D. J., Ford, C. L., and Nichol, L. W., *Arch. Biochem. Biophys. 234*: 15–23 (1984).
56. Nichol, L. W., Owen, E. A., and Winzor, D. J., *Arch. Biochem. Biophys. 239*: 147–154 (1985).
57. Shearwin, K. E., and Winzor, D. J., *Arch. Biochem. Biophys. 260*: 532–598 (1988).
58. Harris, S. J., and Winzor, D. J., *Arch. Biochem. Biophys. 265*: 458–465 (1988).
59. Ogston, A. G., and Winzor, D. J., *J. Phys. Chem. 79*: 2496–2500 (1975).
60. Minton, A. P., *Biophys. Chem. 12*: 271–277 (1980).
61. Williams, J. W., Van Holde, K. E., Baldwin, R. L., and Fujita, H., *Chem. Rev. 58*: 715–806 (1958).
62. Milthorpe, B. K., Jeffrey, P. D., and Nichol. L. W., *Biophys. Chem. 3*: 169–176 (1975).
63. Jeffrey, P. D., Nichol, L. W., Turner, D. R., and Winzor, D. J., *J. Phys. Chem. 81*: 776–781 (1977).
64. Siezen, R. J., and Owen, E. A., *Biophys. Chem. 18*: 181–194 (1983).
65. Shearwin, K. E., and Winzor, D. J., *Eur. J. Biochem. 190*: 523–529 (1990).
66. Van Damme, M.-P. I., Murphy, W. H., Comper, W. D., Preston, B. N., and Winzor, D. J., *Biophys. Chem. 33*: 115–125 (1989).
67. Bergman, D. A., Shearwin, K. E., and Winzor, D. J., *Arch. Biochem. Biophys. 274*: 55–63 (1989).

68. Shearwin, K. E., and Winzor, D. J., *Arch. Biochem. Biophys. 282*: 297–301 (1990).
69. Nichol, L. W., Ogston, A. G., Winzor, D. J., and Sawyer, W. H., *Biochem. J. 143*: 435–443 (1974).
70. Privalov, P. L., *Adv. Protein Chem. 33*: 167–241 (1979).
71. Tanford, C., Buzzell, J. G., Rands, D. G., and Swanson, S. A., *J. Am. Chem. Soc. 77*: 6421–6428 (1955).
72. Buzzell, J. G., and Tanford, C., *J. Phys. Chem. 60*: 1204–1207 (1956).
73. Monod, J., Wyman, J., and Changeux, J.-P., *J. Mol. Biol. 12*: 88–118 (1965).
74. Koshland, D. E., Jr., Némethy, G., and Filmer, D., *Biochemistry 5*: 365–385 (1966).
75. Bergman, D. A., and Winzor, D. J., *J. Theor. Biol. 137*: 171–189 (1989).
76. Baghurst, P. A., and Nichol, L. W., *Biochim. Biophys. Acta 412*: 168–180 (1975).
77. Klotz, I. M., *Arch. Biochem. 9*: 109–117 (1946).
78. Oberfelder, R. W., Barisas, B. G., and Lee, J. C., *Biochemistry 23*: 3822–3826 (1984).
79. Hogg, P. J., Jackson, C. M., and Winzor, D. J., *Biochim. Biophys. Acta 1073*: 609–613 (1991).
80. Holcomb, D. N., and Van Holde, K. E., *J. Phys. Chem. 66*: 1999–2006 (1962).
81. Corbett, R. J. T., and Roche, R. S., *Biochemistry 23*: 1888–1894 (1984).
82. Gerlsma, S. Y., *J. Biol. Chem. 243*: 957–961 (1968).
83. Bello, J., *Biochemistry 8*: 4535–4541 (1969).
84. Gerlsma, S. Y., and Stuur, E. R., *Int. J. Peptide Protein Res. 4*: 377–383 (1972).
85. Hill, T. L., *Introduction to Statistical Thermodynamics,* Addison-Wesley, Reading, Massachusetts (1960).
86. Hill. T. L., *Thermodynamics for Chemists and Biologists,* Addison-Wesley, Reading, Massachusetts (1968).

13

Molecular Basis for Protein Separations

REX E. LOVRIEN, MARK J. CONROY, AND TIMOTHY I. RICHARDSON
University of Minnesota
St. Paul and Minneapolis, Minnesota

I. INTRODUCTION

Solvent–protein interactions govern much of protein separation design and practice. The "solvent" includes organic cosolvents, ligands, and ions that bind to proteins, in addition to the workhorse solvent, water, which sometimes is simply also a cosolvent. Development of newer and older biotechnologies and industrial processes depends on provision of separated enzymes, antibodies, and other classes of proteins such as lectins. Diagnostics devices predominantly depend on purified proteins. Commodity industries' use of enzymes such as proteases, cellulases, and amylases are approaching many thousand tons of enzymes each year [1] perhaps \$1 billion by 1995 [2]. The viability of some of these industries, even of the research and development it takes to determine feasibility, sharply depends on costs.

Costs of refined proteins, if they have to be chromatographed or crystallized, mostly occur in separation costs, not in production of crudes. For example bacterial β-galactosidase from *Escherichia coli* retails for \$20–30 for a single milligram of chromatographed enzyme. These bacteria are very easily grown and are easily induced to produce large quantities of internal β-galactosidase. Getting from the crude (broken cell extracts) to well-purified enzymes usually multiplies costs relative to crudes, however, at least 10 times, often 100 or even 1000 times, because of the labor, time, and required facilities.

Industrial separation expenses comprise four main areas: (a) losses; (b) scalability; (c) time needed to get through current separation steps; (d) dilution—many proteins are excreted or extracted in 0.01–0.2% concentrations as crudes. Such crudes are freighted with salts, metabolites, lipids, other proteins, etc. Separation problems mostly are hybrids of these four underlying problems. Frequently, unwanted enzymes need to be separated out quickly (inexpensively if possible) because they degrade other, sought-after, proteins or polymers. Proteases, especially lysosomal proteases, vigorously nick and cleave other proteins. Ribonucleases (present practically everywhere) split nucleic acids, and DNAases do the same to DNA. Catalase enzymes coexcreted with glucose oxidase upset the use of glucose oxidase in several applications, because catalase efficiently destroys hydrogen peroxide. Similarly, NAD oxidase enzymes interfere with the use of other NAD-utilizing enzymes. Hence, several enzymes, such as glucose oxidase, ribonuclease, proteases, and NAD oxidase, need to be efficiently and quickly separated, not always because they are wanted; rather, they hinder use of other enzymes and sometimes cause outright destruction.

Losses in enzyme separation can be large and expensive. For example, there is much demand for Factor VIII coagulation protein now. Deficiency of Factor VIII causes about 80% of all human hemophilia, and hemophilia is increasing. Nearly 75% of all Factor VIII produced as crude human plasma, however, is lost in separation [3]. Some is lost because plasma has to be sanitized (to get rid of viruses). Until recently, only about 25% of Factor VIII emerged as FDA-approved product from the crude. Hence, much of the need for scale up—and for plasma—occurs from large losses in separations.

Energetic searches are under way in industry and in university laboratories to make protein separation less expensive and faster. Chromatography, both analytical (small-scale) and large-scale chromatography of many kinds, is predominant now in downstream separations. Chromatography, however, despite its elegance and its resolving power, usually is not very scalable except at large expense. It dilutes rather than concentrates, and it is not particularly rapid. The lifetimes of some plasma proteins are comparable with or shorter than the throughput times of chromatographic techniques used to isolate them [4]. How can one chromatograph 5000 liters of anything?

Bringing proteins directly out of solution as precipitates if suitable conditions can be found has increasing appeal, because precipitation methods may be more easily scaled up than chromatography usually can. *Scale up* may sound like an issue for production engineers to contend with, and in fact they do so. Indeed, there has been little input from biophysical chemists, who usually want to keep proteins in solution—"solubilize" them for study. Engineers tend to deal with proteins as if proteins are rigid particles, perhaps spheres characterized by a "size," and a "charge" depending on pH [5]. To be sure, a few proteins behave that way and are remarkably tough. Ribonuclease and lysozyme are examples; they are able to withstand high temperatures, pH extremes, and several kinds of abuse and recover their activity. The trouble comes when proteins are not rugged. They engage in conformation changes depending on several factors, one of which is the nature of the solvent. When proteins unfold, they are susceptible to attack by proteases, oxidizing and/or reducing agents, and several other factors; a few of these can be controlled if it is realized that they need attention. This chapter aims to indicate, mostly with anecdotes but perhaps glimpsing a few principles, how protein conformational motility, solvent–protein interactions, and protein separations design may be brought together. The emphasis is on means for prompting proteins to come out of solution as precipitates, as crystals and cocrystals.

Protein separation design likely needs to take into account that separation and isolation of proteins often involves *reactions*. The course such reactions take involves the conformationally motile nature of proteins, and separation design needs focus on this. Many ligands, including simple salt anions, which rather generally bind to proteins, tighten protein molecules if the proteins were floppy before ligands and salt ions became bound. Tightening protein conformation is a general recipe for protecting proteins. Sometimes tightening, and thereby protection, is fairly easily achieved.

Concerning scale up and industrialization of protein separation, whatever is designed has to be simple. Complicated processes, apparatuses, or chemistries or biochemistries may make for interesting papers. They tend to not be feasible to use, however, especially in scale up. Protein molecular floppiness or the tendency to become floppy, especially in water, is basic to why proteins too often behave not like simple particles in separations, but as reactants. Using organic solvents and cosolvents, and inorganic and organic ligands to tighten proteins simplifies their handling. In their tightened, protected state, proteins may approach the rigid particles envisioned by bioseparations designers. Integration of protein molecule folding $\rightleftharpoons$ unfolding reactions with the protection of proteins has not been a goal of most separation engineering efforts. It is realized, however, that ligands used for chromatograph immobilization of proteins are protective [6]. Ammonium sulfate–protein precipitates almost always are much more stable than the same proteins kept dissolved in water [7].

Protein separation as a practical problem soon may become able to draw on biophysical information concerning protein folding, conformation change and conformation motility. Our approach is the formation of precipitates, coprecipitates, and cocrystals. Protein–solvent interaction is at the same junction: The choice of solvent components such as ligands, soluble inorganic ions, and cosolvents can be made to protect proteins and, so to say, influence water in the right way.

II. PROTEIN REACTIVITY AND CONFORMATION GOVERNANCE IN SEPARATIONS

Plasma albumin (bovine serum albumin, BSA; human serum albumin, HSA) remains the foremost but not an unusual example for one of the main points laid out above: Ligands control protein conformation motility, bearing on how these proteins are protected, separated, and "crystallized" (cocrystallized). There are differences between BSA and HSA with respect to genus, and also between albumins from genetic variants and species. Differences in conformation cause changes in behavior and physical parameters that are small enough to not seriously affect what follows, as far as is presently known. Accordingly, HSA and BSA are both called simply *plasma albumin*(s).

Plasma albumins engage in large conformation changes, seen by hydrodynamic measurements in the acid pH region below pH 4 and at alkaline pHs above 10. Plasma albumin has a more subtle conformation change transition ($N \rightarrow B$ transition) in the pH 7–9 region. The $N \rightarrow B$ transition is not accompanied by much hydrodynamic change. From the ORD (optical rotatory dispersion) measurements of Foster and coworkers [8], however, and from the deuterium exchange rates of some of the slower exchangeable protons from peptide chain amides reported by Benson et al. [9], plasma albumin engages in conformation motility and fluctuations during the $N \rightarrow B$ transition. In the well-expanded region, below pH ~3.5 and above pH ~ 9, there is little nonexchanging core left; populations of peptide hydrogens refusing to exchange after several hours.

The pH-dependent conformational changes of plasma albumin are influenced by ions and ligands upon binding. Organic ions, very polarizable "soft" inorganic ions in millimolar concentrations, and the less polarizable "hard" ions like SO_4^{2-}, Cl^-, and ClO_4^- [10] in 0.1 M to molar concentrations bind to plasma albumin and promote conformational changes. In some pH regions, described in more detail below, these inorganic ions tighten the conformation of plasma albumin to enable separation of the protein. The protein can coprecipitate, and it can cocrystallize. This brings us to a first point of departure from common interpretations of what ions such as chloride and sulfate do, in changing salt concentrations, to proteins in general. On observing salt-dependent behavior, the assumption is usually made that these ions are exerting ionic strength effects in the Debye–Hückel sense. Such assumptions often are valid, but they carry with them the supposition that the electrolyte ions are not bound. Presumably they are mobile and free. Sensitivity to 0.005–0.20M salt is presumably diagnostic for ionic strength control, as in hydrogen ion titration, in viscosity behavior, and in macromolecule electrostatic energy content [11].

Ionic strength control in the Debye–Hückel paradigm—that electrolyte ions are all mobile—has been integrated into a number of hypotheses concerning protein–electrolyte interaction in precipitation and salting out. An interesting example is the Melander–Horvath paper [12] focusing on the solvent cavity surface tension increment, σ. Proteins occupy a cavity in the solvent. If the cavity inner surface tension σ [13] is driven too high by salt and ionic strength increases, proteins are in effect squeezed on and therefore precipitate out. This model, or hypothesis, does not include the likelihood that part of what such salts do is modify protein conformation and perhaps their extent of hydration on binding salt ions directly into the protein. Nevertheless, part of what they do might be exerted through such a σ effect. Matters are far from settled in both regards; binding and conformation control, cavity surface tension increment control, or a mixture of both.

Table 1 shows data on the behavior of plasma albumin, examples both of how it behaves in solution relative to conformation control by organic and inorganic ligands, and how it behaves in its separation and precipitation behavior. Table 2, which deals with plasma albumin and a few other proteins, sketches cocrystallization behavior (often misnamed crystallization) when ligands and ions seem to be

TABLE 1 Plasma Albumin's Conformation Response to Ligands and Ions

Method, tool	Ligand	Response	Reference
Relative viscosity	Various aromatic anions	Decrease relative viscosity from 6 *M* urea unfolded form	15
Intrinsic viscosity	Dodecyl sulfate, submillimolar	Reverse alkaline expansion	16
Optical rotation	Dodecyl sulfate, millimolar	Inhibit urea swelling, reverses swelling	17
Sedimentation, diffusion	Dodecyl sulfate	Restore native sedimentation constant, diffusion constant	17
Proteolysis, three enzymes	Methyl orange, submillimolar	Major protection against proteolysis	18
Allosterism, flow spectrophotometry	Bilirubin, palmitate	Fatty acid, 1–5 mole per mole protein enhances bilirubin association	19
Hydrogen exchange	Dodecyl sulfate	Conformation tightening decreases exchange rates	20
Scanning calorimetry	Palmitate	Increases denaturation temperature	21
Chromogen catalysis, spectrophotometry	Lipids, detergent	Conformational motility decreases, slows conversion rates	22
Tyrosine ionization spectra	Dodecanoate	Reverse tyrosine ionization	23

TABLE 2 Solubility, Coprecipitation, and Cocrystallization of Plasma Albumin

System	Results	Reference
pH-dependent solubility profile in salts dependent on N,F fractions	Behavior assigned to protein microheterogeneity	24
Ammonium sulfate dependence of albumin solubility	High salt precipitates the protein	25
Solubility-dependent on addition, removal of hydrocarbonaceous ligands	Removal of ligands erases apparent microheterogeneity behavior	26
pH 3.0, 0.2–3 *M* KCl	Aggregation to apparent molecular weight values several times larger than monomer	27
C_{10}–C_{18} aliphatic ligands as crystallization aids	Major improvement in crystallization (cocrystallization) by alkanoic ligands	28
Diverse ionic compounds crystallize (cocrystallize) human mercaptalbumin mercury dimer	Ca. 20 each of inorganic cation and inorganic anion complexes; 12 organic ligands in 5–30% organic cosolvents yielded crystals	29

required. The solvent, water, organic cosolvents, ligands, and ions that bind have more than tangential control of cocrystallization. There are major, decisive controls because of what they do in conformation control and folding equilibria.

Some papers referred to in Table 1 were written supposing plasma albumin expands and contracts isotropically, as a sphere. Bloomfield [14] concluded the molecule more likely is a linked series of three domains, each of which might be approximately spherical. They are connected by one polypeptide chain. Plasma albumin's overall shape may be an ellipsoid that expands below pH 3.6. The distinction between a single sphere and a short ellipsoid of subspheres probably does not affect the conclusions drawn from Table 1.

III. THE PLASMA ALBUMIN PROTOTYPE: CONFORMATION BEHAVIOR, REACTIVITY TOWARD LIGANDS, CONSEQUENCES IN COPRECIPITATION, AND COCRYSTALLIZATION

Many authors call plasma albumin a "typical protein." Others say albumin is unique (implying it is not typical) because it has the ability to bind such an unusual variety of compounds. Steinhardt and Reynolds [30] reviewed plasma albumin's binding behavior along with several other typical/atypical proteins in 1969. The synthetic polymer polyvinylpyrrolidone binds nearly as many compounds, however, and as diversely [31] as plasma albumin. So do several of the liver glutathione transferases, called *ligandins* [32], which have broad affinities for toxic compounds. Nevertheless, plasma albumin's response in solution to ions and ligands that propel considerable conformation changes, including shrinkage and tightening, is a close cousin to what happens with many other proteins and enzymes. Regarding bioseparation, crystallization, protection of the protein on binding organic ligands, and tightening the protein macromolecule, plasma albumin in its way is, indeed, typical. Tables 1 and 2 summarize some of the relevant examples. Plasma albumin usually is cited as a carrier protein, not as an enzyme. Some readers may be surprised that albumin can in fact act as a fairly respectable catalyst. It is able, when it is conformationally motile, to catalyze geometry-switching *cis* $\rightleftharpoons$ *trans* isomerization reactions in azo dyes. Geometry-switching chromogen catalysis occurs across the pH scale closely in step with albumin's conformation change properties [22]. Taylor [33] reported albumin accelerated Meisenheimer complex decomposition to rates $\times\ 10^4$ solvent-only rates. This catalysis depended on maintenance of native protein conformation [34].

Plasma albumin's reactivity—conformational reactivity—toward detergents and lipids also may seem opposite from that commonly expected. Detergents like dodecyl sulfate usually are employed as unfolding agents, as in denaturing gel electrophoresis (SDS gels, sodium dodecyl sulfate). Dodecyl sulfate indeed denatures, "normalizes," nearly all proteins in unfolded form. This, however, occurs above the cmc (critical micelle concentration). Below the cmc, dodecyl sulfate and many other hydrocarbonaceous anions, particularly alkane anions, do the opposite. They are protein-compacting and conformation-tightening agents. Detergent and lipid anions of properly adjusted concentrations, $y \sim 1$–20 organic anions per protein molecule, exert rather dramatic effects in these regards. They protect plasma albumin from denaturation while blood plasma is pasteurized (heat shocked) to kill viruses that might be in the plasma. Lipid anions such as caprylate, decanoate, etc., in relatively low concentrations protect against proteolysis and against acid pH extremes (Table 1).

The abilities to tighten conformation and protect plasma albumin clearly show up in albumin's hydrodynamic behavior in solution, correlating with the conditions that produce cocrystallization. Plasma albumin often is cited as a crystalline protein, and also as a precipitable protein. In fact, however, it cocrystallizes and coprecipitates, as noted in Table 2. The molecule strictly requires such ligands, failing to crystallize in homogeneous form unless it gets them. Binding of a few, critical hydrocarbonaceous ligands influences the protein to take on a compact, less hydrated, ordered conformation; this is the prerequisite for cocrystallization. This has been seen with many other proteins, to be discussed below. Because of the large amount of research concerning plasma albumin's conformational and hydrodynamic behavior in solution, one can try to deduce its separation and cocrystallization behavior when it is taken out of solution. Macromolecule tightening and contraction, narrowing into only a few conformations and perhaps only one conformation, supplies the basic need for crystallization. The process requires bound ligands to drive it, and it is actually cocrystallization. Plasma albumin remains a prime example of how these two areas, ligand-driven conformation change in solution (Table 1) and ligand-driven conformation control to achieve cocrystallization (Table 2), are in conjunction.

IV. SALT COUNTERION CONTRACTION OF PROTEINS FROM ACID-EXPANDED CONFORMATION

Acid unfolding and expansion of proteins is the leading example of how low-pH charging of proteins imposes physicochemical changes on the macromolecules. There are some uncoiling constraints, notably disulfide cross-links that prevent complete swelling to random coils. Major expansion occurs, however, with added hydration and penetration by water. Macromolecule expansion relieves excess electrostatic free energy, and repulsion between charged side chains. In low ionic strengths, 0.01–0.15, Coulombic repulsion, and therefore polyelectrolyte expansion, is expected to be enhanced. This is frequently seen, although there are exceptions. Ribonuclease apparently does not acid expand [35]. On the other hand, plasma albumin increases its intrinsic viscosity $[\eta]$ from 3.7 cm^3/g at the isoionic point near $Z_{H^+} \approx 0$, to 8.3 cm^3/g when $Z_{H^+} \approx +60$ and ionic strength is 0.15. The $[\eta]$ value range from 3.7 to 17 cm^3/g at ionic strength 0.01 [36]. Ribonuclease titrated downward to its maximum positive charge, $Z_{H^+} \approx +19$, maintains a virtually constant viscosity, $[\eta] = 3.3$ cm^3/g, over the acid pH range. These kinds of behavior for some time were mostly explained as the expression of a balance that must be struck among Coulombic repulsion forces leading to expansion, and disulfide cross-links and hydrophobic intramolecular forces, which should both lead to contraction and compactness in solution.

Upon addition of even more acid and of additional inorganic salt, several proteins, including plasma albumin, aggregate and precipitate [25,27]. Such aggregation–precipitation by proteins is practically universal upon addition of trichloroacetic and perchloric acid. These acids are standard reagents for protein analysis via precipitation. New insight into these older, still very useful phenomena was recently obtained by Goto, Fink, and collaborators [37,38]. They measured circular dichroism spectra and tryptophan fluorescence of apomyoglobin, β-lactamase, and cytochrome c in the pH region ~2 as more acid and also several inorganic salts were added. The pH 2 expanded proteins were seen to shrink, to contract. Ordinarily, addition of more protons leads to further expansion if any side chains were still available to titrate, or at least maintenance of the expansion already achieved at pH 2. Inorganic anion binding into strongly positively charged proteins, however, did what organic anions and some inorganic anions do to plasma albumin when it is positively charged in acid pH. Anions drew the acid-expanded macromolecules inward in plasma albumin and likewise contracted the three proteins investigated by Goto and Fink. Dependence of this shrinkage on the kind of inorganic anion (sulfate, perchlorate, thiocyanate, iodide, chloride, bromide, acetate, fluoride—not in that order) led Goto, Fink et al. to assign direct anion binding as the reversing force. The order of effectiveness of various inorganic anions approximated the order of selectivity by anion exchange resins for the same anions [38].

The behavior of plasma albumin and of the three proteins studied by Goto, Fink et al. can be summarized as follows, in relation to designing separation methods, described in more detail later, and in relation to inorganic and organic ion binding, particularly anions. Conformational flexibility and motility is key in separation design if such design is to be put on any molecular basis. Simply protein "size" and "charge" are not the only important variables. Protein flexibility raises major obstacles but at the same time provides great, although uncultivated, opportunities. One avenue for controlling protein conformation is to use ions and ligands that decrease motility and tighten macromolecules. Evidence for conformation control is best gotten from biophysical tools such as spectroscopy, which, however, apply to proteins in solution before aggregation. Conformation tightening also squeezes out some water, which helps densify proteins with their bound ions and ligands. Inorganic ions below about 0.1–0.2 ionic strength exert their screening or electrostatic shielding effects over distances roughly predicted by their reciprocal Debye radius $1/\kappa$. In large salt concentrations, preferential exclusion (squeezing out, salting out) occurs [39], sometimes called "high ionic strength effects," but this term and its implications are dubious. Spanning these salt concentrations, a third major effect of inorganic salts bears on protein conformation status, and consequently on protein separations technology; namely, direct binding by inorganic ions. The more polarizable inorganic ions tend to be most effective, and they generate the largest exothermic enthalpies of binding [40].

Organic ions such as organic sulfonates, a number of examples discussed below, promote physicochemical and conformation effects similar to those deriving from the inorganics, but with some added features important in separation technology. They bring organic material, quite a lot of it directly into the protein domain. Such organic material displaces some water, replacing it with lower dielectric constant alkanes, aromatics, etc. This strongly enhances electrostatic Coulombic repulsion and attraction depending on the kinds of charges born on protein side chains and on ligand ionic substituents. Displacement of water and tightening of proteins because of ligand-charged protein side-chain interaction in effect help dry out protein–ligand complexes. Driving some portion of water out of proteins causes them to become less buoyant and more dense than the well-hydrated, open conformation form they may have had in solution. Coprecipitates of this kind are relatively easy to deal with in centrifugation or filtration. Conventional engineering approaches to materials difficult to filter or centrifuge are to build better filters or centrifuges. Another approach, however, quite undeveloped, is to engineer solvent–ligand–protein interactions so that more primitive equipment can do the job.

V. COCRYSTALLIZATION OF PROTEINS WITH INORGANIC AND ORGANIC IONIC LIGANDS

Crystallization of many proteins actually is often cocrystallization; it requires an added ligand. A requirement for a cocrystallizing ligand usually is first seen in experimental screening attempts to get the first crystals and seeds. Ammonium sulfate normally is regarded as a salting-out agent, which includes its behavior as an electrolyte, as an exclusion agent akin to what neutral polymers do, and as a solvent component affecting the activity of water [41]. Part of its overall role may be seen from the fact that lesser amounts of sulfate become site bound to some proteins. This is the first signpost in the hypothesis that cocrystallization and coprecipitation of proteins may be the easier route in separation, as opposed to homogeneous precipitation or crystallization without bound ligands.

Table 3 lists protein cocrystals in which structure determinations and x-ray crystallographic evidence show inorganic ions esconced in repeating, ordered sites. Sulfate is interesting; it is the most nearly universally employed salt "additive and precipitant" for crystallization. That it binds to discrete sites on, or in, proteins has largely been shrugged aside. Friedberg and Bose [50] in 1969, using membrane electrode techniques, found α-chymotrypsin binds two to four SO_4^{2-} and Cl^- in solution in the pH 3–6 range. McPherson [51], summarizing crystallizing conditions for 291 proteins, lists 140 proteins in which ammonium sulfate

TABLE 3 Inorganic Ligands in Protein Cocrystals

Ligand	Number seen	Protein	Reference
Sulfate	3	Designer protein	42
Sulfate	5	Triosephosphate isomerase	43
Sulfate	7	Myoglobin	44
Sulfate	4	Annexin	45
Mg^{2+}, Mn^{2+}, Ca^{2+}	—	Parvalbumin	46
Zn^{2+}	—	aPP hormone	47
Nitrate	3	Lysozyme	48
NH_4^+	1	Actinidin	49

was the principal added salt. Sulfate or other inorganics are not necessarily site bound in all protein crystals so listed. Detailed binding data, pro or con, are not available. When careful analysis is carried out in x-ray crystallography, however, searching the evidence for inorganic ligand binding to proteins in solution [30], one to about five inorganic cations or anions such as sulfate are seen to be bound with appreciable energy [40]. Polarizable, "soft" inorganic ionc usually are most tightly bound [52].

There are good precedents for believing that when inorganic salts are needed for protein "crystallization," the basic need actually is for cocrystallization and the products are cocrystals. Ribonuclease, lysozyme [52], and several proteases and other enzymes are cocrystalline. They contain ligands that help stabilize an ordered conformation so that cocrystallization is preferred. Lysozyme accepts many sorts of inorganic anions with which to cocrystallize. Cocrystals most easily form with the more polarizable soft anions such as SCN^-, thiocyanate [53].

Organic ligands as cocrystallizing agents fall in two broad categories: (a) synthetic ligands such as detergents and dyes, mostly anions; (b) bioactive ligands including cofactors and coenzymes, substrates, and inhibitor-like molecules. That bioactive ligands are able to tighten protein conformation is seen, reviewed at some length [30,54], and intuitively expected. Because bioactive compounds such as NAD^+, FAD, and ATP tighten and protect enzymes for which they are coenzymes, they also cocrystallize them, yielding cocrystalline holo enzymes. Cocrystallization protection of proteins against several kinds of denaturing forces, conformation control, and tightening are all linked. The other side of the coin, however, is that ligand binding, capable of stabilizing the native conformation of proteins, can also stabilize the denatured inactive conformation. Likewise, ligands may stabilize the amorphously insoluble, as opposed to the cocrystalline and insoluble, form of proteins [55]. Solubility versus insolubility, crystallinity versus cocrystallinity, and protected versus unprotected properties should be of high priority in bioseparation technology. That control of such properties can be gotten by judicious use of added ligands is recognized in many basic research papers. It has not, however, been well recognized in a sustained way for large-scale use in industry. Serious activity losses in large-scale chromatography are common; for example, 50% losses occur for tissue plasminogen activator (TPA) protein production [56]. So far as we know, there has been no research and development in coinjecting ligands with input samples to protect key proteins during chromatographic operations. Protein conformation control via organic ligands is not employed in current large-scale separations. Instead, losses are made up by either using more crude, further scale up, or both.

Table 4 lists a small sampling of organic ligands, naturally occurring and synthetic ligands, illustrating their protective and cocrystallizing capacities. During the interval that synthetic ligands are bound to their target proteins and enzymes, the host enzymes usually are inactive. Synthetic ligands such as detergent

TABLE 4 Organic Ligands: Conformation Control, Protection, and Cocrystallization

Enzyme, protein	Denaturing condition	Protective ligand	Cocrystallizing ligand	Reference
ras protein			GDP	60
Lysozyme (mutant)	pH 3, heat	Benzene hydrocarbon		61
Phosphofructokinase	pH < 6	ATP		62
3-Phosphoglycerate kinase			Mg ATP	63
Insulin			Congo Red	64
Phospholipid transfer protein			Phospholipids	65
Ribosome inactivating protein			Adenine sulfate	66
Transcription factor			DNA oligomers	67
Glycogen phosphorylase a			Glucose	68
Chemoreceptor protein			Galactose	69
Glutamine PRP amidotransferase	Oxidative inactivation		AMP	70
Valyl-tRNA synthetase	Reducing conditions	NADP		71
Invertase	Heat, aging	Sorbitol		72
Fibrinogen	Reducing conditions	Ca^{2+}		73
Phosphoglyceromutase	Chymotryptic attack	3-phosphoglycerate		74
Bovine growth hormone	Insolubilizing conditions	bGH fragments		75
PEP carboxylase	Diethylpyrocarbonate deactivator	Phosphoenolpyruvate		76
L-Asparaginase	Heat, proteases	Asparagine, aspartic acid		77
Glutathione reductase	Heat, redox inactivation	Oxidized GSH		78

anions and "matrix ligands," introduced below, are inhibiting while in bound form. Such inhibition usually can be reversed by removal or protecting ligands in about 0.1–1 hour. Enzyme activities are restored upon removal of synthetic ligands by ion exchange or by XAD-type resins [57]. When apoenzymes are isolated or otherwise maneuvered while in inactive conformation, restoration of activity is often feasible by adding back coenzymes such as NAD^+, NADH, etc. Jaenicke [58], and Teipel and Koshland [59] each describe several examples involving dehydrogenases, enolase, fumarase, and aldolase. Table 4, by bringing the two themes of protection and cocrystallization together in one table, may help unify two parts of separation technology not normally seen to be connected.

VI. WATER INSIDE, WATER OUTSIDE PROTEINS

Proteins commonly, almost traditionally, are studied in water. Water maintains its place as an important reactant and product from metabolic turnover [79]. As the exclusive choice or universal solvent, though, water has been oversold. In parallel, exclusion of water from the nonpolar hydrophobic interior of proteins and some of the images thence drawn also have been oversold.

In relation to protein separation design, water in proteins can be seen as a close cousin to water's behavior toward surfaces. If water were repelled by nonpolar surfaces analogous to the protein hydrophobic core, one might expect heats of water adsorption to be relatively small. Yet, on some carbon blacks, water adsorption enthalpies reach 14–16 exothermic kcal/mole water adsorbed in the low coverage regions, roughly one water molecule/100 Å^2 of surface [80]. Water adsorbs onto very polar surfaces with enthalpies of the same magnitude [81]. Polar surfaces such as amorphous cellulose do not produce clearly different energies upon binding water than is the case with nonpolar crystalline cellulosic surfaces. One way around this is to hypothesize that many ostensibly nonpolar surfaces such as graphitized carbon are more polar than perhaps assumed.

Likewise, protein interior regions, so often assumed nonpolar, may indeed be nonpolar, but they are not necessarily dry. Water molecules are increasingly found as discrete single molecules, also in clusters, inside a number of proteins such as trypsin, which has 30 internal water clusters [82]. Series of proteases endowed with polypeptide structural homology all have close to 21 conserved buried water molecules [83]. Hydrophobic (nonpolar) forces are often said to account for protein compactness. Indeed, the dense "knots" inside proteins are very dry [84]. At the same time, though, water molecules in short bridges between chains in protein interiors can act as stabilizers, helping to glue together an overall compact conformation, as in papain [85]. Limitations in water penetration increasingly appear to be set nearly as much by polypeptide steric and packing constraints as the polar versus nonpolar character of protein interiors. Water molecules may act as lubricants in conformation changes involving α-helix rearrangements for 19 proteins [86]. Water usually is thought to be a good solvent for polar compounds, which it usually is, but not good for nonpolar compounds such as benzene and toluene (homologs of phenylalanine side chain). Although benzene is nonpolar in that it has zero dipole moment, benzene is very polarizable, 25.1 $Å^3$ [87]. The polarizability of water equals 1.44 $Å^3$. Benzene is not insoluble in water. It has a molar aqueous solubility (25°) of 0.023 *M* from liquid benzene [88]. The solubility of water in benzene (20°) is 0.007 *M* [89]. Benzene and water molecules hydrogen bond together [90].

Increasingly, water should be looked upon as a cosolvent, as a cross-linking molecule inside small spaces, as a cavity-filling molecule, and as a ligand besides its role as a metabolic reactant and product. Water in small clusters has considerable variability toward its immediate surroundings. If such clusters partly become clathrate within alkane groups in low-temperature icelike fashion, they may not offer many free hydrogen-bonding opportunities. Very low temperature ice at its surface is as nonpolar as Teflon, according to Adamson and colleagues [91]. This likely extends up to temperatures of melting ice. Even liquid water may have a partial hydrogenic surface, nonpolar in character [91].

Densely hydroxylated compounds of other kinds behave similarly; their solid surfaces behave as if they are nonpolar. Crystalline cellulose is an example. Truly crystalline cellulose has all its hydroxyl groups well bonded to one another in the interior and near the subsurface. The external surface, however, is made of methine C-H groups pointing outward [92]. No hydroxyl groups are available at the immediate surface. Accordingly, crystalline cellulose has a critical surface tension of roughly 30–35 ergs/cm^2 [93], about the same as polyethylene. Crystalline cellulose, despite its great hydroxylation, is not at all soluble in water, but it binds alkanes such as *n*-decane and long-chain alkanoic acids on its surface [94]. No doubt water is our leading hydroxylated polar solvent. But because so many water molecules are being found now well inside proteins, evidently for structural and also functional purposes, one questions how dry, therefore how hydrophobic, protein interiors actually are on average. If clusters of water molecules have structures with their hydroxyl groups mostly well hydrogen bonded, the remainder of such clusters may be compatible with, and even add to, internal hydrophobicity.

Bulk water is a leveling solvent. Lower alcohols, methanol and ethanol, also are leveling solvents [95]. Higher alcohols, C_3, C_4, and larger, are differentiating solvents. *Leveling* means solvents interact so strongly with solutes that solute physicochemical behavior (such as ion conductances) and chemical reactivities (such as dissociation constants) are largely governed by hydration and/or solvation. *Differentiating* means weaker solvation such that the intrinsic properties (as in gas phase) of solutes govern most of what they do. For example, in the series Li^+, Na^+, K^+, Rb^+, Cs^+, conductometric ionic mobilities rank in the opposite direction from that expected: The smallest ion is the most slowly moving ion (in water) and ought to be fastest. This occurs because water levels the system and hydrates ions having small radii more than those having large crystallographic radii. Similarly, association—dissociation and ion pairing are often leveled in water, made to behave similarly within a series of different salts and have approximately equal pK values [95]. In differentiating solvents, dissociation pKs spread apart, in step with the intrinsic chemical properties of the solutes.

Dissociation of ion hydrates in equilibria determined by Kebarle [96], represented by

$$M^+(H_2O)_n \rightarrow M^+(H_2O)_{n-1} + H_2O, \qquad n = 1, 2, 3, 4, 5$$

give enthalpies between about 15 and 25 endothermic kcal/single mole of water, for several monovalent cations including NH_4^+, Na^+, K^+, etc. Binding five to six water molecules in vapor to make $M^+(H_2O)_n$, $n \sim 5$ to 6, produces a total of $(-)110$ to $(-)130$ kcal heat/mole central ion from M^+, in gas phase. When hydration occurs in water, the water molecules of course come from bulk water. Routing the hydration process to gas phase requires an enthalpy input of $+9.7$ kcal/mole of water (heat of evaporation from liquid). Hence, five to six water molecules hydrating M^+: Na^+, K^+, etc., in water should produce roughly $-120 + 5 \times 10$ kcal of enthalpy change, about -60 kcal. Common inorganic anions such as Cl^- behave similarly. These enthalpies combined with crystal lattice energies ($+166$ kcal/mole for KCl [97]) give calculated net heats of dissolving crystals in water in good agreement with observations. Net heats of dissolving such salts are tiny compared with the heats of the contributing processes: breaking water molecules out of the liquid, pulling apart the crystal, allowing water molecules to hydrate. For example, the integral heats of dissolving NaCl, KCl, and Na_2SO_4 are $+1.2$, $+4.5$, and -0.5 kcal/mole, respectively. Water's ability to act as a leveling solvent, its small size enabling it to get close to central ions, underlies much of the energetics, at least in the enthalpies and probably in the entropies of such processes.

Interestingly, organic molecules such as dimethoxyethane and dimethyl either bind to K^+ in gas phase, with exothermic heats of $(-)20$–$(-)30$ kcal/mole organic ligand, surpassing water in binding heats of the first molecules to the central ion [98]. Only one or two organic molecules develop such large exothermic enthalpies for the cations, however, perhaps because, unlike water, there is simply not room [98] for more. One of the reasons water may act as a leveling solvent versus organic solvents or cosolvents that "differentiate" is that water molecules are small and can get in close. Part of the perspective of water, as opposed to organic solvents and cosolvents, leveling versus differentiating character, is that water molecules are the smallest solvent molecules. Internal packing possibilities are more diverse for water molecules than for organic solvents in the looser domains of proteins. In very tight domains, as mentioned above, the "knots," are dry, impervious to both classes of solvents.

Figure 1 summarizes some of the aspects of solvents, cosolvents, and ligands that can be maneuvered for protein separation engineering, that is, in precipitating,

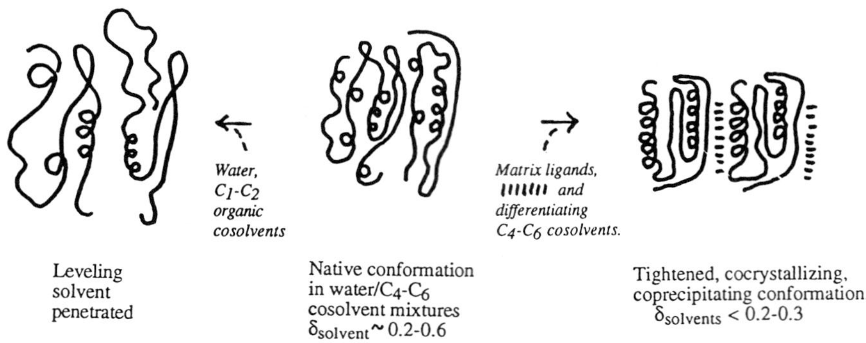

FIGURE 1 Leveling solvents, water and smaller alcohols, penetrate protein molecules and promote somewhat expanded conformation. Organic ion ligands and differentiating cosolvents promote tightened, protected, ordered conformation, able to coprecipitate and sometimes cocrystallize.

coprecipitating, and even cocrystallizing them. Scaling up protein separations likely shall have to head in these directions if current methods that depend on keeping proteins in solution or adsorbed to surfaces continue to have the limitations they evidently have now.

VII. PROTEIN PRECIPITATION FROM FOUR-CARBON COSOLVENT, t-BUTANOL

Salting out proteins with inorganic salts, and frigid solvent precipitation in $-5--20°C$ temperatures using acetone or ethanol are still the most used methods for precipitative isolation from dilute crude protein. At this "upstream" stage, the intent usually is not to try for much purification of the sought-for protein. Rather the intent is to get rid of large volumes of water or fermentation broth ("dewatering") to remove low-molecular-weight compounds, sugars, pigments, metabolites, etc. Inorganic salting out and frigid solvent technique are scalable and simplify dealing with mixed, complex crudes. They are over a century old [99].

When organic solvents are used for precipitating proteins at room temperature and activity yields are the prime criteria, results are very variable. Conventional lore is that organic solvents at room temperature denature most enzymes and other proteins. In contrast, many enzymes have been found to function quite adequately as catalysts in such mixtures. Some of them extend their capacity to catalyze reactions that are not feasible in water, for example, synthesizing esters and peptides from carboxylic acids and alcohols or amines. These syntheses can be performed now in organic solvents, including rather dry solvents. This is reviewed by A. Russell in Chapter 6.

A renewal of interest in nonaqueous solvents and partly aqueous cosolvents for precipitating proteins is at hand because of the opportunities for scaling up isolation of enzymes if they can be kept alive; they regain their activity when redissolved in water. The concept of leveling versus differentiation in organic cosolvent behavior toward proteins led us to try developing some new approaches, particularly for crudes. Crudes are a main concern, because they are the least researched, yet often the most expensive, sector of protein isolation. Research on chromatography, "downstream methods," is particularly active and dominant now. Despite the elegance and resolving power of several chromatographic methods such as antibody affinity chromatography, however, they usually are not very scalable except at extraordinary cost. Chromatographic methods tend to be diluting and time consuming. They need protein concentrates for their inputs. The eluates frequently are much more dilute than the inputs.

Separations in general require more than a series of steps, however advantageous each step may be. Much of the overall problem is to get steps to be compatible with one another. When chromatography is required to get resolution, but large volumes of crude need to be processed, a precipitative technology that gives a concentrate suitable for the next step fits neatly in the sequence.

Based on 1972 research involving seven enzymes in t-butanol cosolvents [100], we concluded t-butanol acts as a differentiating cosolvent in maintenance of enzyme activity in water–cosolvent mixtures. Methanol and ethanol at room temperature usually are not helpful in refolding enzymes. In contrast, t-butanol sometimes promotes refolding of heat-denatured enzymes when the heating step was carried out in water, as in L-glutamate dehydrogenase [100]. Other C_4 and C_5 alcohols, butanols and pentanols, are interesting to use in some cases, but they are of limited solubility in water whereas t-butanol is infinitely soluble in neat water. Hence, t-butanol became the basis for three-phase partitioning (TPP) for crudes using cosolvents for upstream protein isolation. It was first employed with multienzyme cellulases, and subsequently with 14 enzymes including the cellulases, in 1984–1987 [101].

Figure 2 shows how TPP-t-butanol is performed. The aqueous crude is adjusted to a suitable pH, and ammonium sulfate is added. Not enough salt is added to precipitate the sought-for protein, as in conventional salting out. Approximately 20–30% salt concentration is used. Addition of neat t-butanol first saturates the aqueous solution with t-butanol. Further addition to a total of about 1 ml t-butanol per ml of crude causes formation of a second, upper layer, phase. Simultaneously, a third, semisolid, phase of precipitated enzymes or proteins appears, indicated in the figure. The three layers partition cleanly in low-speed centrifugation. The "third phase" of precipitated enzymes, actually a coprecipitate, is easily captured. Kept in t-butanol-coprecipitated form, enzymes tend to remain protected. The coprecipitates usually dissolve quickly in neat water or in aqueous buffer. A few enzymes lose their activity in TPP. Most enzymes examined so far, however, retained their activity, therefore presumably their optimum folding.

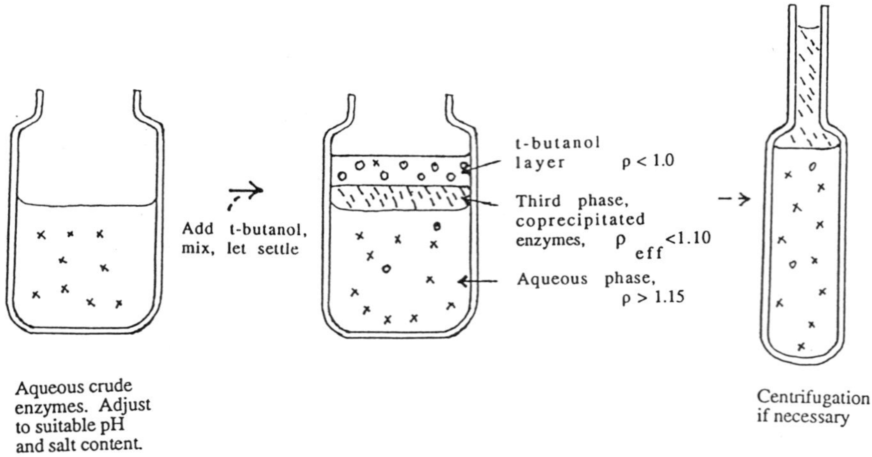

FIGURE 2 Enzyme isolation, TPP (three-phase partitioning) using t-butanol cosolvent. Coprecipitated proteins bind cosolvent, increasing their buoyancy, floating the coprecipitate above the aqueous layer. ρ is the density. Centrifugation is optional.

Some proteins, such as human hemoglobin, are denatured at room temperature by t-butanol, but this has been turned to advantage by Pol et al. [102]. Isolation of enzymes present in small quantities can be troubled by the large quantities of hemoglobin inevitably present in erythrocytes. Rapid denaturation and precipitation of hemoglobin enabled them to retrieve catalase, carbonic anhydrase, and superoxide dismutase free of hemoglobin, in 40–80% yields. Cathepsin L, a protease, has been isolated from liver, using TPP as the first step in a two-step isolation completed in only 6 hours [103]. Useful aspects of TPP via t-butanol for separations biotechnology are the following: (a) It concentrates and dewaters. (b) The organic (upper) layer removes pigments and unwanted apolar compounds. The TPP method is extractive for pigments, etc., as in conventional organic solvent extraction. (c) The TPP procedure usually can be accomplished on a multiliter scale in 5–10 minutes. On a large scale, centrifuging sometimes is not required. Simple settling often suffices. (d) In comparison with protein partitioning via the polymer phase builder technique using polyethylene glycol (PEG) and dextran [104], TPP via t-butanol has certain advantages. The product is precipitated out in TPP, not merely partitioned into another liquid phase. Therefore, TPP-t-butanol is maximally concentrating. Materials used for producing the "third phase" in TPP, such as t-butanol and ammonium sulfate, are easier to strip from the protein product (e.g., by dialysis) because they are low-molecular-weight materials. Polymers such as PEG and dextrans, in conventional polymer partitioning, usually are difficult to completely remove except by chromatography. When chromatography is required for this purpose, labor and time inputs are increased, and overall scalability is decreased.

A question brought out in the TPP-t-butanol technique pertains to other precipitative processes as well. Why does the protein precipitate position itself as it does, between the aqueous lower layer and the t-butanol upper layer? This behavior is useful in processing, because it requires only low-speed centrifugation or even just simple settling. Protein precipitates from conventional aqueous salting out usually sediment to the bottom.

The t-butanol cosolvent must adhere to the protein, decreasing the effective density of the precipitate and increasing its buoyancy. Such behavior is similar to lipoproteins that float, having effective densities lower than water because of their attached lipids. Governance of how such a particle moves is determined by the buoyancy of the protein hydrodynamic domain, which includes ligands and solvents inside the particle. The notation is ϕ' = effective partial specific volume of the entire domain, ρ = density of the solvent, $\bar{v}_p$ = partial specific volume of the dry protein, $\bar{v}_j$ = partial specific volume of components bound to protein inside the domain, $\bar{v}_1$ = partial specific volume of water, δ_1 = specific amount of bound water, δ_j = specific amounts of other components such as ligands that are bound. The relationship is [105]

$$\text{Overall buoyancy of the protein, ligand and included solvent domain} = 1 - \phi'\rho = (1 - \bar{v}_p\rho) + \delta_1(1 - \bar{v}_1\rho) + \sum_j \delta_j(1 - \bar{v}_j\rho)$$

The partial specific volume $\bar{v}_p$ for most proteins is close to 0.73 cm^3/dry g. The reciprocal 1/0.73 = 1.37 $g/cm^3 = \rho_{effective}$ of the dry protein. Ammonium sulfate–buffer systems, both in conventional salting out and in a TPP lower layer, have densities of about 1.10–1.20 g/ml. Hence, a completely dry protein or a moderately hydrated protein precipitate should sink. In conventional salting out, many protein precipitates do so. The TPP-t-butanol-separated proteins, on the contrary, float above the aqueous layer, likely because they bind cosolvent and lesser amounts of water, giving increased buoyancy and lowered effective density. Ligands and cosolvents that tighten and protect proteins are both precipitation agents, "phase builders," and flotation agents. The normal density of t-butanol is 0.79, and so its $\bar{v}$ value is roughly estimated as 1.27 cm^3/g. In normal salting out, also in TPP, which also employs ammonium sulfate, bound salt acts as a "sinker" to decrease buoyancy of proteins. Although a few sulfate ions are more frequently bound to proteins than perhaps formerly thought, it is unlikely that roughly 2–10 SO_4^{2-} ions can affect the $1 - \phi'\rho$ quantity very much by themselves (unless they tightened conformation enough to squeeze out water). The conclusion is that t-butanol's ability to cause protein precipitates to float and become buoyant originates from binding considerable amounts of t-butanol in, or on, the protein.

There is room for considerably increased application of ligand-binding and cosolvent-binding processes, which confer protein conformation control, in industrial processing where high-speed centrifuging is necessary. Conventional practice has been to simply build larger, more powerful centrifuges. Centrifuging costs increase mostly with the products, $V_t\omega^2$, where V_t = throughput volume and ω = angular velocity [104]. The time may be near for choosing ligands and cosolvents to control molecular properties in precipitates and coprecipitates, to either densify or float out desired proteins, thereby decreasing both V_t and ω. This should enable use of less sophisticated, hence less costly, machinery.

VIII. MATRIX COPRECIPITATION BY ORGANIC ION LIGANDS

Organic ligands able to bind to proteins and simultaneously bind parts of the ligands to one another provide another means for isolating proteins by coprecipitating them. Such a ligand–ligand interactive network forms a matrix. The general method, matrix coprecipitation and matrix coprecipitation-cocrystallization (MCC) [52] is scalable. Matrix ligands promote two contiguous reactions, represented in Fig. 3. The notation for the three most important quantities are y = moles ligand added/mole protein (for unknown protein molecular weights, a nominal molecular weight of about 40,000 daltons is assumed); $\nu_{\text{coppt.}}$ = moles ligand bound/mole coprecipitated protein; and Z_{H^+} = charge of the protein relative to the isoionic point from H^+ titration plots of proteins. When such titration data is not at hand, several coprecipitation experiments on a 0.5–2 ml scale are carried out to optimize isolation as a function of pH.

Figure 4 shows structures of three matrix ligands, two of which we synthesized. One of them, Rocellin, was recrystallized from the commercially available parent dye (Aldrich Co.) after ion exchanging its sodium salt to piperidinium. These organic anion ligands bind to a number of cationic proteins when the pH is lower than the isoionic point of the proteins. Similarly, alkyl-substituted acridinium ligands (cation organic ligands) are able to bind to proteins in their macroanion form, in pH ranges alkaline to the pI, as with pepsin [52]. The large organic ligands displace a certain amount of water, a higher dielectric, with their own lower dielectric organic bulk. This reinforces electrostatic attraction between matrix ligands and protein macroions of opposite charge. As binding progresses, the number of ligands becoming bound approaches Z_{H^+} if enough ligand is available to fill the available sites. When that happens, proteins coprecipitate with their ligand counterions. Occasionally, the complexes cocrystallize [52].

Compounds with flat ring systems such as those shown in Fig. 4 are able to stack, draw protein–ligand complexes together, and form the matrix represented by the second reaction in Fig. 3. Azobenzene and azonaphthalene compounds have intense visible absorption spectra, and molar absorption coefficients ϵ ranging

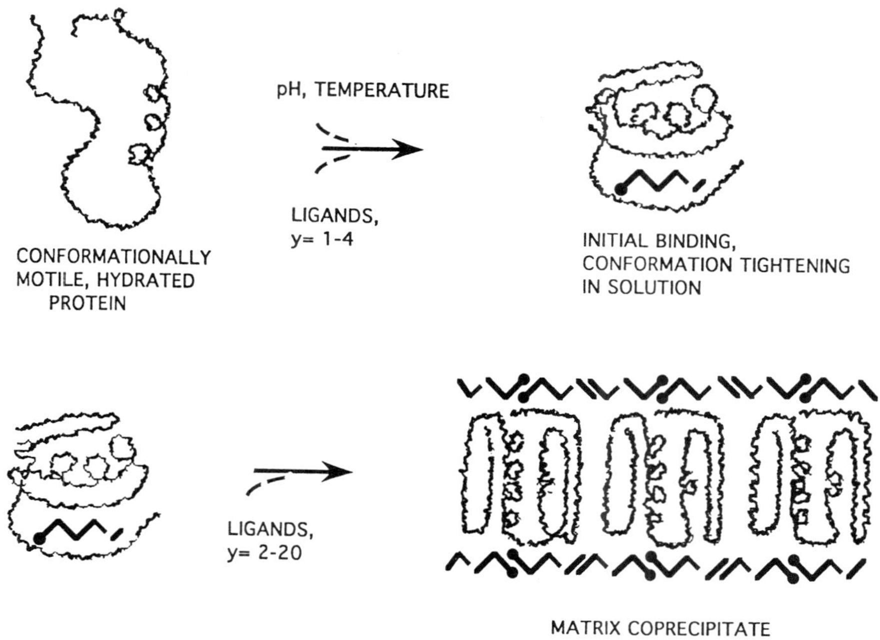

FIGURE 3 Molecular basis of MCC, matrix coprecipitation–cocrystallization. Organic ion ligands force a tightened conformation on proteins in solution to which they bind. Ligands able to stack using their hydrophobic tails form a matrix and pull the reactions from the right by coprecipitating out product. If the system is well ordered, the product may be cocrystalline [52].

from 25,000 to 40,000 M^{-1} cm^{-1}. Their coprecipitates with proteins are easy to analyze for ligand contents, giving $\nu_{\text{coppt.}}$ values, for plotting ligand binding isotherms. Three features borne by such azo ligands are important in their ability to form matrices. Briefly, these features are the coplanarity natural to aromatic diazo compounds, reinforcement of coplanarity by the hydroxyl groups *ortho* to the axo nitrogens, and aliphatic hydrocarbon ring substituents giving additional hydrophobic character and adhesion in the matrix.

Hydrogen atoms of hydroxyl groups *ortho* to diazo nitrogens in azobenzene and azonaphthalene compounds tautomerize. The molecules commonly exist in the hydrazone form, with the hydroxylic hydrogen transferred to the adjacent nitrogen [106]. Similar tautomerism and forcing to coplanarity occur in molecular complexes of such ligands with nucleic acid bases, found in x-ray crystal structure analyses by Ojala et al. [107]. In turn, ligand–ligand stacking, squeezing out water, and hydrophobic association are promoted. Possibly, protein–protein association in the coprecipitates provides some of the drive to make matrices and might

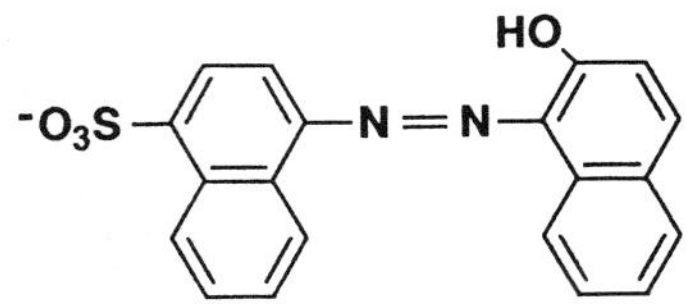

Roccellin

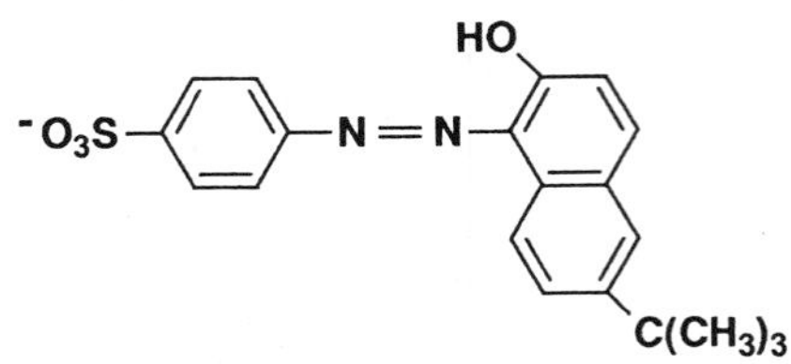

Little Rock Orange

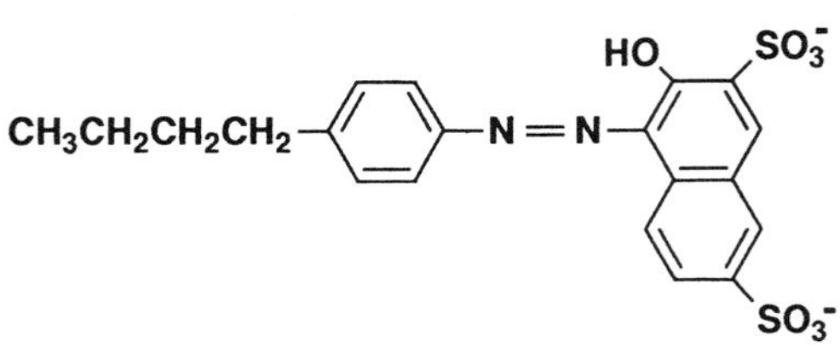

Razorback Red

FIGURE 4 Matrix ligands, azoaromatic sulfonate anions. Organic tails of ligands associate, helping to form matrices with proteins to which ionic heads are bound. Ortho hydroxyl groups of the kinds shown tautomerize with their H atom transferred to azo nitrogen (hydrazone form), enforcing coplanarity, increasing ability to stack, and squeezing out water.

even be dominant in the overall energetics. Most of the proteins we have worked with, however, are very soluble proteins, and matrix ligands are required to initiate the association.

For practical purposes, protection of proteins is an issue nearly as important as isolation of them. Organic matrix ligands integrate protection with coprecipitation in nearly all instances. We examined this using thermal, pH, and prooxidative stress (Fenton reagent hydroxyl radical treatment) on proteases. Figure 5 shows MCC ligand protection of bromelain, a proteolytic enzyme, against a combination of pH and thermal shock. Without the ligand, bromelain becomes almost completely denatured and loses its activity. Protected by a matrix ligand, bromelain survives acidic and heat treatment. After the thermal–pH stress interval, pH was shifted by about 2 units, and anion exchange resin (Dowex 50 in Cl^- form) is added. MCC ligands are trapped out in a few minutes, releasing the enzyme into solution with regain of its activity.

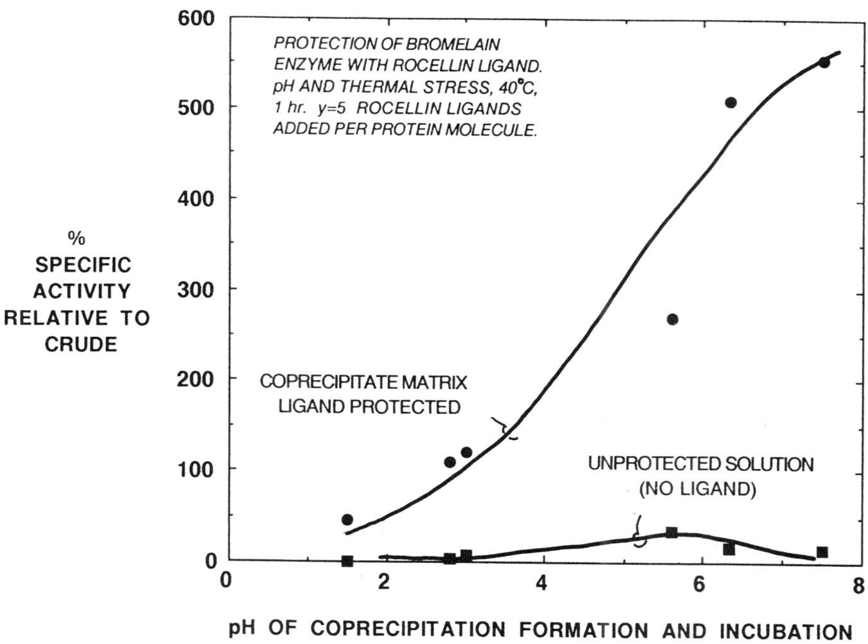

FIGURE 5 Protection of bromelain in coprecipitate form (redissolved after pH and thermal stress for assay) compared with bromelain not coprecipitated and subjected to the same stress. Coprecipitation also sheds some foreign protein, increasing specific activity. Assay conditions: CBZ-lysine-p-nitrophenyl ester substrate, pH 4.6, 25°, acetate buffer.

Matrix ligand coprecipitation of proteins is decisively different from exclusion mechanisms using polyethylene glycol (PEG) polymers and similar neutral osmolytes to precipitate proteins. Such polymers (and related osmolytes) push proteins out of solution; they do not pull them out. Pushing mechanisms, according to Timasheff and colleagues [41], originate from a combination of steric exclusion, preferential hydration, and increases in protein chemical potential. Protein chemical potential increases are relieved and canceled by protein–protein association and precipitation as homogeneous proteins. Matrix ligand coprecipitation (the "co-" prefix is important) pulls proteins out of solution and does not rely on pushing them out. Four characteristics of matrix ligand coprecipitation are noted now to contrast differences—in experimental conditions and in the consequences—between "pulling" proteins out of solution by ligand–matrix technique versus pushing them out by foreign polymer exclusion. (a) Concentrations of matrix ligands necessary to pull out targeted proteins range around 10^{-4}–10^{-5} M ligand. Concentrations of osmolytes and polymers necessary to push out proteins range up to whole molar concentrations. Matrix ligands are far too low in concentration to be expected to have much effect on bulk water structure or water activity. (b) Matrix ligand precipitation scavenges sought-for proteins. MCC quantitatively captures enzymes from 0.01 to 0.10% weight concentrations of proteins, as in pepsin [52] and a number of other enzymes studied more recently, including chymosin, α-chymotrypsin, and cellulases. (c) Matrix ligand behavior is modulated by electrolyte ions and more strongly affected and competed with by the quite polarizable chaotropic ions such as iodide and thiocyanate. Electrolyte concentrations in which this occurs, 10^{-3}–10^{-4} M electrolyte, indicate binding competition between electrolyte ions and ligands for discrete sites in proteins. (d) Binding isotherms of matrix ligands to proteins, plots of $\nu_{\text{coppt.}}$ vs. y in Fig. 6, show that matrix ligands are extraordinarily able to force *coprecipitated* proteins to acquire their full complement of ligands set by the protein's H^+ titration charge Z_{H^+} at the reigning pH.

Figure 6, an isotherm for α-chymotrypsin and Little Rock Orange ligand interaction, shows that when small amounts of ligand are added with levels $y \approx 2$ or 3, coprecipitates form having $\nu_{\text{coppt.}}$ values already near 14–16. Coprecipitation starts early, simply limited by the amount of available ligand, that is, by y. Whatever protein is pulled down by matrix formation, however, the value of $\nu_{\text{coppt.}}$ is close to Z_{H^+}, which in turn is determined by the pH and the protein's H^+ titration character. Continued addition of ligand after the plateau where $\nu_{\text{coppt.}} = Z_{H^+}$ is reached coprecipitatively titrates all enzyme remaining in solution. The plateau for all monoanion sulfonate ligands studied so far in this way with ribonuclease, lysozyme, and various proteases intercepts the ordinate at Z_{H^+}. The general behavior is similar to antigen–antibody coprecipitative titration: The precipitated product has nearly constant composition over the entire range of antigen addition, or precipitating ion addition, respectively.

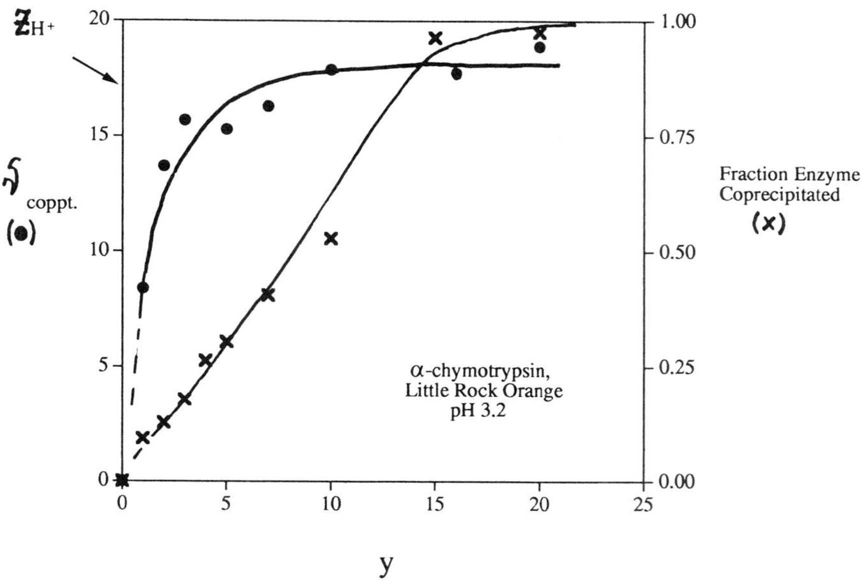

FIGURE 6 Coprecipitate binding isotherm, $\nu_{coppt.}$ vs. y of α-chymotrypsin–Little Rock Orange matrix ligand (left ordinate), coprecipitate yield vs. y, right ordinate. Binding isotherm plateaus, $\nu_{coppt.}$ reaches maximum value near Z_{H^+}, the hydrogen ion titration charge of the protein molecule.

Thus, protein matrix coprecipitation exerts considerable drive, drawing in one ligand anion for each side-chain H^+-titrated cation contributing to protein overall charge. There usually is no need to go to pH extremes, that is, to maximize Z_{H^+}, for the MCC technique. Adjusting the pH so that Z_{H^+} equals $+5-+20$ usually suffices. Accordingly, $\nu_{coppt.}$ values from spectrophotometric analysis show 5–20 of the azo dyes are bound.

Inorganic anions seldom produce such coprecipitation, an exception perhaps being trichloroacetate. Inorganic ions lack an organic group for gluing together a matrix network. Electrostatics and charge–charge interaction initiates, and partly guides, counterion ligand binding. Electrostatic forces alone, however, usually do not provide all that is necessary to bring down most proteins as precipitates or coprecipitates from dilute solution when the solvent is water. Organic matrix ligands get the system over the hump. Once coprecipitation starts, organic ligands drive it quite strongly, ensnaring the sought-for protein or inducing proteins to self-associate and pull themselves down.

In practical use, one is concerned about rates of coprecipitation and of redissolving coprecipitates, steps necessary to remove ligands from coprecipitates.

Not many detailed studies of coprecipitation rates are completed. Most coprecipitates form and settle out in 5–20 minutes, although some are slower. Redissolving coprecipitates is usually easy to do and is rapid. Coprecipitates are given a pH change of 2 to 4 units toward an alkaline pH. Ion exchange resins, often Dowex 50 in the Cl^- form, strip the system of ionic ligands in a few minutes. Enzymes are released in 70–100% total activity yields from their coprecipitates.

IX. INORGANIC AND ORGANIC ION-BINDING THERMOCHEMISTRY

Inorganic salting out, inasmuch as it involves binding inorganic ions, particularly anions, to proteins, is a form of coprecipitation, and when crystals are obtained, of cocrystallization. The MCC technique partly parallels conventional inorganic salting out: use of organic ions instead of inorganic ions for the same ultimate purpose; coprecipitation and cocrystallization to isolate proteins. For both approaches, it would be helpful to have new and improved thermodynamic data.

Heat conduction calorimeters able to use reactants having millimolar and submillimolar concentrations in volumes of roughly 0.5–2 ml provide good prospects for this. There is needed, however, a new, or at least a renewed, emphasis on the status of protein samples used for gathering such data. Here, we briefly review some of the rationale for gathering calorimetric enthalpies and combining the enthalpies with free energies to next determine binding entropies. In order to put such measurements on the same footing, it is especially helpful if mixed resin bed deionization of proteins [16] is the starting point.

Mixed resin bed deionization of proteins usually is rather easy to do. It is effective even with proteins such as plasma albumin, which strongly binds many anions. The reason for emphasizing this is that a large fraction of binding data often is not determined simply by the availability of sites on proteins with assurance that they are open. Instead, such data often is the net sum of at least two reactions: (a) dissociation of a ligand l_1 already bound in the protein; and (b) binding of the new incoming ligand, l_2. The net reaction seen in colorimetry is a displacement or exchange:

$$P \cdot l_1 + l_2 \rightarrow P \cdot l_2 + l_1$$

Probably the most studied such reactions aiming at unraveling the thermodynamics of site binding pertain to albumin. The intent in a Scatchard et al. paper in 1950 [108] was to determine the thermodynamics of filling the first strong site and subsequent sites with inorganic anions such as Cl^- and SCN^-. They concluded that the van't Hoff enthalpy of binding chloride into the first six sites of plasma albumin was nearly zero: $\Delta H^\circ = +430 \pm 540$ calories/mole of bound chloride. This correlated with estimates at that time that binding a number of organic sulfonate anions was nearly athermal also (i.e., gave quite small or nearly zero heats). Therefore, the Gibbs free energies of albumin–anion binding appeared to be entropy driven. In 1964, however, Scatchard and Yap [109], using electrodialyzed plasma albumin redetermined the van't Hoff enthalpies bracketing 0° and 25°C, fitting the data with three classes of sites. There were $n_1 = 1$ strong sites, $n_2 = 4$ intermediate, and $n_3 = 22$ weaker sites. Molar van't Hoff enthalpies for Cl^- binding were estimated as -3, -1.3, and 0 kcal/mole, respectively. The entropies $\Delta S°$ were all positive in sign: 3.2, 4, and 4.6 e.u. respectively. Several authors have used the Scatchard et al. results to interpret their own work, as in Minton and Edelhoch's estimation of the temperature dependence of albumin's net charge [110]. In the meantime, however, calorimetry and careful resin deionization of albumin combined to produce a quite different picture of albumin–inorganic ion binding thermodynamics, especially into the first sites. Mixed bed resin treatment of aqueous plasma albumin forces the protein to give up all its electrolyte counterions. The very strong cation and anion affinities by resin exchangers trap Na^+ and replace it with H^+. The mixed bed resins also replace Cl^-, etc., with OH^-. Hence, all protein counterions are replaced by H_2O. The protein is completely stripped and automatically rendered isoionic as an eluate from the resin exchanger [111].

Such deionization and adjustment to isoionic pH is particularly helpful in both the H^+ titration technique and in protein–ligand calorimetry. It provides a reliable, rather rapid, means to ensure starting with the protein under study in a controlled, reproducible reference state, without foreign ions. All sites are empty. In calorimetry, one mixes the deionized protein with the anionic ligand being studied to directly determine the heat via the Seebeck heat conduction principle [112]. The temperature rise ΔT is small, less than 0.20°C, even with heat production ranging up to 20–40 millicalories of observed heat. Hence, such calorimetric measurements essentially are isothermal measurements. There are negligible heat capacity effects underlying the data, as is otherwise the case in many van't Hoff determinations. The magnitudes of the observed heats and the resolving power of the general method enable determination of enthalpies of ligand binding into the first, individual site or sites, with good assurance that the protein is "empty," with its sites open to begin with. (In the case of plasma albumin, which tenaciously binds one or two moles of fatty acid anion, a "defatting" procedure such as charcoal-acid treatment [113] may be needed.)

This program, utilizing batch mixing calorimetry, has been carried out for plasma albumin [114], hemoglobin [115], and β-lactoglobulin [116]. β-Lactoglobulin is quite insoluble when completely deionized. Such insolubility can be dealt with in two ways: (a) staying below the solubility limit [117] during mixed resin deionization; (b) Carrying out deionization with the protein confined in a dialysis sac, leaving the mixed bed resin outside to drag out the counterions. Hemoglobin binds ATP, adenosine triphosphate, anion and other organic phosphate anions, but it also binds chloride and inorganic phosphate. Without mixed bed resin deionization, organic ligands binding to hemoglobin are forced into competitive

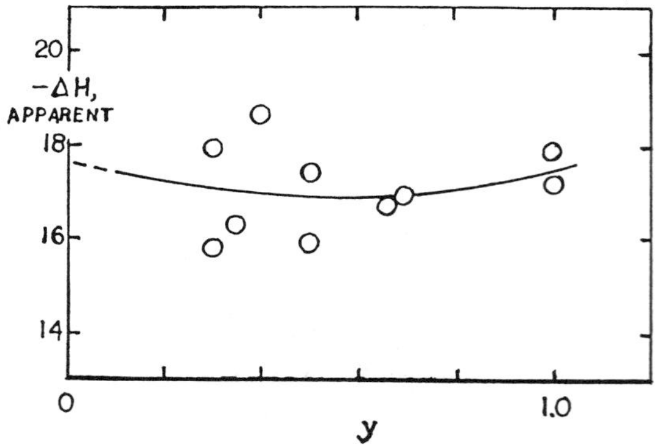

FIGURE 7 Heat conduction calorimetric determination of molar apparent enthalpy of iodide binding to plasma albumin (mixed bed resin deionized protein) in 1.8×10^{-4} *M* protein under condition in which *y*, the relative molar amount of I^- mixed with protein, was 1.0 or less, staying under first site binding levels, 25°.

binding from buffer or electrolyte ions already bound to the protein. As with plasma albumin, such competition decreases observed heats by large amounts, obscuring heats that should be produced from true open site binding.

These techniques provide renewed, and likely improved, estimates of binding enthalpies into primary binding sites on proteins, and thence binding entropies. Once the first site binding thermodynamics are determined, later sites and their heats can be assessed as ligand concentrations are increased. Figure 7 plots molar colarimetric enthalpies of iodide binding to BSA under conditions where *y* (moles I^- added/mole BSA) < 1.0. Extrapolated to the zero limit of *y*, taking into account the binding constants, enthalpies of I^- first single site binding to well-deionized plasma albumin are close to -18 kcal/mole. Entropies of the interaction and related data for chloride and dodecyl sulfate are shown in Table 5, from the original work [114].

In general, association of "hard" anions with proteins into constellations of positive charge are somewhat exothermic for the first, strong sites, namely, with production of -2–-5 kcal heat/bound ion. If the incoming ion has to displace or compete off an already bound anion, the overall reaction may appear to produce a small, even endothermic (positive), heat. Larger, soft, polarizable anions including even iodide have heat productions commonly in the -10–-20 kcal/mole range. These kinds of heats are gotten for many organic ions if they do indeed bind. Such enthalpies contribute enough free energy drive to compensate for the rather large negative entropies that seem rather general, although there are a few exceptions. Albumin-binding reactions are but one example of "enthalpy driven, entropy hindered" reactions. A number of antibody–hapten reactions surveyed by

TABLE 5 Thermodynamics of Anion Binding to Isoionic Plasma Albumin

Anion	Class of site i	Number of bound ions	$-\Delta H_i$ (kcal/mol)	$-\Delta G_i^\circ$ (kcal/mol)	ΔS° (cal deg^{-1} mol^{-1})
Iodide	1	1	18.0	5.4	−42
	2	8	3.2	3.5	+1.1
	3	18	2.0	1.5	−1.6
Chloride	1	1	8.3 ± 3.0	4.6	−12.3
	2	8	3.1 ± 1.0	2.7	−1.2
Dodecyl sulfate	1	1	18 ± 4		

25°C; enthalpies based on heat conduction calorimetry, mixed bed resin stripped protein; Ref. 114.

Wiesinger and Hinz [40] produce very large exothermic enthalpies in binding. Similarly, β-lactoglobulin [116] and many enzymes [40,52] are "enthalpy driven" in organic ligand–protein association in water. Such enthalpies, commonly −5−−20 kcal/mole of ligand bound into the first sites of proteins, are sufficient to compensate for considerable adverse entropies presumably produced in protein conformation tightening, some dehydration, and protection of proteins by bound ligands. Even when the thermodynamic data are reasonably accurate, the connection of thermochemical data to these phenomena, conformation tightening, etc., is not straightforward. Some of the bridges, however, still rather shaky and needing reinforcement with more data, may be discernible. Such bridges connect protein-binding reactions in solution, and what might be maneuvered for better separation technology and separation engineering. We have tried to glimpse how such reactions prepare proteins to precipitate and coprecipitate, with some success in applied results.

ACKNOWLEDGMENTS

This chapter was written with support from the University of Minnesota Agricultural Experiment Station and from AURI (Agricultural Utilization Research Institute).

Permission to use figures from previous publishers and copyright owners is gratefully acknowledged: Figure 2 from Alan Liss, Inc. publications (1987); Figure 7 from American Chemical Society publications (1971), granted by these publishing companies.

REFERENCES

1. I. Godfrey and J. Reichelt (1983). "Industrial Enzymology," Macmillan Pub., Ltd., London, 1–7.
2. P. Layman (1990). *Chem. and Eng. News* (Sept 24), 17–18.
3. V. J. Cabasso (1989). *Plasmapheresis 3*, 17–21.
4. E. R. Jeans, P. J. Marshall, and C. R. Lowe (1985). *Tr. in Biotechnology 3*: 267–271.
5. P. Belter, E. Cussler, and W.-S. Hu (1988). "Bioseparations: Downstream Processing for Biotechnology," Wiley-Interscience, New York, 347–359.
6. J. F. Kennedy and J. S. Cabral (1983), in W. H. Scouten, ed., "Solid Phase Biochemistry,: *Chemical Analysis* series, Vol. 66, 358–391.
7. R. K. Scopes (1987). "Protein Purification," Springer-Verlag, New York, 52.
8. W. J. Leonard, K. K. Vijai, and J. F. Foster (1962). *J. Biol. Chem. 238:* 1984–1988.
9. E. S. Benson, B. E. Hallaway, and R. W. Lumry (1964) *J. Biol. Chem. 239*: 122–129.
10. R. G. Pearson (1973), in R. Pearson, ed., "Hard and Soft Acids and Bases," Benchmark Series; Douden, Hutchinson and Ross, Inc., Stroudsbury, Pennsylvania, 67–78.
11. C. Tanford (1958), in A. Neuberger, ed., "Symposium on Protein Structure," Methuen and Co., London, 1–23.

12. W. Melander and C. Horvath (1977). *Arch. Biochem. Biophys. 183*: 200–215.
13. O. Sinanoglu (1968), in B. Pullman, ed., "Molecular Associations in Biology," Academic Press, New York, 427–445.
14. V. Bloomfield (1966). *Biochemistry 5*: 684–689.
15. E. L. Duggan and J. M. Luck (1948). *J. Biol. Chem. 172*: 205–221.
16. R. Lovrien (1963). *J. Am. Chem. Soc. 85*: 3677–3682.
17. G. Markus, R. Love, and F. Wissler (1964). *J. Biol. Chem. 239*: 3687–3693.
18. G. Markus, D. McClintock, and B. Castellani (1967). *J. Biol. Chem. 242*: 4402–4408.
19. R. G. Reed (1977). *J. Biol. Chem. 252*: 7483–7487.
20. E. Benson and B. Hallaway (1970). *J. Biol. Chem. 245*: 4144–4149.
21. A. Shrake and P. Ross (1988). *J. Biol. Chem. 263*: 15392–15399.
22. R. Lovrien and T. Linn (1967). *Biochemistry 6*: 2281–2293.
23. K. Zakrzewski and H. Goch (1968). *Biochemistry 7*: 1835–1842.
24. H. A. Petersen and J. F. Foster (1965). *J. Biol. Chem. 240*: 2503–2507.
25. K. P. Wong and J. F. Foster (1968), in "The Solution Properties of Natural Polymers," Spec. Pub. No. 23, Chemical Society, London, 25–48.
26. R. H. McMenamy and Y. Lee (1967). *Arch. Biochem. Biophys. 122*: 635–643.
27. M. Kronman, M. Stern, and S. Timasheff (1956). *J. Phys. Chem. 60*: 829–831.
28. E. J. Cohn, W. L. Hughes, and J. H. Weare (1947). *J. Am. Chem. Soc. 69*: 1753–1761.
29. J. Lewin (1951). *J. Am. Chem. Soc.* **73**: 3906–3911.
30. J. Steinhardt and J. A. Reynolds (1969). "Multiple Equilibria in Proteins," Academic Press, New York, 85–128.
31. P. Molyneux and H. P. Frank (1961). *J. Am. Chem. Soc. 83*: 3169–3174, 3175–3180.
32. W. B. Jakoby (1978). *Adv. Enzymol. 46*: 389–395.
33. R. P. Taylor (1976). *J. Am. Chem. Soc. 98*: 2684–2686.
34. R. P. Taylor, S. Berga, V. Chau, and C. Bryner (1975). *J. Am. Chem. Soc. 97*: 1943–1948.
35. J. G. Buzzell and C. Tanford (1956). *J. Phys. Chem. 60*: 1204–1207.
36. C. Tanford, J. Buzzell, D. Rands, and S. Swanson (1955). *J. Am. Chem. Soc. 77*: 6421–6428.
37. Y. Goto, L. Çalciano, and A. L. Fink (1990). *Proc. Natl. Acad. Sci. 87*: 573–577.
38. Y. Goto, N. Takahashi, and A. L. Fink (1990). *Biochemistry 29*: 3480–3488.
39. T. Arakawa, Y. Kita, and L. Narhi (1991), in C. Suelter, ed., *Meth. Biochem. Anal. 35*: 87–126.
40. H. Wiesinger and H.-J. Hinz (1986), in H.-J. Hinz, ed., "Thermodynamic Data for Biochemistry and Biotechnology," Springer-Verlag, Berlin, Chapter 7.
41. S. N. Timasheff (1992). *Biochemistry 31*: 9857–9864.
42. C. P. Hill, D. Anderson, L. Wesson, W. DeGrado, and D. Eisenberg (1990). *Science 249*: 543–546.
43. R. Wierenga, M. Noble, G. Vriend, S. Nauche, and W. Hol (1991). *J. Mol. Biol. 220*: 995–1015.
44. B. P. Schoenborn (1988). *J. Mol. Biol. 201*: 741–749.
45. R. Huber, R. Berendes, A. Burger, M. Schneider, A. Karshikov, and H. Lueke (1992). *J. Mol. Biol. 223*: 683–704.
46. J.-P. Declercq, B. Timant, J. Parello, and J. Rambaud (1991). *J. Mol. Biol. 220*: 1017–1039.
47. J. E. Pitts (1990). *Nature 346*: 113–114.
48. P. Artymiuk and C. Blake (1981). *J. Mol. Biol. 152*: 737–762.

49. E. Baker (1980). *J. Mol. Biol. 141*: 441–484.
50. F. Friedberg and S. Bose (1969). *Biochemistry 8*: 2564–2567.
51. A. McPherson (1982). "Preparation and Analysis of Protein Crystals," Wiley-Interscience, New York, Appendix, 127–159.
52. M. Conroy and R. Lovrien (1992). *J. Crystal Growth 122*: 213–222.
53. M. Ries-Kautt and A. Ducruix (1991). *J. Crystal Growth 110*: 20–25.
54. N. Citri (1973). *Adv. in Enzymol. 37*: 397–648.
55. J. Wyman and S. Gill (1990). "Binding and Linkage," University Science Books, Mill Valley, California, 237–268.
56. R. Datar, T. Cartwright, and C.-G. Rosen (1993). *Bio/Technology 11*: 349–355.
57. M. Knuth and R. Burgess (1987), in R. Burgess, ed., "Protein Purification: Micro to Macro," A. Liss, Inc., New York, 279–305.
58. R. Jaenicke (1987). *Prog. Biophys. Molec. Biol. 49*: 117–237.
59. J. Teipel and D. Koshland (1971). *Biochemistry 10*: 792–805.
60. S.-H. Kim and S. Holbrook (1990). *J. of NIH Resch. 2*: 64–65.
61. A. Eriksson, W. Baase, J. Wogniak, and B. Matthews (1992). *Nature 355*: 371–373.
62. J. Alpers, H. Paulus, and G. Bazylewicz (1971). *Proc. Natl. Acad. Sci. 68*: 2937–2940.
63. G. Davies, S. Gamblin, J. Littlechild, and H. Watson (1992). *J. Mol. Biol. 227*: 1263–1264.
64. W. Turnell and J. Finch (1992). *J. Mol. Biol. 227*: 1205–1233.
65. E. Pebay-Peroula, C. Cohen-Addad, M. Lehmann, and D. Marion (1992). *J. Mol. Biol. 226*: 563–564.
66. M. Miyano, K. Appelt, M. Arita, N. Habuka, J. Kataoka, H. Ago, H. Tsuge, M. Noma, V. Ashford, and N. Xuong (1992). *J. Mol. Biol. 226*: 281–283.
67. T. Hakosha, Y. Teranishi, T. Ohira, K. Suzuki, N. Ogawa, and Y. Oshima (1993). *J. Mol. Biol. 229*: 556–569.
68. S. Sprang, K. Acharya, E. Goldsmith, D. Stuart, K. Varvill, R. Fletterick, N. Madsen, and L. Johnson (1988). *Nature 336*: 215–221.
69. N. Vyas, M. Vyas, and F. Quiocho (1988). *Science 242*: 1290–1295.
70. D. Bernlohr and R. Switzer (1981). *Biochemistry 20*: 5675–5681.
71. S. Black (1986). *Science 234*: 1113–1115.
72. D. Combes, T. Yoovidhya, E. Girbal, and P. Monsan (1987). *Ann. N.Y. Acad. Sci. 501*: 59–62.
73. R. Procyk and B. Blombäck (1990). *Biochemistry 29*: 1501–1507.
74. L. Fothergill-Gilmore and H. Watson (1989). *Adv. in Enzymol. 62*: 271–277.
75. D. N. Brems (1988). *Biochemistry 27*: 4541–4546.
76. A. Sandana (1988). *Biotech. Adv. 6*: 349–466.
77. T. Tosa, R. Sano, K. Yamamoto, M. Nakamura, and I. Chibata (1972). *Biochemistry 11*: 217–222.
78. M. Pinto, A. Mata, and J. Lopez-Barea (1984). *Arch. Biochem. Biophys. 228*: 1–12.
79. F. Franks (1983). "Water," Royal Soc. of Chemistry, London, 1–4.
80. A. Zettlemoyer, F. Micale, and K. Klier (1975), in F. Franks, ed., "Water: A Comprehensive Treatise," Vol. 5, Plenum Press, New York, 249–291.
81. D. S. Reid (1979), in M. N. Jones, ed., "Biochemical Thermodynamics," Elsevier Pub., Amsterdam, 168–183.
82. H. Bartunik, L. Summers, and H. Bartsch (1989). *J. Mol. Biol. 210*: 813–828.

83. U. Sreenivisan and P. Axelsen (1992). *Biochemistry 31*: 12785–12791.
84. R. Gregory and R. Lumry (1985). *Biopolymers 24*: 301–326.
85. J. H. Berendsen (1975), in F. Franks, ed., "Water: A Comprehensive Treatise," Vol. 5, Plenum, New York, 293–349.
86. M. Sundaraligam and Y. Sekharudu (1989). *Science 244*: 1333–1337.
87. A. W. Adamson (1979). "Physical Chemistry," 2nd ed., Academic Press, New York, 84–86.
88. W. May, S. Wasik, and D. Freeman (1978). *Analyt. Chem. 50*: 997–1000.
89. P. Schatzberg (1963). *J. Phys. Chem. 67*: 776–779.
90. W. Klemperer (1992). *Science 257*: 887–888.
91. A. Adamson, L. Dormant, and M. Orem (1967). *J. Coll. Interface Sci. 25*: 206–217.
92. B. Odegaard, P. Anderson, and R. Lovrien (1984). *J. App. Biochem. 6*: 156–183.
93. P. Luner and M. Sandell (1969). *J. Polymer Sci., Part C, 28*: 115–142.
94. F. J. Ehrhardt (1984). Ph.D. thesis, Institute of Paper Chemistry and Lawrence University, Appleton, Wisconsin, 27–36.
95. C. W. Davies (1962). "Ion Association," Butterworths, London, 95–160.
96. P. Kebarle (1977). *Ann. Rev. Phys. Chem. 28*: 445–476.
97. C. Kittel (1986). "Introduction to Solid State Physics," 2nd ed., J. Wiley Pub., New York, 70–72.
98. W. Davidson and P. Kebarle (1976). *Can. J. Chem. 54*: 2594–2599.
99. J. T. Edsall (1979). *Ann. N.Y. Acad. Sci. 325*: 53–74.
100. K. Tan and R. Lovrien (1972). *J. Biol. Chem. 247*: 3278–3285.
101. R. Lovrien, C. Goldensoph, P. Anderson, and B. Odegaard (1987), in R. Burgess, ed., "Protein Purification: Micro to Macro," A. Liss, Inc., New York, 131–148.
102. M. Pol, H. Deutsch, and L. Visser (1990). *Int. J. Biochem. 22*: 179–185.
103. G. Jacobs, R. Pike, and C. Dennison (1989). *Analyt. Biochem. 18*: 169–171.
104. M.-R. Kula (1987), in R. Burgess, ed., "Protein Purification: Micro to Macro," A. Liss, Inc., New York, 99–115.
105. J. Steele, C. Tanford, and J. Reynolds (1978). *Meth. Enzymol. 48*: 11–14.
106. A. Olivieri, R. Wilson, I. Paul, and D. Curtin (1989). *J. Am. Chem. Soc. 111*: 5525–5532.
107. W. H. Ojala, L. K. Lin, W. B. Gleason, T. I. Richardson, and R. Lovrien, *Acta Crystallographica Sec. B* (In press).
108. G. Scatchard, I. Scheinburg, and S. Armstrong (1950). *J. Am. Chem. Soc. 72*: 535–546.
109. G. Scatchard and W. Yap (1964). *J. Am. Chem. Soc. 86*: 3434–3438.
110. A. Minton and H. Edelhoch (1982). *Biopolymers 21*: 451–458.
111. J. Edsall and J. Wyman (1958). "Biophysical Chemistry," Academic Press, New York, 600.
112. J. Sturtevant (1974). *Ann. Rev. Biophys. Bioeng. 3*: 35–51.
113. R. Chen (1967). *J. Biol. Chem. 242*: 173–181.
114. R. Lovrien and J. Sturtevant (1971). *Biochemistry 10*: 3811–3815.
115. B. Hedlund, C. Danielson, and R. Lovrien (1972). *Biochemistry 11*: 4660–4668.
116. R. Lovrien and W. Anderson (1969). *Arch. Biochem. Biophys. 131*: 139–142.
117. J. Treece, R. Sheinson, and T. McMeekin (1964). *Arch. Biochem. Biophys. 108*: 99–108.

Index